Continuum Physics with Application to Cosmology

Other Titles by the Author

Eléments de Dynamique des Solides et Structures Déformables (in French)

A Geometric Approach to Thermomechanics of Dissipating Continua

Covariance and Gauge Invariance in Continuum Physics: Application to Mechanics, Gravitation, and Electromagnetism

Introduction à la Mécanique des Milieux Continus Déformables (in French)

Continuum Physics with Application to Cosmology

Lalaonirina Rakotomanana Ravelonarivo
University of Rennes, France

World Scientific

NEW JERSEY • LONDON • SINGAPORE • BEIJING • SHANGHAI • HONG KONG • TAIPEI • CHENNAI • TOKYO

Published by

World Scientific Publishing Co. Pte. Ltd.

5 Toh Tuck Link, Singapore 596224

USA office: 27 Warren Street, Suite 401-402, Hackensack, NJ 07601

UK office: 57 Shelton Street, Covent Garden, London WC2H 9HE

Library of Congress Cataloging-in-Publication Data
Names: Rakotomanana, Lalao author
Title: Continuum physics with application to cosmology / Lalaonirina Rakotomanana Ravelonarivo, University of Rennes, France.
Description: New Jersey : World Scientific, [2026] | Includes bibliographical references and index.
Identifiers: LCCN 2026001689 | ISBN 9789819828616 hardcover |
ISBN 9789819828623 ebook for institutions | ISBN 9789819828630 ebook for individuals
Subjects: LCSH: Space and time--Mathematics | Field theory (Physics) | Geodesics (Mathematics)
Classification: LCC QC173.59.S65 R35 2026
LC record available at https://lccn.loc.gov/2026001689

British Library Cataloguing-in-Publication Data
A catalogue record for this book is available from the British Library.

For any available supplementary material, please visit
https://www.worldscientific.com/worldscibooks/10.1142/14748#t=suppl

Desk Editor: Shaun Tan Yi Jie

Typeset by Stallion Press
Email: enquiries@stallionpress.com

This book is dedicated to my mother Lalasoa (1932–2022) and to my father Albert (1920–1975). They gave us the best and priceless heritage: the taste for effort and study.

Preface

For centuries, different theoretical models, mostly depending on the choice of length scale observation, exist to describe the numerous and varied physical phenomena in the Universe. Besides the classical theories of Newtonian mechanics or Einstein gravitation, and Maxwell electromagnetism, the best known are quantum mechanics, the standard model of particle physics, loop quantum gravitation, and string theory. These latter two mainly constitute attempts to unify quantum mechanics with relative gravitation.[1] In classical physics, the continuum approach has always been in tension with, but complementing, the corpuscular approach; this has been the case for a long time. The deliberate choice of this work remains, however, within the continuum description, given the very large dimensions involved in relative gravitation and even more so in cosmology.

On the one hand, the usual path to derive theoretical models starts with the description of behaviors of test particles, and then it concentrates on the short-range interactions between these particles before extending them to long-distance interactions possibly going up to galactic and intergalactic distances. On the other hand, both electromagnetism and gravitation can describe long-range interactions. The idea here is rather to develop a simple continuum model. We do not start from particles and their properties but we directly consider the action-at-a-distance field, and even the large galactic scales point of view of physical interactions as gravitation and electromagnetism. *In fine* we are interested in looking more in depth at the vacuum spacetime constitution and its geometrical properties.

[1]Publisher's Note: The author uses the term *relative gravitation* throughout this book. Readers unfamiliar with the term might know it better as *general relativity*.

Spacetime is the quintessential example of a continuous medium. According to Descartes' view, and subsequently Newton and Maxwell, and generally speaking, the view of modern physics on the quantum mechanical approach, spacetime is flat, and therefore physical phenomena are ingredients introduced to "populate" this vacuum spacetime. Later, despite the conception of quantum vacuum, which attributes an energy to each point-event in the spacetime, the scientific community notes the resilience of the fundamental difference between quantum mechanics and relative gravitation, because the latter directly identifies physics in the spacetime geometry, following Einstein's view. The unification of quantum mechanics with relative gravitation still faces this fundamental difference. Along with the correspondence between gravitation and curvature of spacetime, presence of an electromagnetic field in a vacuum naturally calls for the extension of the geometry of vacuum spacetime.

The main area of this book can therefore rightly be considered as the physics of generalized continuum. Indeed, we base, almost exclusively, the entire development of this work on a model whose geometric structure rests on the existence of two fields on a differentiable manifold: metric tensor and affine connection. These two variables simply result from the original topological definition of a continuum as being a compact and connected differential manifold.

This monograph is designed for scientists and graduate students in physics, theoretical physics, mechanics and mathematical physics. It may be considered as a third volume, although self-consistent, following two previous books: *A Geometric Approach to the Thermomechanics of Dissipating Continua*, 2003 dealing with the thermomechanics of generalized continuum and *Covariance and Gauge Invariance in Continuum Physics*, 2018. In the same stream the present book is also essentially based on Riemann-Cartan geometry. The main subject of the present work concerns vacuum spacetime models and their influence on cosmology. The methodology rests on the application of variational calculus which were voluntarily omitted in these previous books and then gives an alternative way to model important issues of the acceleration of the Universe's expansion. Indeed, after the discovery of this accelerated expansion, hypothetical existence of dark matter and dark energy, which would constitute nearly 95% of mass and energy of the Universe, is nowadays commonly admitted. Dark energy as well as dark matter are unknown. An approach based on the generalization of the vacuum geometry seems a valuable tool to better understand this problem, which is probably nowadays the biggest problem in fundamental physics. The present work gives, on the one hand, some insights to link relative gravitation and generalized continuum, and on the other hand, also indicates some possible way to include electromagnetism in the geometry structure of the spacetime. For that purpose, both gravitation and electromagnetism are considered in this monograph particularly for they mutually interact with a vacuum spacetime. A formal link between torsion tensor and electromagnetic potential is mathematically derived by means of variational calculus. Being a generalized continuum, we consider the spacetime as sets of microcosms allowing the incompatibilities of scalar and vector fields on manifold. This formally implies the necessity of introducing independent affine connection in addition to metric tensor. For completeness, the vacuum model is thus defined by a Lagrangian depending on these variables. The general approach is *in fine* applied to relative gravitation, electromagnetism, and then to cosmology for obtaining first the field equations, and second the conservation and constitutive laws.

From a synthetic point of view, the common thread we follow all along this book is to pursue the eternal quest for a vacuum spacetime which includes the following progressive steps for the sake of clarity:

(1) Consider flat *Euclidean spacetime* of Newton's classical mechanics where time and space are absolute and uncoupled to give Galilean spacetime, and for which physical phenomena are added to enrich the overall model (e.g., gravitational potential);
(2) Recall briefly *Minkowski spacetime* which couples time and space, thus allowing compatibility of Newton's mechanics and Maxwell's electromagnetism and by extension requires a formulation of physics in four dimensions while remaining flat (physical interactions are also added ingredients to enrich the model);
(3) Consider *Einstein spacetime*, using Riemannian geometry and where time remains coupled with space, which defines a formal framework to include gravitation as an element of the vacuum geometry;
(4) Generalize to an *Einstein-Cartan spacetime* which allows us to reduce the number of added physical ingredients to a minimum, that is tentative to include the two fundamental interactions (gravitation and electromagnetism) in the vacuum geometry.

Since we are interested in physical interactions occurring in very large domains of the Universe, we have deliberately restricted the subject of study to the continuum physics of the cosmological vacuum (infinitely large). Another step would have been to consider the physical phenomenon that occur on the smallest scales and thus include the weak and strong nuclear forces (infinitely small), therefore at the quantum level. But the current state of physics does not yet allow us to link gravitation and quantum physics in a satisfactory way, hence the reason for limiting ourselves to classical fields and to work with density of variables conforming to a continuum approach. We nevertheless believe that the descent of models towards the small microcosms of the Universe (Cartan path disclosure on Riemann-Cartan manifold), involving linear and surface defects of spacetime, which are embodied by curvature and torsion fields, might bring new insights towards the modeling of loop quantum gravitation. The book includes two parts.

(1) The first part presents some mathematical elements, particularly differential geometry of manifolds, and defines the concept of generalized continuum with the formulation of generic conservation laws.
(2) The second part deals with applications to cosmology including gravitation and electromagnetism. This part gives the field equations of gravito-electromagnetism and provides the formal link between physics (gravitation and electromagnetism) and spacetime geometry (metric, curvature and torsion). An original guess to give a possible explanation of dark energy in cosmology is proposed.

Acknowledgements. First of all, I wish to thank all my colleagues and former students (Polytechnic Institute, University of Antsiranana, Swiss Federal Institute of Technology — Lausanne, University of Lausanne, and University of Rennes) with whom I have worked during these last 45 years. The teachers and professors I had during my studies were the first to shape my vision of science in particular and my way of approaching problems in general. May they be thanked by this grateful testimony. My job as a professor over the last few decades has allowed me to give lectures and discuss with students; this permanent exchange has allowed me to constantly improve my understanding of the areas I have taught. I will never stop sincerely thanking them even if I cannot name them individually. They will recognize themselves. I am grateful to the editing staff at World Scientific for their help and guidance during the writing of this book. Last but not least, I particularly want to thank my wife Oly, and my children Rindra, Herinarivo, and Haga for their understanding and their constant encouragement.

Rennes, 25 December 2025
Lalaonirina Rakotomanana Ravelonarivo

About the Author

Dr Lalaonirina Rakotomanana Ravelonarivo is Professor Emeritus at the University of Rennes, France. Between 2002 and 2013, he was Head of Mechanics at Rennes Institute of Mathematical Research, before becoming Director of the Mathematics Department from 2013 to 2016. He has over 200 peer-reviewed publications. He holds a PhD from Swiss Federal Institute of Technology Lausanne (EPFL), Switzerland.

Contents

PART 1
Generalized Continuum: Mathematical Background

Chapter 1

General Introduction

Throughout its history, humanity had developed concepts to describe the heavens, stars and planets to better understand the Universe as a whole. Babylonian and Egyptian astronomers had particularly developed systems for ordering and understanding the heavens, which significantly enriched the foundations of ancient Greek astronomy. This was continued by Islamic astronomers to refine and complete foundations of the early modern European astronomy.

1.1 Brief Historical Introduction

The cosmology documented by Western civilization draws its scientific foundations from the concepts of Aristotle (384–322 BC) and Ptolemy (≃100–168). If Aristotle had essentially put forward the fundamental models of physics to describe the Universe, Ptolemy on the other hand had gone rather towards the mathematical foundations in order to describe the physical reality describing this Universe.[1] Islamic astronomers produced a

[1]Among a dozen scientific treatises written by Ptolemy, the *Almagest*, originally entitled *Mathematical Treatise*, is considered as the foundational written document on mathematical astronomy and cosmology, although restricted to its main assumption on geocentrism. We pass over the controversies concerning the possible plagiarism of Ptolemy accused of having copied the work of astronomer and mathematician Hipparchus (≃190–120 BC), whose writings had already disappeared. Nevertheless, without the *Almagest*, including 13 books, most of the knowledge of ancient Greece on astronomy would not have reached us. It is thanks to *Almagest* that Hipparchus' stature as one of the greatest astronomers of ancient Greece was recognized. Hipparchus himself probably compiled an amount of Babylonian astronomical observations to write his treatises.

large amount of works between the 8th and the 15th centuries by improving the Plotemaic concept and observations. Afraganus wrote a quite non-mathematical presentation of Ptolemy's *Almagest* around 833 entitled *Elements of Astronomy on the Celestial Motions*. Some other astronomers had also contributed to improve and propose a basis of cosmology, but Abu Mashar probably is the most known since his works played an essential role in translating Aristotle's ideas either in the Islamic world or later in European countries. Among the most important contributions between the 16th and 17th centuries, we must mention the respective works of Copernicus' *On the Revolutions of the Heavenly Spheres* in 1543, Brahe's *Rudolphine Tables* with Kepler, *De Nova Stella* in 1573, and Kepler's *The Cosmographic Mystery* in 1596 and *The Harmony of the World* in 1619. Then came the numerous thinkers (non-exhaustive list) on the cosmology theories who will be presented as they contribute throughout this work.

Like most scientific theories, cosmology, which by definition is the science which studies the origin and evolution of the Universe over time, is based on mathematical modeling and on experimental observations and measurements. These two aspects are of essential and comparable importance. They constitute the two crutches that allow us to advance in the detailed understanding of our Universe. We do not venture to speak of other universes which may exist in parallel, but we will be content to approach only that part which could be accessible to our imagination. In the present book, the object of our study is limited to those portions of the Universe between galaxies or aggregates of galaxies which we call vacuum spacetime in which we note that some physical actions can be transmitted even in the absence of matter.

From the outset, it seems appropriate to mention that four-dimensional spacetime is not just a theoretician's artifice but rather represents the fundamental concept and indispensable tool of physics, particularly in cosmology e.g. [1]. The notion of three-dimensional space, even if it seems very intuitive, does not exist outside of us as observers of physical phenomena, and more precisely this intuitive notion is closely linked to the fact that we implicitly admit that the speed of light is infinite. This three-dimensional space, a primitive concept and cornerstone of Newton's approach to classical mechanics, is indeed an image of the world (Universe) by the asymptotic state at rest e.g. [2]. Indeed, this three-dimensional space exists for a given observer if he admits the fact that the universe around him is visible to him at once in its entirety at a given instant t. The long path of the development of any physical theory is therefore subject to this acceptance

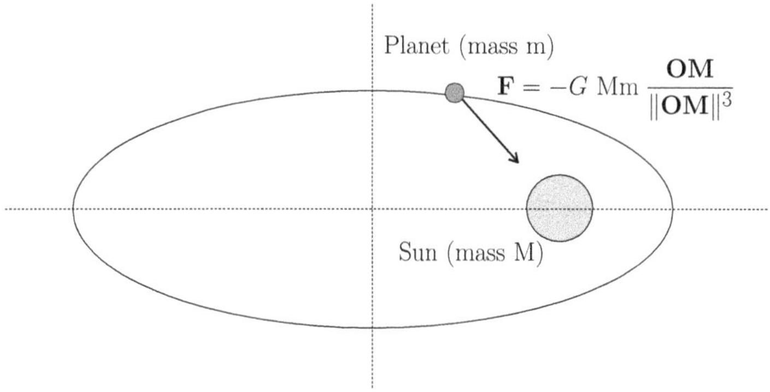

Fig. 1.1: Gravitation: Example of action-at-a-distance force (unmediated) exerted by the Sun to a planet. Newton proposed the universal gravitational force $\mathbf{F}$. The trajectory of the planet is deduced among other actions to this force.

or not of three-dimensional space decoupled from the flow of time. Newton's approach to spacetime, better formalized by Galileo with his group of transformations and inertial frames of reference, corresponds to a static Universe with a succession of sets of an infinite three-dimensional space, but the physical world is in reality in constant evolution in different parts of the Universe and each part has its own evolution.

Therefore, the concept of action-at-a-distance is intimately linked to the concept of spacetime, the latter being the playing medium of the different actions that are present in physics. The study of action-at-a-distance (Fig. 1.1) cannot be dissociated from analytical reflections concerning this playing field that is spacetime. And as we will see throughout this work, the two physical forces that interest us are those which have an infinite radius of action, that is to say gravitation and electromagnetism. From the outset we remain in a classic framework and we deliberately leave aside very important phenomena which occur at very small distances such as the quantum physics of elementary particles.

From different points of view, the notion of space and time has always been the object of interest of scientists for millennia. In the early period of natural philosophy, there was quite a general agreement that excluded the possibility of physical action-at-a-distance without intermediate medium e.g. [3]. Aristotle and Descartes for instance rejected any possibility of any action-at-a-distance. Nevertheless daily observations showed that some physical phenomena already presented exceptions such as light propagation and magnetic phenomena.

One further phenomenon that obeys this principle is the ebb and flow of the tides which already was a matter of concern as early as in 850[2] by Albumashar and by Grossteste in 1227, the latter attributing the tides to light. Of course, planetary motion was a classical battlefield for the pro and contra of the unmediated physical action. In 1609 Kepler was one of the pioneers to propose the existence of unmediated action for describing the movements of planets, and among others showed that Mars's orbit is not a combination of pre-established geometrical circles but rather has an elliptical shape resulting from physical laws. Namely Kepler found confirmation that the mutual attraction of terrestrial and lunar together with the Sun's gravity was responsible for the tides of terrestrial oceans. Eighty years later in 1687, Newton's gravitation was formalized with his *Principia* and opened a broad path for the *spacetime* concept (even for him time is uncoupled from space). Modern physics was born with the Galilean spacetimes, the class of referential frames having constant and uniform translation with respect to each other. In his later career, Newton himself suggested the existence of an aether consisting of particles with very high elasticity to transfer gravitational force between matter. The possibility of action-at-a-distance is closely related to the existence of vacuum spacetime. The debate on the existence of unmediated action still remains nowadays even if the form of the debate has changed and it finds its essence about the construction of the model of the Universe in cosmology.

The construction of a scientific model of cosmology therefore relies not only on mathematics to define the geometric variables which describe this spacetime but also on observations with experimental measurements to corroborate the mathematical models thus designed. Briefly speaking, cosmological models, and to a lesser extent spacetime models, are based on interpretations of observations and measurements of the movements and transformations of planets, galaxies, aggregates of galaxies and superclusters, at different distances and in different past eras of the evolution of the Universe. On the one hand, on a mathematical level, the condition of consistency of the theory is to respect the hypotheses accepted following physical observations while avoiding as much as possible the addition during development of other hypotheses which could contradict those initially admitted. On the other hand, the physical consistency of the theory is to be able to describe real physical phenomena as best as possible using mathematical models. The two "consistencies" are essential to make it an acceptable

[2] *The Great Introduction to the Science of Astrology* (Abu Ma'shar, 850)

theory, which of course will not be immune from questioning following new observations or more detailed experimental measurements which could lead to new concepts. This is the scientific dialectic. This book is written not to claim to build models of cosmology and not even of relative gravitation. Far from it, it provides a view to analyzing all the possible consequences that can be drawn from an initially accepted geometric structure, as simple as possible. The book therefore remains limited to a theoretical study of theoretical consistency and does not claim an experimental confrontation, other than those which already exist in the literature.

1.2 Spacetime and Fields

For a brief recall of the history of gravitation, Newton admitted an absolute space and a time uncoupled from a Cartesian three-dimensional space for deriving the famous Newton's laws in mechanics. Einstein then considered coupled spacetime thanks to his theory of special relativity to reconcile Newton's mechanics and Maxwell's theory of electromagnetism, resulting in a Minkowski spacetime. Basically, the concept of spacetime is tightly related to the concept of fields e.g. [3]. Then to account for gravitational forces, Einstein proposed a space and a time which are always coupled and merged to a spacetime, but which presents geometrical properties other than Euclidean, by considering its curvature. Then came the concept of Einstein-Cartan spacetime curved but also with torsion within the framework of the geometry of Riemann and Cartan, for instance in cosmology theory e.g. [4]. Since then, various types of spacetime have been proposed to solve specific problems, and particularly for quantum physics. Indeed, space and time are the basis of all physical theory and particularly relative gravitation [2]. Complexity of physical phenomenon related to the vacuum spacetime is more or less captured by the introduction of additional mainly scalar but possibly tensorial mathematical variables.

Taking a short path from a geometric point of view, knowing that spacetime is basically defined as a continuum (i.e., a compact and connected manifold), we thus can say that Newton's approach to spacetime is based on Euclidean space equipped with a uniform metric throughout the volume of the Universe. Einstein's approach, on the other hand, considers an Einstein spacetime whose metric is not uniform but depends on the event (point of space and instant). This spacetime admits a metric connection and also a non-zero curvature field. Finally the "natural" extension of Einstein gravitation on a geometric level consists of considering the Einstein-Cartan spacetime by taking a metric and an independent connection compatible with that metric e.g. [5]. Considering an independent connection means introduction of a new variable which is the torsion. This approach using the properties of the manifold can be classified as classical (in the sense of non-quantum), and can be justified by the logic of first constructing the environment where the physical phenomena will take place and second adding to it subsequently the mathematical variables which will define these phenomena. This approach is common, for example, in thermomechanics and physics of continuous media e.g. [6]. This is probably not the only approach, but it is the one we take in this book. In any case, the old problem of defining vacuum spacetime has remained a research motivation on relative gravitation in particular and in relativistic physics in general. Remaining within the classical approach, the consideration of electromagnetic phenomena coupled with gravitation, even if their geometrization draws back more than a century old, still remains relevant, particularly with the discoveries relating to the expansion of the Universe.

Let's have a short digression on the reason we are interested in Riemann-Cartan geometry. The Riemann curvature tensor, which characterizes the Einstein gravitational field, and the Faraday electromagnetic tensor, including the electric and magnetic fields, which measure the electromagnetic field in a vacuum, seem to play different roles in classical physics, limiting ourselves to only the two fundamental forces that can act at a distance.

For a long time, numerous attempts have been made to unify these models, including both phenomena, into a single theoretical model. The first step was to consider the existence of an electromagnetic field in a curved space with non-zero curvature. This method remains a mixed approach that is not entirely geometric.

Indeed, two different points of view emerged. In the first, spacetime is considered only as an arena where physical phenomena take place. Physical variables are then introduced ad hoc once spacetime is defined. For the second approach, spacetime itself becomes the source of all physical phenomena. Matter, electric charges, and possibly magnetic charges reflect manifestations of this spacetime e.g. [7]. For this second approach, gravitation was successfully incorporated into curved spacetime in Einstein's theory of relative gravitation. The incorporation of the electromagnetic field as a manifestation of spacetime still presents some problems; in any case, there is not yet a consensus like there is for gravitation. A theoretical approach that consists of showing that the curvature of spacetime is also sufficient to understand electromagnetic phenomena dates back to the early work of Rainich [8] — contemporaneous with the introduction of Cartan geometry — and in the 1950s with the geometrization of classical physics e.g. [7]. These works seem to conclude that Riemannian geometry is sufficient to account for electromagnetism. This would have effectively been a unification of the two existing actions-at-a-distance. Nevertheless, a closer reading shows that this approach implicitly considers a multiply connected spacetime inspired by Volterra's work on dislocations, and thus is closer to Riemann-Cartan manifolds. The concept of generalized continuous media can actually be appreciated in very different ways [9]. One possible option is to start from the enrichment of the underlying geometry. For so-called second gradient materials, which come from the modeling of continuously distributed defects, the use of compact and connected manifolds leads to Riemann-Cartan manifolds. An interesting link comes from the definition of a medium (spacetime) called Planck-Kleinert crystals [10] which is nothing other than a special case of generalized continuous media e.g. [11] but with a Planck length scale $\ell_P \simeq 10^{-35}$ $[m]$. This is one of the main reasons we immediately place ourselves in a broader geometry by considering spacetime as a Riemann-Cartan manifold, and thus an Einstein-Cartan spacetime: a continuum equipped with a metric and an independent connection with non-zero torsion.

Soon after the field equations of relative gravitation were developed in 1915, Einstein added a parameter Λ, called *cosmological constant*, to ensure static spacetime solutions of his field equations. The Standard Model of cosmology named FLRW spacetime model with evolving but uniform curvature admitting isotropy and homogeneity of the Universe was developed independently by Friedmann (1922), Lemaître (1927), Roberston (1929), and Walker (1935) to account for the continuous expansion of the Universe conforming to astronomical observations of Hubble [12]. In 1932, Einstein and de Sitter also proposed a spacetime model of a continuously expanding Universe with $\Lambda \neq 0$. Theoretically, the constant Λ may have positive or negative values; however, this value is very small $\Lambda \simeq 10^{-52}[m^2]$. Behind this concept of field was hidden implicitly but effectively the very delicate notion of empty space. The transmission of gravitational forces (and at the same time of electromagnetic waves) without the existence of intermediate matter had remained for millennia a fundamental problem for physicists e.g. [3]. The empty spacetime becomes more and more a vacuum spacetime, meaning the existence of the energy of emptiness are not exactly known at the present time, although recent scientific advances in the field of quantum physics, particularly quantum electrodynamics, have been able to determine the fluctuations in the electromagnetic field of what is called quantum vacuum e.g. [13, 14].

There are many hypotheses ranging from relative gravitation to quantum mechanics and physics of particles related to the so-called cosmological constant e.g. [15, 16]. Using opto-electronic sampling, these theoretical and experimental discoveries provide access to zero-point fluctuations in vacuum spacetime. This side of research opens the way to link macroscopic electromagnetism with quantum physics within the framework of quantum electrodynamics e.g. [17]. At the particle level, transmission of gravitational and electromagnetic forces are associated with gravitons (although it remains hypothetical to a large extent) with mass $m = 0$ and spin 1, and photons with mass $m = 0$ and spin 2. The range of action of gravitational and electromagnetic forces is infinite, contrary to strong and weak nuclear forces, respectively $10^{-15}[m]$ and $10^{-18}[m]$. At the human length scale level, only the gravitational and electromagnetic forces can be considered as action-at-a-distance. It should be mentioned that quantum effects are usually relevant at small distances as per the Planck length $\ell_P \simeq 10^{-35}[m]$; at larger length scales, they may be significant but only in certain limits at low temperature. Throughout this book, the main motivation will be to find a link between electromagnetism, gravitation and the cosmological constant.

1.3 Cosmological Constant, Dark Matter and Dark Energy

The discovery that the Universe is not static but expanding, first shown by Hubble, and that this expansion is accelerating, as demonstrated by Riess and coworkers [18], and independently by Perlmutter *et al.* [19], constitutes a major turning point in the theory of cosmology and in our understanding of vacuum spacetime (see review in [15]). What is actually observed is that the repulsive gravitational force generating the accelerating expansion is greater than the attractive gravitational force due to the presence of matter. Several decades after the observational discovery of the accelerated expansion of the Universe, the use of the cosmological constant Λ still remains relevant despite being essentially based on the simplifying hypotheses of isotropy and homogeneity of spacetime, which remain the fundamental principles of cosmology. This tool Λ, which has the advantage of having a gravitational interpretation as a force of repulsion, nevertheless remains phenomenological and has raised several questions including its extremely low value close to zero; the search for its physical origin still remains and has sparked several great projects e.g. [20]. If all these concepts are correct, there are three major contributions in cosmology: (a) the regular "visible" baryonic matter, made of electron and quarks, (b) the dark matter of unknown composition and (c) the dark energy which could be assumed to be relics of photons. Although postulation of this dark energy is convenient to give an explanation of the accelerating expansion, the exact nature of this dark energy (and dark matter to a lesser extent) is a matter of speculation. Its existence, in whatever form, is needed to reconcile measurements of the Universe's expansion with its matter contents.

On the one hand, the observations show that the magnitude of the gravitational attraction requires the introduction of dark matter (attractive force), and on the other hand, the expansion is generally attributed to the hypothetical existence of a dark energy that causes this expansion with acceleration (repulsive force) e.g. [21, 22]. A physical quantity which plays a central role in the quest for dark matter and dark energy (e.g., [23]) is again the cosmological constant, denoted Λ (e.g., [24]). It constitutes a central piece in the construction of the gravitational theories of cosmology. Beyond the possible but legitimate question on the merits of the conceptual hypotheses of dark matter and dark energy which seem to be unanimously accepted by contemporary cosmology researchers, we must also ask ourselves some subsidiary questions on the real nature of this dark matter and dark energy and what are the interrelations between these two entities. The question on the existence of a positive cosmological constant (contraction of the Universe, be cautious with the sign convention), or better on the simultaneous existence of regions in contraction with regions in expansion (inhomogeneous Universe), also remains relevant e.g. [25]. The relationship between the cosmological constant Λ and, more generally, dark matter and dark energy with quantum physics is at the forefront of efforts to explain phenomena at the level of quantum physics. The relevance of current models, even those remaining within non-quantum general relativity but becoming increasingly complex, needs to be clarified, particularly with regard to their coupling to electromagnetism, including the notion of magnetic monopoles (e.g., [26, 27]), whose theoretical existence in terms of the magnetic charge was first proposed in 1931 within the quantum domain by the founding work of Dirac [28].

The search for cosmological models capable of explaining the evolution of the Universe and conforming to experimental measurements is still, and probably will remain, in progress knowing, however, that we nowadays face a very challenging problem due to this accelerating expansion. It is an integral part of the concept of spacetime e.g. [16] and shown as early as the

1960s to involve also quantum physical particles [29]. Recent results have demonstrated the relevance of the hypothesis of fluctuations in the electromagnetic vacuum energy, not only theoretically but also supported by experimental measurements. These results were obtained through electro-optic sampling techniques within the framework of quantum electrodynamics; see [13] for the theoretical basis and [17] for applications. This provides direct access to zero-point fluctuations, which is the basic concept of vacuum quantum energy. Roughly speaking, the cosmological constant Λ arises from the vacuum spacetime energy. In some sense, the vacuum is not an empty spacetime but it has its proper energy independently on the baryonic matter evolving within the vacuum. This point is central, although indirectly, for the motivation of the present work.

The cosmological constant problem comes from the fact that the theoretical expectations exceed experimental measurement limits by some 120 orders of magnitude. At least five approaches are possible to solve this problem: the tools of supersymmetry and superstrings, the anthropic model (which is more a state-of-mind), the adjustment mechanisms, the methods of modified gravity theories, and the quantum cosmology approach; see review with details of each method in e.g. [30]. All of them might be developed on their own and constitute branches of advanced theories of cosmology with a special mention to the quantum approach according to this review.

From these discoveries, the only evidence of dark matter and dark energy is *a priori* observed by their influence on cosmological gravitation. It then seems natural, even if this approach may seem elementary compared to the approach with quantum particle physics, that the first step would be to assume that their origins could be estimated with a phenomenological approach e.g. [23]. This has motivated numerous researchers to consider the cosmological constant as an integrative variable to model the phenomena. In this framework, a coupled study of the gravitational field combined with the electromagnetic field as a spacetime background could be chosen as a starting point. Indeed, the widely accepted, so-called ΛCDM (Cold Dark Matter) model is obtained by adding the cosmological constant to the standard Friedmann-Robertson-Lemaître-Walker (FLRW) model in cosmology. Therefore, the concept of the cosmological constant constitutes a cornerstone for the putative existence of dark energy and dark matter which drive the accelerating expansion of the Universe e.g. [21]. However, recent models, notably what is called $CCC+TL$ which is an extension of the standard model that introduces Constant Covarying Coupling in such a way that the

FLRW metric is weighted by a time-dependent scalar function with exponential behavior proposed by e.g. [31], seems to make it possible to dispense with the notions of dark energy and dark matter while maintaining consistency with experimental data. The confrontation of different models and their long-term unification constitutes probably one of the numerous challenges to be met in the coming years.

1.4 Spacetimes vs. Geodesics and Autoparallels

In accordance with Newton's formulation of the laws of mechanics, Painlevé issued in 1889 the *Axiomes de la Mécanique* by categorizing them into three groups. The first gives the axioms on the topology of three-dimensional (Cartesian) space and time (a continuous linear one-dimensional space) which is synthesized into a concept of four-dimensional spacetime later in relativity. The second group defines the axioms concerning the fundamental transformations on the relative displacements/movements of spacetimes, thus defining the concept of inertial systems which reduces to the group of Galileo transformations (inertial systems of references are in relative translation with respect to each other) in Newtonian mechanics. The geodesic curves in these spacetimes without "external forces" are then straight lines. Finally, the last group includes the specifically mechanical axioms which are summarized in the notion of force (concretized by Newton's second law in classical mechanics of the point) when the trajectory in spacetime (world line) no longer follows these geodesics. It is the introduction of this force in Galilean spacetime which makes it possible to measure the deviation of a point in the Universe with respect to a world line, associated with the group of Galilean transformations with 10 parameters in Newtonian mechanics.

Therefore, it is common to analyze spacetime by the characteristics of its geodesics, the paths traveled by the hypothetical test particles without external forces applied. Indeed, given an arbitrary spacetime, the analysis of geodesics and autoparallels, also called world lines, constitutes a common method to define the geometrical and physical interpretation of this spacetime as an extension of the well-known geodesic deviation equation which was first derived by Singe [32] and independently by Levi-Civita [33] for Einstein spacetime in relative gravitation. Since then, the analysis of the geodesic deviation equation has been generalized to include other physical phenomena, such as electromagnetism e.g. [34] and also towards more complicated spacetimes of larger dimensions e.g. [35].

The present work is far from pretending to provide an ambitious model for cosmology. It only considers a background spacetime, which we call vacuum spacetime, meaning an absence of matter, or even of particles. Cosmological models are mostly admitting the presence of fluid to explain dark energy and dark matter e.g. [15, 36]. In the present work, we restrict ourselves to a macroscopic analysis, assuming an extended geometric structure in the form of Einstein-Cartan spacetime e.g. [37–39], and avoid the aspects which call upon quantum physics. It should also be highlighted that the two physical fields that interest us in this book concern gravitation and electromagnetism which, it should be remembered, are interactions whose radiuses of action are theoretically infinite, unlike strong and weak nuclear fundamental forces with radii of action of the order of $10^{-15}[m]$ and $10^{-18}[m]$. Theory and practice of these fields mostly consist in analyzing the trajectories and movement of test particles which are subject only to the influences of these fields; trajectories are called geodesics or in some cases autoparallels e.g. [34, 40–45], and many other references.

Analyses of geodesics and their deviation remain an intensive field of research from both theoretical aspects and for practical applications such as the development of methods for following planets e.g. [45] or any flying objects and in the Global Positioning System e.g. [46]. In such a case, the knowledge of geodesic deviation may be a valuable tool in astronomy. For macroscopic objects, often considered as test particles, two fields are usually present in nature: gravitation and electromagnetism. Depending on the required precision and speed of the test particle, gravitational fields can be classified as Newtonian, Einsteinian, and even Einstein-Cartan gravitation in some sense. The use of the latter has gained interest in cosmology e.g. [4].

Over the years, torsion, mainly introduced by Cartan [47], becomes more and more suggested to intervene in the modeling of continuum physics at various length scales (see recent reviews e.g. [48]) not only in continuum thermomechanics e.g. [49] but also in continuum physics e.g. [50–53]. Extension of relative gravitation from Riemann geometry to Riemann-Cartan geometry therefore arose quite naturally. The use of Einstein-Cartan spacetime is increasingly used in cosmology for different reasons, notably the inclusion of electromagnetism, spin fluids, and even more recently in the geometric modeling of energy. Role of torsion in the modeling of dark matter and dark matter may be found in e.g. [54]. The determination of the torsion field in certain studies nevertheless rests on purely geometric restrictive hypotheses which assume that the geodesics must be confused with the autoparallels. This aspect deserves to be better clarified because it is ultimately not based on any physical principle, as shown in the case of an electromagnetic field coupled with gravitation e.g. [34], where it is necessary to introduce an additional physical source other than gravitation.

1.5 Gravitation and Electromagnetism

As well as special relativity, the history of the theory of relative gravitation is closely linked to that of electromagnetism. Despite the uncertainties as to the exact origin of electromagnetic fields in the Universe, primordial or coming from plasmas, these fields play an essential role in the evolution of the Universe, in particular for its structuring into galaxies, clusters, and filaments (see for review e.g. [55, 56]). The introduction of electromagnetic fields in addition to gravitation constitutes a key point of this book, in addition to the extension of the spacetime geometry. In short, the geometrization of gravitation linked to electromagnetism has historically been associated with the names of Einstein, Reichenbach e.g. [57, 58] and Weyl to varying degrees e.g. [59]. Indeed, the advent of Maxwell's theory of electromagnetism led to Einstein's special relativity. Namely, the finiteness of the speed of light signals, a particular case of electromagnetic waves, was one of the main motivations for the development of special relativity. This also led to the genesis of the theory of relative gravitation to unify the two theories well before quantum particle physics. Newton's gravitation theory is based on the idea that all masses generate an attractive gravitational force instantaneously. Gravitational force in Newton's theory can travel faster than the speed of light. Most advances in astronomy are obtained by Newtonian mechanics, including the tracking of satellites or planets with acceptable accuracy. For instance, complete solutions of the geodesic structure in this context have been largely obtained e.g. [43]. Einstein's theory of gravitation replaces forces by curvature of spacetime where it propagates with a certain speed as waves. Gravitational waves of finite speed are vibrations of spacetime generated by the acceleration of objects in the Universe. Their existence was theoretically predicted more than 100 years ago by Albert Einstein but the measurement confirming their definite existence was announced in [60]. The accounting of fluid spin distributed within spacetime suggested the introduction of torsion tensor in the gravitation models e.g. [61]. Furthermore, there is evidence that spacetime geometry may include additional features such as curvature and torsion, particularly when both gravitational and electromagnetic fields are present, as suggested in cosmological vacuum spacetime e.g. [62].

The influence of electromagnetism on gravitation, both theoretically and experimentally, cannot be overlooked and must be incorporated into the model, including for applications in quantum physics. Staying at the macroscopic level, the first step is to consider electromagnetism and particularly waves since most if not all information for analyzing the evolution of the Universe involving gravitation are obtained by means of electromagnetic waves e.g. [63, 64]. Electromagnetic signals are known to play a fundamental role in relativity and particularly when measuring distance in spacetime e.g. [65]. What is commonly accepted is that magnetic fields have played and still play a role in structuring the Universe e.g. [66]. A review of the mechanism of production of these fields before and during the decoupling of photons with matter can be seen in e.g. [56] and can be considered a *Galactic Dynamo* or *Primordial* process. Roughly speaking, a dynamo process consists of a conversion of turbulent kinetic energy from electrically charged conducting fluids to magnetic energy. The primordial existence of magnetic fields results directly from a pre-existing field which gets adiabatically compressed when the galactic cloud collapses. These pre-galactic fields are not necessarily produced in the early Universe. Regardless of its origin, it is widely accepted that magnetic fields play an important role in the structuring of the Universe. Their significance, which increases over time due to amplification, justifies considering them in cosmological models of spacetime e.g. [67]. Investigating the respective roles of the initial distribution of magnetic fields — whether homogeneous or stochastic — and their helicity in the evolution of the primordial magnetic fields, particularly in relation to large-scale structure formation, remains meaningful e.g. [66].

There are also speculations concerning the appearance of primordial magnetic helicities. Different positions exist, notably those who think that helicities may not have existed from the origin of the Universe but came later through mechanisms such as the Biermann battery effect. Some theories attribute their emergence to shocks during the early stages of the Universe, to the electroweak or quantum electrodynamics phase transitions, or simply due to the initial inflation of the Universe. One thing is almost certain: it was not possible to have a homogeneous electromagnetic field at the genesis of the Universe and certain inhomogeneity structures were already there. These uncertainties add a lot of vagueness as to the existence or not of primordial magnetic helicities, and the role of the Cherns-Simons current is advocated e.g. [68]. True to the perspective of this book, we lean towards a more geometric approach to what empty spacetime could be, the basis of all theories concerning astrophysics and cosmology. In any case, electromagnetic waves, propagating in a vacuum, a zone devoid of matter, have an invariant characteristic which can be approached using the concept of the photon wave function e.g. [69]. This invariant is optical helicity, which includes magnetic helicity and electric helicity, and which allows the first bridges to be built between Maxwell classical electromagnetism and the notion of zero-point energy fluctuation of quantum vacuum e.g. [70–72] by rewriting Maxwell's equations in the form of a Schrödinger equation e.g. [73–75]. In an Euclidean spacetime and even an Einstein spacetime, the invariance of optical helicity was established; the question is the validity of this result within an extended spacetime. For a thorough study of empty spacetime, the temptation to immediately apply the quantum approach remains valid when considering regions with increasingly smaller dimensions. For this, one could add external physical fields, namely particles to describe quantum physics, but it is known that to complete the quantum ontology, taking into account the electromagnetic field is sufficient to arrive at this approach e.g. [77], at least for the objective we have set ourselves. The gravitational field together with electromagnetic field are known to be influenced by this geometry structure including metric and affine connection of the spacetime e.g. [78]. One of the most important consequences of the gravitation-electromagnetism interaction is the bending of a light beam in an intense gravitational field. This phenomenon is one of the key experimental confirmations of Einstein's theory of relative gravitation and was observed by Dyson, Eddington, and Davidson [79]. It is now understood that the curvature of vacuum spacetime, which represents the gravitational field, is responsible for this bending.

Sixteen years after the theoretical confirmation of light deflection in an intense gravitational field, a publication by Euler showed that photons can scatter off each other [80]. His work with Heisenberg demonstrated that light, which is an electromagnetic wave, can interact with a strong magnetic field within the framework of quantum physics [81]. They suggested as a theoretical basis for the vacuum polarization what is now called the Euler-Heisenberg Lagrangian. What is curious then is that the electromagnetic field, like the gravitational field, also influences the propagation of light beams. Transverse magnetic fields modify the light propagation in a vacuum spacetime, as discovered experimentally by [83]. This was done with the objective of detecting if a photon has a magnetic moment. This aspect is less popular than the deflection of a light beam by gravitation and is called vacuum birefringence e.g. [84], which is playing a role in cosmology e.g. [71, 85] and more accurately on the similacrum of dark energy e.g. [86, 87], where the trace of Maxwell stress-energy tensor deduced from nonlinear electrodynamics based on Euler-Heisenberg Lagrangian was explicitly linked to dark energy. One of the questions that the present study therefore asks is the geometric counterpart of spacetime which would contribute to this phenomenon. The role of torsion via the spin of fluid has been studied for many years e.g. [61]. Such a model is necessary in the presence of electrically charged spinning fluids and the torsion fields they generate. Considering torsion in the absence of fluid may also be of interest, as it allows the separation of torsion intrinsic to spacetime from torsion generated by matter-spin and from torsion arising from minimal fermion coupling e.g. [38]. The Einstein–Cartan spacetime provides a valuable geometric framework, as it naturally incorporates other phenomena related to torsion, including fermionic and bosonic fields, by accounting for Dirac fermion fields e.g. [88].

The interweaving of Riemann-Cartan geometry, electromagnetism and gravitation via cosmological constant Λ increasingly shows the close interrelation between these three domains e.g. [4, 38, 62]. Interestingly, dark matter is suggested to be generated by torsion in the Friedmann cosmology e.g. [89]. Torsion is also related to the possible emergence of magnetic monopoles, which may be considered as defect fields of spacetime e.g. [90]. On one hand, it is established that the torsion field on a manifold is a fundamental geometrical entity closely associated with defects in an otherwise uniform continuum e.g. [49], a concept that can be extended to defect fields within the spacetime continuum e.g. [90], particularly in cosmological models of the Universe e.g. [4]. On the other hand, the torsion field, representing distributed defects within spacetime, can also be linked to the existence of magnetic monopoles e.g. [90, 91]. This leads naturally to the conclusion that the torsion field is closely interconnected with the magnetic monopole field e.g. [54, 92]. A recent review may be found in e.g. [26], whose authors discuss the developments of theory and experimental/observational aspects of magnetic monopoles either for colliders or in the cosmos. Their main conclusion was that magnetic monopoles can potentially play a role as a dark matter candidate. From a rather simple example, the role of magnetic monopoles on the anti-de Sitter spacetime might be also found in [6]. The role of magnetic monopoles, advocated by Dirac in quantum physics [28], may then provide some new interesting insights even at the phenomenological length scale.

Another aspect linked to the Universe expansion comes from what is called the *Hubble tension*, which can be explained in a very simplified manner by the determination of the Hubble constant H_0, which differs significantly depending on the measurement methods used: (a) that obtained by the Cosmic Microwave Background, and (b) that obtained by measurements on supernovae (or similar methods) e.g. [15]. An attenuation of Hubble tension could also be achieved by the presence of diffuse large-scale magnetic fields in galaxies and extra-galaxies, [93]. This increasingly illustrates, during recent studies in cosmology, the more and more significant role of electromagnetic fields in the dynamic expansion of the Universe e.g. [64, 93] (these authors argue that weak magnetic fields in the early Universe would lead to the accelerating expansion seen today), [94]. At this step, the choice of Einstein-Cartan spacetime used extensively in this work will be argued as the text progresses, but it already seems judicious to recall the notion of defects distributed in the Universe which are supposed sources of torsion. In this context, twist and curvature model line defects and surface defects, and even grain defects, analogous to those in condensed matter physics. The analogy between the formation of defects during the thermomechanical evolution of condensed matter and the formation of defects in spacetime — leading to microcosms [2] or local spacetime grains — deserves to be mentioned in this book. An integrated approach including electromagnetism and Eintein-Cartan spacetime to relative gravitation may then provide new insights for modeling the evolution of the Universe. In particular, it offers a relatively simple framework for analyzing the cosmological constant Λ and the phenomenological notions of dark matter and dark energy without directly involving quantum mechanics.

1.6 Cosmology, Electromagnetism, and Geodesics

There are more to be known about dark energy and specifically the cosmological constant (Λ) problem. The seminal paper of [29] was a starting point to relate the cosmological constant with quantum physics, which was consistent with the concept of zero-point (vacuum spacetime) fluctuations resulting from the developments of quantum electrodynamics in the mid-to-late 1920s, for which the electromagnetic field is treated as a collection of quantized harmonic oscillators (each quantized harmonic oscillator has "zero-point" energy). The torsion of spacetime has been also suggested by numerous studies as being a possible source of dark energy. This dark energy is commonly represented by the cosmological constant Λ e.g. [38], and it can even take negative values [95]. Some studies have further related dark energy to a background electromagnetic field without proposing additional new physical interpretations e.g. [62].

Beyond the oversimplified introduction of the cosmological constant in the Lagrangian to account for the expansion of the Universe, it should be pointed out that most enigmas of modern physics are related to this constant; the recent extended review in [15] highlights the wide range of physics domains involved with this seemingly simple constant: particle physics, astrophysics, quantum physics, to name but a few. Geodesic and autoparallel curves are thus of definite interest, at least because they represent some observables, if only for the analysis of the trajectories of planets or satellites in the distant horizons of the Universe where it is generally admitted that dark energy plays an important role e.g. [15]. Over the years, the role of electromagnetism in defining the cosmological constant, dark matter and dark energy has been discussed in the literature e.g. [96, 97], and so has quantum electromagnetism e.g. [98]. The interest of Einstein-Cartan spacetime is then accordingly obvious, even though its influence might be modest because torsion is often associated with the electromagnetic field e.g. [6, 38, 50, 99, 100].

Detection of gravitational waves also illustrates a well-known application of the geodesic deviation equation e.g. [101]. The measuring of the separation of two neighbored geodesic curves, which are the trajectories of two test particles in a Einstein spacetime, may be evaluated by means of the separation acceleration that we briefly recall in this work. Some previous studies have extended this deviation equation to include the relativistic top moving in a gravitational field e.g. [101], or to reformulate the geodesic deviation in terms of teleparallel gravity [102] (in this approach the torsion field is considered to engender the gravity field, instead of the curvature). A generalized Jacobi equation was also derived in the presence of curvature and torsion e.g. [103], and e.g. [104], where a deviation operator was proposed using covariant derivative and Lie differential operator, and also e.g. [105], including the electrical field.

The geodesic deviation equation describes the relative motion of two structureless particles determined by a spacetime geometry with non-zero curvature. When particles are charged electrically, the deviation equation should account for the electric charge q e.g. [34] where geodesic curves are called autoparallel curves or worldlines in which higher-order deviations are also studied e.g. [105].

1.7 Goal of the Present Book

In view of all of these numerous aspects, such as cosmological constant, gravitation, electromagnetism and accelerating expansion of the Universe (with defect distribution), the main motivation of the present book is to search for the link between them while avoiding particle physics, despite their preeminent role in cosmology, but rather focusing on the concept of continuum fields occurring in a defined spacetime, without any consideration of spinning fluids embodying the vacuum spacetime.

To define the framework of this book, the present work makes some basic assumptions to define the geometrical background of the continuum spacetime and the physical events occurring within the spacetime:

(1) the spacetime has a structure of four-dimensional manifold endowed with a metric;
(2) the manifold is endowed with a linear connection;
(3) only gravitation and electromagnetism are considered as physical fields.

The general methodology consists of exploiting the geometric structure of spacetime by progressively giving some recalls on Galilean, Minkowski, Einstein and finally Einstein-Cartan spacetimes. The idea is to reduce the phenomena of gravitation and electromagnetism to the geometric variables characterizing these spacetimes, without going through the theory of quantum physics which goes beyond the scope of this book. Spacetime torsion will be a matter of concern all along this work, and implicitly we consider that spacetime is more and more assimilated with an infinitely small set of microcosms, as is for defected materials [39, 49].

This book includes seven chapters and a Final Remarks chapter. After a general introduction in Chapter 1, we revisit in Chapter 2 the mathematical background common to all continuum physics which are the differentiable manifolds. The spacetime concept is a particular case of a continuum whose two basic characteristics are the compactness (existence of a metric structure) and the connectedness (existence of a connection, maybe independent of the metric).

The geometric structure of spacetime is closely linked to a Lagrangian $\mathscr{L}$ on the continuum whose arguments are generally the metric and its partial derivatives as well as the connection and its partial derivatives, originally proposed by e.g. [58]. Chapter 3 rigorously discusses the necessary and sufficient condition for a spacetime model thus defined to be covariant, in particular the Einstein-Hilbert Lagrangians in relative gravitation and that of Yang-Mills of electromagnetism in this framework.

Chapter 4 of this book concerns the development of conservation laws in the case of Einstein-Cartan spacetime by using intensively the Principle of General Covariance. It is similar and gives new results compared to the results obtained in curved and twisted spacetimes e.g. [4]. Similarity with behavior of continuum with irreversible deformation is given by analogy.

Chapter 5 is devoted to classical relative gravitation within the framework of Einstein's theory in a pseudo-Riemanian space. The spacetimes of Schwarzschild and those of de Sitter are summoned, together with the Jacobi equation for Riemanian geodesics. In this chapter, we are more interested in the link that exists between relative gravitation and electromagnetism using the actions of Hilbert-Einstein and that of Yang-Mills.

Chapter 6 constitutes the heart of this work, devoted to coupling of relative gravitation and electromagnetism in the Einstein-Cartan spacetime. Field equations and examples of coupling are treated. The role of the geometric structure combined with gravitation and electromagnetism is highlighted with the derivation of field equations, the influence of the Chern-Simmons current being fundamental in this coupling effect, namely that of optical helicity e.g. [107]. We are also interested in developing the extension of the deviation equation for geodesics to autoparallel curves where the spacetime is curved with torsion, without *a priori* considering electric charge. In view of the previous results where the torsion tensor of the spacetime is identified to be directly related to spin angular momentum of the electromagnetic field e.g. [6, 108, 109], it is necessary to investigate the influence of the torsion on the deviation of autoparallel curves. For geodesics analysis, the contribution of this chapter would be slightly different since we are interested in developing the extension of the deviation equation for geodesics to autoparallel curves where the spacetime is curved and distorted, after developing a geometric approach for the cosmological constant. Same steps will be used for Newton gravitation, Einstein gravitation, and Einstein-Cartan gravitation: search of the geodesics or autoparallel curves for each spacetime, constructing the deviation equation (Jacobi equation) and solving for the equation for specific cases as illustration.

For all three cases, the idea is that a particle test with a relatively small mass is subjected to a gravitational field which is not modified by this particle. Using the variational procedure, the close link between the torsion field and the electromagnetic field in spacetime is also highlighted. Last but not least, a mathematical equation is formally derived relating torsion field and electromagnetic field within the Einstein-Cartan spacetime. Even if such a link has been proposed for a long time, our approach here is more systematic with only the hypothesis that spacetime is characterized by the sum of the actions of Einstein-Palatini and Yang-Mills. Knowing that each of the actions are in any case valid separately in relative gravitation and in electromagnetism, we can consider that our hypothesis is truly minimal.

Chapter 7 is an application of the Einstein-Cartan gravitation and electromagnetism to cosmology. Gravitation and a standard approach to spacetime in cosmology are addressed, namely the FLRW spacetime. In the process, Friedmann equations are then generalized to account for the presence of a diffuse magnetic monopole field in the cosmic Universe. The magnetic monopole, via the torsion field, is clearly shown to contribute to the acceleration of the expansion of the Universe. The influence of the volume-form is deeply analyzed in this chapter where the field equations are derived accordingly depending on the choice of volume-form together with the necessity of its compatibility with metric and connection. The consequences of this choice combined with the presence of the electromagnetic field are sketched for the occurrence of Universe anisotropy. This might induce questions about the validity of cosmological principles: isotropy and homogeneity. Concluding remarks are given in the last chapter.

Chapter 2

Elements of Geometry for Spacetime Models

Elements of geometry are given in this chapter concerning differentiable manifolds endowed with a metric and an independent affine connection. We recall basics on the torsion and curvature tensor fields on a manifold. The covariant derivatives (to ensure coordinate system independence) and Lie derivatives (to obtain intrinsic tool for the formulation of conservation laws or other physical laws) on scalars, vectors and 1-forms as well as tensors of arbitrary order are revisited. Spacetime geometry is defined by means of a continuum: compacted and connected manifold. Necessary differential operations of tensor fields on continuum are introduced, with a special mention to the calculus of torsion and curvature variations in view of deriving equations for spacetime models.

2.1 Manifolds and Group of Transformations

By extending the original ideas of astronomers of ancient Greece, it is implicitly assumed that there is the existence of a fixed underlying space, and then a background spacetime, in which planets and stars move, subject essentially to gravitational forces. In short, the stars with their planets, the aggregates of stars and the interstellar matter forming the galaxies, and the galaxies themselves forming the clusters of galaxies (superclusters) form the observable Universe. But all of them are found to move in an assumed empty background space, which is Euclidean and therefore flat. In the 1920s and 1930s, however, various physicists dealing with cosmology had demonstrated through observations that galaxies and galaxy clusters kept more or less relatively their mutual distances fixed (probably due to the gravitational interactions connecting them) while the background space itself is expanding [12]. The mutual interaction between matter and the properties of space (*in extenso* spacetime) following the findings of Einstein's work on relative gravitation breaks the flatness of space (resp. spacetime) and leads to generalizing its geometry to Riemann, and even Riemann-Cartan, manifolds.

Therefore, continuous bodies and spacetimes are often described by three- or four-dimensional continua, respectively. Mathematical models of continuum mechanics and relativistic gravitation then dwell upon the concept of neighborhoods and continuous mappings. This naturally leads to the notion of manifold e.g. [5], [110]. Namely, in Galilean view, space is a three-dimensional manifold and time is a one-dimensional manifold; they exist separately. From a Minkowski point of view, for relativistic physics, spacetime is a four-dimensional manifold, where time is coupled with space. In both theories, manifolds are differentiable. After introducing manifolds which is the background of continuum geometry and kinematics, two variables constitute the basic characteristics of spacetime and continuous material models: a metric tensor and an independent connection. Figuratively speaking, for both relative gravitation and quantum physics, there is no empty space (*in extenso* spacetime). On the one hand, for relative gravitation, vacuum spacetime is not empty in the sense that it is not flat because inherent gravitational forces are related to its curvature (and we will also see that the electromagnetic forces will be related to torsion). On the other hand, for quantum physics, vacuum spacetime is not empty in the sense that it is filled with quantum activities of elementary particles and fluctuations of localized energies (mainly below the Planck length $\ell_P = 10^{-35}[m]$). In accordance with the objective of this work, we remain at the length scale within the continuum approach for relative gravitation.

2.1.1 *Manifolds, Tangent Space, Cotangent Space*

We recall some basic elements of manifolds for self-consistency of the book. Extensive developments may be seen elsewhere e.g. [5]. Briefly, we fix once and for all the n-dimensional vector space E (reference set, for example $\mathbb{R}$). An n-dimensional manifold $\mathscr{B}$, embedded onto E, is a point set which is covered completely by a countable set of neighborhoods $U_1, U_2, \ldots$, such that each point $P \in \mathscr{B}$ belongs to at least one of these neighborhoods. U_k is also called coordinate patch. The manifold $\mathscr{B}$ is said to be differentiable if

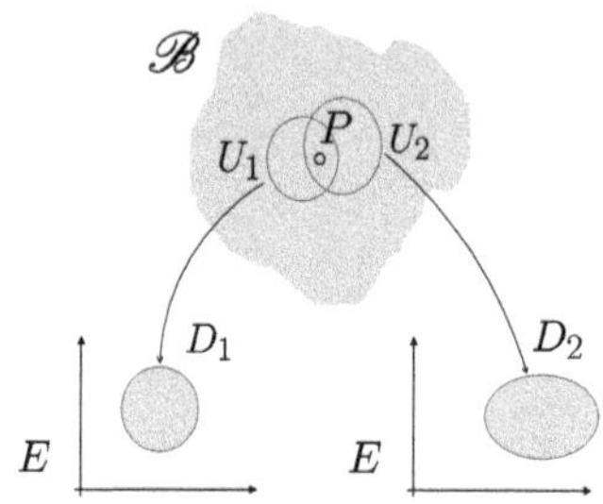

Fig. 2.1: Overlapping of coordinate patches. The point $P \in \mathscr{B}$ is reported as being associated with two overlapping coordinate patches U_1 and U_2. In the overlap, there will be points that have coordinates from each patch. Sets $D_1 := \varphi_1(U_1)$ and $D_2 := \varphi_2(U_2)$ belong to E.

there is at least one covering of $\mathscr{B}$ by a set of coordinate patches $\{U_k,\}$ such that in each overlap of the patches, the coordinates are mutually related by differentiable functions. For two given manifolds, say $\mathscr{B}_1$ and $\mathscr{B}_2$, a diffeomorphism is an invertible mapping $\varphi : \mathscr{B}_1 \to \mathscr{B}_2$, such that φ and its inverse φ^{-1} are both C^∞, meaning that the mappings are continuous and infinitely differentiable. For the case where $\mathscr{B}_1 = \mathscr{B}_2 = \mathscr{B}$, any mapping φ is invertible and provided that $\varphi \in C^\infty$, then it is a diffeomorphism.

A coordinate system is defined on each U_k such that one may assign in a unique manner n real numbers $x^1, \ldots, x^n$ to each point $P \in U_k$. As P ranges over U_k, the corresponding numbers $x^1, \ldots, x^n$ range over an open domain D_k of E. There thus exists a one-to-one mapping of each U_k onto D_k, which will be assumed continuous. The numbers $x^1, \ldots, x^n$ are called the coordinates of P. In the case where $U_1 \cap U_2 \neq \emptyset$ there are two sets of coordinates associated to a same point $P \in U_1 \cap U_2$. In sum, the manifold $\mathscr{B}$ is then provided with a family of pairs $\{U_k, \varphi_k\}$ such that for any two coordinate patches U_n and U_m such that $U_n \cap U_m \neq \emptyset$, the map $\varphi_n \left(\varphi_m\right)^{-1} : \varphi_m \left(U_n \cap U_m\right) \to \varphi_n \left(U_n \cap U_m\right)$ is infinitely differentiable. φ_n are called coordinate functions (or coordinates), and for any point $P \in U_k$, represented by $\left(x^1(P), \ldots, x^n(P)\right)$. Practically, manifold differentiability may be approached by means of a change of coordinate system. Calculus on manifolds of dimension n starts with the definition of function and tangent space. A function on manifold $\mathscr{B}$ denoted $f : \mathscr{B} \to \mathbb{R}$ is a (smooth) map from $\mathscr{B}$ to real $\mathbb{R}$. It is represented in Figure 2.2.

Let us now consider a curve on the manifold $\mathscr{B}$ defined by the function $\mathscr{C} : [a, b] \in \mathbb{R} \to \mathscr{B}$ and a function $f : \mathscr{B} \to \mathbb{R}$. Say a real $t = 0 \in [a, b]$. The tangent vector of the curve at $\mathscr{C}(0)$ is the directional derivative of a

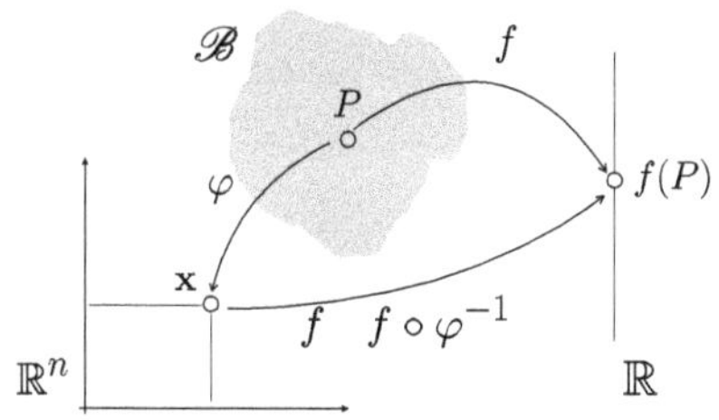

Fig. 2.2: Function f on manifold $\mathscr{B}$. The function $f : P \in \mathscr{B} \to \mathbb{R}$ is defined by its coordinate presentation $f\varphi^{-1} : (x^1, \ldots, x^n) \in \mathbb{R}^n \to f(P) \in \mathbb{R}$.

function $f[\mathscr{C}(t)]$ along the curve $\mathscr{C}$ at the point $t = 0$:

$$\frac{d}{dt} f[\mathscr{C}(t)]\,|_{t=0} = \frac{\partial\left(f\varphi^{-1}(x^\mu)\right)}{\partial x^\alpha}\,\frac{dx^\alpha\{\mathscr{C}(t)\}}{dt}|_{t=0} \tag{2.1}$$

where the "partial derivative" of f in the composition of functions is practically expressed in terms of the partial derivative of its coordinate presentation. It is worth defining the tangent vector $\mathbf{u}$ at P of $\mathscr{B}$ as:

$$u^\alpha := \frac{dx^\alpha\{\mathscr{C}(t)\}}{dt} \tag{2.2}$$

and use a synthetic expression of the relation (2.1):

$$\frac{d}{dt} f[\mathscr{C}(t)]\,|_{t=0} = \mathbf{u}[f] = u^\alpha \frac{\partial f}{\partial x^\alpha} \tag{2.3}$$

Remark 2.1. This is the directional derivative of $f(\mathbf{x})$ along $\mathbf{u}$.

Definition 2.1 (Tangent space). All the equivalent classes of curves $\mathscr{C}_k$, and by extension all tangent vectors $\mathbf{u}_k$ at point P, form a vector space called the tangent space of $\mathscr{B}$ at point P, denoted $T_P\mathscr{B}$.

The base of the tangent space of dimension n is denoted by:

$$\mathbf{f}_k := u_k^\alpha \frac{\partial}{\partial x^\alpha}, \quad \alpha, k = 1, \ldots, n \tag{2.4}$$

where k stands for the labelling vectors and α for coordinates.

Definition 2.2 (Cotangent space). The cotangent space at P of the manifold $\mathscr{B}$ is the set of linear function from $T_P\mathscr{B} \to \mathbb{R}$. It is denoted $T_P^*\mathscr{B}$.

An arbitrary element of the cotangent space $\omega \in T_P^*\mathscr{B}$ is also called dual vector. A basic example of a dual vector is the differential df of a real function:

$$df = \frac{\partial f}{\partial x^\alpha} dx^\alpha \tag{2.5}$$

Then the action of the linear map $df \in T_P^*\mathscr{B}$ on any vector $\mathbf{u} \in T_P\mathscr{B}$ is defined by the inner product:

$$(df, \mathbf{u}) = \mathbf{u}\,[f] = u^\alpha \frac{\partial f}{\partial x^\alpha} \tag{2.6}$$

The base of the cotangent space is denoted here $\{dx^\alpha\}$, allowing us to decompose any 1-form ω on the manifold $\mathscr{B}$ as $\omega = \omega_\beta dx^\beta$, where ω_β are the covariant components of the 1-form ω.

2.1.2 *Coordinate Systems and Group of Transformations*

Some confusion may appear between change of reference frames and (passive) change of coordinate systems e.g. [111]. Coordinate systems have no physical meaning whereas the frame of reference is a fundamental concept in physics theory (principle of relativity). There are two different concepts of transformations in physics, and basically for the kinematics of a point. The first is coordinate transformations (CT). Taking the description of a given trajectory of a material point, described in one coordinate system, CT changes to its description in another coordinate system. The second is object transformations (OT). Taking a given trajectory of a material point, described in a given coordinate system, and the OT changes it into a different trajectory, described in the same coordinate system as the original trajectory. We deal with coordinate transformations in this section. The accounting for time as coordinate induces an additional problem which should be solved.

2.1.2.1 *Change of Coordinate System (Passive Diffeomorphism)*

We consider the point $P \in \mathscr{B}$ of Figure 2.1 belonging to the differentiable manifold $\mathscr{B}$. We consider a change of coordinate system (passive diffeomorphism) on the same manifold (spacetime). Both coordinate systems describe $\mathscr{B}$. Of course, any physical theory can be made invariant under change of coordinate system (see Figure 2.3) because physical phenomena do not care which coordinate system you use to describe it.

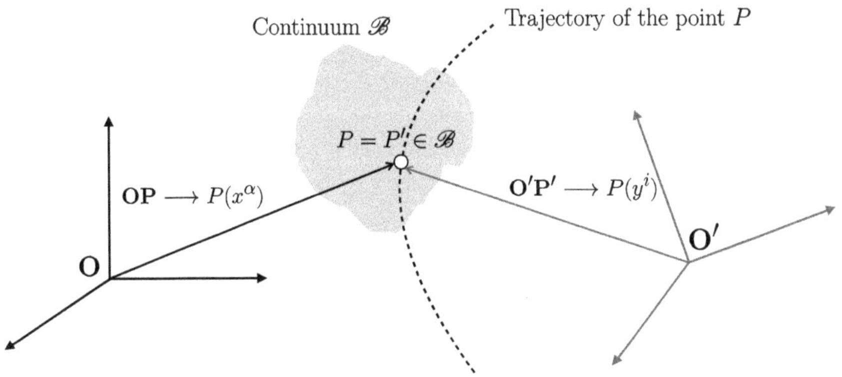

Fig. 2.3: Passive diffeomorphism: Change of spacetime coordinate system $x^\alpha \to y^i(x^\mu)$. The interpretation is a "passive diffeomorphism" on this figure, where the material point $P = P' \in \mathscr{B}$ moves with respect to the reference system Σ_x — along the dotted trajectory — with coordinates (y^i) with respect to another reference Σ_y (in gray) (time origin change, translation, and rotation).

Let (y^i) *(Latin indices)* and (x^α) *(Greek indices)* correspond to coordinate systems respectively in U_1 and U_2. Latin indices and Greek indices will be used to distinguish two different coordinate systems. The transformation between coordinates (y^i) and (x^α) is diffeomorphism, more precisely a passive diffeomorphism:

$$y^i = y^i(x^\alpha), \quad x^\alpha = x^\alpha(y^i) \tag{2.7}$$

$$J_i^\alpha = \frac{\partial x^\alpha}{\partial y^i}, \quad J_{ij}^\alpha = \frac{\partial J_i^\alpha}{\partial y^j} = \frac{\partial^2 x^\alpha}{\partial y^i \partial y^j}, \quad J_{ijk}^\alpha = \frac{\partial J_{ij}^\alpha}{\partial y^k} = \frac{\partial^3 x^\alpha}{\partial y^i \partial y^j \partial y^k}$$

We also have that for any coordinate system (y^i), and for any permutation $\sigma \in \mathfrak{S}_n$

$$\frac{\partial^n}{\partial y^1 \cdots \partial y^n} = \frac{\partial^n}{\partial y^{\sigma(1)} \cdots \partial y^{\sigma(n)}}$$

meaning that the passive transformations we consider are $\mathscr{C}^\infty$ diffeomorphisms.

Remark 2.2. When a lower- and uppercase index such as $j, k, l, \ldots$ appear twice in a term then summation over that index is applied. The range of summation is $1, \ldots, n$, where the letter n is excluded from the summation.

We have

$$\frac{\partial x^\alpha}{\partial y^i}\frac{\partial y^i}{\partial x^\beta} = J_i^\alpha A_\beta^i = \delta_\beta^\alpha, \qquad \frac{\partial y^i}{\partial x^\alpha}\frac{\partial x^\alpha}{\partial y^j} = A_\alpha^i J_j^\alpha = \delta_j^i$$

with summation (Einstein) over i and α from 1 to n.

In elementary linear algebra and vector calculus, the tensor concept is concerned mainly by linear transformations of coordinates and their consequences on the definition of vector, form and tensors, meaning that the matrices A_α^i and J_i^α do not depend on the coordinates y^i and x^α. In such a case, components of vector $\mathbf{u}$, form ω and tensor $\mathbf{T}$ are respectively defined by the rule of transformations:

$$u^i = A_\alpha^i\, v^\alpha, \quad \omega_j = J_j^\alpha\, \pi_\alpha, \quad T_k^{ij} = A_\alpha^i A_\beta^j J_k^\gamma\, T_\gamma^{\alpha\beta} \tag{2.8}$$

where u^i and v^α, ω_j and π_α, and T_k^{ij} and $T_\gamma^{\alpha\beta}$ are the components in the actual and initial coordinate systems respectively. When the coordinate changes are nonlinear, these definitions of rule transformations do not change except that the matrices A_α^i and J_i^α depend on the coordinates.

$$u^i = A_\alpha^i\,(x^\mu)\, v^\alpha, \quad \omega_j = J_j^\alpha\,(x^\mu)\,\pi_\alpha, \quad T_k^{ij} = A_\alpha^i\,(x^\mu)\, A_\beta^j\,(x^\mu)\, J_k^\gamma\,(x^\mu)\, T_\gamma^{\alpha\beta} \tag{2.9}$$

where the dependence of matrices on the coordinates are explicitly written for the sake of clarity. This will be rigorously defined as such in the framework of tensor fields on differentiable manifold.

2.1.2.2 *Examples of Coordinate Transformations*

Both the continuum medium and the vacuum spacetime are described by manifolds: in this work, the continuum of dimension three, and the spacetime of dimension four e.g. [112], [113]. Here we present some fundamental examples for coordinate system changes where change of coordinates may include change of reference frames e.g. [111]. For the sake of simplicity, let us denote $x^0 := ct$ the time variable and $x^i = (x^1, x^2, x^3)$ the space coordinates. The spacetime coordinates hold as $x^\alpha = (x^0, x^1, x^2, x^3)$.

(1) One of the most common examples is given by a change of coordinate system, from spherical (x^0, x^1, x^2, x^3) to Cartesian (y^0, y^1, y^2, y^3):

$$\begin{cases} y^0 = x^0 \\ y^1 = x^1 \sin x^2\ \cos x^3 \\ y^2 = x^1 \sin x^2\ \sin x^3 \\ y^3 = x^1 \cos x^2 \end{cases} \tag{2.10}$$

The matrix A^i_α is straightforward to obtain. This is not the case for the inverse J^α_i, although it is not out of reach:

$$A^i_\alpha = \begin{pmatrix} \sin x^2 \cos x^3 & x^1 \cos x^2 \cos x^3 & -x^1 \sin x^2 \sin x^3 \\ \sin x^2 \sin x^3 & x^1 \cos x^2 \sin x^3 & x^1 \sin x^2 \cos x^3 \\ \cos x^2 & -x^1 \sin x^2 & 0 \end{pmatrix}, \quad i, \alpha = 1, 2, 3$$

whereas $A^0_0 = 1$, $A^i_0 = 0$, and $A^0_\alpha = 0$ for the other indices. Formally, the inverse is calculated using the definition $A^i_\alpha J^\alpha_j = \delta^i_j$.

(2) As it is an important case in special relativity theory, we recall the so-called Lorentz transformations which connect the coordinates of an event with respect to two inertial frames of reference in uniform rectilinear motion relative to each other at velocity u. This is obtained by assuming that a test particle moving at the speed of light c in such a frame must move at this speed in all these frames. *Lorentz transformations* are given by the famous relations:

$$\begin{cases} y^0 \,:\, \gamma\,(x^0 - \beta\, x^1) \\ y^1 = \gamma\,(x^1 - \beta\, x^0) \\ y^2 = x^2 \\ y^3 \,:\, x^3 \end{cases} \tag{2.11}$$

where

$$\beta := \frac{u}{c} \quad \text{and} \quad \gamma := \sqrt{1 - \beta^2} \tag{2.12}$$

We simplify by choosing a translation along the x^1-axis. The term u is interpreted as the relative velocity between the two sets of coordinates, and c is the speed of light. The coupling of time x^0 and space x^1 is the genesis of the name *spacetime*.

(3) The *Leibniz group* is the group of transformations:

$$y^j := R^j_i(x^0) x^i + u^j(x^0) \quad \text{for } i, j = 1, 2, 3, \quad \text{and} \quad y^0 := x^0 + u^0 \tag{2.13}$$

where $R^j_i(x^0)$ is an orthogonal transformation depending on time variable, and $u^\alpha = (u^0, u^1, u^2, u^3)$. These are transformations between rigid Euclidean reference frames with arbitrary rotation and arbitrary acceleration with respect to each other e.g. [114].

(4) The second is the *Maxwell group* represented by the transformations:

$$y^j := R^j_i\, x^i + u^j(x^0) \quad \text{for } i, j = 1, 2, 3, \quad \text{and} \quad y^0 := x^0 + u^0 \tag{2.14}$$

where R_i^j is a uniform orthogonal transformation, independent of the time variable. The Maxwell group includes fixed rotating frames but arbitrary accelerating frames. This is expressed by the dependence of u^j on x^0. In the case where the dependence of u^j is linear with respect to x^0, we recover the group of Galilean transformations.

(5) The third example consists of the *Newton-Cartan group*:

$$y^j := R_i^j \, x^i + v^j x^0 + u^j, \quad \text{for } i,j = 1,2,3, \quad \text{and} \quad y^0 := x^0 + u^0 \tag{2.15}$$

where $v^j = (v^1, v^2, v^3)$ is a constant velocity. This is the *Galilean* group (or Galilean Lie group [115]) with 10 parameters. The basic root of the design of Galilean transformation comes from the "intuitive" observation that the velocity of a material point as seen by observers at rest in two reference frames differs and depends on the relative constant velocity of the two frames, linearly.

The role of these groups with regard to constitutive laws may be found in e.g. [49] when considering the principle of material indifference. An extension of the Newton-Cartan group is the local Galilean group $R_i^j(x^k)$ and $u^j(x^k)$, where the linear transformation and the translation depend on space coordinates but not on the time variable x^0.

(6) An important group is that of *internal transformations*. We can attach numerous coordinate systems to one reference frame. For that purpose, it is worth considering the internal transformations e.g. [116]:

$$y^0 = y^0(x^0, x^1, x^2, x^3), \quad y^j = y^j(x^1, x^2, x^3), \quad j = 1,2,3 \tag{2.16}$$

where x^0 and y^0 correspond to the time variable in each coordinate system. Leibniz and Maxwell groups are particular cases. The first equation expresses a synchronization of the time measure (clock) within the frame of reference, whereas the second equation expresses a change of coordinate system within the three-dimensional space of the spacetime.

Remark 2.3. Internal transformations are nonlinear extensions of Galilean transformations, but they do not allow us to derive the adequate transformations of spacetime. Moreover, if we assume that inertial forces are present in a given frame, they cannot be removed by means of internal transformations. This basically constitutes the subtle difference between coordinate systems and frames of reference.

2.2 Tensor Field on Manifold

Let $T_P\mathscr{B}$ be the tangent space of $\mathscr{B}$ at point P and $T_P\mathscr{B}^*$ its dual. Let $\{\mathbf{e}_1, \ldots, \mathbf{e}_n\}$ be a base of $T_P\mathscr{B}$ (contravariant vectors with lower index) and its dual base $\{\mathbf{f}^1, \ldots, \mathbf{f}^n\}$ in $T_P\mathscr{B}^*$ (covariant vectors with upper index) such that $\mathbf{f}^i(\mathbf{e}_j) \equiv \delta^i_j$.

2.2.1 *Forms and Tensors*

Definition 2.3 (Tensor). Let $\{\mathbf{u}_1, \ldots, \mathbf{u}_q\} \in T_P\mathscr{B}$ and $\{\mathbf{v}^1, \ldots, \mathbf{v}^p\} \in T_P\mathscr{B}^*$ be some arbitrary vectors. A p-contravariant and q-covariant tensor field $\mathcal{T}$ on $\mathscr{B}$ is a multilinear form defined at each point $P \in \mathscr{B}$ by $\mathcal{T} : (\mathbf{v}^1, \ldots, \mathbf{v}^p, \mathbf{u}_1, \ldots, \mathbf{u}_q) \in (T_P\mathscr{B}^*)^p \times (T_P\mathscr{B})^q \longrightarrow \mathcal{T}(\mathbf{v}^1, \ldots, \mathbf{v}^p, \mathbf{u}_1, \ldots, \mathbf{u}_q) \in \mathbb{R}$. The sum $(p+q)$ is called the rank of the tensor field. The couple (p, q) is its type.

For any two tensors $\mathcal{T}$ and $\mathcal{S}$, respectively q and q' covariant on $\mathscr{B}$, we define the tensor product $\mathcal{T} \otimes \mathcal{S}$ as a linear application such that:

$$\mathcal{T} \otimes \mathcal{S}(\mathbf{u}_1, \ldots, \mathbf{u}_q, \mathbf{w}_1, \ldots, \mathbf{w}_{q'}) \equiv \mathcal{T}(\mathbf{u}_1, \ldots, \mathbf{u}_q)\, \mathcal{S}(\mathbf{w}_1, \ldots, \mathbf{w}_{q'}) \quad (2.17)$$

This definition may be extended to contravariant and mixed tensors e.g. [49].

Definition 2.4. A differential form of order q, or a q-form, is a totally antisymmetric tensor of type $(0, q)$.

A type $(1, 0)$ tensor, denoted $\mathbf{u}$, is called vector field and a type $(0, 1)$ tensor, denoted ω, is a co-vector field or a 1-form field on the manifold $\mathscr{B}$. Let us give some examples of tensor products of vectors and 1-forms:

(1) $\mathbf{e}_{i_1} \otimes \mathbf{e}_{i_2}, \mathbf{e}_{i_1} \otimes \mathbf{e}_{i_2} \otimes \mathbf{e}_{i_3}, \ldots, \mathbf{e}_{i_1} \otimes \mathbf{e}_{i_2} \otimes \mathbf{e}_{i_3} \otimes \cdots \otimes \mathbf{e}_{i_r}$
(2) $\mathbf{f}^{j_1} \otimes \mathbf{f}^{j_2}, \mathbf{f}^{j_1} \otimes \mathbf{f}^{j_2} \otimes \mathbf{f}^{j_3}, \ldots, \mathbf{f}^{j_1} \otimes \mathbf{f}^{j_2} \otimes \mathbf{f}^{j_3} \otimes \cdots \otimes \mathbf{f}^{j_r}$
(3) $\mathbf{e}_{i_1} \otimes \mathbf{f}_{j_1}, \ldots, \mathbf{e}_{i_1} \otimes \mathbf{f}^{j_1} \otimes \mathbf{f}^{j_2} \otimes \cdots \mathbf{e}_{i_r}$

where the indices $i_1, i_2, \ldots, i_r$ and $j_1, i_2, \ldots, j_r$ may take any value of $1, 2, \ldots, n$. The number of these indices depends on the type (p, q).

Definition 2.5. Let $\{\mathbf{f}^{i_1}, \ldots, \mathbf{f}^{i_r}\}$ be a set of r one-forms on the manifold $\mathscr{B}$ of dimension n, with $r \leq n$. Indices $(i_1, \ldots, i_r)$ take the values $(1, \ldots, n)$. Consider all permutations (even or odd) $\sigma : (i_1, \ldots, i_r) \rightarrow (\sigma(i_1), \ldots, \sigma(i_r))$

with its sign:

$$\text{sign}(\sigma) := \epsilon_{i_1,\ldots,i_r}^{\sigma(i_1),\ldots,\sigma(i_r)} = \begin{cases} +1 & \text{for even permutations} \\ -1 & \text{for odd permutations} \end{cases}$$

The wedge product is defined by the totally antisymmetric tensor product:

$$\mathbf{f}^{i_1} \wedge \mathbf{f}^{i_2} \wedge \cdots \wedge \mathbf{f}^{i_r} := \sum_{\sigma} \text{sign}(\sigma)\ \mathbf{f}^{\sigma(i_1)} \otimes \mathbf{f}^{\sigma(i_2)} \otimes \cdots \otimes \mathbf{f}^{\sigma(i_r)}$$

where summation runs over all permutations σ.

A typical example of differential form of order 1 on an open subset of $\mathbb{R}^n$ is given as follows. Consider a scalar $\mathscr{C}^1$ function $\varphi : (\xi^1, \ldots, \xi^n) \in \mathbb{R}^n \to \varphi(\xi^1, \ldots, \xi^n) \in \mathbb{R}$, then its total differential (cf. exterior derivative in appendix) is a differential 1-form:

$$d\varphi = \partial_i \varphi\ d\xi^i$$

where the set $\{d\xi^1, \ldots, d\xi^n\}$ constitutes covariant independent 1-forms. Let us give some examples of wedge product of 1-forms:

(1) $d\xi^{i_1} \wedge d\xi^{i_2} = d\xi^{i_1} \otimes d\xi^{i_2} - d\xi^{i_2} \otimes d\xi^{i_1}$
(2) $d\xi^{i_1} \wedge d\xi^{i_2} \wedge d\xi^{i_3} = d\xi^{i_1} \otimes d\xi^{i_2} \otimes d\xi^{i_3} + d\xi^{i_2} \otimes d\xi^{i_3} \otimes d\xi^{i_1} + d\xi^{i_3} \otimes d\xi^{i_1} \otimes d\xi^{i_2} - d\xi^{i_1} \otimes d\xi^{i_3} \otimes d\xi^{i_2} - d\xi^{i_2} \otimes d\xi^{i_1} \otimes d\xi^{i_3} - d\xi^{i_3} \otimes d\xi^{i_2} \otimes d\xi^{i_1}$
(3) $\ldots, d\xi^{i_1} \wedge d\xi^{i_2} \wedge d\xi^{i_3} \wedge \cdots \wedge d\xi^{i_r}$, with $r \leq n$

where the indices $i_1, i_2, \ldots, i_r$ may take any value of $1, 2, \ldots, n$.

Remark 2.4. In an Euclidean space, it is typical to represent a 1-form by a "vector". However, it may induce confusion namely when we consider large deformations of a continuum $\mathscr{B}$ (manifold) undergoing an arbitrary transformation even if it is a diffeomorphism $\varphi : \mathscr{B} \to \varphi(\mathscr{B})$. Indeed, let's consider the linear tangent transformation $d\varphi$ (whose components are triads F^i_μ) with the Cartesian components $\partial_\mu \varphi^i$ (also usually called deformation gradient in continuum mechanics).

Figure 2.4 displays an illustration of tangent and cotangent space. During the deformation of a square to a parallelogram, the two vectors $\mathbf{u}_1$ and $\mathbf{u}_2$ become $d\varphi(\mathbf{u}_1)$ and $d\varphi(\mathbf{u}_2)$ respectively, while the 1-forms ω_1 and ω_2 are transformed into $d\varphi^*(\omega_1)$ and $d\varphi^*(\omega_2)$ where the components of φ^* are $\text{Det}(F^j_\nu)\left(F^i_\mu\right)^{-1}$. For a tridimensional Euclidean manifold endowed with metric $g_{\mu\nu}$ and a volume-form ω_0 (determinant), 1-forms ω_1 and ω_2 may be represented by vectors (homeomorphism between vectors and

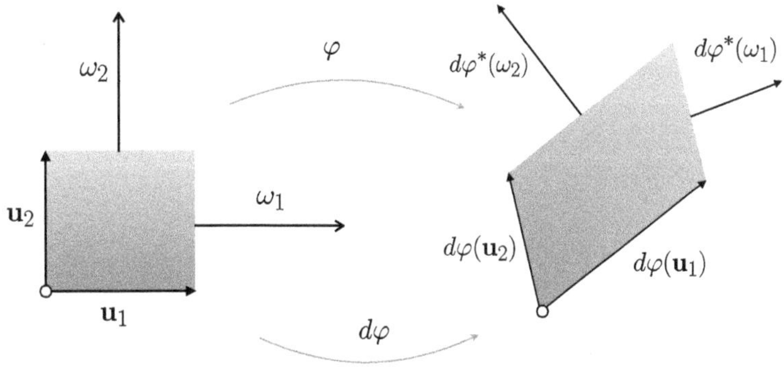

Fig. 2.4: Illustration of dual vectors. Transformation of a square to a parallelogram (in a tridimensional Euclidean space). Vectors $\mathbf{u}_1$ and $\mathbf{u}_2$ are sides of the square whereas 1-forms ω_1 and ω_2 are related to the normal vectors of the square.

normal vectors of surface). Deformation of vectors and those of 1-forms are different.

Definition 2.6 (Components of tensor). If $\mathcal{T}$ is a tensor field of type (p, q) on a manifold $\mathscr{B}$ then the scalars $\mathcal{T}^{l_1 \cdots l_p}_{j_1 \cdots j_q}$ are the components of $\mathcal{T}$ projected onto the base formed by $\{\mathbf{e}_{j_1}, \ldots, \mathbf{e}_{j_q}\}$ and $\{\mathbf{f}^{l_1}, \ldots, \mathbf{f}^{l_p}\}$, defined by $\mathcal{T}^{l_1 \cdots l_p}_{j_1 \cdots j_q} := \mathcal{T}\left(\mathbf{f}^{l_1}, \ldots, \mathbf{f}^{l_p}, \mathbf{e}_{j_1}, \ldots, \mathbf{e}_{j_q}\right)$.

Any (p, q) tensor $\mathcal{T}$ may thus be decomposed along the tensor base according to the tensor product (2.17):

$$\mathcal{T} := \mathcal{T}^{j_1 \cdots j_q}_{l_1 \cdots l_p}\, \mathbf{f}^{l_1} \otimes \cdots \otimes \mathbf{f}^{l_p} \otimes \mathbf{e}_{j_1} \otimes \cdots \otimes \mathbf{e}_{j_q}$$

In the present book, a tensor will be assimilated to its components, as soon as the vector bases are defined. If $\mathcal{T}^{h_1 \cdots h_r}_{k_1 \cdots k_s}$ constitute the components of a type (r, s) tensor, then under the transformation (2.7)

$$\mathcal{T}^{j_1 \cdots j_r}_{l_1 \cdots l_s} = A^{j_1}_{\alpha_1} \cdots A^{j_r}_{\alpha_r}\, J^{\beta_1}_{l_1} \cdots J^{\beta_s}_{l_s}\, \mathcal{T}^{\alpha_1 \cdots \alpha_r}_{\beta_1 \cdots \beta_s} \tag{2.18}$$

and the corresponding inverse formulation

$$\mathcal{T}^{\alpha_1 \cdots \alpha_r}_{\beta_1 \cdots \beta_s} = J^{\alpha_1}_{j_1} \cdots J^{\alpha_r}_{j_r}\, A^{l_1}_{\beta_1} \cdots A^{l_s}_{\beta_s}\, \mathcal{T}^{j_1 \cdots j_r}_{l_1 \cdots l_s} \tag{2.19}$$

Theorem 2.1. *According to the previous definitions:*

(1) *A scalar Φ is a type $(0, 0)$ tensor which has the same form in any coordinate system: Φ in (y^i), $\overline{\Phi}$ in (x^α) and $\Phi = \overline{\Phi}$.*
(2) *If all tensor components vanish in a coordinate system then they vanish in any other coordinate system.*

2.2.2 *Metric and Galilean Structures*

Considering that the purpose of this work is to limit itself to the spacetimes of relative gravitation, we give in this subsection the bare minimum on the metric tensor and the Galilean structure, in particular to just introduce the three different vectors in relativistic physics that are timelike, lightlike, and spacelike.

Definition 2.7 (Metric). Let $\mathscr{B}$ be a differentiable manifold. A metric $\mathbf{g}$ is a type $(0,2)$ tensor on $\mathscr{B}$ which satisfies the following axioms at each point $P \in \mathscr{B}$:

(1) $\mathbf{g}(\mathbf{u},\mathbf{v}) = \mathbf{g}(\mathbf{v},\mathbf{u})$, $\forall\ \mathbf{u},\mathbf{v} \in T_P\mathscr{B}$
(2) $\mathbf{g}(\mathbf{u},\mathbf{u}) \geq 0$ where the equality holds only when $\mathbf{u} = 0$; $\forall\ \mathbf{u} \in T_P\mathscr{B}$

Projected onto a base $\{\mathbf{e}_1, \ldots, \mathbf{e}_n\}$ of $T_P\mathscr{B}$, components of $\mathbf{g}$ are g_{ij}, and components of its inverse $\mathbf{g}^{-1}$ are g^{ij} with $g^{ik}g_{kj} = \delta^i_j$. Since $g_{ij} = g_{ji}$ the eigenvalues of the matrix are real and strictly positive (or negative depending on the convention) and their eigenvectors orthogonal. We say that $\mathbf{g}$ is a pseudo-Riemann metric if it satisfies the first axiom and the second is replaced by $\mathbf{g}(\mathbf{u},\mathbf{v}) = 0$ for any $\mathbf{u} \in T_P\mathscr{B}$, then $\mathbf{v} = 0$. For pseudo-Riemann metric, some of the eigenvalues are negative. Let j be the number of negative (or positive depending on the convention) eigenvalues.

Definition 2.8 (Riemann manifold). A differential manifold $\mathscr{B}$ endowed with a (pseudo-)metric $(\mathscr{B}, \mathbf{g})$ is a (pseudo-)Riemann manifold.

In the particular Riemann case where one can find a global coordinate system $(x_1, \ldots, x_n)$ on the whole manifold such that the components of the metric tensor $\mathbf{g}$ are given by $g_{ij} = \delta_{ij}$, then $\mathscr{B}$ is a *Euclidean manifold.*

For $j = 1$, a pseudo-Riemann four-dimensional manifold $(\mathscr{B}, \mathbf{g})$ is called Lorentzian. On such a manifold, the tangent 4-vectors $\mathbf{u} \in T_P\mathscr{B}$ can be divided into three classes (with the convention $j = 1$ being the number of positive eigenvalues):

(1) $g_{\alpha\beta}u^\alpha u^\beta < 0$, vector $\mathbf{u}$ is spacelike
(2) $g_{\alpha\beta}u^\alpha u^\beta = 0$, vector $\mathbf{u}$ is lightlike
(3) $g_{\alpha\beta}u^\alpha u^\beta > 0$, vector $\mathbf{u}$ is timelike

in the framework of relativistic mechanics.

Definition 2.9 (Galilean structure). Let $\mathscr{B}$ be a four-dimensional differentiable manifold, endowed with a $(0,2)$ tensor field $\mathbf{g}$ and a nonsingular

1-form τ. The triplet $(\mathscr{B}, \mathbf{g}, \tau)$ constitutes a Galilean structure provided the two following conditions are satisfied:

(1) Tensor $\mathbf{g}$, called space metric, is symmetric and degenerate of rank 3 (three non-zero eigenvalues).
(2) There exists a vector field $\mathbf{u} \neq 0$ on $\mathscr{B}$ such that $\mathbf{g}(\mathbf{u}, \mathbf{v}) \equiv 0$ for any vector field $\mathbf{v} \in T_{\mathbf{x}}\mathscr{B}$, and $\tau(\mathbf{u}) = 1$.

This definition introduces the lightlike vector field $\mathbf{u} \in T_{\mathbf{x}}\mathscr{B}$, which is an element of the Galilean structure. The 2-covariant symmetric tensor $\tau \otimes \tau$, degenerate of rank 1 (one non-zero eigenvalue) is called the time metric. The spacetime metric is therefore defined as $\hat{\mathbf{g}} := \tau \otimes \tau - \mathbf{g}$ (hereafter, the hat on the spacetime metric will be dropped if there is no ambiguity). In mechanics, it is worth introducing a local coordinate system $(x^0 := ct, x^1, x^2, x^3)$ in the neighborhood of any event (spacetime point) in such a way that $\tau := cdt$ and for any index i, $\tau(\partial_i) = 0$, meaning that the space coordinates x^i are independent variables with respect to time t e.g. [117]. c denotes the speed of light.

2.2.3 *Volume-Form and Orientation of Manifold*

Governing equations of continuum physics mainly express the conservation laws of some physical quantities such as mass, energy and information, to name but a few. The majority of the models we will work on are based on Lagrangian functions whose actions are integrals of these functions over a finite (or infinite) volume of continuum. The main uses of differential forms concern the integration of physical extensive quantities either within manifolds $\mathscr{B}$ or at their boundary $\partial\mathscr{B}$. For that purpose, a coordinate-free theory of integration is necessary on manifolds.

Definition 2.10 (Volume-form). A volume-form on a manifold $\mathscr{B}$ of dimension n is a nowhere vanishing n-form ω_n on $\mathscr{B}$.

A manifold $\mathscr{B}$ of dimension n is orientable if there exists an n-form, denoted ω_n, which vanishes nowhere e.g. [110]. Accordingly, a volume-form on an n-dimensional spacetime $\mathscr{B}$ is a nowhere vanishing n-form on $\mathscr{B}$. A particular n-form is obtained as $\mathbf{f}^1 \wedge \mathbf{f}^2 \cdots \wedge \mathbf{f}^n$ where $\{\mathbf{f}^1, \mathbf{f}^2, \ldots, \mathbf{f}^n\}$ is the dual base of $\{\mathbf{e}_1, \mathbf{e}_2, \ldots, \mathbf{e}_n\}$. We then have: $\left(\mathbf{f}^1 \wedge \mathbf{f}^2 \cdots \wedge \mathbf{f}^n\right)(\mathbf{e}_1, \mathbf{e}_2, \ldots, \mathbf{e}_n) = 1$. For any spacetime continuum, a volume-form may thus be designed as $\omega := f(\mathbf{x})\, \omega_n$ provided any strictly positive function $f : \mathbf{x} \in \mathscr{B} \to f(\mathbf{x}) \in \mathbb{R}_+^*$. Given the n-form $\mathbf{f}^1 \wedge \mathbf{f}^2 \cdots \wedge \mathbf{f}^n \neq 0$, we can define other volume-forms

$\omega_n := f\ \mathbf{f}^1 \wedge \mathbf{f}^2 \cdots \wedge \mathbf{f}^n$, where $f(\mathbf{x}) \in \mathbb{R}$ is a positive non-vanishing C^∞ scalar function. There is an infinity of volume-forms on a manifold. For (pseudo-)Riemann manifold $(\mathscr{B}, \mathbf{g})$, there exists a natural volume-form which is invariant under coordinate transformation, say in a coordinate system $(x^1, \ldots, x^n)$ e.g. [5].

Definition 2.11. Let $\mathscr{B}$ be a Riemann manifold of dimension n endowed with a metric $\mathbf{g}$, and with a coordinate system $(x^1, \ldots, x^n)$. We define the Riemann volume-form as the n-form:

$$\boxed{\overline{\omega}_n := \sqrt{|\mathrm{Det}\mathbf{g}|}\ dx^1 \wedge dx^2 .. \wedge dx^n} \tag{2.20}$$

Remark 2.5. The volume form $\omega_n := \sqrt{|\mathrm{Det}\mathbf{g}|}\ dx^1 \wedge dx^2 .. \wedge dx^n$ is invariant under the passive diffeomorphism (2.7). In terms of the y-coordinates, the volume-form holds as (the metric is written more explicitly) the extension of the area form:

$$\omega_n = \sqrt{|\mathrm{Det}\left(J_i^\alpha J_j^\beta g_{\alpha\beta}\right)|}\ dy^1 \wedge dy^2 .. \wedge dy^n \tag{2.21}$$

Given that $dy^i := A_\alpha^i dx^\alpha$, we obtain the volume-form in terms of y-coordinates $\omega_n := \pm\sqrt{|\mathrm{Det}\mathbf{g}|}\ dy^1 \wedge dy^2 .. \wedge dy^n$ where $\sqrt{|\mathrm{Det}\mathbf{g}|}$ is the determinant with the coordinates y^i. The sign $\pm$ may be dropped if we consider a passive diffeomorphism with strictly positive Jacobian everywhere.

Remark 2.6. Two volume-forms ω_n and ω_n' on $\mathscr{B}$ are said to be equivalent if $\omega_n = f\ \omega_n'$ for some scalar function on $\mathscr{B}$, say $f \in \mathscr{C}^\infty$ with $f > 0$ at every point $M \in \mathscr{B}$. For example $dx \wedge dy \wedge dz$ is not equivalent to $dy \wedge dx \wedge dz$, and $dx \wedge dy \wedge dz$ (Cartesian coordinate system) is not equivalent to $dr \wedge d\varphi \wedge d\theta$, where (r, θ, φ) is the standard spherical coordinate system. Conversely, $dx \wedge dy \wedge dz$ is equivalent to $dr \wedge d\theta \wedge d\varphi$ for $r \neq 0$, and $\theta \in]0, \pi[$ since $dx \wedge dy \wedge dz = r^2 \sin\theta dr \wedge d\theta \wedge d\varphi$.

Definition 2.12. On an oriented metric manifold $\mathscr{B}$ of dimension n, the Lagrangian (n-form) is defined by Lagrangian density $\mathscr{L}$ multiplied by the volume form ω_n, and thus defines the action:

$$\mathscr{S} := \int_{\mathscr{B}} \mathscr{L}\ \omega_n \tag{2.22}$$

This allows us to integrate the Lagrangian $\mathscr{L}$ over the oriented manifold $\mathscr{B}$. The scalar field $\mathscr{L}$ behaves like a tensor density e.g. [110] and not as a scalar (tensor), but if we include the term $\sqrt{|\mathrm{Det}\mathbf{g}|}$ in the Lagrangian density, we can define a scalar (tensor) field $\mathscr{L}\sqrt{|\mathrm{Det}\mathbf{g}|} \to \mathscr{L}$ where the factor $\sqrt{|\text{Det}\mathbf{g}|}$ is included in order to have the proper weight. The presence of this factor is essential to recover the Einstein field equations in vacuo. The Lagrangian density cannot explicitly depend on the coordinates (x^μ) by virtue of the Lorentz translational invariance as shown by Kibble in e.g. [118].

2.3 Affine Connection

Connection was already implicit in Newton's work (1687) concerning particle acceleration when it was necessary to calculate the difference between two velocities of a same point located in two different places. The intuition of parallel transport was already present. Subsequently, Gauss (1827) and Riemann (1854) considered vector transport on surfaces equipped with metric tensors, based essentially on geodesics. Christoffel (1869) highlighted the relationships between connection coefficients and derivatives of metric by comparing the value of metric at two neighboring points. Later, Ricci and Levi-Civita (1900) formally laid more conceptual foundations of connection by considering the invariance of the Laplace-Beltrami equation [119]. As far as we are concerned, for electromagnetism Levi-Civita (1917) and Weyl (1918) derived the equations in relation to connections. Closer to us, it was the work of Elie Cartan which made it possible to separate metric and connection on a manifold (1923, 1924) in relativity.

2.3.1 *Intuitive Introduction*

The concept of affine connection may be seen in Figure 2.5 where each material point is represented by continuum grains (rigorously containing infinite number of particles) surrounded by an infinitesimal volume of matter. Two adjacent grains may be moved to be superimposed by a transformation not possibly reduced to an isometry. For instance, grain interfaces are sources of dislocation during plastic flow of polycrystalline solids. Dislocations may multiply within grains by a multiple cross-slip process.

Geometrically, the word *connection*, whose mathematical symbol is ∇, comes from a geometric origin e.g. [2]. Consider a vector field $\mathbf{v}$ on the manifold $\mathscr{B}$, and if ∇ is a connection on the tangent bundle (set of all tangent spaces over the manifold), then one defines the vector field $\nabla_{\mathbf{u}}\mathbf{v}$ which is assumed to describe the change of the vector field along the flow of the vector field $\mathbf{u}$. Of course, the problem is that the vector field $\mathbf{v}$ is considered here at two different points of the manifold and then belongs

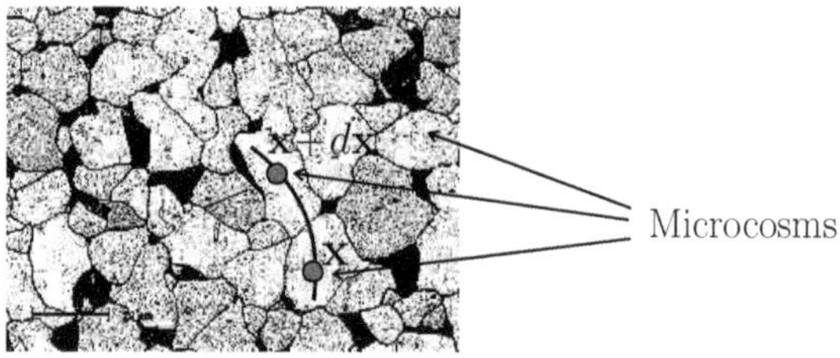

Fig. 2.5: Two connected material points x^μ and $x^\mu + dx^\mu$ along a material line. At this length scale $\ell \simeq \mu m$, the metric alone is not sufficient for modeling the shape change due to relative motions of grains.

to different tangent spaces. The connection is the operation which relates them. The same dissection of spacetime holds relativistically to obtain local inertial microcosms — attached to local flat inertial frames of reference. From these microcosms, in which the special relativity theory holds, the global structure of the gravitational field is reconstructed after introducing curvature to obtain a continuous bound of the whole.

2.3.2 *Affine Connection, Affinely Connected Manifold*

A connection specifies how tensors are transported along a curve on a manifold. To start with, a local tangent base is associated with a coordinate system: the contravariant base $\{\mathbf{e}_\alpha\}$ is associated with the system $\{y^\alpha\}$.

Definition 2.13 (Affine connection). An affine connection ∇ on a manifold $\mathscr{B}$ is a map defined by ∇: $(\mathbf{u}, \mathbf{v}) \in T_P\mathscr{B} \times T_P\mathscr{B} \longrightarrow \nabla_\mathbf{u}\mathbf{v} \in T_P\mathscr{B}$ which satisfies the following conditions (λ and μ are scalars, ϕ is the scalar field)

(1) $\nabla_{\lambda\mathbf{u}_1+\mu\mathbf{u}_2}\mathbf{v} = \lambda\nabla_{\mathbf{u}_1}\mathbf{v} + \mu\nabla_{\mathbf{u}_2}\mathbf{v}$
(2) $\nabla_\mathbf{u}(\lambda\mathbf{v}_1 + \mu\mathbf{v}_2) = \lambda\nabla_\mathbf{u}\mathbf{v}_1 + \mu\nabla_\mathbf{u}\mathbf{v}_2$
(3) $\nabla_{\phi\mathbf{u}}\mathbf{v} = \phi\nabla_\mathbf{u}\mathbf{v}$
(4) $\nabla_\mathbf{u}(\phi\mathbf{v}) = \phi\nabla_\mathbf{u}\mathbf{v} + \mathbf{u}(\phi)\mathbf{v}$

For convenience, we define a local base whenever necessary without losing the generality of our purpose. Connection coefficients Γ^c_{ab} are implicitly defined by $\nabla_{\mathbf{e}_a}\mathbf{e}_b := \Gamma^c_{ab}\mathbf{e}_c$. $\nabla_\mathbf{u}$ represents a covariant derivative along the direction $\mathbf{u} \in T_P\mathscr{B}$. It generalizes the derivative of tensor fields on manifolds. The covariant derivative of a type (p, q) tensor is a type $(p, q+1)$ tensor. For example, the covariant derivative of a scalar field ϕ and a vector field $\mathbf{w}$ along the vector $\mathbf{e}_k$ (a vector of the tangent space $T_P\mathscr{B}$) may be expressed in terms of their components on the base $\{\mathbf{e}_a\}$ associated to coordinate system (y^a): $\nabla_{\mathbf{e}_k}\phi = \partial_k\phi$, and $\nabla_{\mathbf{e}_k}\mathbf{w} = (\partial_k w^a + \Gamma^a_{kc} w^c)\,\mathbf{e}_a$. By definition, the covariant derivative of a scalar field is a 1-form $\nabla\phi := \partial_k\phi\ \mathbf{e}^k = \nabla_k\phi\ \mathbf{e}^k$. The covariant derivative of a vector field is a $(1,1)$ type tensor:

$$\begin{aligned}\nabla\mathbf{w} &= \nabla_{\mathbf{e}_k}(w^a\mathbf{e}_a)\otimes\mathbf{e}^k = [(\nabla_{\mathbf{e}_k}w^a)\,\mathbf{e}_a + w^c\nabla_{\mathbf{e}_k}\mathbf{e}_c]\otimes\mathbf{e}^k\\ &= (\partial_k w^a + \Gamma^c_{kc} w^c)\,\mathbf{e}_a\otimes\mathbf{e}^k := \nabla_k w^a\ \mathbf{e}_a\otimes\mathbf{e}^k\end{aligned}$$

For tensor components, we adopt whenever needed the notations $\partial_k(..) = (..)_{,k}$ and $\nabla_{\mathbf{e}_k}(..) = (..)_{|k}$. This allows us to define the components of the covariant derivative of a vector $\mathbf{w}$, and that of a 1-form ω on a manifold $\mathscr{B}$, as: $\nabla_k w^a := \partial_k w^a + \Gamma^a_{kc}\ w^c$, and $\nabla_k\omega_a := \partial_k w_a - \Gamma^c_{ka}\ \omega_c$ by using the reciprocity relation $\omega_a(\mathbf{w}^b) := \delta^b_a$ for the two dual fields.

Definition 2.14 (Affinely connected manifold). A differential manifold $\mathscr{B}$ endowed with an affine connection ∇ is an affinely connected manifold.

Definition 2.15 (Metric-compatible connection). On a manifold $\mathscr{B}$ endowed with connection ∇, and metric $\mathbf{g}$, ∇ is metric-compatible if and only if $\nabla\mathbf{g} \equiv 0$.

Definition 2.16 (Metric-affine manifold). A manifold $\mathscr{B}$ endowed with a metric $\mathbf{g}$ and an affine connection ∇ is called metric-affine (connected) manifold.

Connection also plays a central role in continuum physics since it represents the evolution of material manifold topology during non-holonomic deformation e.g. [49]. It supports dynamical concepts in four-dimensional spacetime, such as reference frames e.g. [112].

2.3.3 *Transformation Rules for Connection*

We derive these laws in component-form. Consider a passive diffeomorphism, that is a change of coordinate system on $\mathscr{B}$:

$$x^\alpha \to \tilde{x}^\mu = \tilde{x}^\mu(x^\alpha) \quad \text{and its inverse} \quad x^\alpha = x^\alpha(\tilde{x}^\mu) \tag{2.23}$$

Elementary formulations of the (same ∇) connection coefficients with respect to these two coordinate systems are given by:

$$\frac{\partial\mathbf{e}_\beta}{\partial x^\alpha} := \Gamma^\gamma_{\alpha\beta}\ \mathbf{e}_\gamma \quad \frac{\partial\tilde{\mathbf{e}}_\nu}{\partial\tilde{x}^\mu} := \tilde{\Gamma}^\lambda_{\mu\nu}\ \tilde{\mathbf{e}}_\lambda \quad \tilde{\mathbf{e}}_\nu = \frac{\partial x^\beta}{\partial\tilde{x}^\nu}\ \mathbf{e}_\beta \tag{2.24}$$

where $\mathbf{e}_\beta := \partial_{x^\beta}$ and $\tilde{\mathbf{e}}_\nu := \partial_{\tilde{x}^\nu}$ are the two local vector bases associated to the two coordinate systems (x^β) and $(\tilde{x}^\nu)$ respectively. We deduce:

$$\frac{\partial}{\partial \tilde{x}^\mu}\left(\frac{\partial x^\beta}{\partial \tilde{x}^\nu}\,\mathbf{e}_\beta\right) = \tilde{\Gamma}^\lambda_{\ \mu\nu}\left(\frac{\partial x^\sigma}{\partial \tilde{x}^\lambda}\,\mathbf{e}_\sigma\right)$$

which gives, by using appropriate indices to have the projection onto vector $\mathbf{e}_\gamma$:

$$\frac{\partial^2 x^\gamma}{\partial \tilde{x}^\mu \partial \tilde{x}^\nu}\,\mathbf{e}_\gamma + \frac{\partial x^\beta}{\partial \tilde{x}^\nu}\,\frac{\partial x^\alpha}{\partial \tilde{x}^\mu}\,\Gamma^\gamma_{\ \alpha\beta}\,\mathbf{e}_\gamma = \tilde{\Gamma}^\lambda_{\ \mu\nu}\,\frac{\partial x^\gamma}{\partial \tilde{x}^\lambda}\,\mathbf{e}_\gamma$$

By multiplying each side of the equation with $\partial\tilde{x}^\sigma/\partial x^\gamma$ we deduce the transformation rules of the connection coefficients following a passive diffeomorphism:

$$\boxed{\tilde{\Gamma}^\sigma_{\ \mu\nu} = \frac{\partial x^\alpha}{\partial \tilde{x}^\mu}\,\frac{\partial x^\beta}{\partial \tilde{x}^\nu}\,\frac{\partial \tilde{x}^\sigma}{\partial x^\gamma}\,\Gamma^\gamma_{\ \alpha\beta} + \frac{\partial^2 x^\gamma}{\partial \tilde{x}^\mu \partial \tilde{x}^\nu}\,\frac{\partial \tilde{x}^\sigma}{\partial x^\gamma}} \tag{2.25}$$

This is a classical result and the presence of the second term on the right-hand-side of (2.25) shows that *the connection is not a tensor* e.g. [5]. In the chapter on Einstein's relative gravitation, we will give more details on the relation of connection to the inertial frame of reference.

2.3.4 *Practical Formula for Covariant Derivative*

There are numerous index conventions for the component formulation of the covariant derivative. We recall the covariant derivative of general tensor $\mathbf{T} = (T^{\alpha_1\cdots\alpha_p}_{\beta_1\cdots\beta_q})$:

$$\begin{aligned}\nabla_\gamma T^{\alpha_1\cdots\alpha_p}_{\beta_1\cdots\beta_q} &= \partial_\gamma T^{\alpha_1\cdots\alpha_p}_{\beta_1\cdots\beta_q} + \sum_{s=1}^{s=p} \Gamma^{\alpha_s}_{\gamma\mu} T^{\alpha_1\cdots\alpha_{s-1}\ \mu\ \alpha_{s+1}\cdots\alpha_p}_{\beta_1\cdots\beta_q} \\ &\quad - \sum_{s=1}^{s=q} \Gamma^{\mu}_{\gamma\beta_s} T^{\alpha_1\cdots\alpha_p}_{\beta_1\cdots\beta_{s-1}\ \mu\ \beta_{s+1}\cdots\beta_q}\end{aligned} \tag{2.26}$$

where the index γ is the first covariant index of connection coefficients for the second $\Gamma^{\alpha_s}_{\gamma\mu}$ and third $\Gamma^{\mu}_{\gamma\beta_s}$ terms. Some authors use the transpose covariant derivative and thus put the γ index at the second place e.g. [110]. For illustration, the covariant derivative of tensor $T^{\alpha_1\alpha_2}$, $T_{\beta_1\beta_2}$, $T^{\alpha_1}_{\beta_1\beta_2}$, and

$T^{\alpha_1}{}_{\beta_1\beta_2\beta_3}$ are respectively:

$$\begin{aligned}
\nabla_\gamma T^{\alpha_1\alpha_2} &= \partial_\gamma T^{\alpha_1\alpha_2} + \Gamma^{\alpha_1}_{\gamma\mu} T^{\mu\alpha_2} + \Gamma^{\alpha_2}_{\gamma\mu} T^{\alpha_1\mu} \\
\nabla_\gamma T_{\beta_1\beta_2} &= \partial_\gamma T_{\beta_1\beta_2} - \Gamma^{\mu}_{\gamma\beta_1} T_{\mu\beta_2} - \Gamma^{\mu}_{\gamma\beta_2} T_{\beta_1\mu} \\
\nabla_\gamma T^{\alpha_1}{}_{\beta_1\beta_2} &= \partial_\gamma T^{\alpha_1}{}_{\beta_1\beta_2} + \Gamma^{\alpha_1}_{\gamma\mu} T^{\mu}{}_{\beta_1\beta_2} - \Gamma^{\mu}_{\gamma\beta_1} T^{\alpha_1}{}_{\mu\beta_2} - \Gamma^{\mu}_{\gamma\beta_2} T^{\alpha_1}{}_{\beta_1\mu} \\
\nabla_\gamma T^{\alpha_1}{}_{\beta_1\beta_2\beta_3} &= \partial_\gamma T^{\alpha_1}{}_{\beta_1\beta_2\beta_3} + \Gamma^{\alpha_1}_{\gamma\mu} T^{\mu}{}_{\beta_1\beta_2\beta_3} - \Gamma^{\mu}_{\gamma\beta_1} T^{\alpha_1}{}_{\mu\beta_2\beta_3} \\
&\quad - \Gamma^{\mu}_{\gamma\beta_2} T^{\alpha_1}{}_{\beta_1\mu\beta_3} - \Gamma^{\mu}_{\gamma\beta_3} T^{\alpha_1}{}_{\beta_1\beta_2\mu}
\end{aligned}$$

2.3.5 *Levi-Civita Connection and Christoffel's Symbols*

Metric and connection need not depend on each other. However, the metric alone determines a particular connection $\overline{\nabla}$. Let's consider a particular and important connection deduced from the metric and for which the two lower indices can be interverted (symmetric), and denoted $\overline{\nabla}$. This is called Levi-Civita connection.

Proposition 2.1 (Christoffel's symbols). *Let's consider a connection $\overline{\nabla}$ on the metric manifold $\mathscr{B}$. Assume that this connection is compatible with the metric* $\mathbf{g}$ *and such that its coefficients are symmetric with respect to the lower indices. Then its coefficients are given by:*

$$\boxed{\overline{\Gamma}^{\gamma}_{\alpha\beta} = \frac{1}{2} g^{\gamma\mu} \left(\partial_\beta g_{\alpha\mu} + \partial_\alpha g_{\mu\beta} - \partial_\mu g_{\alpha\beta}\right)} \tag{2.27}$$

Proof. It is classical to consider the coefficients on a base. First, let's consider the metric compatibility:

$$\overline{\nabla}_\mu g_{\alpha\beta} = \partial_\mu g_{\alpha\beta} - \overline{\Gamma}^{\gamma}_{\mu\alpha}\, g_{\gamma\beta} - \overline{\Gamma}^{\gamma}_{\mu\beta}\, g_{\alpha\gamma} = 0$$

This allows us to write by permuting indices:

$$\begin{aligned}
\partial_\mu g_{\alpha\beta} &= \overline{\Gamma}^{\gamma}_{\mu\alpha}\, g_{\gamma\beta} + \overline{\Gamma}^{\gamma}_{\mu\beta}\, g_{\alpha\gamma} && \text{(a)} \\
\partial_\alpha g_{\mu\beta} &= \overline{\Gamma}^{\gamma}_{\alpha\mu}\, g_{\gamma\beta} + \overline{\Gamma}^{\gamma}_{\alpha\beta}\, g_{\mu\gamma} && \text{(b)} \\
\partial_\beta g_{\alpha\mu} &= \overline{\Gamma}^{\gamma}_{\beta\alpha}\, g_{\gamma\mu} + \overline{\Gamma}^{\gamma}_{\beta\mu}\, g_{\alpha\gamma} && \text{(c)}
\end{aligned}$$

Second, consider (c) + (b) – (a) and using the symmetry of the metric and the connection (for the lower indices), we easily obtain:

$$\overline{\Gamma}^{\gamma}_{\alpha\beta} = \frac{1}{2} g^{\gamma\mu} \left(\partial_\beta g_{\alpha\mu} + \partial_\alpha g_{\mu\beta} - \partial_\mu g_{\alpha\beta}\right)$$

□

Connection coefficients (2.27) are called Christoffel's symbols. The discovery of Christoffel's symbols in 1869 was really a starting point of tensor analysis mainly on Riemann manifolds e.g. [119]. They had an important role in the development of Einstein's relative gravitation.

Parallel transport of Levi-Civita. Say a vector field $\mathbf{v}$ on the Riemann manifold $\mathscr{B}$, defined along a curve $\mathscr{C}$ with parameter $t \in \mathbb{R}$, such that $x^\mu = x^\mu(t)$ for any point of the curve. The derivative of the vector field with respect to the parameter is defined, as usual, as:

$$\frac{d\mathbf{v}}{dt} := \lim_{\epsilon \to 0} \frac{1}{\epsilon} \left[\mathbf{v}(t+h) - \mathbf{v}(t)\right] \tag{2.28}$$

At each point $\mathbf{x} \in \mathscr{C} \subset \mathscr{B}$, the tangent to the curve $\mathscr{C}$ is locally defined by its tangent vector:

$$\mathbf{u} := \frac{d\mathbf{x}}{dt} \tag{2.29}$$

Consider a base $(\mathbf{f}_1, \ldots, \mathbf{f}_n)$ on tangent space $T_{\mathbf{x}}\mathscr{B}$. Thanks to the properties of connection $\overline{\nabla}$, the covariant derivative of the vector field $\mathbf{v}$ along the curve $\mathscr{C}$ is thus defined by:

$$\overline{\nabla}_{\mathbf{u}}\mathbf{v} = \overline{\nabla}_{(u^\mu\ \mathbf{f}_\mu)}\ (v^\gamma\ \mathbf{f}_\gamma) = u^\mu\ (\partial_\mu v^\gamma + \overline{\Gamma}^\gamma_{\mu\beta}\ v^\beta)\ \mathbf{f}_\gamma$$

where $\mathbf{u}$ is the tangent vector along the curve. This allows us to write transport (or displacement) of the vector along the curve:

$$\overline{\nabla}_{\mathbf{u}}\mathbf{v} = \left(\frac{dv^\gamma}{dt} + \overline{\Gamma}^\gamma_{\alpha\beta}\ u^\alpha v^\beta\right) \mathbf{f}_\gamma \tag{2.30}$$

The transport of the vector $\mathbf{v}(t)$ along the curve $\mathscr{C}$ is called the parallel displacement of Levi-Civita e.g. [120]. As we shall see in the general case, there are an infinite number of distinct connections on a manifold $\mathscr{B}$. For a Riemann manifold, we have the following theorem.

Theorem 2.2. *Let $\mathscr{B}$ be a Riemann manifold endowed with a metric $\mathbf{g}$. Then, there is a unique symmetric connection that satisfies the Leibnitz rule:*

$$\frac{d}{dt}\left[\mathbf{g}(\mathbf{v}, \mathbf{w})\right] = \mathbf{g}\left(\frac{\nabla \mathbf{v}}{dt}, \mathbf{w}\right) + \mathbf{g}\left(\mathbf{v}, \frac{\nabla \mathbf{w}}{dt}\right) \tag{2.31}$$

for any pair of vector fields $(\mathbf{v}, \mathbf{w})$ on $\mathscr{B}$ defined along a parametrized curve $\mathscr{C}$, and this connection is the Levi-Civita connection $\nabla = \overline{\nabla}$, defined by (2.27).

Proof. It is obvious by considering two vector fields defined on the neighborhood of the curve $\mathscr{C}$. By hypothesis, we have:

$$\frac{\partial}{\partial x^\gamma}\left(g_{\alpha\beta}\ v^\alpha w^\beta\right) \equiv g_{\alpha\beta}\left(\frac{\partial v^\alpha}{\partial x^\gamma} + \Gamma^\alpha_{\gamma\mu}\ v^\mu\right) w^\beta + g_{\alpha\beta}\ v^\alpha\left(\frac{\partial w^\beta}{\partial x^\gamma} + \Gamma^\beta_{\gamma\nu}\ w^\nu\right)$$

$$= \frac{\partial g_{\alpha\beta}}{\partial x^\gamma}\ v^\alpha w^\beta + g_{\alpha\beta}\ \frac{\partial v^\alpha}{\partial x^\gamma}\ w^\beta + g_{\alpha\beta}\ v^\alpha\ \frac{\partial w^\beta}{\partial x^\gamma}$$

Since it is assumed that this relation holds for any vector fields $\mathbf{v}$ and $\mathbf{w}$, we deduce:

$$\frac{\partial g_{\alpha\beta}}{\partial x^\gamma} = g_{\mu\beta}\ \Gamma^\mu_{\gamma\alpha} + g_{\alpha\mu}\ \Gamma^\mu_{\gamma\beta}$$

By solving the coefficients, we see that it is exactly the same equation as for the Levi-Civita connection. This shows that the coefficients of the guess symmetric connection ∇ reduce to Christoffel's symbols (2.27):

$$\Gamma^\gamma_{\alpha\beta} = \overline{\Gamma}^\gamma_{\alpha\beta} = \frac{1}{2} g^{\gamma\mu}\left(\partial_\beta g_{\alpha\mu} + \partial_\alpha g_{\mu\beta} - \partial_\mu g_{\alpha\beta}\right)$$

This solution is accordingly unique. □

Remark 2.7. Thanks to locally calculated connection coefficients, we can extend by defining in each atlas of the manifold (coordinate system) $\Gamma^\gamma_{\alpha\beta} = \overline{\Gamma}^\gamma_{\alpha\beta}$. The uniqueness we have proven shows that the locally defined covariant derivative in the atlases agrees in each overlap and thus implies the connection globally.

2.3.6 *Torsion and Curvature*

Say a manifold $\mathscr{B}$ is endowed with an affine connection ∇. Covariant derivative of a vector field $\mathbf{w}$ with components (w^a) (considered as $(1,0)$ tensor) is a $(1,1)$ tensor field with components $\nabla_k w^a$. The second covariant derivative leads to a $(1,2)$ tensor with components $\nabla_l \nabla_k w^a$ which may be different to $\nabla_k \nabla_l w^a$ even if $w^a(\mathbf{x})$ is twice differentiable with respect to components x^b. Torsion and curvature fields are introduced for that purpose.

Definition 2.17. Let $(\mathscr{B}, \nabla)$ be an affinely connected manifold. Lie-Jacobi brackets and Cartan coefficients of structure are defined by: $[\mathbf{u}, \mathbf{v}](\Phi) := \mathbf{uv}(\Phi) - \mathbf{vu}(\Phi)$, and then $[\mathbf{e}_\alpha, \mathbf{e}_\beta] := \aleph^\gamma_{0\alpha\beta}\ \mathbf{e}_\gamma$, where Φ is a scalar field on $\mathscr{B}$, with $\mathbf{u}$ and $\mathbf{v}$ vector fields (Fig. 2.6).

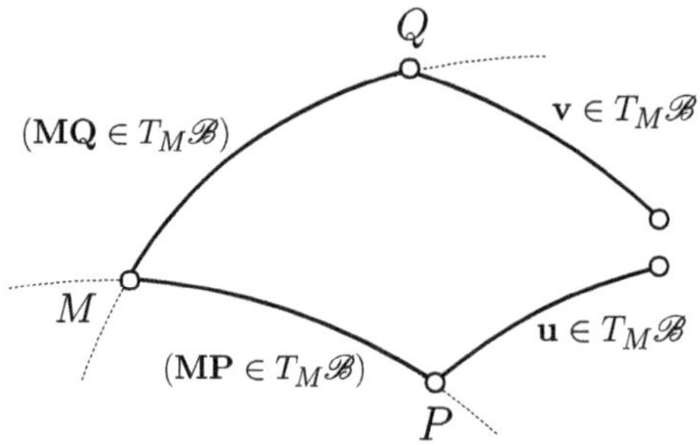

Fig. 2.6: Cartan parallelogram. Let M be a point on the manifold $\mathscr{B}$ and (MP) and (MQ) two curves intersecting at M. Both $\mathbf{MP}$ and $\mathbf{MQ}$ are assumed infinitesimal and constitute two vectors of the tangent space $T_M\mathscr{B}$. Vectors $\mathbf{u}$ and $\mathbf{v}$ are the images of $\mathbf{MQ}$ and $\mathbf{MP}$ by parallel transport along (MP) and (MQ) respectively. The torsion measures the disclosure of the parallelogram made up of infinitesimal vectors $\mathbf{MP}$ and $\mathbf{MQ}$ and their parallel transport $\mathbf{v}$ and $\mathbf{u}$.

Considering two vector bases on the tangent space $T_M\mathscr{B}$, there are two possibilities: (a) Case where $(\mathbf{e}_1, \ldots, \mathbf{e}_n)$ is a base associated to a global coordinate system $(x^1, \ldots, x^n)$, then $\aleph^{\gamma}_{0\alpha\beta} \equiv 0$; (b) Case where $(\mathbf{e}_1, \ldots, \mathbf{e}_n)$ is a base which cannot be associated to a global coordinate system, then $\aleph^{\gamma}_{0\alpha\beta} \neq 0$. The basis $\{\mathbf{e}_\alpha\}$ is said to be anholonomic when the coefficients of structure do not vanish; the frames $\{\partial_\alpha\}$ are holonomic because its members commute to each other.

Definition 2.18 (Torsion). The torsion tensor $\aleph$ is a type $(1,2)$ tensor

$$\begin{cases} \aleph(\mathbf{f}^\gamma, \mathbf{e}_\alpha, \mathbf{e}_\beta) = \mathbf{f}^\gamma \left(\nabla_{\mathbf{e}_\alpha}\mathbf{e}_\beta - \nabla_{\mathbf{e}_\beta}\mathbf{e}_\alpha - [\mathbf{e}_\alpha, \mathbf{e}_\beta]\right) \\ \aleph^{\gamma}_{\alpha\beta} = \Gamma^{\gamma}_{\alpha\beta} - \Gamma^{\gamma}_{\beta\alpha} - \aleph^{\gamma}_{0\alpha\beta} \end{cases} \tag{2.32}$$

where $\mathbf{e}_\alpha$ and $\mathbf{f}^\gamma$ are vector and 1-form on $\mathscr{T}_M\mathscr{B}$, respectively.

To highlight the role of torsion on a manifold, consider a scalar field $\Phi\left(x^\mu\right)$ twice differentiable on $\mathscr{B}$ $(\mathscr{C}^2)$. Consider the difference between second covariant derivatives, assuming that we use a coordinate vector base:

$$\begin{aligned} \nabla_\alpha\nabla_\beta\Phi - \nabla_\beta\nabla_\alpha\Phi &= (\partial_\alpha\partial_\beta\Phi - \partial_\beta\partial_\alpha\Phi) - (\Gamma^{\gamma}_{\alpha\beta} - \Gamma^{\gamma}_{\beta\alpha})\partial_\gamma\Phi \\ &= -\aleph^{\gamma}_{\alpha\beta}\,\partial_\gamma\Phi = -\aleph^{\gamma}_{\alpha\beta}\nabla_\gamma\Phi \end{aligned} \tag{2.33}$$

showing that on a manifold with torsion, the commutativity of second covariant derivatives is not ensured even for a smooth $\mathscr{C}^2$ scalar field. This demonstrates that torsion is a characteristic of the manifold rather than of the field Φ. Even for a smooth field Φ on the continuum, the

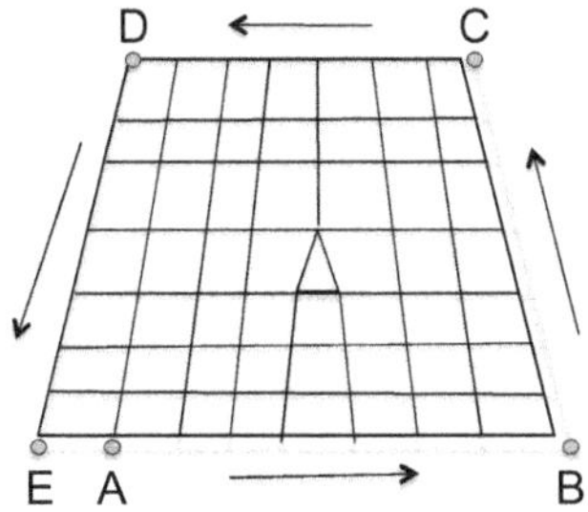

Fig. 2.7: Cartan parallelogram in a discrete crystal lattice. The non-closure of the path $ABCDE$ measures the number of defects inside the parallelogram. Only one defect (triangle) is represented here. The opening **EA** defines the so-called Bürgers vector.

second (covariant) derivative may be non-commutative. If $\aleph^{\gamma}_{\alpha\beta}$ vanish in some coordinate system, they will vanish in any other coordinate system. The skew symmetry does not depend on the choice of coordinate system. Torsion tensor may be associated with defects in a continuum and crystal lattice. It is well known that defects are often due to the non-integrability of displacement and of gradient of displacement as sketched on Figure 2.7. Torsion measures the density of defects on this figure, which reports the discrete lattice version of the continuum Cartan parallelogram disclosure. The Cartan disclosure may be associated with defects in matter, namely for crystalline solids with dislocations e.g. [11], but also for spacetime when considering geometrization of quantum effects e.g. [121]. The Bürgers vector is not topologically invariant conversely to torsion and might not be used to model these defects [122].

Definition 2.19 (Curvature). The curvature tensor $\mathfrak{R}$ is a type $(1,3)$ tensor (Fig. 2.8)

$$\begin{cases} \mathfrak{R}(\mathbf{f}^k, \mathbf{e}_i, \mathbf{e}_j, \mathbf{e}_l) = \mathbf{f}^k(\nabla_{\mathbf{e}_i}\nabla_{\mathbf{e}_j}\mathbf{e}_l - \nabla_{\mathbf{e}_j}\nabla_{\mathbf{e}_i}\mathbf{e}_l - \nabla_{[\mathbf{e}_i,\mathbf{e}_j]}\mathbf{e}_l) \\ \mathfrak{R}^k_{ijl} = (\Gamma^k_{jl,i} + \Gamma^m_{jl}\Gamma^k_{im}) - (\Gamma^k_{il,j} + \Gamma^m_{il}\Gamma^k_{jm}) - \aleph^m_{0ij}\Gamma^k_{ml} \end{cases} \tag{2.34}$$

where $\mathbf{e}_i$ and $\mathbf{f}^i$ are vector and 1-form on $\mathscr{B}$, respectively.

Curvature tensor is also a characteristic of the manifold $\mathscr{B}$, and is skew-symmetric in its first pair of lower indices (i, j). In the following, we assume that the base is always associated to a coordinate system, say $\aleph^m_{0ij} \equiv 0$. In a n-dimensional manifold there are $n^2(n^2-1)/12$ independent components. Because of its symmetry properties, only one tensor can be constructed

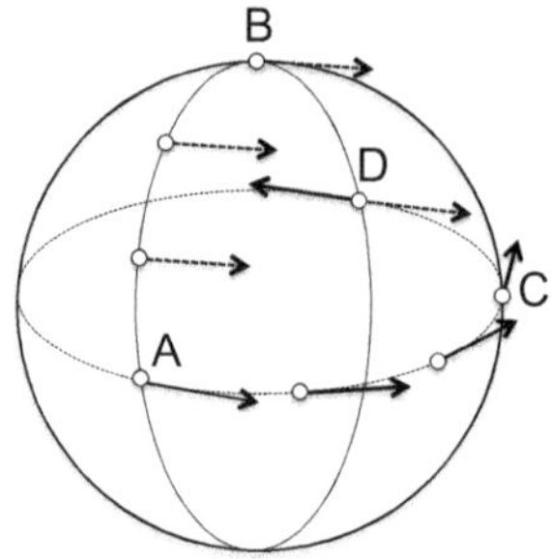

Fig. 2.8: Curvature of a sphere. A vector $\mathbf{w}$ is defined at point A. This vector is (parallel) transported along the path $\widehat{ABD}$ of the vertical circle by maintaining $\mathbf{w}$ parallel to itself and represented by a dotted arrow. The same vector $\mathbf{w}$ is transported along the path $\widehat{ACD}$ along the horizontal circle by maintaining $\mathbf{w}$ tangent to the circle. At point D the two transported vectors, initially the same, are opposite. This happens due to the non-vanishing curvature of the sphere.

from the curvature tensor, namely the Ricci curvature tensor $\Re_{\beta\lambda} := \Re^{\alpha}_{\alpha\beta\lambda}$ with six independent components on a three-dimensional manifold e.g. [5].

Remark 2.8. The existence of torsion and curvature tensors does not require the existence of a metric on the manifold $\mathscr{B}$. The theorem of Frobenius, e.g. [5], allows us to simplify the expression of torsion and curvature projected onto a coordinate base.

Remark 2.9. To a connection $\Gamma^{\gamma}_{\alpha\beta}$ on a differentiable manifold $\mathscr{B}$ we have associated torsion $\aleph^{\gamma}_{\alpha\beta}$ and curvature $\Re^{\gamma}_{\alpha\beta\lambda}$ tensors and their respective higher-order covariant derivatives. Conversely, the knowledge of torsion and curvature on a neighborhood of each point of the manifold also allows us to determine the connection, at least locally.

Remark 2.10. Let $(\mathbf{u}, \mathbf{v}, \mathbf{w})$ be three arbitrary vectors belonging to the tangent space $T_{\mathbf{x}}\mathscr{B}$. Torsion and curvature may be also defined as operators on vectors by writing:

$$\begin{cases} \aleph(\mathbf{u}, \mathbf{v}) := \nabla_{\mathbf{u}}\mathbf{v} - \nabla_{\mathbf{v}}\mathbf{u} - [\mathbf{u}, \mathbf{v}] \\ \Re(\mathbf{u}, \mathbf{v}, \mathbf{w}) := \nabla_{\mathbf{u}}\nabla_{\mathbf{v}}\mathbf{w} - \nabla_{\mathbf{v}}\nabla_{\mathbf{u}}\mathbf{w} - \nabla[\mathbf{u}, \mathbf{v}]\mathbf{w} \end{cases} \tag{2.35}$$

By using the definition of the connection ∇, it is easy to show that for three scalar real functions (f, g, h) on the manifold $\mathscr{B}$, we have $\Re(f\mathbf{u}, g\mathbf{v}, h\mathbf{w}) = fgh\Re(\mathbf{u}, \mathbf{v}, \mathbf{w})$ proving that $\Re$ is a multilinear object. So is the case for $\aleph$.

Theorem 2.3. *Let's consider a vector field* $\mathbf{u}$ *defined on a connected manifold* $\mathscr{B}$, *where* $\{\mathbf{e}_1, \ldots, \mathbf{e}_n\}$ *and* $\{\mathbf{f}^1, \ldots, \mathbf{f}^n\}$ *are vector bases of* $T_M\mathscr{B}$ *and of dual* $T_M^*\mathscr{B}$ *respectively. Then the non-commutativity of the second covariant derivatives of the vector field holds:*

$$\nabla_\alpha \nabla_\beta u^\lambda - \nabla_\beta \nabla_\alpha u^\lambda = -\aleph^\gamma_{\alpha\beta} \nabla_\gamma u^\lambda + \Re^\lambda_{\alpha\beta\gamma} u^\gamma \tag{2.36}$$

Proof. It suffices to write first its definition and then its distributive property:

$$\begin{aligned}(\nabla_\alpha \nabla_\beta u^\lambda - \nabla_\beta \nabla_\alpha u^\lambda)\mathbf{e}_\lambda &:= \nabla_{\mathbf{e}_\alpha} \nabla_{\mathbf{e}_\beta} (u^\lambda \mathbf{e}_\lambda) - \nabla_{\mathbf{e}_\beta} \nabla_{\mathbf{e}_\alpha} (u^\lambda \mathbf{e}_\lambda) \\ &= \left[\nabla_{\mathbf{e}_\alpha} \left(\nabla_{\mathbf{e}_\beta} u^\lambda\right) \mathbf{e}_\lambda + \nabla_{\mathbf{e}_\alpha} \left(\nabla_{\mathbf{e}_\beta} \mathbf{e}_\lambda\right) u^\lambda\right] \\ &\quad - \left[\nabla_{\mathbf{e}_\beta} \left(\nabla_{\mathbf{e}_\alpha} u^\lambda\right) \mathbf{e}_\lambda + \nabla_{\mathbf{e}_\beta} \left(\nabla_{\mathbf{e}_\alpha} \mathbf{e}_\lambda\right) u^\lambda\right] \\ &= \left(-\aleph^\gamma_{\alpha\beta} \nabla_\gamma u^\lambda + \Re^\lambda_{\alpha\beta\gamma} u^\gamma\right) \mathbf{e}_\lambda\end{aligned}$$

thanks to the definition of torsion and curvature, together with (2.33). This relation is valid in any coordinate base. □

For a compatible connection, it is obvious to derive the relation:

$$\nabla_\alpha \nabla_\beta u_\lambda - \nabla_\beta \nabla_\alpha u_\lambda = -\aleph^\gamma_{\alpha\beta} \nabla_\gamma u_\lambda + g_{\lambda\mu} g^{\gamma\nu} \Re^\mu_{\alpha\beta\gamma} u_\nu$$

Remark 2.11. Consider $\{\mathbf{e}_1, \ldots, \mathbf{e}_n\}$ and $\{\mathbf{f}^1, \ldots, \mathbf{f}^n\}$ which are coordinate basis and coordinate dual basis. It should be mentioned that even components $u^\lambda(x^\alpha) \in \mathscr{C}^2$ are twice differentiable; the non-commutativity of second covariant derivatives comes from the twisting and the curving of the manifold $\mathscr{B}$.

Remark 2.12. In view of the property (2.36) the curvature and the torsion tensors may induce independently the non-commutativity of covariant derivatives of a vector field $\mathbf{u}$ on a curved and twisted manifold $\mathscr{B}$.

Torsion and curvature tensors are intrinsic properties of a connection, and then of the metric-affine continuum manifold $(\mathscr{B}, \mathbf{g}, \nabla)$. In relativity, affine connection may be interpreted from two different point of views. Torsion and curvature may represent force field (usually the field for the curvature) and the geodesics of connections define the classical inertial path of free particles in standard relativity. The second utilization of affine connection is the curved (and/or twisted) spacetime where the material body is evolving e.g. [123]. For a torsionless metric-affine manifold $\mathscr{B}$, the curvature tensor $\Re^\gamma_{\alpha\beta\lambda} \equiv 0$ identically vanishes if and only if the manifold is flat.

It means that the Cartesian coordinate system with $g_{\alpha\beta} = \delta_{\alpha\beta}$ (Kronecker symbols) and connection coefficients $\Gamma^{\gamma}_{\alpha\beta} = 0$ can be introduced throughout $\mathscr{B}$. It also means that the parallel transport of vectors on $\mathscr{B}$ is independent of path, and the covariant derivatives commute.

2.3.7 *Bianchi's Identities*

Bianchi's identities have a role in the development of conservation laws for relative gravitation. Although Bianchi's identities had already been published in 1902 [124], Einstein was not aware of their contracted form when he developed his relative gravitation theory in 1915.

2.3.7.1 *First Identities of Bianchi*

We report in this subsection the two Bianchi identities on a Riemann-Cartan manifold, which are purely mathematical relations e.g. [49]. The sign $\sum_{(\alpha\beta\gamma)}$ means a cyclic permutation of the indices (α, β, γ) for the two identities. The component-free form of the first identities are (details may be found in a previous book):

$$\sum_{(\alpha\beta\gamma)} \Re\left(\mathbf{u}_\alpha, \mathbf{u}_\beta, \mathbf{u}_\gamma, \omega^\lambda\right) = \sum_{(\alpha\beta\gamma)} \left(\nabla_{\mathbf{u}_\alpha}\aleph\right)\left(\mathbf{u}_\beta, \mathbf{u}_\gamma, \omega^\lambda\right) + \aleph\left[\aleph(\mathbf{u}_\alpha, \mathbf{u}_\beta), \mathbf{u}_\gamma, \omega^\lambda\right] \tag{2.37}$$

where $\mathbf{u}_\mu$ are elements of the tangent space $T_{\mathbf{x}}\mathscr{B}$, and ω^λ of the cotangent space $T_{\mathbf{x}}\mathscr{B}^*$. Projected onto a base, the first identity takes the form of:

$$\sum_{(\alpha\beta\gamma)} \left(\Re^{\lambda}_{\alpha\beta\gamma} - \nabla_\alpha \aleph^{\lambda}_{\beta\gamma} - \aleph^{\sigma}_{\alpha\beta}\, \aleph^{\lambda}_{\sigma\gamma}\right) = 0, \quad \lambda = 1, n \tag{2.38}$$

In relative gravitation we have (α, β, γ) = cyc. perm. of $(0, 1, 2, 3)$, and $\lambda = 0, 1, 2, 3$. Identity (2.38) holds in any Riemann-Cartan manifold, and thus for an Einstein-Cartan spacetime.

Remark 2.13. For a Riemann manifold, the torsion vanishes and the first Bianchi's identity holds:

$$\sum_{(\alpha\beta\gamma)} \Re^{\lambda}_{\alpha\beta\gamma} = 0$$

For relative gravitation the same remark on indices holds, and application will be for Einstein spacetime.

2.3.7.2 *Second Identities of Bianchi*

The Bianchi identities provide constraints on the Riemann curvature tensor, torsion tensor, and its derivatives. Particularly, the Bianchi identities play a key role in the derivation of the theory of relative gravitation. For instance, they ensure the derivation of local conservation of energy and momentum in the presence of a curvature field. Their implications on the extended theory of Einstein-Cartan spacetime highlight the increasing importance of geometric structures in cosmology e.g. [4]. Again, starting from the general definition of curvature the component-free form of the second Bianchi's identities is given by e.g. [49]:

$$\sum_{(\alpha\beta\gamma)} (\nabla_{\mathbf{u}_\gamma} \Re) \left(\mathbf{u}_\alpha, \mathbf{u}_\beta, \mathbf{v}, \omega^\lambda\right) + \sum_{(\alpha\beta\gamma)} \Re \left(\aleph(\mathbf{u}_\alpha, \mathbf{u}_\beta), \mathbf{u}_\gamma, \mathbf{v}, \omega^\lambda\right) = 0 \tag{2.39}$$

where $\mathbf{v}$ is also an arbitrary vector field on the tangent space $T_{\mathbf{x}}\mathscr{B}$. The projection onto a base is obtained if we can introduce $\mathbf{v} = \mathbf{u}_\mu$:

$$\sum_{(\alpha\beta\gamma)} \nabla_\gamma \Re^\lambda_{\alpha\beta\mu} + \sum_{(\alpha\beta\gamma)} \aleph^\sigma_{\alpha\beta} \, \Re^\lambda_{\sigma\gamma\mu} = 0 \tag{2.40}$$

As with the first identity, the second identity is simplified for Riemann manifolds:

$$\sum_{(\alpha\beta\gamma)} \nabla_\gamma \Re^\lambda_{\alpha\beta\mu} = 0$$

Remark 2.14. All proofs and developments with details of Bianchi's identities may be found in [49].

2.3.7.3 *Application of Bianchi's Second Identity to Relative Gravitation*

Consider a Riemann manifold (Einstein spacetime). In order to recover the Einstein field equations for vacuum spacetime, we can rewrite:

$$g^{\beta\lambda} \overline{\nabla}_\alpha \mathcal{R}_{\beta\lambda} = g^{\beta\lambda} \overline{\nabla}_\nu \left(\Re^\gamma_{\gamma\beta\lambda}\right) \tag{2.41}$$

where $\overline{\nabla}$ is the Levi-Civita connection. To derive a system of 10 equations to determine the 10 components of the metric, let's recall Bianchi's first identity:

$$\overline{\nabla}_\nu \Re^\gamma_{\alpha\beta\lambda} + \overline{\nabla}_\beta \Re^\gamma_{\nu\alpha\lambda} + \overline{\nabla}_\alpha \Re^\gamma_{\beta\nu\lambda} = 0 \tag{2.42}$$

in the case of Riemann manifold without torsion. Putting $\alpha = \gamma$ gives us:

$$\overline{\nabla}_\nu \mathcal{R}_{\beta\lambda} - \overline{\nabla}_\beta \mathcal{R}_{\nu\lambda} - \overline{\nabla}_\alpha \Re^\alpha_{\nu\beta\lambda} = 0 \tag{2.43}$$

where we have used the skew-symmetry property of the curvature with respect to the two first covariant indices, and where we have introduced the definition of the Ricci curvature. Multiplying by the inverse of the metric $g^{\beta\lambda}$ allows us to put in high position indices and also to recover the Ricci curvature tensor again for the third term:

$$\overline{\nabla}_\nu \mathcal{R} - \overline{\nabla}_\beta \mathcal{R}^\beta_\nu - \overline{\nabla}_\alpha \mathcal{R}^\alpha_\nu = 0 \tag{2.44}$$

Finally, by applying the trace operation, we find the following equations of mixed components:

$$\overline{\nabla}_\beta G^\alpha_\beta = 0, \quad G^\alpha_\beta := \mathcal{R}^\alpha_\beta - \frac{\mathcal{R}}{2}\, \delta^\alpha_\beta \tag{2.45}$$

which have exactly the shape of the Einstein field equations e.g. [125].

Remark 2.15. However, a distinction must be made between these two equations: the Einstein field equation is derived from a variational formulation (the action is the integral of the scalar curvature) which is a special case of a physical vacuum spacetime model, while the Bianchi equation is a purely geometric result in Riemanian space which remains valid in any other Riemann space. For Einstein-Cartan, we do not arrive at the same field equation by following the two paths: the one with the same action shape, the other with the appropriate second equation of Bianchi for Riemann-Cartan space. The coincidence of the form of the equations is only observed for Rieman manifolds.

Remark 2.16. It seems that the role of the volume-form ω_n has often been minimized in the construction of relative gravitation theory. Studies have generally focused on the metric $\mathbf{g}$ and the connection ∇, compatible or not. Throughout the following chapters, we gradually show its subtle but almost confusing role in certain respects.

2.4 Lie Derivative and Applications

The invariance of the shape of the equations following a coordinate transformation is an obvious necessary condition for any mathematical modeling of a physical phenomena. They are called *passive diffeomorphisms*. Contrary to these, *active diffeomorphisms* are the transformations that connect different continua, knowing that passive transformations only allow us to represent the same spacetime but with another coordinate system. Now we consider translations (of both the space and the time) as groups with smooth continuous parameters denoted ξ^{μ}, called *Lie groups.* They constitute active diffeomorphisms which are essential to derive the conservation laws. Another main interest of the Lie derivative is that it is not defined from the geometric structures of manifolds but is based only on the differentiation property of the used variables e.g. [126].

2.4.1 *Lie Derivative of a Vector, and* 1*-Form*

For Lorentz transformations, there are six transformations that generate the entire Lorentz group $\mathscr{O}(1,3)$; three of these generators are spatial rotations, and the other three generators are time-space operators, called boosts. Boosts are related to the change of an observer's reference frame from another's. Mathematically, because Lorentz boosts and spacetime translations do not necessarily return a system to its original state, the Poincaré group is non-compact. Figure 2.9 presents the elements for defining the Lie derivative. On a manifold $\mathscr{B}$, the mapping $\varphi(s,\mathbf{x}) := \varphi_s(\mathbf{x})$ $(s \in \mathbb{R})$ transforms the point $\mathbf{x} \to \mathbf{y} = \varphi_\epsilon(\mathbf{x})$ for an infinitesimal value of the parameter $s = \epsilon$. Conversely we have $\mathbf{x} = \varphi_{-\epsilon}(\mathbf{y})$. The tangent space $T_{\mathbf{x}}\mathscr{B}$ at $\mathbf{x}$ is transformed by the linear tangent mapping $d\varphi_\epsilon$ at $\mathbf{x}$ to the tangent space $T_{\mathbf{y}}\mathscr{B}$ at the point $\mathbf{y}$. Then the vector $\mathbf{u}$ is transformed as $\mathbf{v} = d\varphi_\epsilon\,[\mathbf{u}]$, since the two tangent spaces are not the same. For comparing $\mathbf{u}$ and $\mathbf{v}$, we must pull back the vector $\mathbf{v}$ to the vector $\mathbf{v}' \in T_{\mathbf{x}}\mathscr{B}$, calculated as $\mathbf{v}' := d\varphi_{-\epsilon}\,[\mathbf{v}]$ where the linear tangent mapping is defined at $\mathbf{y}$. Lie derivative of vector $\mathbf{u}$ along vector field ξ generator of the mapping φ_s is defined as:

$$\mathcal{L}_\xi \mathbf{u} := \lim_{\epsilon \to 0} \frac{1}{\epsilon} \left\{ d\varphi_{-\epsilon} \left[\mathbf{u} \left(\varphi_\epsilon(\mathbf{x}) \right) \right] - \mathbf{u}(\mathbf{x}) \right\} \tag{2.46}$$

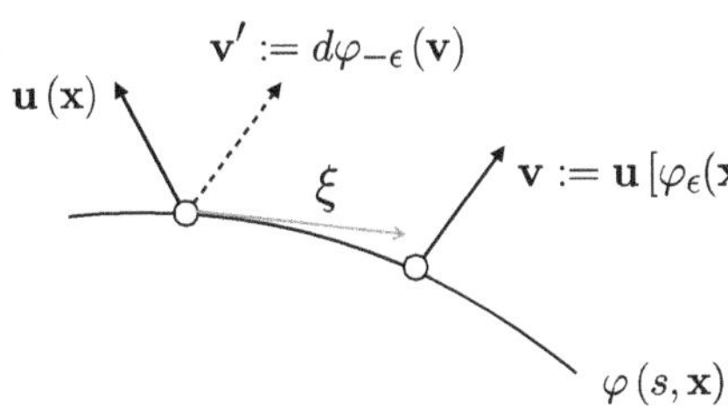

Fig. 2.9: Lie derivative of a vector field along the curve. Definition of the difference $\mathbf{v}' - \mathbf{u}$, where $s = \epsilon$ for the calculus.

where the generator vector of the mapping φ_ϵ is defined as:

$$\xi := \frac{d\varphi(s, \mathbf{x})}{ds}$$

which is tangent to the the curve.

It is necessary to introduce the Lie derivative since we cannot merely take the difference between components of the vector fields $\mathbf{u}$ and $\mathbf{v}$. These vectors do not operate on the same tangent space, so it is necessary to pull back $\mathbf{v}$ to $\mathbf{v}'$. Now, consider a metric-affine manifold $\mathscr{B}$ (three-dimensional material continuum or four-dimensional spacetime continuum), and spacetime local translation. Say an (active) diffeomorphism which is a smooth invertible map $\varphi : \mathscr{B} \to \mathscr{B}$. Let θ be a scalar field on $\mathscr{B}$, and a new scalar field $\tilde{\theta}$ defined as $\tilde{\theta}(\mathbf{x}) := \theta\left[\varphi(\mathbf{x})\right]$, meaning that the initial scalar field is pushed forward to a new manifold, using nevertheless the same coordinate values. For instance the transformed 1-form under an active diffeomorphism (particular case of Poincaré transformations e.g. [127] of the form $\tilde{\mathbf{x}} := \mathbf{x} - \xi\ d\lambda$; the "small" term $d\lambda \in \mathbb{R}$ is introduced here to mean an infinitesimal diffeomorphism) holds:

$$\begin{aligned}
\tilde{u}_\mu\left(\mathbf{x}\right) &= J_\mu^\alpha\ u_\alpha\left[\mathbf{x} + \xi\left(\mathbf{x}\right) d\lambda\right] \\
&= u_\mu\left(\mathbf{x}\right) + \left[\xi^\alpha\left(\mathbf{x}\right) \partial_\alpha u_\mu\left(\mathbf{x}\right) + u_\alpha\left(\mathbf{x}\right) \partial_\mu \xi^\alpha\left(\mathbf{x}\right)\right] d\lambda
\end{aligned}$$

since $J_\mu^\alpha\left(\mathbf{x}\right) = \delta_\mu^\alpha + \partial_\mu \xi^\alpha\left(\mathbf{x}\right) d\lambda$, and where we might drop the dependence on $\mathbf{x}$ for the simplicity of notation. With the help of formulae and definition, we obtain the Lie derivative of scalar, covariant and contravariant vectors, respectively:

$$\mathcal{L}_\xi \phi(\mathbf{x}) = \xi^\alpha(\mathbf{x}) \partial_\alpha \phi(\mathbf{x}) \tag{2.47}$$

$$\mathcal{L}_\xi \omega_\mu\left(\mathbf{x}\right) = \xi^\alpha\left(\mathbf{x}\right) \partial_\alpha \omega_\mu\left(\mathbf{x}\right) + \omega_\alpha\left(\mathbf{x}\right) \partial_\mu \xi^\alpha\left(\mathbf{x}\right) \tag{2.48}$$

$$\mathcal{L}_\xi u^\mu\left(\mathbf{x}\right) = \xi^\alpha\left(\mathbf{x}\right) \partial_\alpha u^\mu\left(\mathbf{x}\right) - u^\alpha\left(\mathbf{x}\right) \partial_\alpha \xi^\mu\left(\mathbf{x}\right) \tag{2.49}$$

Remark 2.17. From (2.47), (2.48), and (2.49) which are valid for any coordinate system, we can rewrite Lie derivatives as (their intrinsic definitions):

$$\mathcal{L}_\xi \Phi(\mathbf{x}) = \xi\left[\Phi\right], \quad \mathcal{L}_\xi u^\mu\left(\mathbf{x}\right) = \left[\xi, \mathbf{u}\right]$$

where the first term is reduced to the directional derivative of a scalar Φ. Once again, it should be emphasized that the Lie derivatives (2.47), (2.48), and (2.49) do not depend on the geometric structure of the manifold.

2.4.2 *Lie Derivative of a Tensor*

Let ξ be a given vector field on $\mathscr{B}$. Let $\mathbf{T}(T^{\alpha_1 \cdot \alpha_p}_{\beta_1, \cdot \beta_q})$ be a tensor of type (p, q) (i.e., contravariant of order p and covariant of order q) on the same manifold. By extending relations (2.48) and (2.49), Lie derivative of $\mathbf{T}$ with respect to (or along the field) ξ is the tensor of type (p, q):

$$\mathcal{L}_\xi T^{\alpha_1 \cdots \alpha_p}_{\beta_1 \cdots \beta_q} := \xi^\gamma \partial_\gamma T^{\alpha_1 \cdots \alpha_p}_{\beta_1 \cdots \beta_q} - T^{\gamma \alpha_2 \cdots \alpha_p}_{\beta_1 \cdots \beta_q} \partial_\gamma \xi^{\alpha_1} - \cdots - T^{\alpha_1 \cdots \alpha_{p-1} \gamma}_{\beta_1 \cdots \beta_q} \partial_\gamma \xi^{\alpha_p}$$
$$+ T^{\alpha_1 \cdots \alpha_p}_{\gamma \beta_2 \cdots \beta_q} \partial_{\beta_1} \xi^\gamma + \cdots + T^{\alpha_1 \cdots \alpha_p}_{\beta_1 \cdots \beta_{q-1} \gamma} \partial_{\beta_q} \xi^\gamma \tag{2.50}$$

For Riemann manifold, partial derivatives can be replaced by covariant derivatives, whereas for Riemann-Cartan manifold, some additional calculus should be undertaken. It is straightforward to verify that this definition with coordinate is equivalent to the intrinsic definition.

Theorem 2.4. *Lie derivative verifies the following properties (which could be in turn considered as definitions of the Lie derivative):*

(1) *for any real numbers α and β, and any two vector fields $\mathbf{u}$ and $\mathbf{v}$ on $\mathscr{B}$, we have:*

$$\mathcal{L}_\xi (\alpha\ \mathbf{u} + \beta\ \mathbf{v}) = \alpha\ \mathcal{L}_\xi \mathbf{u} + \beta\ \mathcal{L}_\xi \mathbf{v} \tag{2.51}$$

(2) *for any scalar field φ and any vector field $\mathbf{u}$ on $\mathscr{B}$, we have:*

$$\mathcal{L}_\xi (\varphi\ \mathbf{u}) = \varphi\ \mathcal{L}_\xi \mathbf{u} + \mathcal{L}_\xi \varphi \otimes \mathbf{u} \tag{2.52}$$

(3) *for any two real numbers α and β, and for any two tensor fields $\mathbf{T}$ and $\mathbf{S}$ of the same type (p, q), the following holds:*

$$\mathcal{L}_\xi (\alpha\ \mathbf{T} + \beta\ \mathbf{S}) = \alpha\ \mathcal{L}_\xi \mathbf{T} + \beta\ \mathcal{L}_\xi \mathbf{S} \tag{2.53}$$

(4) *the Leibniz rule applies for any two tensors $\mathbf{T}$ and $\mathbf{S}$:*

$$\mathcal{L}_\xi (\mathbf{T} \otimes \mathbf{S}) = \mathcal{L}_\xi \mathbf{T} \otimes \mathbf{S} + \mathbf{T} \otimes \mathcal{L}_\xi \mathbf{S} \tag{2.54}$$

(5) *for any p-form ω^p and q-form ω^q, we have:*

$$\mathcal{L}_\xi (\omega^p \wedge \omega^q) = \mathcal{L}_\xi \omega^p \wedge \omega^q + \omega^p \wedge \mathcal{L}_\xi \omega^q \tag{2.55}$$

Lie derivative of identity $\mathbf{I}$ (and symbol of Krönecker) vanishes and Lie derivative commutes with the process of contraction.

Proof. It can be found in classical books or textbooks on differential geometry e.g. [5], [128]. It could also be shown by using the coordinate version; this way does not present any particular difficulty but is quite lengthy. □

Covariant expressions of Lie derivatives. For illustrating the covariant forms of Lie derivatives, relations (2.47), (2.48), and (2.49) can be rewritten by means of covariant derivatives as follows:

$$\mathcal{L}_\xi \phi(\mathbf{x}) = \xi^\alpha(\mathbf{x}) \nabla_\alpha \phi(\mathbf{x}) \tag{2.56}$$

$$\mathcal{L}_\xi \omega_\mu(\mathbf{x}) = \xi^\alpha(\mathbf{x}) \nabla_\alpha \omega_\mu(\mathbf{x}) + \omega_\alpha(\mathbf{x}) \nabla_\mu \xi^\alpha(\mathbf{x}) + \xi^\alpha(\mathbf{x}) \aleph^\rho_{\alpha\mu}(\mathbf{x}) \omega_\rho(\mathbf{x}) \tag{2.57}$$

$$\mathcal{L}_\xi u^\mu(\mathbf{x}) = \xi^\alpha(\mathbf{x}) \nabla_\alpha u^\mu(\mathbf{x}) - u^\alpha(\mathbf{x}) \nabla_\alpha \xi^\mu(\mathbf{x}) - \xi^\alpha(\mathbf{x}) \aleph^\mu_{\alpha\rho}(\mathbf{x}) u^\rho(\mathbf{x}) \tag{2.58}$$

where ∇ and $\aleph$ are connection and its associated torsion respectively. This is a rewriting of the variations/Lie derivatives which have been *a priori* defined independently of any connection on the manifold $\mathscr{B}$.

Remark 2.18. Lie derivative commutes with exterior differentiation and does not need the existence of an affine connection on the manifold. Lie derivative $\mathcal{L}_\xi$ is different from covariant derivative along the vector ξ, say ∇_ξ, since the first does not require any affine connection on the manifold. Moreover, Lie derivative involves the derivative of the vector field ξ while covariant derivative does not.

The main reason to utilize Lie derivative is to compare two values of a tensor, say at point x^μ and point $x^\mu + \xi^\mu d\sigma$ along a flow (evolution curve). It is not possible to make operation as difference between the values of the components at these two points $\mathbf{x}$ and $\mathbf{x} + \xi d\sigma$ since the tangent spaces $T_{\mathbf{x}}\mathscr{B}$ and $T_{\mathbf{x}+\xi d\sigma}\mathscr{B}$ at these two points are not the same along the integral curve of the vector field ξ.

Theorem 2.5 (Hadamard's Lemma). *Let $\mathbf{u}(\mathbf{x})$ be a vector field on the manifold $\mathscr{B}$. Let $f(\mathbf{x})$ be a continuously differentiable function in a neighborhood $\mathscr{U}$ of $\mathbf{x} \in \mathscr{B}$. Then for sufficiently small values of $t \in \mathbb{R}$, there is a function g_t, continuously differentiable in t and points near $\mathbf{x}$, such that:*

$$g_0 = \mathbf{u}(\mathbf{x})\,[f], \quad \text{and} \quad f\left(\phi_t(\mathbf{x})\right) = f(\mathbf{x}) + t\; g_t(\mathbf{x}) \tag{2.59}$$

that is

$$f \circ \phi_t = f + t\; g_t \tag{2.60}$$

where ϕ_t is the flow associated to $\mathbf{u}$.

Proof. See textbooks e.g. [120] (p 127). □

A base (or frame) of vector fields $(\mathbf{f}_1, \ldots, \mathbf{f}_n)$ in a region $\mathscr{U}$ of a manifold $\mathscr{B}$ consists of n independent vectors in $\mathscr{U}$. A coordinate frame is a base where $\mathbf{f}_\alpha = \partial_\alpha$ for some coordinate system $(x^1, \ldots, x^n)$ in $\mathscr{U}$. It may be assumed that $\mathscr{U} = \mathscr{B}$ for the sake of simplicity.

Theorem 2.6. *A base* $(\mathbf{f}_1, \ldots, \mathbf{f}_n)$ *is locally a coordinate frame if and only if:*

$$[\mathbf{f}_\alpha, \mathbf{f}_\beta] = 0, \quad \forall \alpha, \beta = 1, \ldots, n \tag{2.61}$$

In other words, this means that the Cartan constants of structure are identically equal to zero $\aleph^{\gamma}_{0\alpha\beta} \equiv 0, \forall \alpha, \beta, \gamma$.

Remark 2.19. Consider a two-dimensional manifold with the polar coordinate system (r, θ). Let $(\mathbf{e}_r, \mathbf{e}_\theta)$ the usual orthonormal local base. It is not a coordinate base. To check, it suffices to start from the derivation of an arbitrary function Φ on $\mathscr{B}$ by a non-normal base:

$$\begin{cases} \mathbf{f}_r [\Phi] = \frac{\partial \Phi}{\partial r} = \mathbf{e}_r [\phi] \\ \mathbf{f}_\theta [\Phi] = \frac{\partial \Phi}{\partial \theta} = r\mathbf{e}_\theta [\phi] \end{cases} \Longrightarrow \begin{cases} \mathbf{e}_r [\Phi] = \frac{\partial \Phi}{\partial r} \\ \mathbf{e}_\theta [\Phi] = \frac{1}{r} \frac{\partial \Phi}{\partial \theta} \end{cases}$$

The Lie-Jacobi bracket applied to any function Φ gives:

$$[\mathbf{e}_r, \mathbf{e}_\theta] = \mathbf{e}_r [\mathbf{e}_\theta [\phi]] - \mathbf{e}_\theta [\mathbf{e}_r [\phi]] = -\frac{1}{r^2} \frac{\partial \Phi}{\partial \theta}$$

This bracket is not equal to zero except for functions depending only on r. The orthonormal base $(\mathbf{e}_r, \mathbf{e}_\theta)$ is not a coordinate frame. Conversely $(\mathbf{f}_r, \mathbf{f}_\theta)$ is a coordinate frame. This remark is more general and can be applied for many usual local orthonormal bases such as spherical and cylindrical bases; their use requires special attention because in such a case, the constants of structure $\aleph^{\gamma}_{0\alpha\beta}$ used for calculating torsion and curvature are not zero e.g. [5]. The use of orthonormal bases is very reassuring because the components of tensors on these bases are called *physical components* of the tensors. However, the equations risk becoming cumbersome due to the presence of these constants.

2.4.3 *Elements for Integral Invariants (Poincaré)*

Theory of integral invariants was founded by H. Poincaré and appeared in his book on *Méthodes Nouvelles de la Méchanique Céleste*, tome *III*. During their works, Liouville and Boltzmann had already recognized the existence of particular invariant integrals, but it was Poincaré who founded the theory in its general formulation. Conservation laws of continuum physics are essentially based on the variation of integral quantities calculated on the continuum $\mathscr{B}$. Consider a p-form ω_p with $p \leq n$ on the continuum and a diffeomorphism φ from $\mathscr{B}$ into $\mathscr{B}$. The p-form is integral invariant by the application φ if we have:

$$\int_{\mathscr{B}} \omega_p = \int_{\varphi(\mathscr{B})} \omega_p \tag{2.62}$$

Consider now the vector field ξ which generates a local group of transformations φ_t. The field of p-form ω_p is an integral invariant of ξ, if it is integral invariant by each map φ_t. We recall two important results using the Lie derivatives of forms and proofs may be found in differential geometry textbooks e.g. [5], [128].

Lemma 2.1. *Let $\mathscr{U}$ be an open set of $\mathbb{R}^n$ and ω_p a p-form with $p \leq n$. If*

$$\int_c \omega_p = 0, \quad \text{for any } p\text{-cube of } \mathscr{U} \tag{2.63}$$

then $\omega_p = 0$.

This Lemma which is attributed to Haar plays a central role in the localization of conservation laws from integral formulation.

Theorem 2.7. *Let $\varphi \in \mathrm{Diff}\,(\mathscr{B})$ be diffeomorphism of the manifold $\mathscr{B}$ and ω_p be a p-form. Then ω_p is an integral invariant if and only if:*

$$d\varphi^* \, \omega_p = \omega_p \tag{2.64}$$

Let $(\mathbf{u}_{10}, \ldots, \mathbf{u}_{p0})$ be an arbitrary set of p-uplet of vectors of the tangent space $T_{\mathbf{x}}\mathscr{B}$ at the point $\mathbf{x}$ of the manifold $\mathscr{B}$, let ω_p be a p-form and φ a diffeomorphism on $\mathscr{B}$. Say $d\varphi$ the differential of the diffeomorphism, then we have the relations:

$$\mathbf{u}_\alpha := d\varphi\,(\mathbf{u}_{\alpha 0}), \quad (d\varphi^* \omega_p)\,(\mathbf{u}_{10}, \ldots, \mathbf{u}_{p0}) = \omega_p\,(d\varphi(\mathbf{u}_{10}, \ldots, \mathbf{u}_{10})) \tag{2.65}$$

The third theorem allows us to formulate the invariance by means of Lie derivatives e.g. [128] and allows us to write the Principle of General Covariance e.g. [126], [129], [130].

Theorem 2.8. *Let ξ be a vector field on the manifold $\mathscr{B}$ and let ω_p be a p-form field on $\mathscr{B}$. Then ω_p is integral invariant if and only if:*

$$\boxed{\mathcal{L}_\xi \omega_p = 0} \tag{2.66}$$

This last theorem (2.66) is used for deriving physics laws intrinsically. In this book, we also use a particular version by introducing the so-called $\mathscr{B}$-derivative, i.e., time derivative of tensors following a continuum motion e.g. [6].

2.4.4 *Compatibility of Connection and Volume-Form*

Metric compatibility of affine connection is the condition for defining Riemann-Cartan manifolds. In addition, compatibility of connection with the volume form is also an essential condition if we want to apply the *divergence theorem* e.g. [131], [132]. For that purpose, let us express Lie derivatives of metric and volume-form in terms of covariant derivatives. From the definition (2.50), we obtain for metric:

$$\begin{aligned}\mathcal{L}_\xi\, g_{\alpha\beta} &= \xi^\gamma\, \nabla_\gamma g_{\alpha\beta} + g_{\gamma\beta}\, \nabla_\alpha \xi^\gamma + g_{\alpha\gamma}\, \nabla_\beta \xi^\gamma \\ &\quad + \xi^\gamma \left(g_{\alpha\nu}\, \aleph^\nu_{\gamma\beta} + g_{\nu\beta}\, \aleph^\nu_{\gamma\alpha} \right),\end{aligned} \tag{2.67}$$

and for the 4-volume-form:

$$\begin{aligned}\mathcal{L}_\xi\, \omega_{n\ \alpha\beta\mu\lambda} &= \xi^\gamma \nabla_\gamma \omega_{n\ \alpha\beta\mu\lambda} + \omega_{n\ \gamma\beta\mu\lambda} \nabla_\alpha \xi^\gamma + \omega_{n\ \alpha\gamma\mu\lambda} \nabla_\beta \xi^\gamma \\ &\quad + \omega_{n\ \alpha\beta\gamma\lambda} \nabla_\mu \xi^\gamma + \omega_{n\ \alpha\beta\mu\gamma} \nabla_\lambda \xi^\gamma + \xi^\gamma \omega_{n\ \nu\beta\mu\lambda}\, \aleph^\nu_{\gamma\alpha} \\ &\quad + \xi^\gamma \omega_{n\ \alpha\nu\mu\lambda}\, \aleph^\nu_{\gamma\beta} + \xi^\gamma \omega_{n\ \alpha\beta\nu\lambda}\, \aleph^\nu_{\gamma\mu} + \xi^\gamma \omega_{n\ \alpha\beta\mu\nu}\, \aleph^\nu_{\gamma\lambda}\end{aligned} \tag{2.68}$$

It is reminded that the coefficients of an affine connection can be decomposed uniquely into Christoffel's symbols and contortion tensor itself decomposed according to a skew-symmetric and a symmetric part $\mathcal{T}^\gamma_{\alpha\beta}$ e.g. [6]:

$$\Gamma^\gamma_{\alpha\beta} := \overline{\Gamma}^\gamma_{\alpha\beta} + \mathcal{T}^\gamma_{\alpha\beta} \tag{2.69}$$

where $\mathcal{T}^\gamma_{\alpha\beta} := \Omega^\gamma_{\alpha\beta} + D^\sigma_{\alpha\beta}$:

$$\Omega^\gamma_{\alpha\beta} = \frac{1}{2}\, \aleph^\gamma_{\alpha\beta}, \quad D^\sigma_{\alpha\beta} = \frac{1}{2} g^{\sigma\gamma} \left(g_{\alpha\mu}\, \aleph^\mu_{\gamma\beta} + \aleph^\mu_{\gamma\alpha}\, g_{\mu\beta} \right)$$

since the constants of structure are zero for a coordinate base [5]. In sum, the contortion tensor is the difference between the affine connection $\Gamma^\gamma_{\alpha\beta}$ and the Levi-Civita connection $\overline{\Gamma}^\gamma_{\alpha\beta}$ which is symmetric. However, contortion tensor $\mathcal{T}^\gamma_{\alpha\beta}$ is not necessarily skew-symmetric because it includes a symmetric part. When torsion is zero, then the contortion tension is necessary zero. Sometimes, contortion tensor is used instead of torsion tensor for spacetime modeling, namely when accounting for spins e.g. [37],

Remark 2.20. For Riemann manifold, with zero torsion, metric compatibility of connection induces:

$$\mathcal{L}_\xi \, g_{\alpha\beta} = \nabla_\alpha \xi_\beta + \nabla_\beta \xi_\alpha \tag{2.70}$$

which is the usual expression of the deformation rate in classical continuum mechanics e.g. [133]. We observe that the second row of (2.67) reduces to the symmetric part of the contortion tensor.

Let $\mathscr{B}$ be a Riemann-Cartan manifold endowed with a metric $g_{\alpha\beta}$ and an independent connection $\Gamma^\gamma_{\alpha\beta}$.

Definition 2.20. A volume-form on the manifold $\mathscr{B}$ is compatible with the affine connection if, for any vector field ξ on the manifold, e.g. [132]:

$$\boxed{\mathcal{L}_\xi \omega_n = (\nabla_\alpha \xi^\alpha)\, \omega_n} \tag{2.71}$$

For a Riemann manifold, torsion is equal to zero, so it is common to consider the Levi-Civita connection $\overline{\nabla}$ obtained from metric derivatives and Riemann volume-form $\overline{\omega}_n := \sqrt{|\mathrm{Det}\mathbf{g}|}\; dx^1 \wedge \cdots dx^n$. It is easy to verify that the Christoffel symbols $\overline{\Gamma}^\gamma_{\alpha\beta}$ are compatible with this volume-form. For an independent affine connection, we then have to ensure two conditions.

(1) The metric compatibility is met whenever we deal with Riemann-Cartan manifold $\nabla_\gamma g_{\alpha\beta} \equiv 0$;
(2) The connection compatibility with volume-form is satisfied if we have $\mathcal{L}_\xi \omega_n = (\nabla_\alpha \xi^\alpha)\, \omega_n$ for any vector field ξ.

2.4.5 *Volume-Form for Riemann-Cartan Manifold*

Connection ∇, with non-zero torsion, can be compatible with a volume-form under certain conditions e.g. [132]. For a Riemann-Cartan manifold, the compatibility condition between $\Gamma^\gamma_{\alpha\beta}$ and a volume-form like $\omega_n = f(\mathbf{x})\sqrt{|\mathrm{Det}\mathbf{g}|}\; dx^1 \wedge \cdots dx^n$ is that the 1-form $\aleph^\alpha_{\alpha\beta} dx^\beta$ is closed. Saa showed that in this case the volume-form can be written [132]:

$$\omega_n = \mathrm{e}^\vartheta \sqrt{|\mathrm{Det}\mathbf{g}|}\; dx^1 \wedge \cdots \wedge dx^n, \quad \partial_\beta \vartheta := \aleph^\alpha_{\alpha\beta} \tag{2.72}$$

This volume-form ω_n is the "closest" to the volume-form of the Riemann manifold $\overline{\omega}_n = \sqrt{|\mathrm{Det}\mathbf{g}|}\; dx^1 \wedge \cdots \wedge dx^n$. In our case, what matters is the ability to write the divergence theorem and to ensure that such a volume-form exists, close or not to that of Riemann. Compatibility of ω_n with ∇ was studied for Riemann-Cartan manifold in e.g. [132], [134], and related with the gauge invariance of electromagnetic fields in e.g. [132], [135], and also recently in e.g. [131]. For the sake of self-consistency, we re-derive the proof with a direct calculus. For this purpose, let's express the Lie derivative of the components:

$$\begin{aligned}\mathcal{L}_\xi \omega_{\alpha_1,\ldots,\alpha_n} &= \xi^\mu \nabla_\mu \omega_{\alpha_1,\ldots,\alpha_n} + (\nabla_{\alpha_1}\xi^\mu)\,\omega_{\mu,\alpha_2,\ldots,\alpha_n} \\ &\quad + (\nabla_{\alpha_2}\xi^\mu)\,\omega_{\alpha_1,\mu,\alpha_3,\ldots,\alpha_n} + \cdots + (\nabla_{\alpha_n}\xi^\mu)\,\omega_{\alpha_1,\ldots,\alpha_{n-1},\mu} \\ &\quad + \xi^\mu \aleph^\nu_{\mu\alpha_1}\ \omega_{\nu,\alpha_2,\ldots,\alpha_n} + \cdots + \xi^\mu \aleph^\nu_{\mu\alpha_n}\ \omega_{\alpha_1,\ldots,\alpha_{n-1},\nu}\end{aligned}$$

Since there is only one component $\Omega := \omega_{\alpha_1<\alpha_2\cdots<\alpha_n}$, the Lie derivative reduces to:

$$\mathcal{L}_\xi \Omega = \xi^\mu \nabla_\mu \Omega + (\nabla_\mu \xi^\mu)\,\Omega + \tau_\mu\ \xi^\mu\ \Omega, \quad \text{with } \tau_\mu := \aleph^\nu_{\mu\nu} = \sum_{\nu=0}^{\nu=3} \aleph^\nu_{\mu\nu} \tag{2.73}$$

Imposing the compatibility of ∇ with ω_n induces the relation:

$$\xi^\mu \nabla_\mu \Omega + \tau_\mu\ \xi^\mu = 0 \tag{2.74}$$

By definition Ω is always positive, so it is then possible to make a change of variable $\Omega := e^\vartheta$ to obtain (ξ^μ being arbitrary):

$$\nabla_\mu \vartheta = -\tau_\mu \tag{2.75}$$

Coming back to the projection of the volume-form onto an arbitrary tensor base $dx^1 \wedge \cdots dx^n$, we may write:

$$\omega_n = e^\vartheta\ \omega_{12\cdots n}\ dx^1\wedge\cdots\wedge dx^n = e^\vartheta\ \mathrm{Det}(\partial_1, \partial_2, \ldots, \partial_n)\ dx^1\wedge\cdots\wedge dx^n \tag{2.76}$$

Particularly, the compatible volume-form with ∇ can be suggested:

$$\omega_n = e^\vartheta\ \overline{\omega}_n, \quad \vartheta = \vartheta(\mathbf{g}, \aleph) \tag{2.77}$$

Introduction of determinant is necessary to ensure invariance during a passive diffeomorphism. It was also worth assuming the dependence $\vartheta = \vartheta(\mathbf{g}, \aleph)$ to ensure covariance of the function ϑ (see Cauchy-Weyl theorem e.g. [49]).

2.4.6 *Basic Relations for Conservation Laws*

Generalized continua are manifolds $\mathscr{B}$ which have compactness and connectedness [136], respectively associated to a metric $\mathbf{g}$ with components $g_{\alpha\beta}$ and a connection ∇ with coefficients $\Gamma^\gamma_{\alpha\beta}$. Introducing independent connections extends the covariant derivative of fields on the manifold and accounts for torsion $\aleph$ with components $\aleph^{\gamma}_{\alpha\beta}$ and curvature $\Re$ with components $\Re^{\gamma}_{\alpha\beta\lambda}$. Covariant derivative of tensor type (p,q) denoted $\mathbf{T}$ is given by:

$$\nabla_\gamma T^{\alpha_1\cdots\alpha_p}_{\beta_1\cdots\beta_q} = \partial_\gamma T^{\alpha_1\cdots\alpha_p}_{\beta_1\cdots\beta_q} + \sum_{s=1}^{s=p} \Gamma^{\alpha_s}_{\gamma\mu} T^{\alpha_1\cdots\alpha_{s-1}\ \mu\ \alpha_{s+1}\cdots\alpha_p}_{\beta_1\cdots\beta_q}$$

$$-\sum_{s=1}^{s=q} \Gamma^{\mu}_{\gamma\beta_s} T^{\alpha_1\cdots\alpha_p}_{\beta_1\cdots\beta_{s-1}\ \mu\ \beta_{s+1}\cdots\beta_q}$$

Recall that integration of scalar $f(x^1,\ldots,x^n)\in\mathbb{R}$ on $\mathscr{B}$ requires a volume-form:

$$\mathscr{S} = \int_{\mathscr{B}} f(\mathbf{x})\ \omega_n$$

For a density function $\rho(\mathbf{x})$, we obtain the local mass of matter surrounding the point $\mathbf{x}$ as $\rho\omega_n$, which is a form on the tangent space at $\mathbf{x}$. Compatibilities of metric with connection, and the connection with volume-form, are expressed by:

$$\nabla_\gamma g_{\alpha\beta} \equiv 0, \quad \mathcal{L}_\xi \omega_n \equiv (\nabla_\alpha \xi^\alpha)\,\omega_n$$

where ξ is an arbitrary vector. For a Riemann manifold, compatibilities of metric, Levi-Civita connection and Riemann volume-form are:

$$\overline{\nabla}_\gamma g_{\alpha\beta} \equiv 0, \quad \mathcal{L}_\xi \overline{\omega}_n \equiv \left(\overline{\nabla}_\alpha \xi^\alpha\right)\omega_n$$

with Christoffel's symbols and the Riemann volume-form as:

$$\overline{\Gamma}^{\sigma}_{\alpha\beta} = \frac{1}{2} g^{\sigma\gamma}\left(\partial_\alpha g_{\beta\gamma} + \partial_\beta g_{\gamma\alpha} - \partial_\gamma g_{\alpha\beta}\right), \quad \overline{\omega}_n = \sqrt{|\mathrm{Det}\mathbf{g}|}\ dx^1\wedge\cdots\wedge dx^n$$

Invariance of integral along a vector field is given by the Lie derivative e.g. [49]:

$$\mathscr{S} := \int_{\mathscr{B}} \omega \quad \text{integral invariant} \quad \Longleftrightarrow \quad \mathcal{L}_\xi \omega = 0$$

This property is essential for deriving conservation laws in most domains of continuum physics. It is therefore mandatory to define a volume-form for formulating conservation equations.

Remark 2.21. For a Riemann-Cartan manifold endowed with a metric $g_{\alpha\beta}$ and an independent connection (with torsion) $\Gamma^{\gamma}_{\alpha\beta}$, we have:

$$\mathcal{L}_\xi \overline{\omega}_n = \left(\overline{\nabla}_\alpha \xi^\alpha\right)\overline{\omega}_n = \left(\nabla_\alpha \xi^\alpha - \aleph^{\alpha}_{\alpha\beta}\ \xi^\beta\right)\overline{\omega}_n \neq \left(\nabla_\alpha \xi^\alpha\right)\overline{\omega}_n$$

considering that for a Riemann-Cartan manifold, metric compatibility $\nabla_\gamma g_{\alpha\beta} \equiv 0$ holds. Application of divergence theorem requires some caution e.g. [132], [137].

2.5 Spacetime Models and Continuum Physics

Introduction of four-dimensional spacetime was made in physics with the advent of relativistic theories. The conceptualization of four-dimensional spacetime was a long process lasting several millennia.[1] It seems that the first approach to the fourth dimension in physics is due to Nicole Oresme in the 14th century who used the word "4^{am} *dimensionem*". By introducing what are now called Galilean transformations, after his name (1594–1642), Galileo can be considered the father of modern physics. His approach explicitly considers a four-dimensional spacetime: three for space and one for time. This step is fundamental in the development of both classical and relativistic mechanics theories. Schematization of the observable world is effectively achieved by a geometric structure including the combination of these two entities, Euclidean or not. A contemporary of Newton, Henry More had succeeded in his quest for the fourth dimension but in theology: the three dimensions concern matter and the fourth the spirit (1682). In mechanics, a fourth dimension had been formally written by D'Alembert in 1754 and Lagrange in 1797 (configuration space), translated in these terms [138]: "*Since the position of a point in space depends upon three rectangular coordinates, these coordinates in the problems of mechanics are conceived as being functions of the time t. Thus we may regard mechanics as a geometry of four dimensions, and mechanical analysis as an extension of geometrical analysis.*" Whether for classical or for relativistic mechanics, the use of a four-dimensional spacetime is required.

[1] A passage from Emile Borel's book seems very appropriate to us concerning the general process for modeling spacetime in physics. We report here the English translation of the original French version of this text, by the author: "To measure lengths and to measure time, we will have to start by constructing the whole theory of physics and here we are, from the outset, stopped, since we cannot think, it seems, of observing physical phenomena, if we do not already know how to measure lengths and time. This difficulty is great, but it is far from being new in the history of science; it is by successive approximations that we arrive at knowing nature: *the essential thing is that these approximations converge.*" (L'espace et le temps, Borel, Presse Universitaire de France, 1949)

2.5.1 *Space and Time Background*

Prior to any description of continuum motion, it is sound to recall some notions of material points in Newtonian mechanics. However, even if Newtonian mechanics is essentially based on the existence of an absolute time and an absolute space which are not influenced by the physical phenomena which occur there, we will see as we progress in reading this work that the vacuum spacetime geometry cannot be constructed in isolation from gravitation and electromagnetism.

2.5.1.1 *Referential and Motion*

Newtonian mechanics admits the existence of an absolute time and an absolute space. This approach has a hint of Aristotle's scheme stipulating that there exists an immutable and underlying geometric substrate-ambient space independent of any physical phenomenon. The concept of absolute may be considered as an asymptotic and idealization of time and space.

Newtonian spacetime. Two hypotheses stand for Newtonian mechanics:

(1) There exist a non-deformable referential solid (space $\mathscr{E}$) and a referential periodic phenomenon (time $\mathscr{T}$).
(2) Time and space are homogeneous (translation, addition), isotropic (rotation, reflection), Euclidean (measurable), continuous (indefinitely sub-divisible), and of infinite size.

Consequently, it is possible to define a *referential* defined by a rigid body and a clock (Figure 2.10). The rigid body is defined by an origin (point) O and a Cartesian vector base $(\mathbf{e}_1, \mathbf{e}_2, \mathbf{e}_3)$. Time is represented by $t \in \mathbb{R}$ defining a universal instant (simultaneously standard), throughout the absolute space.

Motion of a material point is defined by the function φ:

$$\varphi : t \in \mathbb{R} \to \mathbf{OM}\,(t) = \varphi\,(M, t) \in \mathcal{E} \tag{2.78}$$

where $\mathcal{E}$ is the vector space underlying the affine space of geometric spatial points $\mathscr{S}$. To each point M of $\mathscr{S}$ is associated one and only one vector

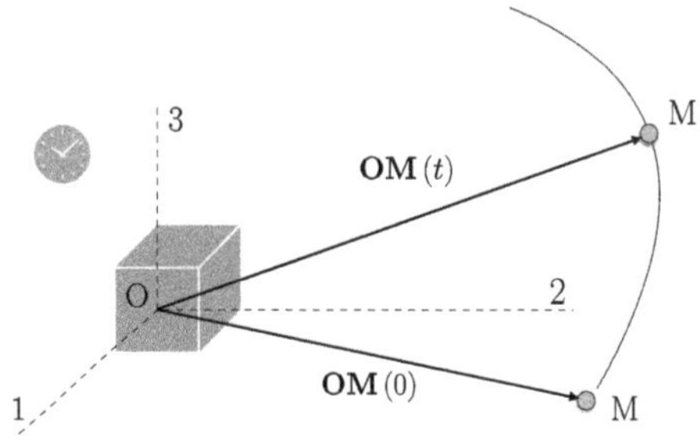

Fig. 2.10: Referential: A material point M located at $\mathbf{OM}(t)$ and reference body (rigid solid). To the reference body is attached an axis system with origin O. The reference body and the clock constitute the *referential* for analyzing the motion of the material point M located initially at $\mathbf{OM}(0)$ for $t = 0$.

$\mathbf{u} \in \mathcal{E}$ such that $\mathbf{OM} := \mathbf{u}$. As it is *a priori* admitted, space is decoupled from time in such a theory.

Remark 2.22. Newton's two hypotheses are not without contradiction in their conception. Indeed, they implicitly define absolute translational or rotational motion, as a motion relative to this absolute space and described relative to an absolute clock. It was Leibniz who raised this apparent contradiction of Newton's original approach e.g. [139]. If absolute space, with absolute time, exists then body motions with respect to this absolute space should be measurable, which is not the case. Descriptions require frames of reference. This led physicists to introduce the concept of invariance with respect to the choice of referential frame [140], which led to Galilean spacetime.

2.5.1.2 *Length and Angle Measures in Space*

Newtonian absolute space (also called ambient space) is modeled by an Euclidean manifold. It is endowed with a Euclidean metric to measure vector length, or distance between two points of the space, and angle of two vectors. The metric is obtained from a bilinear form (scalar product): $\hat{\mathbf{g}} : \mathbf{u}, \mathbf{v} \in \mathcal{E} \to \hat{\mathbf{g}}(\mathbf{u}, \mathbf{v}) := \mathbf{u} \cdot \mathbf{v} \in \mathbb{R}$ where $\mathcal{E}$ is the vector space underlying the affine ambient space (space of points) such that:

$$\begin{cases} \hat{\mathbf{g}}(\mathbf{u}, \mathbf{v}) = \mathbf{g}(\mathbf{v}, \mathbf{u}) \\ \hat{\mathbf{g}}(\mathbf{u}, \mathbf{u}) \geq 0 \\ \hat{\mathbf{g}}(\mathbf{u}, \mathbf{u}) = 0 \quad \text{if and only if} \quad \mathbf{u} = 0 \end{cases}$$

We recall that the metric allows us to calculate:

(1) the length of any vector $\mathbf{u}$ of the space $\mathcal{E}$: $\|\mathbf{u}\| = \sqrt{\mathbf{u} \cdot \mathbf{u}}$;
(2) the angle between two vectors $\mathbf{u}$ and $\mathbf{v}$ of the space $\mathcal{E}$: $\cos(\mathbf{u}, \mathbf{v}) = \mathbf{u} \cdot \mathbf{v} / \|\mathbf{u}\| \|\mathbf{v}\|$.

The existence of a Euclidean metric permits us to define an orthonormal base $(O\mathbf{e}_1\mathbf{e}_2\mathbf{e}_3)$ within and attached to a reference solid body. To any point of the affine space can be attributed its Cartesian coordinates: $\mathbf{OM} = x^1\, \mathbf{e}_1 + x^2\, \mathbf{e}_2 + x^3\, \mathbf{e}_3$ with $\mathbf{e}_i \cdot \mathbf{e}_j := \delta_{ij}$ where δ_{ij} is the Kroenecker symbol: $\delta_{ij} = 1$, for $i = j$ otherwise it is equal to zero. The Euclidean ambient space of classical mechanics has a uniform metric tensor not depending on the point. Mechanics lies upon the analysis of body movement with respect to a referential (spacetime).

Remark 2.23. Admitting an absolute space equipped with a Euclidean metric, with components $g_{ij} = \delta_{ij}$ over the whole space, is not sufficient to construct a model of point mechanics, and even less the mechanics and *in extenso* the continuum physics. We already mentioned that the movements of bodies are not measurable in absolute terms, they are relative. It is therefore necessary to introduce one (and possibly even more) additional mathematical tool, other than the metric, which links the different frames of reference that can be used in theory and practice with respect to which the movement is described. As an illustration, we recall in the following subsection this aspect in a particular case that is classic but important in the physics of continuous media.

2.5.1.3 *Principle of Material Indifference*

Although this is not directly related to the scope of this book, we raised the notion of invariance with respect to the choice of the reference frame by evoking Leibniz's position on the apparent contradiction of Newton's approach. We now recall the Principle of Material Indifference which frames invariance of description when we deal with a set of points continuously distributed in space.

Given a reference body $\mathscr{R}$ on which we can define a frame $(O, \mathbf{e}_1, \mathbf{e}_2, \mathbf{e}_3)$ as well as a clock $\mathscr{T}$ measuring the elapsed duration of time t, the position of a particle M is denoted $\mathbf{OM}(t)$. For another referential $\mathscr{R}'$, the position of the same particle M is denoted $\mathbf{O'M'}$. Since a particle is located in a unique position $M = M'$ with respect to an absolute space, we can write the vector relation (this point is crucial to understand that we truly face a passive difeomorphism):

$$\mathbf{O'M'} = \mathbf{O'O} + \mathbf{OM'} = \mathbf{O'O} + \mathbf{OM} \tag{2.79}$$

Decomposition of each vector onto the bases may be written as follows[2]:

$$\mathbf{O'M'} = y^i\,(t')\,\mathbf{e}'_i, \quad \mathbf{O'O} = u^i\,(t)\,\mathbf{e}'_i, \quad \mathbf{OM} = x^i\,(t)\,\mathbf{e}_i = x^j\,(t)\,Q^i_j\,(t)\,\mathbf{e}'_i \tag{2.80}$$

A change of (rigid) frame of reference (an example of passive diffeomorphism) is then defined by the relations:

$$\begin{cases} y^i(t') = u^i(t) + Q^i_j(t)\ x^j(t) \\ \quad t' = t + t_0 \end{cases} \tag{2.81}$$

with the new frame of reference $(\mathscr{R}', \mathscr{T}')$ with the new base $(O', \mathbf{e}'_1, \mathbf{e}'_2, \mathbf{e}'_3)$ and new time t', and in which $\mathbf{Q}$ is an orthogonal transformation and t_0 a scalar representing a time shift. Now assuming that the material point $M \in \mathscr{B}$ is a constituent point of a continuous medium $\mathscr{B}$, then we see by evidence that relation (2.81) takes a more familiar form in the framework of continuum physics e.g. [141]:

$$\begin{cases} \varphi'(M, t') = \mathbf{u}(t) + \mathbf{Q}(t)\ \varphi\,(M, t) \\ \qquad t' = t + t_0 \end{cases} \tag{2.82}$$

where $\mathbf{u}(t) := \mathbf{O'O}(t)$, $\varphi'(M, t') := \mathbf{O'M'}$ and $\varphi\,(M, t) := \mathbf{OM}$. We stress that M stands for the material point and not its position in the spacetime. For the transformation characterizing the change of reference frame (2.82) in continuum mechanics, we simply need to make a change of notation by defining the spatial coordinates of $\varphi'(M, t')$ as (y^1, y^2, y^3) and the time $y^0 := t'$ as well as the coordinates of $\varphi\,(M, t)$ as (x^1, x^2, x^3) and $x^0 := t$. We thus directly identify it with transformations belonging to the Leibniz group. As a reminder, the material frame indifference asserts as a fundamental principle of classical physics that material properties are independent of the frame of reference (time and space). We therefore can state the following principle for continuum physics:

Principle of Material Indifference. Constitutive equations of continuum physics must be invariant under the Leibniz transformations group,

[2]It should be stressed that the position of the origin O' and the orientation of the new base $(\mathbf{e}'_1, \mathbf{e}'_2, \mathbf{e}'_3)$ are given as functions of the time t.

given by:

$$y^j := Q^j_i(x^0)\, x^i + u^j(x^0) \quad \text{for } i,j = 1,2,3, \quad \text{and} \quad y^0 := x^0 + u^0 \tag{2.83}$$

in the four-dimensional spacetime formulation.

In extenso, scalar, vector, and second-order tensor fields $(\Phi, \mathbf{v}, \mathbf{T})$ are frame-indifferent (or also objective) if they are transformed as (in a three-dimensional form):

$$\Phi' = \Phi, \quad \mathbf{v}' = \mathbf{Q}\,\mathbf{v}, \quad \mathbf{T}' = \mathbf{Q}\mathbf{T}\mathbf{Q}^{\mathrm{T}} \tag{2.84}$$

Leibniz group transformations are passive diffeomorphisms in the sense that they change the (possibly local) system of coordinates, but not change the physical objects. As an illustration in continuum mechanics, given the velocity field $\mathbf{v}$ on a continuum $\mathscr{B}$, an objective tensor is the deformation rate $\mathbf{D} := 1/2(\overline{\nabla}\mathbf{v} + \overline{\nabla}\mathbf{v}^{\mathrm{T}})$ with $\overline{\nabla}$ the Levi-Civita connection in the Euclidean space, whereas the skew-symmetric tensor $\Omega := 1/2(\overline{\nabla}\mathbf{v} - \overline{\nabla}\mathbf{v}^{\mathrm{T}})$ is not objective e.g. [133]. This brief reminder about material indifference, for Newtonian mechanics, aims to warn us about the choice of mathematical variables that can appear as arguments of Lagrangian $\mathscr{L}$ of a physical system. In the next chapter, we will generalize the Principle of Material Indifference (here limited to the Leibniz group) to the case of all groups of diffeomorphisms, while remaining passive.

2.5.2 *Spacetime Structure in Classical Mechanics*

The space structure is a three-dimensional Euclidean geometry. The spacetime is a four-dimensional manifold defined as a product of a monodimensional Euclidean time with a three-dimensional Euclidean space $\mathscr{T} \times \mathscr{E}$. Absolute space $\mathscr{E}$ and absolute time $\mathscr{T}$ do not depend upon physical events happening inside. The spacetime coordinates can be written as (t, x^1, x^2, x^3). Both of them have their proper metric for measuring the flow of time $|t - t_0|$ and the distance $\|\mathbf{MM}_0\|^2 := (x^1 - x^1_0)^2 + (x^2 - x^2_0)^2 + (x^3 - x^3_0)^2$ between points respectively and the connection which relates the values of the local field, namely the velocity field $\mathbf{v}(t, M)$, at two arbitrary neighboring points $M(t, x^i)$ and $\tilde{M}(t, x^i + dx^i)$ of this spacetime. The notion of an orthonormal frame attached to, or otherwise underlying, a rigid solid — which becomes a reference system — is an abstraction (implicitly) linked to the reduction of the geometry of space to its Euclidean structure. But it is clear that a perfectly rigid body is only a point of view of the mind and remains an ideal but physically unrealizable notion. There is no need to mention the effectiveness of Newtonian mechanics for body speeds well below the speed of light. It should be remembered, however, that the notion of spacetime as a rigid whole remains an asymptotic point of view. The contribution of the theorists of relativity is not so much to have introduced four-dimensional spacetime but above all to have introduced the notion that spacetime cannot be treated as globally rigid, only locally. Reviews on classical mechanics neglecting relativity may be found in e.g. [142].

2.5.2.1 *Euclidean Spacetime and Isometries*

An Euclidean spacetime is a quadruplet, called affine fiber bundle, $(\mathscr{B}, \mathbf{g}, T, \tau)$, where $\mathscr{B}$ is a four-dimensional differentiable manifold endowed with a metric $\mathbf{g}$, whose elements $(x^\alpha, \alpha = 0, 1, 2, 3) \in \mathscr{B}$ are called events. T is a one-dimensional ordered Euclidean space and having underlying vector space on the real set $\mathbb{R}$. Projection map $\tau : \mathscr{B} \to T$ characterizes the absolute time $\tau(x^\alpha) = ct$ of the event $x^\alpha \in \mathscr{B}$ e.g. [113]. At this step, there is no need of a supplementary structure such as affine structure. There is an absolute time τ and a absolute three-dimensional space. Hence the manifold $\mathscr{B}$, set of the events, may be considered as the Cartesian product of the sets of instants T and the set of all space points $\mathscr{S} \to \mathscr{B}_t$, say $\mathscr{B} := T \times \mathscr{B}_t$ e.g. [1]. Any instantaneous subspace $\mathscr{B}_t$ is then implicitly assumed to be Euclidean 3-spaces (Fig. 2.11).

Remark 2.24. The term c is the light speed in order to introduce the space-component $x^0 := ct$. In non-relativistic mechanics, c plays no role since light is assumed to propagate at an infinite speed.

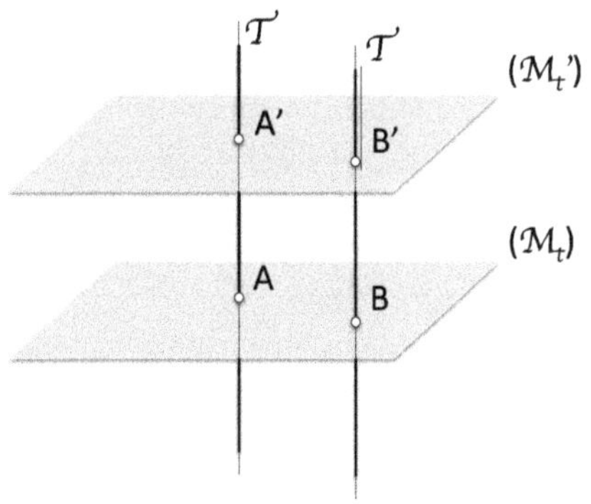

Fig. 2.11: Euclidean spacetime. The $3D$ affine Euclidean (sub)-spaces $\mathscr{B}_t := \{x^\alpha \in \mathscr{B}, \text{such that } \tau(x^\alpha) = ct\}$, $t \in T$ are instantaneous spaces. The structure group of the bundle is the group of isometries of a three-dimensional Euclidean space. Time T remains the same for all points of the spacetime (absolute time).

2.5.2.2 *Galilean Spacetime and Newtonian Spacetime*

Admitting an absolute spacetime to obtain four-dimensional Euclidean spacetime is not sufficient to model classical point mechanics. The reason is simple: all relative motions of material points are rigorously described in the framework of Newtonian mechanics as long as the frames of reference move in uniform translation relative to each other. This is the Galilean symmetry group:

$$\begin{cases} y^i = R^i_j \, x^j + u^i \, t + x^i_0 \\ t' = t + t_0 \end{cases} \tag{2.85}$$

in which t_0 is the time shift, R^i_j a rotation independent of time, u^i a relative translation velocity independent of time, and x^i_0 a shift of the origin, also independent of time. Although the satisfaction of the conditions inherent to this symmetry group is mandatory, its practical use remains limited. It is then more judicious to formulate it via differential geometry by introducing an affine structure. Roughly speaking, all the referential frames should be connected to each other. Therefore, a connection should be added to complete the geometric structure for deriving Newton's laws of motion. They are fundamentally based on the connecting of instantaneous spaces to calculate the acceleration e.g. [113], [143].

Remark 2.25. The affine structure, geometrically embodied by a connection, is designed to distinguish the straight lines (inertial motions) in Figure 2.13 from curved ones (i.e., accelerated motions).

Galilean connection and compatibilities. Instead of taking a rather abstract theoretical path, we prefer here a heuristic approach by taking from the outset a coordinate system and considering the passage from a point (with its volume element) to its neighbor. This was the historical and convenient approach to introduce affine connection ∇ on a differentiable manifold e.g. [2]. Theoretically, a vector is a differentiation operation on a differential manifold. Introduction of a coordinate system is then conveniently accompanied by a vector base:

$$(x^0, x^1, x^2, x^3) \quad \rightarrow \quad \{\partial_0, \partial_1, \partial_2, \partial_3\} \quad \rightarrow \quad \{\nabla_0, \nabla_1, \nabla_2, \nabla_3\}$$

which is associated with a coordinate set. Partial derivatives are however not covariant, in the sense that it does not keep the same formulation when choosing another coordinate system, such as spherical coordinates. We use covariant derivative using connection ∇ instead. For either classical or relativistic mechanics, there is a correspondence between the set of reference frames of spacetime and the set of compatible affine connections e.g. [113].

The use of connection has been exploited to define spacetimes of Leibniz, Galileo, and Newton in a systematic way e.g. [114]. For completeness, we just recall here, although it does not constitute the main object of this book, the notion of Galilean connection. The existence of an infinite number of reference frames is related to the existence of an infinite number of affine connections on a manifold. For Galilean structures (see definition below), the metric compatibility of the connection takes the form of e.g. [111], [117], [144]:

$$\nabla_\gamma h_{\alpha\beta} \equiv 0, \quad \nabla_\beta \tau_\alpha \equiv 0 \tag{2.86}$$

where $\mathbf{h}$ and τ are space metric and 1-form on the manifold $\mathscr{B}$ respectively, which will be defined later. This allows us to define:

Definition 2.21 (Galilean connection). Consider a Galilean structure $(\mathscr{B}, \mathbf{h}, \tau)$ where $\mathbf{h}$ and τ are metric and 1-form on the manifold $\mathscr{B}$ respectively. An affine connection ∇ is a Galilean connection if and only if it is compatible with the metric: $\nabla_\gamma h_{\alpha\beta} \equiv 0$, and $\nabla_\beta \tau_\alpha \equiv 0$.

However, the connection coefficients $\Gamma^\gamma_{\alpha\beta}$ of ∇ are not uniquely determined by the metric compatibility (2.86). From the compatibility of the 1-form τ, we deduce that the exterior derivative vanishes $d\tau \equiv 0$, meaning an integrability of the distribution associated to τ. First, the maximal integral manifolds are the spatial sections of constant time. Second, there is no unique symmetric connection on a Galilean structure (manifold). Any connection with coefficients $\Gamma^\gamma_{\alpha\beta} + 2\left(\tau_\alpha \omega_{\beta\lambda} - \tau_\beta \omega_{\alpha\lambda}\right) h^{\lambda\gamma}$, where $\Gamma^\gamma_{\alpha\beta}$ is Galilean and $\omega_{\alpha\beta}$ are components of arbitrary closed 2-form ($d\omega \equiv 0$), is also Galilean e.g. [144].

Neo-Newtonian or Galilean spacetime. Spacetime geometry for classical (called here neo-Newtonian) physics is described by a symmetric tensor $g^{\alpha\beta}$, a 1-form τ_α, and a symmetric connection $\Gamma^\gamma_{\alpha\beta}$ satisfying e.g. [111], [117], [145]:

$$\begin{cases} g^{\alpha\beta}\ \tau_\beta = 0 \\ \nabla_\gamma g^{\alpha\beta} = 0 \quad \text{and} \quad \nabla_\alpha \tau_\beta = 0 \\ \mathfrak{R}^\gamma_{\alpha\beta\lambda} = 0 \end{cases} \tag{2.87}$$

respectively meaning orthogonality condition of space and time, connection-metric compatibility, and absence of forces (spacetime flatness).

Metric $g^{\alpha\beta}$ is required to be (negative) semi-definite, and of matrix rank 3. We intentionally introduce the contravariant components of the metric since it is related to relativistic metrics. These conditions induce that there is a family of coordinate systems in which:

$$\begin{cases} g^{\alpha\beta} = \text{diag}\{0,-1,-1,-1\} \quad \text{as} \quad \left(= \lim\limits_{c\to\infty} \text{diag}\left\{1/c^2,-1,-1,-1\right\}\right) \\ \tau_\alpha = \text{diag}\{1,0,0,0\} \\ \Gamma^\gamma_{\alpha\beta} \equiv 0 \end{cases}$$

Coordinate systems are related to one another by a group of Galilean transformations (Fig. 2.12). They are identified as inertial frames of references. (2.87) and particularly metric compatibility are not sufficient to define a unique inertial frame but rather a family of frames. Compatibility (2.87) also ensures existence of a scalar function t which satisfies $\tau_\alpha := \partial_\alpha t$, meaning that the Newtonian spacetime possesses an absolute time.

Remark 2.26. Spacetime flatness (third line of 2.87) $\Re^\gamma_{\alpha\beta\lambda}(x^\mu) \equiv 0$ prohibits relativistic rotation between inertial frames. This condition also imposes at most linearity in the time dependence of translation between inertial frames; the relativistic acceleration between them is thus excluded

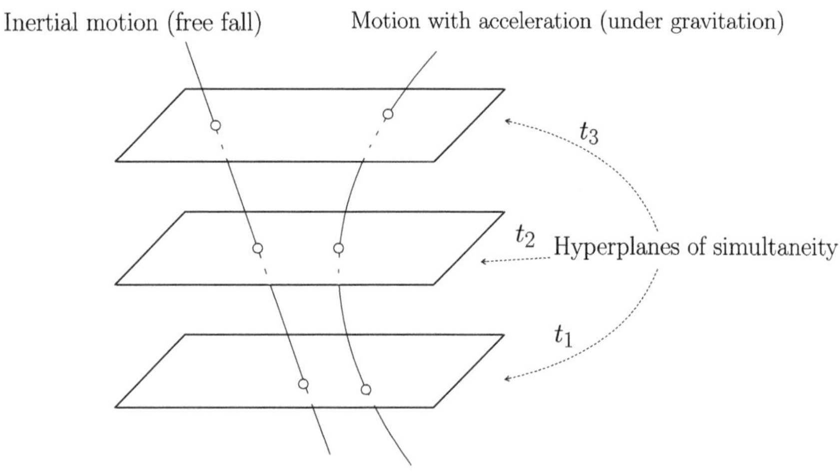

Fig. 2.12: Galilean spacetime. The three-dimensional affine Galilean (sub)-spaces $\mathscr{B}_t := \{x^\alpha \in \mathscr{B}, \text{such that } \tau(x^\alpha) = ct\}$, $t \in T$ are instantaneous spaces (hyperplanes of simultaneity at times t_1, t_2, and t_3). Time $t_i, i = 1,2,3,\ldots$ remains the same for all points of the spacetime (absolute time).

e.g. [111]. This allows us to define the Galilean group of transformations associated with classical mechanics e.g. [111]. The non-relativistic metric is not as simple since it introduces two independent quantities $\mathbf{g}$ and τ.

2.5.3 *Spacetime Structure in Special Relativity*

The advent of the theory of special relativity is due to the discovery of Maxwell electromagnetism and not so much to a purely mechanical consideration that wanted to extend the model to a four-dimensional spacetime. The genesis of the theory essentially comes from the problem of simultaneity and the consequence of the constancy of the light speed c, a special case of electromagnetic waves. Classically, the group of transformations that guarantees the invariance of the wave propagation formula is that of Lorentz defined by:

$$\begin{cases} y^1 = \gamma\,(x^1 - v\,t) \\ t' = \gamma\left(t - \dfrac{v\,x^1}{c^2}\right) \end{cases} \quad \text{where } \gamma := \frac{1}{\sqrt{1-(v^2/c^2)}} \tag{2.88}$$

in which v is the relative translation speed between two frames of reference. We illustrate with only relative uniform translation along the 1-axis for the sake of simplicity. This group is exactly the symmetry group of Maxwell's wave propagation equations. It is understandable that a four-dimensional spacetime is the most appropriate. This transformation guarantees the compatibility of the mechanics of special relativity and electromagnetism. The Lorentz group reduces to the Galileo group when $v \ll c$. Otherwise, time is not decoupled from space.

2.5.3.1 *Minkowski Spacetime and Metric*

Special relativity lies upon two cornerstones: (a) the extended relativity principle which assumes first the existence of inertial or Galilean frames $\mathscr{G}$, just as in classical mechanics, and second that with respect to these frames, all physics laws should be invariant; (b) the light speed axiom assuming that the same value of light speed holds in all Galilean frames, irrespective of the emission properties of the source. Usually, $c :=$ 299,792,458 [km s^{-1}] may be taken as the light speed. Implicitly it then assumes, as in classical mechanics, that there is an absolute spacetime, a set of events with spacetime coordinates (x^μ) interpreted as coordinates of a four-dimensional affine space. Space and time merge into a spacetime which is described as a four-dimensional continuum where any event can be described by coordinates $(x^0 = ct, x^1, x^2, x^3)$ of $\mathbb{R}^4$; index 0 stands for time.

Definition 2.22 (Minkowski spacetime). A Minkowski spacetime $\mathscr{B}$ is the vector space $\mathbb{R}^4$ endowed with an indefinite inner product given by:

$$< \mathbf{x}, \mathbf{y} > := x^0 y^0 - x^1 y^1 - x^2 y^2 - x^3 y^3 \tag{2.89}$$

where x^μ are the coordinates of $\mathbf{x} \in \mathbb{R}^4$, and y^ν the coordinates of $\mathbf{y} \in \mathbb{R}^4$.

To avoid lengthy equations including the light speed c, it is sometimes worthwhile to define the spacetime coordinates $(x^0 := ct, x^1, x^2, x^3)$, and therefore to have the previous metric $\hat{g}_{\alpha\beta} := \operatorname{diag}\{+1, -1, -1, -1\})$ for Minkowski spacetime $\mathscr{B}$. For ambient spacetime, a uniform metric holds for the entire Euclidean space $\hat{g}_{ij} := -\operatorname{diag}\{+1, +1, +1\}$ and $\hat{g}_{\alpha\beta} := \operatorname{diag}\{+1, -1, -1, -1\})$ for Minkowski spacetime.[3] Reference manifold (a four-dimensional "spacetime continuum") is *a priori* assumed e.g. [113] for classical mechanics. The particular choice of reference body with Euclidean (resp. Minkowskian) metric assumes the existence of an implicit flat configuration and constitutes an underlying constitutive assumption.

2.5.3.2 *Hilbert's Causality Principle*

Causality principle constitutes a principle that should be satisfied by any physics model. For general relativity, this means that each event (point) x^μ of the spacetime $\mathscr{B}$ should admit an intrinsic notion of past, present, and future. They do not depend on the mathematical description (invariant notion). When considering dynamical evolution of a metric on a manifold or change of spacetime coordinates, Hilbert's conditions must be satisfied to respect the causality e.g. [146]. Arbitrariness of coordinate change (or dynamics of metric evolution) should be limited if we want to stand on the point of view that two points of the same timeline are in a cause-effect relationship. Four inequalities must be satisfied besides the condition $\det[g_{\alpha\beta}] < 0$ e.g. [123], [146]:

$$g_{00} > 0, \quad g_{11} < 0, \quad \det\begin{bmatrix} g_{11} & g_{12} \\ g_{21} & g_{22} \end{bmatrix} > 0, \quad \det\begin{bmatrix} g_{11} & g_{12} & g_{13} \\ g_{21} & g_{22} & g_{23} \\ g_{31} & g_{32} & g_{33} \end{bmatrix} < 0 \tag{2.90}$$

Minkowski spacetime is a four-dimensional spacetime used to describe the consequences of special relativity theory on the finiteness of light speed,

[3] It is worth noting that covariant components of events hold as $(x_0 = ct, x_1 = -x^1, x_2 = -x^2, x_3 = -x^3)$ in this flat Minkowskian spacetime $\mathscr{B}$.

and for the time-like worldline $g_{\alpha\beta}\, u^\alpha u^\beta := c^2 dt - g_{ij} dx^i dx^j > 0$ while for the light-like worldline $g_{\alpha\beta}\, u^\alpha u^\beta := c^2 dt - g_{ij} dx^i dx^j = 0$. These conditions ensure that change of spacetime coordinates will respect the flow of time $x^0 := ct$ (they allow us to define a "proper time coordinate"). Application of Hilbert's causality conditions provides some restriction on the formation of black holes in cosmology after analysis of the Schwarzschild solutions to Einstein's equations of relative gravitation e.g. [147]. Broadly speaking, Hilbert's causality conditions involve the causal ordering and the univocal determination of metric components. They are strongly tight with the invariance of Lagrangian function and its arguments e.g. [146], and even beyond the covariance, related to the search of invariance group of spacetimes e.g. [148], [149].

Definition 2.23 (Isometry). Let $(\mathscr{B}, \mathbf{g})$ be a (pseudo-)Riemann manifold. A diffeomorphism $\mathbf{f} : \mathscr{B} \to \mathscr{B}$ is an isometry if it preserves the metric:

$$A^i_\alpha\, A^j_\beta\, g_{ij}\left(y^k\right) = g_{\alpha\beta}(x^\mu) \tag{2.91}$$

where y^k and x^μ are the coordinates of $\mathbf{f}(M) \in \mathbf{f}(\mathscr{B})$, and $M \in \mathscr{B}$ respectively.

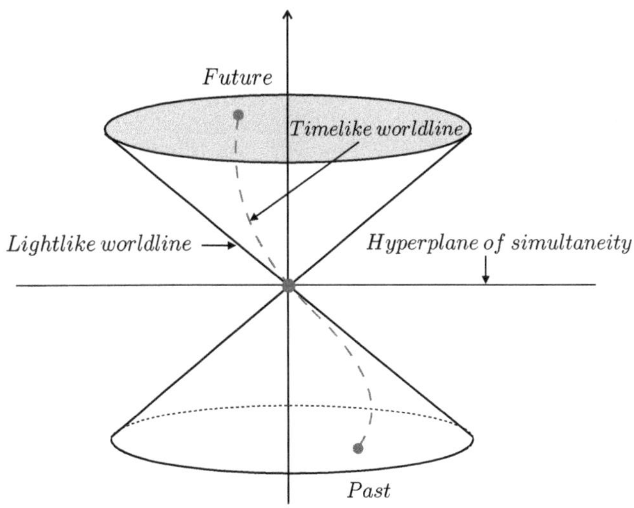

Fig. 2.13: Minkowski spacetime. Diagram showing the light cone for a worldline in Minkowski spacetime. The observer is at the junction between two light cones, the past and the future. Space is represented by an hypersurface of simultaneity.

The orthogonality group $\mathcal{O}(n), n = 3, 4$ is the group of $n \times n$ real matrices whose transpose is equal to their inverse. In an Euclidean n-dimensional manifold, the linear transformation of (tangent) vectors as $\hat{\mathbf{e}}_j = R_{jh}\mathbf{e}_h$, provided that the coefficients R_{jh} satisfy the orthogonality condition $R_{jh}R_{kh} \equiv \delta_{jk}$, implies that Det $\mathbf{R} = \pm 1$. These passive diffeomorphisms are orthogonal transformations, and for the subgroup Det $\mathbf{R} = 1$ rotations (proper orthogonal). Isometries form a group which includes the identity. Since an isometry preserves the length of a (tangent) vector, it characterizes a rigid motion in continuum mechanics. For special relativity, the background is the Minkowski spacetime, a four-dimensional real vector space $\mathbb{R}^{(1,3)}$ endowed with an inner product, which is a non-degenerate bilinear form with signature $(1, 3)$, where x^0 is the time. But the associated metric is not positive-definite. Adopting the notation that $\mathcal{O}(n) := \mathcal{O}(0, n)$, the orthogonal group of the Minkowski spacetime is denoted $\mathcal{O}(1, 3)$. This group $\mathcal{O}(1, 3)$ is called the Lorentz group of transformations.

Definition 2.24 (Lorentz group). The Lorentz group $\mathcal{O}(1, 3)$ is the group of linear transformations that preserve the Minkowski spacetime inner product (2.89) on $\mathbb{R}^4$.

Let us now recall briefly the necessary and sufficient condition for a 4×4 matrix — representing a linear transformation — $\mathbf{A} = (A^{\alpha}_{\beta})$ to leave the inner product (2.89) of any two 4-vectors $\mathbf{u}$ and $\mathbf{v}$ invariant. Suppose the transformed vectors:

$$u'^{\mu} = A^{\mu}_{\alpha} u^{\alpha}, \quad u'^{\nu} = A^{\nu}_{\beta} u^{\beta}, \quad \mathbf{u}' \cdot \mathbf{v}' = g_{\mu\nu} \; A^{\mu}_{\alpha} u^{\alpha} \; A^{\nu}_{\beta} u^{\beta} \equiv g_{\alpha\beta} u^{\alpha} u^{\beta}$$

In order for the condition $\mathbf{u}' \cdot \mathbf{v}' \equiv \mathbf{u} \cdot \mathbf{v}$ to hold for arbitrary vectors $\mathbf{u}$ and $\mathbf{v}$, we deduce the necessary and sufficient condition (then a definition) of Lorentz transformations, which is similar to (2.91):

$$g_{\mu\nu} \; A^{\mu}_{\alpha} \; A^{\nu}_{\beta} \equiv g_{\alpha\beta} \tag{2.92}$$

These transformations constitute a group (closure, associativity, identity and inverse conditions). This group defined as such is as large as the group defined hereafter (boosts and rotations). The subgroup which preserves the time orientation is denoted $\mathcal{O}^{+}(1, 3)$. The subgroup which preserves both the time and the space orientation is the group of proper Lorentz transformations $\mathcal{SO}^{+}(1, 3)$.

2.5.3.3 *Minkowski Spacetime and Lorentz Transformations*

Let $(\mathscr{B}, \mathbf{g})$ be a (pseudo)-Riemann manifold, and consider the kinematics-based Lorentz transformation, where the velocity v is along the axis x^1 for simplifying. The metric is defined by $g_{\alpha\beta} := \mathrm{diag}(1, -1, -1, -1)$, to give the Minkowski line element $ds^2 = g_{\alpha\beta}\, dx^\alpha dx^\beta = (dx^0)^2 - (dx^1)^2 - (dx^2)^2 - (dx^3)^2$. The group of proper Lorentz transformations $\mathscr{O}(1,3)$ is generated by six types of transformations that are three simple spatial rotations, and three time-space operators called boosts.

By introducing the dimension of light velocity, term v is replaced by v/c, and we recover the Galilean group of transformations when $v/c \to 0$ e.g. [1]. Field equations of Newtonian mechanics are invariant under any member of the Galilean group, whereas those of special relativity are invariant under the Lorentz group $v/c \neq 0$. Lorentz transformations built with change of reference (boosts and rotations) together with time reversal and space inversion equals the entire Lorentz group.

Remark 2.27. Components of metric $\{g_{ij}\}$ constitute a symmetric matrix; the eigenvalues are real. For a Riemann manifold, all the eigenvalues are strictly positive. For a pseudo-Riemann manifold, some of them may be strictly negative. The index of the metric is the pair (i, j) where i is the number of positive eigenvalues and j the number of negative eigenvalues. For $i = 1$, the metric is called Lorentzian.

Causality of spacetime events and invariance with respect to the Lorentz group are strongly tight. Consider a Minkowski spacetime $\mathscr{B}$ endowed with the metric $g_{\alpha\beta}$. In 1964, Zeeman's theorem states that the most general point (event) transformation of $\mathscr{B}$ to itself that preserves Hilbert's causality relation within the spacetime is an 11-parameters group, called $\mathscr{G}$ generated by the orthochronous Lorentz transformations, the spacetime translation, and dilations [148]. In other words, the Lorentz group naturally appears as the symmetry group of spacetime $\mathscr{B}$ if a single principle of causality is admitted *a priori*.

2.5.3.4 *Global Poincaré Transformations*

Indeed, common continuous symmetries of special relativistic theory is Lorentz invariant, meaning that the dynamics is the same in any Lorentz frame. The group of Lorentz transformations (which is not compact) can be decomposed into two parts: (a) Boosts, where we go from one Lorentz frame to another, i.e., we change the velocity v; and (b) Rotations, where we change the orientation of the coordinate frame. Lorentz transformations are not the only elements of the symmetry group of the Minkowski spacetime. For special relativity with metric $\hat{g}_{\alpha\beta} = \text{diag}\{+1, -1, -1, -1\}$, the Poincaré-Minkowski symmetry group of the spacetime is obtained by adding translation in space and time $u^\nu, \nu = 0, 1, 2, 3$:

$$y^\mu = A^\mu_\nu \, x^\nu + u^\nu, \quad \hat{g}_{\alpha\beta} \, A^\alpha_\mu \, A^\beta_\nu = \hat{g}_{\mu\nu} \tag{2.93}$$

This constitutes the symmetry group of the physics laws.[4] In the language of group theory, we have the definition.

Definition 2.25. Poincaré's group is $\mathcal{SO}^+(1,3) \ltimes \mathbb{R}^{1,3}$, the semi-direct product of the group of translations of the Minkowski spacetime $\mathbb{R}^{1,3}$ and the proper Lorentz transformations $\mathcal{SO}^+(1,3)$.

Zeeman [148] showed that the causality group of a Minkowski spacetime $\mathscr{B}$ was the group generated by the orthochronous Lorentz group, the translation of $\mathscr{B}$ and dilations of $\mathscr{B}$. Later, Williams [149] extended this result to show that the group of automorphisms $\mathscr{G}$ of $\mathscr{B}$ that preserved the norms of timelike vectors was in fact the complete Poincaré group. In that way, it is needless to use the principle of relativity since the invariance of the light velocity c leads to the invariance with respect to the Poincaré group e.g. [149]. The Poincaré group is a larger group of symmetries than the Lorentz group, obtained when we add translations to the set of symmetries (space and time translation). This is the main idea behind Einstein's theory of special relativity. The set of conserved quantities associated with the Poincaré group is larger. Translational and boost invariance implies conservation of four-momentum, and rotational invariance implies conservation of angular momentum.

2.5.4 *Riemann-Cartan Manifolds and Loops*

Before going further, we have to come back to the conceptualization of spacetime. Newtonian classical mechanics, quantum mechanics, and special relativity all rely on a fundamental postulate (among many others) about

---

[4]Given the two groups $\mathbb{R}^{1,3}$ and $\mathcal{SO}^+(1,3)$ such that $\mathbb{R}^{1,3} \cap \mathcal{SO}^+(1,3) = \{\mathbb{I}\}$, the semi-direct product of $\mathbb{R}^{1,3}$ and $\mathcal{SO}^+(1,3)$ is a group denoted with its element:

$$\mathcal{SO}^+(1,3) \ltimes \mathbb{R}^{1,3} := \left\{g = l\,k \in \mathcal{SO}^+(1,3) \ltimes \mathbb{R}^{1,3}, \text{ where } k \in \mathcal{SO}^+(1,3), l \in \mathbb{R}^{1,3}\right\}.$$

the existence of a spacetime background in which physical phenomena are supposed to take place. General relativity admits a radically different postulate since the metric constitutes a dynamical variable of the theory. In the continuity of relative gravitation, admitting nucleation of evolving defects, during the initial inflation after the Bing Bang, in the Universe leads us to postulate that besides the metric, a connection *a priori* independent of the metric can be present in the geometric structure of spacetime.

Another big challenge of modern physics would be to investigate the quantum aspects. String theory and loop quantum gravity have attempted to reconcile quantum mechanics and general relativity. Briefly, loop quantum gravity postulates that the structure of spacetime is composed of finite loops woven into an extremely fine fabric or network (spin networks). The evolution of such a spin network seems to be relevant at a length scale above the Planck length $\ell_{\mathrm{P}} \simeq 10^{-35}[m]$, then smaller scales are assumed to be meaningless. Therefore, for matter, spacetime has more or less a kind of atomic structure (discrete rather than continuous). However, the present work remains on a continuum physics approach.

Cartan circuit and loops. Nowadays, the basic trend is to develop theory allowing us to mathematically describe the (discrete) quantum spacetime by considering these so-called spin networks. Research has mainly evolved in two directions: the canonical loop quantum gravity, and the covariant loop quantum gravity. Schematically, the latter reassembles to the concept of microcosms according to Gonseth [2], and the affine connection is the main tool to link all these microcosms. At an intermediate length scale level, the consideration of field discontinuity has been used to link defect distribution within a continuum and the torsion and the curvature. The spacetime model which is investigated in this book remains at relative gravitation and electromagnetic fields but the accounting of these spacetime defects (defined on a Cartan parallelogram if adopting the geometric langage) is reconsidered in this chapter and might be hopefully expected to give insights for extended continuum models of relative gravitation.

Spacetime and loops. The common thread of this subsection is based on the formal link between the *concept of loops* at different length scales within a continuum and the torsion and curvature fields within Einstein-Cartan spacetime.

Consider a metric-affine manifold $\mathscr{B}$. Let θ be a scalar field on $\mathscr{B}$. If the variation of θ from one point to another depends on the path then the field presents discontinuity. This may be also applied to vector field $\mathbf{w}$ [6]. The generalized continuum model presented here is based on Cartan's circuit concept where the continuum $\mathscr{B}$ does not have any lattice structure. A continuum $\mathscr{B}$ is said to be affinely equivalent to the Euclidean continuum $\mathscr{E}$ if and only if the torsion and curvature tensor fields are identically null at every point of $\mathscr{B}$. From a topological point of view, an extended model of spacetime may be defined by a locally connected and locally compact manifold, which is formally a Generalized Continuum. The manifold is no longer Euclidean like Cauchy's classical continuum. This class of Generalized Continuum is nevertheless metrizable if it has a one-to-one continuous image in a metric after Proizvolov's theorem e.g. [150] (resp. semi-metric) space (Euclidean ambient space) in classical continuum mechanics (resp. Lorentzian spacetime for relativistic mechanics). This is one of the reasons we remain in the framework of Riemann-Cartan geometry.

Circuit, torsion and curvature. The path dependence is illustrated by the Cartan circuit which is more recently renamed loop together with discontinuities of fields but this intuition may be confusing as we will show in the following.

We do not build the theory on the discrete atom picture, but rather we stay on a continuum approach. To have a compromise between the different scales, between discrete description and Einstein spacetime continuum, we relax the geometric variables to include a discontinuity of fields. Going back to the definition of torsion and curvature, it is worth noting that the proof of the discontinuity theorem requires a circuit in the metric-affine manifold which is a loop (Figure 2.14). Considering each crack individually is akin to modeling point particles, which does not quite fit within a continuum physics approach. It is more rigorous to consider a density of defects, as we will discuss below. Indeed, depending on the loop dimensions in Figure 2.14, many aspects of physics may be captured. When dimensions are of the order of Planck length ℓ_{P}, the quantum effects are relevant and the curvature and the torsion cannot be considered as continuum fields on the manifold and may be represented by discrete entities.

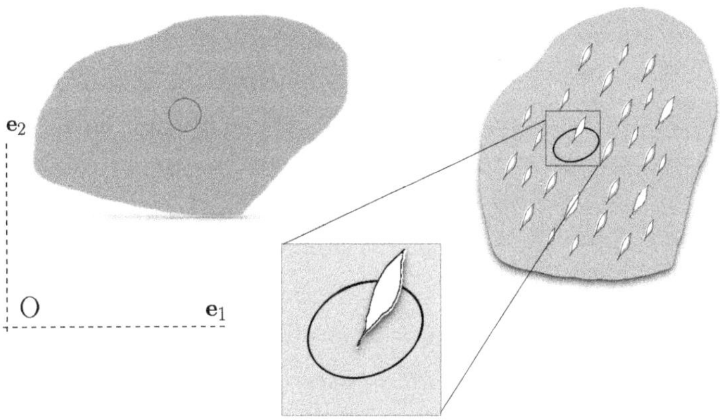

Fig. 2.14: Spacetime with loops. (Left) Smooth spacetime at macroscopic length scale; (right) spacetime with distribution of loops; (center) enhanced view of a loop surrounding a discontinuity between two lips (boundaries between microcosms). Theorem 2.9 expresses the interweaving between loops (or Cartan circuit) and the torsion and curvature fields.

Theorem 2.9. *Say an affinely connected manifold $\mathscr{B}$. Let's consider two arbitrary vectors $\mathbf{f}_1$ and $\mathbf{f}_2$ at a point M; they define two paths of length ϵ_1 and ϵ_2. Non-zero torsion and curvature fields induce the following relationships [11]:*

$$\begin{cases} \lim\limits_{(\epsilon_1,\epsilon_2)\to 0} (\theta' - \theta'')/\epsilon_1\epsilon_2 = \aleph(\mathbf{f}_1, \mathbf{f}_2)[\theta] \\ \lim\limits_{(\epsilon_1,\epsilon_2)\to 0} [\mathbf{w}'' - \mathbf{w}']/\epsilon_1\epsilon_2 = \Re(\mathbf{f}_1, \mathbf{f}_2, \mathbf{w}) - \nabla_{\aleph(\mathbf{f}_1,\mathbf{f}_2)}\mathbf{w} \end{cases} \tag{2.94}$$

where $\theta(\mathbf{x})$, and $\mathbf{w}(\mathbf{x})$ are respectively a scalar field and a vector field on $\mathscr{B}$. We notice $\theta' := \theta(M')$, and $\theta'' := \theta(M'')$; and $\mathbf{w}' := \mathbf{w}(M')$, and $\mathbf{w}'' := \mathbf{w}(M'')$.

Proof. Some parts of the proof are given in [11]. Complete proof including two steps is given here.

(1) For the first equation, consider the two portions MM_1 and M_1M' of the path MM'. Consider a scalar field $\theta(M)$, we have respectively the relations:

$$\begin{cases} \theta(M_1) = \theta(M + \varepsilon_1\ \mathbf{f}_1). = \theta(M) + \nabla_{\varepsilon_1\mathbf{f}_1(M)}\theta(M) \\ \theta(M') = \theta(M_1 + \varepsilon_2\mathbf{f}_2(M + \varepsilon_1\mathbf{f}_1)) = \theta(M_1) + \nabla_{\varepsilon_2\mathbf{f}_2(M+\varepsilon_1\mathbf{f}_1)}\ \theta(M_1) \end{cases}$$

where we have:

$$\varepsilon_2\mathbf{f}_2\left(M+\varepsilon_1\mathbf{f}_1\right)=\varepsilon_2\mathbf{f}_2\left(M\right)+\varepsilon_2\varepsilon_1\nabla_{\mathbf{f}_1(M)}\mathbf{f}_2(M)$$

For the sake of simplicity, the term (M) of the right-hand side of the equation can be dropped since all terms are now dependent on the point M. By using the properties of the affine connection ∇, we obtain the value of the scalar field at M' in terms of its value at M, by noticing $\theta\left(M'\right)=\theta'$ and $\theta\left(M\right)=\theta$:

$$\begin{aligned}\theta' &= \theta+\varepsilon_1\nabla_{\mathbf{f}_1}\theta+\varepsilon_2\nabla_{\mathbf{f}_2}\theta+\varepsilon_2\varepsilon_1\nabla_{\nabla_{\mathbf{f}_1}\mathbf{f}_2}\theta+\varepsilon_2\varepsilon_1\nabla_{\mathbf{f}_2}\nabla_{\mathbf{f}_1}\theta\\&\quad+\varepsilon_2\varepsilon_1^2\nabla_{\nabla_{\mathbf{f}_1}\mathbf{f}_2}\nabla_{\mathbf{f}_1}\theta\end{aligned}$$

Consider now the two portions MM_2 and M_2M'' of MM''. With the same method, we easily obtain the value of the field at M'' by exchanging index 1 with 2:

$$\begin{aligned}\theta'' &= \theta+\varepsilon_2\nabla_{\mathbf{f}_2}\theta+\varepsilon_1\nabla_{\mathbf{f}_1}\theta+\varepsilon_1\varepsilon_2\nabla_{\nabla_{\mathbf{f}_2}\mathbf{f}_1}\theta+\varepsilon_1\varepsilon_2\nabla_{\mathbf{f}_1}\nabla_{\mathbf{f}_2}\theta\\&\quad+\varepsilon_1\varepsilon_2^2\nabla_{\nabla_{\mathbf{f}_2}\mathbf{f}_1}\nabla_{\mathbf{f}_2}\theta\end{aligned}$$

The gap between the values at θ' and θ'' is given by:

$$\begin{aligned}\theta'-\theta'' &= \varepsilon_1\varepsilon_2\left[\nabla_{\nabla_{\mathbf{f}_1}\mathbf{f}_2}\theta-\nabla_{\nabla_{\mathbf{f}_2}\mathbf{f}_1}\theta+\nabla_{\mathbf{f}_2}\nabla_{\mathbf{f}_1}\theta-\nabla_{\mathbf{f}_1}\nabla_{\mathbf{f}_2}\theta\right]\\&\quad+\text{higher-order terms}\end{aligned}$$

where the higher-order terms are proportional to $\varepsilon_2\varepsilon_1^2$ or $\varepsilon_1\varepsilon_2^2$ and can be neglected when calculating the limit. Let's introduce the following definitions and properties of the covariant derivatives of a scalar field: $\nabla_{\mathbf{u}}\theta:=\mathbf{u}\left[\theta\right]$, then it is straightforward to deduce the relation:

$$\theta'-\theta''=\varepsilon_1\varepsilon_2\left\{\left(\nabla_{\mathbf{f}_1}\mathbf{f}_2-\nabla_{\mathbf{f}_2}\mathbf{f}_1\right)\left[\theta\right]-\left(\mathbf{f}_1\ \mathbf{f}_2-\mathbf{f}_2\ \mathbf{f}_1\right)\left[\theta\right]\right\}$$

where we pointed out the Lie-Jacobi bracket $(\mathbf{f}_1\ \mathbf{f}_2-\mathbf{f}_2\ \mathbf{f}_1)$. Finally, the density of the discontinuity of a scalar field per surface area is obtained as:

$$\lim_{\epsilon_1\epsilon_2\to 0}\frac{\theta'-\theta''}{\varepsilon_1\varepsilon_2}=\aleph\left(\mathbf{f}_1,\mathbf{f}_2\right)\left[\theta\right]$$

considering the definition of torsion operator and by neglecting higher-order terms.

(2) For the second equation, we proceed with the same method by considering the two paths term by term, the difference being that we calculate the values of a vectorial field $\mathbf{w}\left(M'\right)$ and $\mathbf{w}\left(M''\right)$ as a function of its

value at M. It suffices to replace the scalar field θ by $\mathbf{w}$ to obtain respectively:

$$\mathbf{w}' = \mathbf{w} + \varepsilon_1 \nabla_{\mathbf{f}_1} \mathbf{w} + \varepsilon_2 \nabla_{\mathbf{f}_2} \mathbf{w} + \varepsilon_2 \varepsilon_1 \nabla_{\nabla_{\mathbf{f}_1} \mathbf{f}_2} \mathbf{w} + \varepsilon_2 \varepsilon_1 \nabla_{\mathbf{f}_2} \nabla_{\mathbf{f}_1} \mathbf{w}$$
$$+ \varepsilon_2 \varepsilon_1^2 \nabla_{\nabla_{\mathbf{f}_1} \mathbf{f}_2} \nabla_{\mathbf{f}_1} \mathbf{w}$$

and

$$\mathbf{w}'' = \mathbf{w} + \varepsilon_2 \nabla_{\mathbf{f}_2} \mathbf{w} + \varepsilon_1 \nabla_{\mathbf{f}_1} \mathbf{w} + \varepsilon_1 \varepsilon_2 \nabla_{\nabla_{\mathbf{f}_2} \mathbf{f}_1} \mathbf{w} + \varepsilon_1 \varepsilon_2 \nabla_{\mathbf{f}_1} \nabla_{\mathbf{f}_2} \mathbf{w}$$
$$+ \varepsilon_1 \varepsilon_2^2 \nabla_{\nabla_{\mathbf{f}_2} \mathbf{f}_1} \nabla_{\mathbf{f}_2} \mathbf{w}$$

As for the scalar field, the gap of the vector field divided by the area after neglecting the higher-order terms gives:

$$\lim_{\epsilon_1 \epsilon_2 \to 0} \frac{\mathbf{w}'' - \mathbf{w}'}{\varepsilon_1 \varepsilon_2} = \nabla_{\mathbf{f}_1} \nabla_{\mathbf{f}_2} \mathbf{w} - \nabla_{\mathbf{f}_2} \nabla_{\mathbf{f}_1} \mathbf{w} + \nabla_{\nabla_{\mathbf{f}_2} \mathbf{f}_1} \mathbf{w} - \nabla_{\nabla_{\mathbf{f}_1} \mathbf{f}_2} \mathbf{w}$$

Due to the properties of ∇ and the definition $\nabla_{\mathbf{f}_2} \mathbf{f}_1 - \nabla_{\mathbf{f}_1} \mathbf{f}_2 = -\aleph(\mathbf{f}_1, \mathbf{f}_2) - [\mathbf{f}_1, \mathbf{f}_2]$ the last two terms may be rearranged to obtain:

$$\lim_{\epsilon_1 \epsilon_2 \to 0} \frac{\mathbf{w}'' - \mathbf{w}'}{\varepsilon_1 \varepsilon_2} = \nabla_{\mathbf{f}_1} \nabla_{\mathbf{f}_2} \mathbf{w} - \nabla_{\mathbf{f}_2} \nabla_{\mathbf{f}_1} \mathbf{w} - \nabla_{[\mathbf{f}_1, \mathbf{f}_2]} \mathbf{w} - \nabla_{\aleph(\mathbf{f}_1, \mathbf{f}_2)} \mathbf{w}$$

By using the definition of the curvature operator, we then have shown the second equation of the theorem:

$$\lim_{\epsilon_1 \epsilon_2 \to 0} \frac{\mathbf{w}'' - \mathbf{w}'}{\varepsilon_1 \varepsilon_2} = \Re(\mathbf{f}_1, \mathbf{f}_2) \mathbf{w} - \nabla_{\aleph(\mathbf{f}_1, \mathbf{f}_2)} \mathbf{w}$$

which expresses the density of the vector field discontinuity per unit area by neglecting all higher-order terms.

Only terms factor of the area $\epsilon_1 \epsilon_2$ are retained since all others tend to zero. □

Remark 2.28. The system (2.94) is the continuum extension of the discrete dislocation loop which is characterized by the Burgers vector $\mathbf{b} := [\mathbf{u}]$, defined by the discontinuity of the displacement field within continuum $\mathbf{b} := -\oint_{\mathcal{C}} d\mathbf{u}$. The integral is calculated along a closed loop $\mathcal{C}$ around the isolated dislocation in the limit case but it should not be forgotten that the proof (2.94) hightlights that the right members are density per surface of area $\varepsilon_1\,\varepsilon_2$. Torsion $\aleph$ and curvature $\Re$ are physically interpreted as densities of discontinuities at point $\mathbf{x} \in \varphi(\mathscr{B})$.

Alternatively, for multi-valued fields, the concept of generalized continuum allows us to define locally connected (and locally compact) manifolds rather than directly accounting for discontinuity of scalar and vector fields, $\theta(\mathbf{x})$ and $\mathbf{w}(\mathbf{x})$. If the curvature vanishes, the vector field is however necessarily not single-valued as shown by the second equation (2.94). These relations show the interdependence of translation and rotation dislocations and therefore their practical co-existence. The twisted and curved continuum manifold offers a representation of a material (resp. spacetime) manifold that is everywhere dislocated (resp. singular or with continuously distributed spins). Moreover, these relations may be read from right to left direction, and state that whenever $\aleph$ or $\Re$ are not equal to zero then any scalar and vector fields on the affinely connected continuum $\mathscr{B}$ are necessarily multi-valued fields.

Remark 2.29. For Riemann continua, only vector fields may be discontinuous, whereas for Weitzenböck continua e.g. [151], only scalar fields and some particular vector fields may be discontinuous, since $\Re \equiv 0$. We consider here arbitrary vector fields and not only displacement.

2.5.5 *Spacetime Structure in Relative Gravitation*

In Newtonian mechanics, gravitation is commonly associated with the presence of some matter within an absolute flat spacetime. In relative gravitation, spacetime and matter are strongly tight. Contrary to Galilean and Minkowski spacetimes, relative gravitation requires the introduction of a metric function of the event x^μ.

2.5.5.1 *Spacetime Classification and Hierarchy*

Modeling spacetime as a generalized continuum imposes the choice of a geometric structure: a *continuum* is a compact and connected differentiable manifold. It is therefore provided with a metric $\mathbf{g}$ and a connection ∇, which we assume affine and *a priori* independent of the metric [136]. Thanks to compactedness and connectedness, one summarizes the hierarchical order of spacetime geometric structures:

$$\begin{array}{c} \text{Metric} \\ \text{affine} \\ \text{geometry} \end{array} \quad \nabla\mathbf{g} \equiv 0 \quad \begin{array}{c} \text{Riemann/} \\ \text{Cartan} \\ \text{geometry} \end{array} \left\{ \begin{array}{ccc} \aleph \equiv 0 & \begin{array}{c} \text{Riemann} \\ \text{geometry} \end{array} & \Re \equiv 0 \\ & & \\ \Re \equiv 0 & \begin{array}{c} \text{Weitzenboeck} \\ \text{geometry} \end{array} & \aleph \equiv 0 \end{array} \right\} \begin{array}{c} \text{Euclidean/} \\ \text{Minkowski} \\ \text{geometry} \end{array}$$

The metric compatibility is an essential condition of a continuum to survive as a continuum after non-holonomic transformations of the continuum e.g. [152]. From right to left, the geometric structure describes a strongly continuous body e.g. [153], a body with rotational and/or translational dislocations, and most general continua e.g. [11]. The analogy with spacetime holds. The non-compatibility $\nabla_\gamma g_{\alpha\beta} \neq 0$ leads to a more extended version of continuum. Weyl developed the electromagnetism by means of this non-compatibility but seemed to face many reactions e.g. [57]. In sum, a metric-affine manifold $(\mathscr{B}, \mathbf{g}, \nabla)$ with the most general metric-compatible connection is called Riemann-Cartan manifold, also denoted U_4. If the torsion $\aleph^{\gamma}_{\alpha\beta}$ vanishes, a U_4 becomes a Riemannian manifold, denoted V_4 in general relativity theory. If alternatively curvature $\Re^{\lambda}_{\alpha\beta\mu}$ vanishes, then a U_4 manifold becomes Weitzenböck's teleparallel manifold, denoted T_4 (Tele Parallel Gravity theory). Finally, condition of zero curvature $\Re^{\lambda}_{\alpha\beta\mu} = 0$ transforms a V_4 into a Minkowski spacetime M_4, and zero torsion $\aleph^{\gamma}_{\alpha\beta} = 0$ transforms a T_4 into an M_4.

2.5.5.2 *Spacetime Structure in Einstein Gravitation*

For Einstein gravitation, metric tensor and connection with zero torsion and curvature hold. The metric satisfies Hilbert causality and connection with the metric compatibility:

$$g_{\alpha\beta} \quad \nabla_\gamma g_{\alpha\beta} = 0 \quad \Re^{\gamma}_{\alpha\beta\lambda} \neq 0 \tag{2.95}$$

where again Hilbert's causality principle given by (2.90) is satisfied for the metric, the second equation means metric compatibility of connection, and the third inequality expresses that a gravitation field exists.

Torsionless but curved spacetime $(\mathscr{B}, \mathbf{g}, \nabla)$ is slightly more general than Newton-Cartan spacetime. Such a model is considered as a "structureless continuum" if the torsion vanishes. The spacetime is summarized by the Riemannian structure with metric $g^{\alpha\beta}(x^\mu)$, torsion $\aleph^{\gamma}_{\alpha\beta}$, and curvature $\Re^{\delta}_{\alpha\sigma\gamma}(x^\mu)$:

$$\nabla_\gamma g_{\alpha\beta} = 0, \quad \aleph^{\gamma}_{\alpha\beta} \equiv 0, \quad \Re^{\delta}_{\alpha\sigma\gamma} \neq 0 \tag{2.96}$$

Remark 2.30. In relations (2.95), both metric and curvature depend on x^μ meaning that they are variable with respect to time and space.

2.5.5.3 *Spacetime Structure for Einstein-Cartan Gravitation*

In addition to these two variables, the metricity of the connection might be also included in the list but we voluntarily let this one at first step. Riemann spacetime is extended to Riemann-Cartan spacetime by considering non-zero torsion. It is known that the test of spacetime with torsion can be only conducted with particles or continuum with internal structure, such as intrinsic spin e.g. [154], [155] or fluid with vortices e.g. [156]. It was shown that only intrinsic spin and not global angular rotation of matter couples with spacetime e.g. [157]. Let us consider a curved spacetime manifold with torsion $(\mathscr{B}, \mathbf{g}, \nabla)$ with the metric $g^{\alpha\beta}(x^\mu)$, the torsion $\aleph^{\gamma}_{\alpha\beta}(x^\mu)$, and the curvature $\Re^{\delta}_{\alpha\sigma\gamma}(x^\mu)$:

$$\nabla_\gamma g^{\alpha\beta} = 0, \quad \aleph^{\gamma}_{\alpha\beta} \neq 0, \quad \Re^{\delta}_{\alpha\sigma\gamma} \neq 0 \tag{2.97}$$

Remark 2.31. In relative gravitation, Riemann-Cartan manifold corresponds to Einstein-Cartan spacetime. There are numerous approaches to extend the Einstein spacetime to include more complex physical phenomenon, the simplest being the electromagnetic field. Some are based on a spacetime with dimension higher than four; we choose in this book the extension by using an analogous (but not exact) approach to a Taylor expansion of the metric: metric, torsion and curvature.

2.5.5.4 *Torsion and Curvature Variation for Riemann-Cartan Manifold*

The connection is not a tensor conversely to torsion and curvature. During the transformations of a manifold $\mathscr{B}$, the connection evolves and consequently, torsion and curvature also change. This subsection is largely inspired by the theory of "*weakly continuous media*", named after previous works which concern the irreversible deformation of generalized continua e.g. [11], [49]. For determining the Euler variation[5] of a tensor $\mathbf{A}$, function of ∇, we use the directional derivative:

$$\delta\mathbf{A}(\nabla) := \lim_{\eta\to 0} \frac{1}{\eta}\left[\mathbf{A}(\nabla + \eta\delta\nabla) - \mathbf{A}(\nabla)\right] \tag{2.98}$$

[5]This should not be confused with the Lie variation $\mathcal{L}_\xi \mathbf{A}$ as we have seen earlier, which is a variation along the evolution of the manifold.

along the variation $\delta\nabla$. We apply to torsion and curvature which are solely defined by means of the connection without introducing the metric.

Variation of torsion. Consider any three vectors $\mathbf{u}_\alpha$, $\mathbf{u}_\beta$, and $\mathbf{u}_\lambda$ of the tangent space $T_\mathbf{x}\mathscr{B}$, and a 1-form $\omega^\gamma \in T_\mathbf{x}\mathscr{B}^*$ of its dual. Given a connection ∇ which constitutes a field on $\mathscr{B}$, the torsion is a 2-covariant 1-contravariant tensor defined by the relation:

$$\aleph\left(\mathbf{u}_\alpha, \mathbf{u}_\beta, \omega^\gamma\right) := \omega^\gamma\left(\nabla_{\mathbf{u}_\alpha}\mathbf{u}_\beta - \nabla_{\mathbf{u}_\beta}\mathbf{u}_\alpha - [\mathbf{u}_\alpha, \mathbf{u}_\beta]\right) \tag{2.99}$$

which gives in component-form:

$$\aleph^\gamma_{\alpha\beta} = \Gamma^\gamma_{\alpha\beta} - \Gamma^\gamma_{\beta\alpha} - \aleph^\gamma_{0\alpha\beta} \tag{2.100}$$

with the Lie-Jacobi bracket $[\mathbf{u}_\alpha, \mathbf{u}_\beta] = \aleph^\gamma_{0\alpha\beta}\,\mathbf{u}_\gamma$. Cartan's constants of structure $\aleph^\gamma_{0\alpha\beta}$ are assumed not to change during the variation. It is checked that the torsion is a tensor. Torsion is skew-symmetric with respect to the two lower indices. Its variation is easy to calculate:

$$\boxed{\delta\aleph^\gamma_{\alpha\beta} = \delta\Gamma^\gamma_{\alpha\beta} - \delta\Gamma^\gamma_{\beta\alpha}} \tag{2.101}$$

Remark 2.32. Recall here that the variation is merely obtained by using Frechet (directional) derivative with respect to the direction $\delta\Gamma^\gamma_{\alpha\beta}$ of the connection.

Variation of curvature. Given a connection ∇, curvature is a 3-covariant 1-contravariant tensor defined by the relation:

$$\Re\left(\mathbf{u}_\alpha, \mathbf{u}_\beta, \mathbf{u}_\lambda, \omega^\gamma\right) := \omega^\gamma\left(\nabla_{\mathbf{u}_\alpha}\nabla_{\mathbf{u}_\beta}\mathbf{u}_\lambda - \nabla_{\mathbf{u}_\beta}\nabla_{\mathbf{u}_\alpha}\mathbf{u}_\lambda - \nabla_{[\mathbf{u}_\alpha,\mathbf{u}_\beta]}\mathbf{u}_\lambda\right) \tag{2.102}$$

which gives the component-form:

$$\Re^\gamma_{\alpha\beta\lambda} = (\partial_\alpha\Gamma^\gamma_{\beta\lambda} + \Gamma^\gamma_{\alpha\mu}\,\Gamma^\mu_{\beta\lambda}) - (\partial_\beta\Gamma^\gamma_{\alpha\lambda} + \Gamma^\gamma_{\beta\mu}\,\Gamma^\mu_{\alpha\lambda}) - \aleph^\mu_{0\alpha\beta}\Gamma^\gamma_{\mu\lambda} \tag{2.103}$$

It is checked that the curvature is a tensor. Curvature is skew-symmetric with respect to the first two lower indices. Its variation gives on a coordinate base:

$$\begin{aligned}\delta\Re^\gamma_{\alpha\beta\lambda} = {} & (\partial_\alpha\delta\Gamma^\gamma_{\beta\lambda} + \delta\Gamma^\gamma_{\alpha\mu}\,\Gamma^\mu_{\beta\lambda} + \Gamma^\gamma_{\alpha\mu}\,\delta\Gamma^\mu_{\beta\lambda}) \\ & - (\partial_\beta\delta\Gamma^\gamma_{\alpha\lambda} + \delta\Gamma^\gamma_{\beta\mu}\,\Gamma^\mu_{\alpha\lambda} + \Gamma^\gamma_{\beta\mu}\,\delta\Gamma^\mu_{\alpha\lambda})\end{aligned}$$

Let's calculate the covariant derivatives of tensors $\delta\Gamma^{\gamma}_{\beta\lambda}$ and $\delta\Gamma^{\gamma}_{\alpha\lambda}$:

$$\nabla_{\alpha}\delta\Gamma^{\gamma}_{\beta\lambda} = \partial_{\alpha}\delta\Gamma^{\gamma}_{\beta\lambda} + \Gamma^{\gamma}_{\alpha\mu}\,\delta\Gamma^{\mu}_{\beta\lambda} - \Gamma^{\mu}_{\alpha\beta}\,\delta\Gamma^{\gamma}_{\mu\lambda} - \Gamma^{\mu}_{\alpha\lambda}\,\delta\Gamma^{\gamma}_{\beta\mu}$$
$$\nabla_{\beta}\delta\Gamma^{\gamma}_{\alpha\lambda} = \partial_{\beta}\delta\Gamma^{\gamma}_{\alpha\lambda} + \Gamma^{\gamma}_{\beta\mu}\,\delta\Gamma^{\mu}_{\alpha\lambda} - \Gamma^{\mu}_{\beta\alpha}\,\delta\Gamma^{\gamma}_{\mu\lambda} - \Gamma^{\mu}_{\beta\lambda}\,\delta\Gamma^{\gamma}_{\alpha\mu}$$

We deduce the Palatini relation within an Riemann-Cartan manifold e.g. [158]:

$$\boxed{\delta\Re^{\gamma}_{\alpha\beta\lambda} = \nabla_{\alpha}(\delta\Gamma^{\gamma}_{\beta\lambda}) - \nabla_{\beta}(\delta\Gamma^{\gamma}_{\alpha\lambda}) + \aleph^{\mu}_{\alpha\beta}\,\delta\Gamma^{\gamma}_{\mu\lambda}} \tag{2.104}$$

In the same way, curvature is associated with rotational dislocations in mechanical continuum mechanics. From a slightly different point of view, curvature is not only associated with rotational dislocations, but also to non-smoothness of a vector field e.g. [11].

Remark 2.33. (2.104) generalizes the Palatini relation on the variation of curvature. It plays a crucial role in the development of the Einstein-Maxwell equations which couple gravitation and electromagnetism within Einstein-Cartan spacetime. It also makes it possible to find the Chern-Simons current that links torsion with electromagnetism, which we will develop.

Remark 2.34. All along this chapter, the compatibility of the connection and the metric was naturally imposed. For Einstein-Cartan spacetime, the compatibility of the connection with the volume-form will additionally play an important role for the derivation of field equations and also for the integration of any form on the spacetime manifold.

2.6 General Comment on Spacetime Geometry

In sum, it is essential to return to the history of the geometrization of classical physics, namely gravitation and electromagnetism. Without wishing to be exhaustive, the role of spacetime in physics is more than primordial. It is fundamental to be aware of their classification e.g. [1]. To cite only the original work of Rainich [8] and the more recent work of Misner and Wheeler, an interesting approach that highlights that the sole use of Einstein spacetime (via the Riemann curve manifold) to model gravitation as well as electromagnetic phenomena, point masses, non-quantum electric (or even magnetic) charges, seems sufficient, e.g. [7].

We note, however, that the Riemannian manifolds they used were not purely Riemannian but implicitly introduced the notion of multiple connexity of the manifold, thus in the presence of "holes", which *in fine* amounted to considering loops and discontinuities of vector and scalar fields within the vacuum spacetime. In this chapter, the essential point is therefore to consider the notion of loops identified with Cartan circuits which are closely linked to densities of dislocations and disclinations e.g. [49]. Beyond these characteristics, it should also be retained that for any metric-affined spacetime, metric and connection play an essential role for the definition of referential frames e.g. [159]. Details will be given in the second part of the book treating relative gravitation.

Spacetime as Generalized Continuum. To summarize this chapter, it follows by deduction the necessity of using a generalized continuum (Riemann-Cartan manifold) as a geometric background instead of Riemannian, and then by extension to mainly consider the Einstein-Cartan spacetime for the rest of this book. This allows us to include defect fields (linear and surface) directly in the vacuum spacetime continuum model without needing to introduce them *ad hoc*. In the event that other simpler geometric models of spacetime are used (for some illustrations), we will explicitly mention it.

Chapter 3

Covariance: Form Invariance of Equations on Generalized Continua

We extend in this chapter the Frame Indifference Principle when using Einstein-Cartan spacetime. It is necessary to specify even more the role of invariance. Namely, the differentiation between passive and active diffeomorphisms is essential in the formulation of physical theories. What we explore in this chapter concerns the invariance of mathematical equations of physics phenomena, which brings us to *covariance*, that is, in our vocabulary, invariance with respect to passive diffeomorphisms. We particularly recall the elements necessary to formulate the invariance of Lagrangian formulations with respect to a change of coordinate system.

3.1 Introduction

The notion of covariance has its own history and controversies in relative gravitation e.g. [160]; however, this problem also has its counterpart in classical continuum mechanics, particularly in the theory of constitutive equations e.g. [161]. In parallel with the debate that has arisen over the notion of covariance in relativity, a debate at least as long-standing has also taken place within the framework of what is called Material Frame Indifference. Without going into too much detail, which may be found in Frewer's review, e.g. [162], this parallel is nevertheless worth recalling because it sheds light on a particular point.

3.1.1 *Material Frame Indifference and Covariance*

Basically, the idea is that coordinates are only chosen artifices used in describing physical phenomenon, and therefore cannot play a conceptual role in the formulation of physical laws, and *a fortiori* for either continuum physics or relative gravitation. As an example, in the context of elastic media, the response of the material must be the same regardless of the observers (say under change of frame of references and accordingly associated coordinates). In any physical theory, mathematical variables, which represent geometric and physical quantities, must transform under an appropriate manner, that is, under any element of a defined group of coordinate transformations between admissible frames of reference.

Material Frame Indifference (Truesdell and Noll). Two formulations of the Material Frame Indifference that appeared equivalently in the framework of classical continuum mechanics were proposed in [141]:

(1) The constitutive laws of a continuum must be invariant when the medium undergoes a rigid motion superimposed on its own motion (Hooke, 1678, Poisson, 1829, Cauchy, 1829);
(2) The constitutive laws must be invariant under an arbitrary change of the observer with respect to which the motion is described (Zaremba, 1903, Jaumann, 1906).

After a long debate of almost 60 years on the applicability of the material frame indifference, the most adequate and altogether natural formulation would be that the constitutive laws governing the internal interactions of the different parts of a continum should not depend on the external frames of reference which serve to describe the continuum movement e.g. [162].

The lessons learned from this long debate in classical continuum mechanics nevertheless allow us to arrive at two conclusions: the more efficient use of four-dimensional spacetime instead of a three-dimensional space $\times$ one-dimensional time. This aspect is important to facilitate the application of invariance of scalars, vectors, and tensors. And the second immediate conclusion is to distinguish what we will call passive transformations (which corresponds to the Zaremba-Jaumann formulation) from active transformations (Hooke-Poisson-Cauchy formulation). And therefore, the frame of reference can be defined in a four-dimensional formulation, embedded in the spacetime manifold, as a coordinate system (local in relative gravitation). Passive transformations are merely changes of coordinate system.

For the sake of simplicity, the present chapter mainly deals with a four-dimensional spacetime.

The *Principle of Material Frame Indifference* has been widely exploited from the beginning in continuum mechanics, particularly for the derivation of constitutive laws of complex materials. Taking this principle into account quickly proved essential for the rigorous obtention of the equations governing these media in nonlinear regimes and in large transformations e.g. [141]. The original applications of this principle are generally linked to the search for invariant equations of constitutive laws under rigid transformations of the body reference. In this chapter, we move on to the generalization of this principle to covariance or *form invariance* of equations. Different names have been proposed to evoke the invariance of the equation shape governing relative gravitation with respect to the coordinate system, namely passive covariance, passive diff invariance, formal general covariance, or simply covariance (name chosen for the title of this chapter), knowing that all these names mean *in fine* background independence of the formulation with respect to the choice of coordinate system e.g. [163].

What we call *covariance* in the present chapter, also called diffeomorphism invariance, consists of the invariance of the mathematical form of physical laws under arbitrary (passive) differentiable coordinate transformations, and not limited to rigid body transformations.[1] Practically, what we call passive diffeomorphism invariance refers to the invariance following a change of coordinate system, *i.e.*, the same mathematical object is represented in different coordinate systems. On the contrary, an active diffeomorphism relates distinct states of the spacetime geometries (namely metric, torsion, and curvature) and of other physical fields but may be expressed in the same coordinate system. The term passive and active transformations was originally proposed by Bargmann [164]. We will see in the next chapter that this concept requires the notion of Lie derivatives of various fields, either geometrical or physical. The distinction avoids some misconceptions of this principle e.g. [165].

The concept of covariance has been the subject of dispute in relative gravitation for decades [160] and probably for many more years, if not more decades [166]. In our case, we limit ourselves to the covariance of Lagrangian $\mathscr{L}$ with certain well-defined arguments, which facilitates the formulation.

[1]We generalize the group of Leibniz transformations (2.81) used for the Principle of Material Indifference for continuum mechanics to the group of diffeomorphisms which is more appropriate for gravitation and electromagnetism.

That clearly means invariance of equation shape during a change of coordinate system. Without limiting the generality of our remarks too much, we can say that two types of covariance are mentioned: formal covariance and substantive covariance [166].

Remark 3.1. Classical mechanics admits *a priori* the existence of a space and a time that is immutable independently of the physical phenomena happening there. Special relativity admits the same hypothesis except that time is coupled with space and we speak rather of four-dimensional spacetime. In both cases, a change of four-dimensional coordinate system can be formulated by the transformations $y^i = y^i(x^\alpha)$ in the same spacetime. On the other hand, in relative gravitation, such a change of coordinates also acts on the metric tensor because the tensor $\mathbf{g}$ is part of the dynamic variables of the system to be studied, gravitation depending on the metric of spacetime. For a modified theory of gravitation, the connection of spacetime also comes into play. This peculiarity has been a source of confusion regarding the perception of what is called relative gravitational covariance, which leads to distinguishing passive from active diffeomorphism. The latter considers the evolution of the physical system generated by a flow φ, with the tangent vector field ξ, see Figure 3.1.

3.1.2 *Some General Purpose*

In this book, covariance consists of invariance of Lagrangian during an arbitrary change of coordinate systems. On the one hand, during his quest to unify gravitation and electromagnetism in vacuum, a Lagrange function of spacetime was proposed originally by Reichenbach to depend on a metric, an affine connection independent of the metric, and its partial derivatives [58]: $\mathscr{L} = \mathscr{L}(g_{\alpha\beta}, \Gamma^\gamma_{\alpha\beta}, \partial_\lambda \Gamma^\gamma_{\alpha\beta})$. This Lagrangian is covariant if during any change of coordinates we have:

$$\mathscr{L} = \mathscr{L}(g_{ij}, \Gamma^k_{ij}, \partial_\ell \Gamma^k_{ij}) = \mathscr{L}(g_{\alpha\beta}, \Gamma^\gamma_{\alpha\beta}, \partial_\lambda \Gamma^\gamma_{\alpha\beta}), \quad \text{with} \quad y^i = y^i(x^\alpha) \tag{3.1}$$

where the variables y are defined with a new coordinate system. This condition imposes the same shape of $\mathscr{L}$ by analogy with the frame indifference principle for constitutive laws in continuum mechanics e.g. [133]. On the other hand, in our particular case, the substantive covariance (or general covariance, or sometimes gauge invariance) would consist of the invariance of the action during an active diffeomorphism, notably through a flow

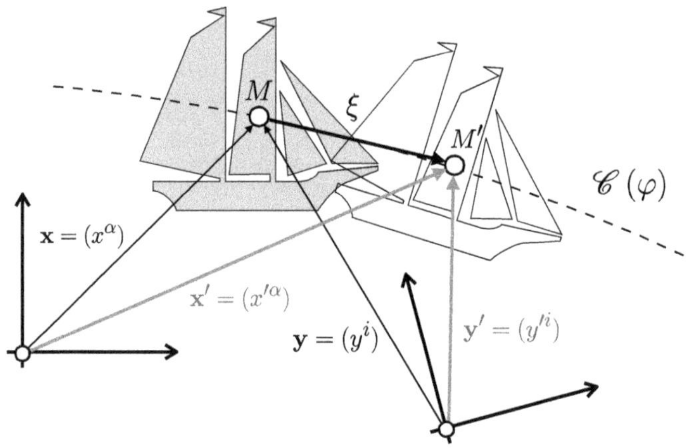

$$M(\mathbf{x}) \to M'(\mathbf{x}') \qquad \varphi(s,\mathbf{x}) \longrightarrow \xi := \frac{\partial}{\partial s}\varphi(s,\mathbf{x})$$

$$M(\mathbf{y}) \to M'(\mathbf{y}') \qquad \varphi(s,\mathbf{y}) \longrightarrow \xi := \frac{\partial}{\partial s}\varphi(s,\mathbf{y})$$

Fig. 3.1: Passive and active transformations. For the point M, passive transformations are characterized by change of coordinate systems $M(x^\alpha)$ and $M(y^i)$ respectively associated with two frames of reference. Active transformations are generated by the flow $\varphi(s,\mathbf{x})$ or $\varphi(s,\mathbf{y})$ depending on which coordinate system is used.

characterized by a vector field ξ (with some specific conditions) which gives us the Lagrangian. Starting with the flow, we have the variations:

$$x^\alpha \to x^\alpha + \xi^\alpha, \quad g_{\alpha\beta} \to g_{\alpha\beta} + \mathcal{L}_\xi g_{\alpha\beta}, \quad \Gamma^\gamma_{\alpha\beta} \to \Gamma^\gamma_{\alpha\beta} + \mathcal{L}_\xi \Gamma^\gamma_{\alpha\beta}$$

The formulation of the substantive invariance, also named Principle of General Covariance [129, 167][2]:

$$\mathscr{S} := \int_{\mathscr{B}} \mathscr{L}(g_{\alpha\beta}, \Gamma^\gamma_{\alpha\beta}, \partial_\lambda \Gamma^\gamma_{\alpha\beta})\omega_n, \quad \text{then} \quad \mathcal{L}_\xi \mathscr{S} = 0$$

Details about the Principle of General Covariance e.g. [130] will be given in the next chapter where the derivation of the conservation laws is the main motivation, especially for Einstein-Cartan spacetimes due to their recent relevance in cosmology e.g. [4]. We focus on the formal covariance (use of passive diffeomorphisms) in the present chapter.

[2]As soon as 1929, this form of action was initially proposed by Reichenbach to include electromagnetism [58], and is called Einstein-Palatini action, for which the connection is independent of the metric tensor. Accordingly, the variation of the metric and that of the connection are independent.

3.1.3 *Place of Covariance in Physics*

Gravitation is clearly associated with the spacetime curvature. As we will observe in the chapter on Einstein-Cartan spacetime, the electromagnetic fields are formally and tightly linked to the torsion field of the manifold. By anticipation, we thus consider the geometric structure of the manifold as arguments: metric, torsion, and curvature. Gravitational field and electromagnetic field in vacuum are therefore assumed to be integral parts of the spacetime geometry. Coming back to physics, this chapter[3] is therefore devoted to the "axiom of general invariance of Hilbert" e.g. [146]. In 1905, Hilbert proposed two axioms for setting the basis for the covariant derivation of both relative gravitation and electromagnetism.

Hilbert's axioms. The two axioms of Einstein gravitation including electromagnetism were proposed as follows:

(1) (Mie axiom on Lagrangian) The Lagrangian density $\mathscr{L}$ depends on the spacetime metric $g_{\mu\nu}$ (10 components), their first and second derivatives, and the electromagnetic potential A_μ (four components) and their first derivatives;
(2) (Covariance) The shape of the Lagrangian function $\mathscr{L}$ is invariant with respect to arbitrary transformations of the point-event coordinates x^μ.

The present chapter deals with the second Hilbert's axiom. For historical context, we recall the Cartan theorem. Let $\mathscr{B}$ be a four-dimensional manifold endowed with a metric tensor $g_{\alpha\beta}(x^\nu)$. Assume that the manifold is orientable with a volume-form $\omega_n(x^\nu)$. It is worth revisiting the theorem of Cauchy (1850)-Weyl (1939) stating that for any scalar-valued function $\mathscr{L}$ with vector arguments $\mathbf{e}_\alpha := F^i_\alpha \hat{\mathbf{e}}_i$, $\alpha = 1, n$, elements of the tangent space $T_{\mathbf{x}}\mathscr{M}$, the Lagrangian function $\mathscr{L}$ is invariant under the rotation group $\mathcal{O}^+$ if and only if it exists a representation of the function $\tilde{\mathscr{L}}$ such that: $\mathscr{L} = \tilde{\mathscr{L}}(g_{\alpha\beta}, \det(g_{\alpha\beta}))$. Proof with details may be found in e.g. [49] (annex E), slightly extending the version of the theorem of [141]. This allows us to assume *a priori* the metric components as arguments of the Lagrangian density of the action: $\mathscr{S} := \int_{\mathscr{B}} \mathscr{L}\, \omega_n$, where $\mathscr{L}$ depends on the fields and their derivatives (of finite order), and ω_n a volume-form on $\mathscr{B}$. We study in the following section the invariance of $\mathscr{L}$ under the group of

[3]This chapter was written in part with Nirmal Antonio Tamarasselvame PhD.

diffeomorphisms Diff($\mathscr{B}$). Three theorems are particularly involved for the covariance analysis.

3.2 Some Basic Theorems

We report here three theorems that are useful to the derivation of the Lagrangian covariance: Cartan theorem in the framework of relativistic gravitation, Lovelock's theorem extending the Cartan theorem to nonlinear Lagrangian in the framework of Riemann geometry, and the Quotient theorem necessary for establishing the main result of this chapter.

3.2.1 *Theorem of Cartan*

After the introduction of Minkowski spacetime, the second major modification of the spacetime is recognizing that gravity is not an *a priori* existing force field but an aspect of geometry which is captured by the metric and by the affine connection. Derivation of the gravitation equations consists in searching for 2-covariant tensor $T_{\alpha\beta}$ with the conditions: first, $T_{\alpha\beta}$ is linear with respect to the second derivatives of the metric components; and second, the tensor should be divergence-free for satisfying the conservation laws. In 1922, Cartan derived the Einstein equations of gravity on affinely connected manifolds. He showed in [168] the uniqueness of the Einstein tensor "$G_{\alpha\beta} := \Re_{\alpha\beta} - (1/2)\mathfrak{R}\, g_{\alpha\beta}$", assessing the necessary (and sufficient) condition to use Einstein's equations of gravity within usual assumptions.

Theorem 3.1 (Cartan, 1922). *Let tensor* $T_{\alpha\beta}(g_{\alpha\beta}, \partial_\gamma g_{\alpha\beta}, \partial_\lambda \partial_\gamma g_{\alpha\beta})$ *be linear with respect to the second derivatives of* $g_{\alpha\beta}$*, and covariant in the sense that it has a form invariant with respect to all changes of coordinates* $y^\mu = y^\mu(x^\nu)$*. Then the unique solution takes the form of:*

$$T_{\alpha\beta} = a\ \Re_{\alpha\beta} + b\ \mathfrak{R}\ g_{\alpha\beta} + c\ g_{\alpha\beta} \tag{3.2}$$

where $\Re_{\alpha\beta}$ *and* $\mathfrak{R}$ *are the Ricci and the scalar curvatures respectively.* a *and* b *are real arbitrary constants and* c *is the scalar field.*

Proof. Details may be found in Cartan's original work [168]. □

Steps of proof are analogous to those of Lovelock & Rund and details may be found in [169]. The previous theorem shows that at any point P of a pseudo-Riemannian (and Lorentzian) manifold $\mathscr{M}$, it is possible to find a coordinate system (x^μ) such that the metric tensor writes $g_{\alpha\beta}$ and the first derivatives of the components $\partial_\gamma g_{\alpha\beta}$, with respect to the chosen coordinates, are zero.

However, certain combinations of second-order derivatives $\partial_\lambda \partial_\gamma g_{\alpha\beta}$ which form the Riemann curvature $\Re^\gamma_{\alpha\beta\lambda}$ cannot be eliminated. This is a fundamental result in gravitation theory and constitutes the mathematical basis for gravitation equations $\chi\ T_{\alpha\beta} := \Re_{\alpha\beta} - (1/2)\ \mathcal{R}\ g_{\alpha\beta} + \Lambda\ g_{\alpha\beta}$, where $\chi := 8\pi G/c^4$ is the cosmological constant introduced to be compatible with the universe expansion according to the finding of Hubble in 1929. $G \simeq 6.67 \times 10^{-28}[m^3 kg^{-1} s^{-2}]$ and $c \simeq 3 \times 10^8 [ms^{-1}]$ are the gravity constant and the light velocity respectively. Cartan theorem states that at every point of the four-dimensional Riemannian manifold (spacetime), among the 100 second derivatives of the metric $\partial_\lambda \partial_\gamma g_{\alpha\beta}$, the invariance with respect to passive diffeomorphisms allows us to eliminate 80 of them to retain only 20, corresponding to the components of curvature $\Re^\gamma_{\alpha\beta\lambda}$. The Cartan theorem was extended in the 1970s by using a scalar field rather than a 2-covariant tensor [110]. Since then numerous geometrized theories of gravitation including nonlinear dependence on curvature have been proposed, called fourth-order gravitation theory. In short, they were mainly introduced to prevent the Big Bang singularity in relativistic gravitation, to account for the inflationary cosmological model, and most interestingly to attempt to unify gravitation and quantization of matter fields e.g. [170].

Remark 3.2. The Cartan theorem can be applied to both relative gravitation theory and some classes of strain gradient continuum theory. Following the idea of Cartan, Anderson provided the structure of divergence-free contravariant $(2, 0)$ type tensor [171]. In this paper, the separation of the influence of the metric components and their second partial derivatives was shown.

3.2.2 *Theorem of Lovelock*

Consider a set of fields $\Phi^i(x^\mu)$ on $\mathscr{B}$. Let's assume that the Lagrangian is a function of $\Phi^i(x^\mu)$ and their first and second derivatives with respect to x^μ. Metric components $g_{\alpha\beta}(x^\mu)$ play the role of $\Phi^i(x^\mu)$ in gravitation theory. The associated Euler-Lagrange equations are expected to be of fourth order. However, in the general theory of relativity, the Einstein field equations for gravitation are of second order in terms of metric components. Lovelock investigated the case of Lagrange density which is a function of the metric and its first and second derivatives with respect to x^μ. The following theorem was shown.

Theorem 3.2 (Lovelock, 1973). *Let $\mathscr{M}$ be a four-dimensional torsionless manifold endowed with a metric $g_{\alpha\beta}$, and a metric-compatible affine connection $\Gamma^{\gamma}_{\alpha\beta}$ with a curvature $\mathfrak{R}^{\gamma}_{\alpha\beta\lambda}$. The only second-order Euler-Lagrange equations which can arise from a scalar density $\mathscr{L}(g_{\alpha\beta}, \partial_\gamma g_{\alpha\beta}, \partial_\lambda \partial_\gamma g_{\alpha\beta})$ are:*

$$\mathfrak{R}_{\alpha\beta} - \frac{\mathcal{R}}{2}\, g_{\alpha\beta} + \Lambda g_{\alpha\beta} = 0 \tag{3.3}$$

i.e., the Einstein gravitational equations with the cosmological constant Λ.

Proof. Details may be found in the original paper [172]. □

We will consider in this chapter an extension of such a theorem in the framework of Riemann-Cartan manifolds. It should be remembered that the theorem of Lovelock constitutes one of the major results on the uniqueness of the Einstein equations on manifolds with a metric tensor with signature $(+,-,-,-)$, and also for Einstein-Maxwell equations [173]. In the same way, motivated by the fact that gravitational field equations should be second-order equations in terms of metric components, these equations being obtained from a Lagrangian density $\mathscr{L}(g_{\alpha\beta}, \partial_\gamma g_{\alpha\beta}, \partial_\lambda \partial_\gamma g_{\alpha\beta})$, it was also shown in [174] that in a four-dimensional manifold, the Lagrangian density must take the form of: $\mathscr{L} = a\, \mathcal{R} + b$ where a and b are constants. The resulting Euler-Lagrange equations are deduced and can be rearranged to be exactly (3.3). This result is obtained under very general conditions without the assumption that the Lagrangian is linear in curvature $\mathcal{R}$. In [174], Anderson extended this result to include the interaction of gravitational fields and external sources due to matter immersed in the spacetime $\mathscr{M}$.

3.2.3 *Theorem of Quotient*

For self-consistency, we recall some technical theorems in tensor analysis [110]. The most important is the quotient law saying that a set of real numbers ($\mathbb{R}$) form the components of a tensor of a certain rank, if and only if its scalar product with another arbitrary tensor is again a tensor (practically, we attempt to obtain a scalar by an appropriate choice). This can be used as a criterial test of whether a set of numbers form a tensor or not.

Lemma 3.1. *Locally at point P, let both $(n \times n)$ quantities Σ^{ij} and $\overline{\Sigma}^{ij}$ then both $(n \times n \times n)$ quantities $\Sigma^{ij,k}$ and $\overline{\Sigma}^{ij,k}$ (the comma "," does not* *represent partial derivative). If, for any symmetric type* $(0, 2)$ *tensor* $\mathbf{h}$, *for* i, j, $k = 1, \dots, n$:

$$\Sigma^{ij} h_{ij} + \Sigma^{ij,k} h_{ij|k} = \overline{\Sigma}^{ij} h_{ij} + \overline{\Sigma}^{ij,k} h_{ij|k} \tag{3.4}$$

then

$$(\Sigma^{ij} + \Sigma^{ji}) = (\overline{\Sigma}^{ij} + \overline{\Sigma}^{ji}) \tag{3.5}$$

and

$$(\Sigma^{ij,k} + \Sigma^{ji,k}) = (\overline{\Sigma}^{ij,k} + \overline{\Sigma}^{ji,k}) \tag{3.6}$$

Proof. The equality (3.4) being valid for any symmetric tensor $\mathbf{h}$, it is thus valid for a non-null constant tensor (locally). The covariant derivative vanishes and from (3.4) we have the following equality $(\Sigma^{ij})h_{ij} = (\overline{\Sigma}^{ij})h_{ij}$. The term h_{ij} cannot simplify because of the summation in i and j. For i and j fixed, we choose $h_{ij} = h_{ji} = 1$ and the other components null. For i and j range over $\{1, \dots, n\}$, we obtain

$$\begin{cases} \Sigma^{11} = \overline{\Sigma}^{11} \\ \Sigma^{12} + \Sigma^{21} = \overline{\Sigma}^{12} + \overline{\Sigma}^{21} \\ \Sigma^{13} + \Sigma^{31} = \overline{\Sigma}^{13} + \overline{\Sigma}^{31}, \qquad \dots \end{cases} \tag{3.7}$$

thus we deduce (3.5). In the same way, locally at point P, a null tensor $\mathbf{h}$ with a non-null constant covariant derivative can be chosen too. In such a case from (3.4), we get $(\Sigma^{ij,k})h_{ij|k} = (\overline{\Sigma}^{ij,k})h_{ij|k}$. The term $h_{ij|k}$ cannot be simplified because of the summation in i, j and k. For i, j and k fixed, we choose $h_{ij|k} = h_{ji|k} = 1$ and the other components null. For i, j and k range over $\{1, \dots, n\}$, we obtain

$$\begin{cases} \Sigma^{11,1} = \overline{\Sigma}^{11,1} \\ \Sigma^{11,2} = \overline{\Sigma}^{11,2} \end{cases}, \quad \begin{cases} \Sigma^{12,1} + \Sigma^{21,1} = \overline{\Sigma}^{12,1} + \overline{\Sigma}^{21,1} \\ \Sigma^{12,2} + \Sigma^{21,2} = \overline{\Sigma}^{12,2} + \overline{\Sigma}^{21,2} \\ \Sigma^{13,1} + \Sigma^{31,1} = \overline{\Sigma}^{13,1} + \overline{\Sigma}^{31,1} \\ \Sigma^{13,2} + \Sigma^{31,2} = \overline{\Sigma}^{31,2} + \overline{\Sigma}^{31,2}, \qquad \dots \end{cases} \tag{3.8}$$

thus we obtain (3.6). □

We have also the version with the partial derivative:

Lemma 3.2. *Locally at point P, let both* $(n \times n)$ *quantities* Σ^{ij} *and* $\overline{\Sigma}^{ij}$ *then both* $(n \times n \times n)$ *quantities* $\Sigma^{ij,k}$ *and* $\overline{\Sigma}^{ij,k}$ *(the comma "," does not* *represent partial derivative). If, for any symmetric type* $(0,2)$ *tensor* $\mathbf{h}$, *for* $i,\ j,\ k = 1,\ldots,n$,*:*

$$\Sigma^{ij} h_{ij} + \Sigma^{ij,k} h_{ij,k} = \overline{\Sigma}^{ij} h_{ij} + \overline{\Sigma}^{ij,k} h_{ij,k} \tag{3.9}$$

then

$$(\Sigma^{ij} + \Sigma^{ji}) = (\overline{\Sigma}^{ij} + \overline{\Sigma}^{ji}) \tag{3.10}$$

and

$$(\Sigma^{ij,k} + \Sigma^{ji,k}) = (\overline{\Sigma}^{ij,k} + \overline{\Sigma}^{ji,k}) \tag{3.11}$$

Now the following quotient theorem holds e.g. [110].

Theorem 3.3 (Quotient law). *Locally at point* P, *if the* $(n \times n)$ *quantities* Σ^{ij} *and the* $(n \times n \times n)$ *quantities* $\Sigma^{ij,k}$ *(the comma "," does not represent the partial derivative) are such that the quantities* $\Sigma^{ij} h_{ij} + \Sigma^{ij,k} h_{ij|k}$ *represent a scalar field for any symmetric type* $(0,2)$ *tensor* $\mathbf{h}$, *then the quantities* $(\Sigma^{ij} + \Sigma^{ji})$ *and* $(\Sigma^{ij,k} + \Sigma^{ji,k})$ *respectively represent the components of a tensor type* $(2,0)$ *and the components of a type* $(3,0)$ *tensor.*

Proof. Let's define the scalar

$$\psi = \Sigma^{ij} h_{ij} + \Sigma^{ij,k} h_{ij|k}$$

and

$$\overline{\psi} = \Sigma^{\alpha\beta} h_{\alpha\beta} + \Sigma^{\alpha\beta,\gamma} h_{\alpha\beta|\gamma}$$

in the system (y^i) and (x^α) respectively for any symmetric tensor type (0,2) $\mathbf{h}$. The equality $\psi = \overline{\psi}$ becomes

$$\Sigma^{ij} h_{ij} + \Sigma^{ij,k} h_{ij|k} = \Sigma^{\alpha\beta}\ A^i_\alpha A^j_\beta\ h_{ij} + \Sigma^{\alpha\beta,\gamma}\ A^i_\alpha A^j_\beta A^k_\gamma\ h_{ij|k} \tag{3.12}$$

According to Lemma 3.1

$$\Sigma^{ij} + \Sigma^{ji} = \Sigma^{\alpha\beta}\ A^i_\alpha A^j_\beta + \Sigma^{\alpha\beta}\ A^j_\alpha A^i_\beta \tag{3.13}$$

and

$$\Sigma^{ij,k} + \Sigma^{ji,k} = \Sigma^{\alpha\beta,\gamma}\ A^i_\alpha A^j_\beta A^k_\gamma + \Sigma^{\alpha\beta,\gamma}\ A^j_\alpha A^i_\beta A^k_\gamma \tag{3.14}$$

then a permutation between i and j gives

$$\begin{cases} \Sigma^{ij} + \Sigma^{ji} = \left(\Sigma^{\alpha\beta} + \Sigma^{\beta\alpha}\right) A^i_\alpha A^j_\beta \\ \Sigma^{ij,k} + \Sigma^{ji,k} = \left(\Sigma^{\alpha\beta,\gamma} + \Sigma^{\beta\alpha,\gamma}\right) A^i_\alpha A^j_\beta A^k_\gamma \end{cases} \tag{3.15}$$

Therefore $(\Sigma^{ij} + \Sigma^{ji})$ and $(\Sigma^{ij,k} + \Sigma^{ji,k})$ are respectively components of type $(2,0)$ tensor and $(3,0)$ according to the definitions (2.18) and (2.19). □

Remark 3.3. For a continuum and *a fortiori* spacetime, the two variables which characterize the geometry are respectively the metric and the connection, independent of the metric in the general case. It is therefore necessary to consider the invariance of the Lagrangian when applying a diffeomorphism for which these two entities can be transformed.

3.3 Invariance with Respect to the Metric

Consider a scalar field $\mathscr{L}$ depending on the metric components and their partial derivatives $\mathscr{L} = \mathscr{L}(g_{ij}, \partial_k g_{ij}, \partial_l \partial_k g_{ij})$ and $\mathscr{L} = \mathscr{L}(g_{\alpha\beta}, \partial_\gamma g_{\alpha\beta}, \partial_\lambda \partial_\gamma g_{\alpha\beta})$ respectively in coordinate system (y^i) and (x^α). The metric components $g_{\alpha\beta}(x^\mu)$ ($g_{ij}(y^k)$ in terms of new coordinates) are the unknown functions. We consider general covariance by imposing the invariance of Lagrangian shape under coordinate transformations $y^i(x^\alpha)$. The corresponding partial derivatives are denoted

$$\Lambda^{ij} = \frac{\partial \mathscr{L}}{\partial g_{ij}}, \qquad \Lambda^{ij,k} = \frac{\partial \mathscr{L}}{\partial g_{ij,k}}, \qquad \Lambda^{ij,kl} = \frac{\partial \mathscr{L}}{\partial g_{ij,kl}}. \tag{3.16}$$

They should not be confused with the hypermomenta or currents, which are physical reactions due to the variations of kinematics, and deduced from the variation of the Lagrangian function. Care should be taken since relativistic gravitation and gradient continuum models need hypermomenta and currents to be a viable theory e.g. [175]. Due to the symmetry of $\mathbf{g}$, we have the major properties of symmetry

$$\Lambda^{ij} = \Lambda^{ji}, \qquad \Lambda^{ij,k} = \Lambda^{ji,k}, \qquad \Lambda^{ij,kl} = \Lambda^{ji,kl} = \Lambda^{ij,lk}. \tag{3.17}$$

Minor symmetry property $\Lambda^{ij,kl} = \Lambda^{kl,ij}$ is also satisfied but is not used here. The covariance of $\mathscr{L}$ (which is invariance with respect to passive diffeomorphisms e.g. [141]) takes the form of

$$\mathscr{L}(g_{\alpha\beta}, \partial_\gamma g_{\alpha\beta}, \partial_\lambda \partial_\gamma g_{\alpha\beta}) = \mathscr{L}(g_{ij}, \partial_k g_{ij}, \partial_l \partial_k g_{ij}). \tag{3.18}$$

In 1922, Cartan showed the uniqueness of the Einstein equations of relativistic gravity [168]. Result on covariance we are now proving extends Cartan's result to a nonlinear dependence of Lagrangian on metric second derivatives $\partial_\lambda \partial_\mu g_{\alpha\beta}$. The corresponding Euler-Lagrange equations deduced from the covariant Lagrangian can be used for that purpose. Recall that because no observer is distinguished, laws in physics have to be observer-invariant (frame indifference). The key point is to define the group of invariance. Each physics theory has its invariance group. For covariance, Lagrangian functions $\mathscr{L}$ should have at least the same shape for arbitrary coordinate systems. Interestingly, Ryskin also gave an accurate formulation of objectivity (a more recent term defining the material frame indifference) as *"Any physical law must be expressible in a form independent of the coordinate system"* [165]. This is covariance in the Einstein language. Further, it was extended to four-dimensional formalism, thus including the time variable, and suggests that every physical law verifying the diffeomorphism invariance or form invariance [169] seems also material frame indifferent [161].

3.3.1 *Transformation Rules for the Metric and its Derivatives*

In terms of dimensional analysis, curvature and torsion have respectively the dimensions of second and first derivatives of the metric tensor. According to the definitions (2.18) and (2.19), the metric components and its first and second derivatives satisfy the following rules of transformations

$$\left\{\begin{aligned} g_{ij} &= J_i^\alpha J_j^\beta g_{\alpha\beta} \\ g_{ij,k} &= (J_{ik}^\alpha J_j^\beta + J_i^\alpha J_{jk}^\beta) g_{\alpha\beta} + J_i^\alpha J_j^\beta J_k^\gamma g_{\alpha\beta,\gamma} \\ g_{ij,kl} &= (J_{ikl}^\alpha J_j^\beta + J_{ik}^\alpha J_{jl}^\beta + J_{il}^\alpha J_{jk}^\beta + J_i^\alpha J_{jkl}^\beta) g_{\alpha\beta} \\ &\quad + (J_{ik}^\alpha J_j^\beta J_h^\gamma + J_i^\alpha J_{jk}^\beta J_l^\gamma + J_{il}^\alpha J_j^\beta J_k^\gamma + J_i^\alpha J_{jl}^\beta J_k^\gamma + J_i^\alpha J_j^\beta J_{kl}^\gamma) g_{\alpha\beta,\gamma} \\ &\quad + (J_i^\alpha J_j^\beta J_k^\gamma J_l^\lambda) g_{\alpha\beta,\gamma\lambda} \end{aligned}\right. \tag{3.19}$$

According to the symmetry $J_{pq}^\mu = J_{qp}^\mu$ (the transformation is assumed of class C^2), one can write $J_{pq}^\mu = (1/2)(J_{pq}^\mu + J_{qp}^\mu)$, which induces $(\partial J_{ij}^\alpha / \partial J_{pq}^\mu) = (1/2)\delta_\mu^\alpha(\delta_i^q \delta_j^p + \delta_j^q \delta_i^p)$, $(\partial J_{ijk}^\alpha / \partial J_{pq}^\mu) = 0$ and $(\partial J_i^\alpha / \partial J_{pq}^\mu) = 0$. If one does not consider the symmetric part then there is a loss of some terms in the derivation. This is necessary. Now, introducing the expressions (3.19) into the equality (3.18), and differentiating with respect to J_{pq}^μ give

$$\begin{aligned} 0 = \Lambda^{ij,k} &\left[\delta_\mu^\alpha(\delta_k^q \delta_i^p + \delta_i^q \delta_k^p) J_j^\beta + \delta_\mu^\beta(\delta_k^q \delta_j^p + \delta_j^q \delta_k^p) J_i^\alpha\right] g_{\alpha\beta} \\ &+ \Lambda^{ij,kl}\left[J_i^\alpha J_j^\beta \delta_\mu^\gamma(\delta_k^q \delta_l^p + \delta_l^q \delta_k^p) + J_i^\alpha J_k^\gamma \delta_\mu^\beta(\delta_j^q \delta_l^p + \delta_l^q \delta_j^p)\right] g_{\alpha\beta,\gamma} \\ &+ \Lambda^{ij,kl}\left[J_j^\beta J_k^\gamma \delta_\mu^\alpha(\delta_i^q \delta_l^p + \delta_l^q \delta_i^p) + J_i^\alpha J_l^\gamma \delta_\mu^\beta(\delta_j^q \delta_k^p + \delta_k^q \delta_j^p)\right] g_{\alpha\beta,\gamma} \\ &+ \Lambda^{ij,kl}\left[J_j^\beta J_l^\gamma \delta_\mu^\alpha(\delta_i^q \delta_k^p + \delta_k^q \delta_i^p)\right] g_{\alpha\beta,\gamma} \\ &+ \Lambda^{ij,kl}\left[J_{jk}^\beta \delta_\mu^\alpha(\delta_i^q \delta_l^p + \delta_l^q \delta_i^p) + J_{il}^\alpha \delta_\mu^\beta(\delta_j^q \delta_k^p + \delta_k^q \delta_j^p)\right] g_{\alpha\beta} \\ &+ \Lambda^{ij,kl}\left[J_{jl}^\beta \delta_\mu^\alpha(\delta_i^q \delta_k^p + \delta_k^q \delta_i^p) + J_{ik}^\alpha \delta_\mu^\beta(\delta_j^q \delta_l^p + \delta_l^q \delta_j^p)\right] g_{\alpha\beta} \end{aligned}$$

Previous equation is valid for arbitrary coordinate transformations, in particular for the identity transformation: $x^\alpha = y^i$, $J_i^\alpha = \delta_i^\alpha$, $J_{ij}^\alpha = 0$. In such a case, we simplify

$$
\begin{aligned}
0 = \Lambda^{ij,k} &\left[\delta_\mu^\alpha(\delta_k^q\delta_i^p + \delta_i^q\delta_k^p)\delta_j^\beta + \delta_\mu^\beta(\delta_k^q\delta_j^p + \delta_j^q\delta_k^p)\delta_i^\alpha\right] g_{\alpha\beta} \\
&+ \Lambda^{ij,kl}\left[\delta_i^\alpha\delta_j^\beta\delta_\mu^\gamma(\delta_k^q\delta_l^p + \delta_h^t\delta_k^p) + \delta_i^\alpha\delta_k^\gamma\delta_\mu^\beta(\delta_j^q\delta_l^p + \delta_l^q\delta_j^p)\right] g_{\alpha\beta,\gamma} \\
&+ \Lambda^{ij,kl}\left[\delta_j^\beta\delta_k^\gamma\delta_\mu^\alpha(\delta_i^q\delta_l^p + \delta_l^q\delta_i^p) + \delta_i^\alpha\delta_l^\gamma\delta_\mu^\beta(\delta_j^q\delta_k^p + \delta_k^q\delta_j^p)\right] g_{\alpha\beta,\gamma} \\
&+ \Lambda^{ij,kl}\left[\delta_j^\beta\delta_l^\gamma\delta_\mu^\alpha(\delta_i^q\delta_k^p + \delta_k^q\delta_i^p)\right] g_{\alpha\beta,\gamma}
\end{aligned}
$$

Further simplifications and symmetry of Λ induce[4]

$$
2\Lambda^{q\beta,\gamma p} g_{\mu\beta,\gamma} + 2\Lambda^{p\beta,\gamma q} g_{\mu\beta,\gamma} + \Lambda^{\alpha\beta,pq} g_{\alpha\beta,\mu} + \Lambda^{p\beta,q} g_{\mu\beta} + \Lambda^{q\beta,p} g_{\mu\beta} = 0. \tag{3.20}
$$

In the particular case of a normal coordinate system, these reduce to:

$$
\Lambda^{p\mu,q} + \Lambda^{q\mu,p} = 0
$$

According to the symmetry of $\Lambda^{ij,k}$, for arbitrary indices i, j, k, we have

$$
\begin{aligned}
\Lambda^{ji,k} + \Lambda^{ki,j} &= 0 \\
\Lambda^{kj,i} + \Lambda^{ki,j} &= 0 \quad (i \longleftrightarrow k) \\
\Lambda^{ji,k} + \Lambda^{kj,i} &= 0 \quad (i \longleftrightarrow j)
\end{aligned}
$$

then $\Lambda^{ji,k} = -\Lambda^{ki,j} = \Lambda^{kj,i} = -\Lambda^{ji,k}$, and finally

$$
\Lambda^{ji,k} = \Lambda^{ij,k} = 0. \tag{3.21}
$$

Nevertheless (3.21) is only valid in normal coordinate systems.

3.3.2 *Introduction of Dual Variables*

Introducing the expressions (3.19) into (3.18) and differentiating with respect to $g_{\alpha\beta,\gamma\lambda}$, $g_{\alpha\beta,\gamma}$, $g_{\alpha\beta}$, allow us to write

$$
\begin{cases}
\Lambda^{\alpha\beta,\gamma\lambda} = \Lambda^{ij,kl} J_i^\alpha J_j^\beta J_k^\gamma J_l^\lambda \\
\Lambda^{\alpha\beta,\gamma} = \Lambda^{ij,kl} \dfrac{\partial g_{ij,kl}}{\partial g_{\alpha\beta,\gamma}} + \Lambda^{ij,k} \dfrac{\partial g_{ij,k}}{\partial g_{\alpha\beta,\gamma}} \\
\Lambda^{\alpha\beta} = \Lambda^{ij,kl} \dfrac{\partial g_{ij,kl}}{\partial g_{\alpha\beta}} + \Lambda^{ij,k} \dfrac{\partial g_{ij,k}}{\partial g_{\alpha\beta}} + \Lambda^{ij} \dfrac{\partial g_{ij}}{\partial g_{\alpha\beta}}
\end{cases} \tag{3.22}
$$

[4]Latin and Greek indices mix since the transformation is the identity.

The first equation shows that $\Lambda^{ij,kl}$ are components of a type $(4,0)$ tensor. Conversely the two other equations show that $\Lambda^{ij,k}$ and Λ^{ij} are not components of the tensor. Thus we should introduce two tensorial quantities instead of $\Lambda^{ij,k}$ and Λ^{ij}. Let $\mathbf{h}$ be an arbitrary symmetric type $(0,2)$ tensor (i.e., $\mathbf{h}$ follows the same rule of transformation (3.19) as $\mathbf{g}$), and Π^{ij} and $\Pi^{ij,k}$ (the comma "," does not represent partial derivative) two unknown quantities that verify the following equation

$$\Lambda^{ij,kl}h_{ij,kl} + \Lambda^{ij,k}h_{ij,k} + \Lambda^{ij}h_{ij} = \Lambda^{ij,kl}h_{ij|k|l} + \Pi^{ij,k}h_{ij|k} + \Pi^{ij}h_{ij} \tag{3.23}$$

Covariant derivatives are introduced into (3.23), that reduce to an equality without covariant derivative terms. It depends on $\{\Gamma^k_{ij},\ \Gamma^k_{ij,l}\}$, $\{h_{ij},\ h_{ij,k},\ h_{ij,kl}\}$, $\{\Lambda^{ij},\ \Lambda^{ij,k},\ \Lambda^{ij,kl}\}$ and $\{\Pi^{ij},\ \Pi^{ij,k}\}$. Then, Lemma 3.2 is applied to (3.23) in order to identify the coefficients of $h_{ij,k}$ and h_{ij} (coefficients of $h_{ij,kl}$ are the same on both sides of the equation). Some permutations between the indices are necessary. Recall that Λ^{ij}, $\Lambda^{ij,k}$ and $\Lambda^{ij,kl}$ have symmetry properties. This is not necessarily the case for Π^{ij} and $\Pi^{ij,k}$. Thus we obtain the following equations

$$\left\{\begin{aligned}
\Pi^{ij,k}_{(S)} &= \Lambda^{ij,k} + 2\Gamma^i_{al}\Lambda^{aj,kl} + 2\Gamma^j_{al}\Lambda^{ia,kl} + \Gamma^k_{bl}\Lambda^{ij,bl}\\
\Pi^{ij}_{(S)} &= \Lambda^{ij} + \Gamma^i_{ak,l}\Lambda^{aj,kl} + \Gamma^j_{ak,l}\Lambda^{ia,kl}\\
&\quad - \Gamma^b_{al}\Gamma^i_{bk}\Lambda^{aj,kl} - \Gamma^b_{cl}\Gamma^j_{bk}\Lambda^{ic,kl}\\
&\quad - \Gamma^i_{bl}\Gamma^j_{ck}\Lambda^{bc,kl} - \Gamma^j_{bl}\Gamma^i_{ck}\Lambda^{bc,kl}\\
&\quad - \Gamma^b_{kl}\Gamma^i_{cb}\Lambda^{cj,kl} - \Gamma^b_{kl}\Gamma^j_{cb}\Lambda^{ci,kl}\\
&\quad + (1/2)\Gamma^i_{ak}(\Pi^{aj,k} + \Pi^{ja,k}) + (1/2)\Gamma^j_{ak}(\Pi^{ia,k} + \Pi^{ai,k})
\end{aligned}\right. \tag{3.24}$$

where $\Pi^{ij,k}_{(S)} = (1/2)(\Pi^{ij,k} + \Pi^{ji,k})$ and $\Pi^{ij}_{(S)} = (1/2)(\Pi^{ij} + \Pi^{ji})$.

Lemma 3.3. *$F := \Lambda^{ij,kl}h_{ij,kl} + \Lambda^{ij,k}h_{ij,k} + \Lambda^{ij}h_{ij}$ is a scalar field.*

Proof. Let us notice $F = \Lambda^{ij,kl}h_{ij,kl} + \Lambda^{ij,k}h_{ij,k} + \Lambda^{ij}h_{ij}$ and $\overline{F} = \Lambda^{\alpha\beta,\gamma\lambda}h_{\alpha\beta,\gamma\lambda} + \Lambda^{\alpha\beta,\gamma}h_{\alpha\beta,\gamma} + \Lambda^{\alpha\beta}h_{\alpha\beta}$ respectively in coordinate systems (y^i) and (x^α). According to (3.22), we get

$$\Lambda^{\alpha\beta,\gamma\lambda}h_{\alpha\beta,\gamma\lambda} = \left[\Lambda^{ij,kl}J^\alpha_i J^\beta_j J^\gamma_k J^\lambda_l\right]h_{\alpha\beta,\gamma\lambda}$$

$$\Lambda^{\alpha\beta,\gamma}h_{\alpha\beta,\gamma} = \left[\Lambda^{ij,kl}\frac{\partial g_{ij,kl}}{\partial g_{\alpha\beta,\gamma}} + \Lambda^{ij,k}\frac{\partial g_{ij,k}}{\partial g_{\alpha\beta,\gamma}}\right]h_{\alpha\beta,\gamma}$$

$$\Lambda^{\alpha\beta}h_{\alpha\beta} = \left[\Lambda^{ij,kl}\frac{\partial g_{ij,kl}}{\partial g_{\alpha\beta}} + \Lambda^{ij,k}\frac{\partial g_{ij,k}}{\partial g_{\alpha\beta}} + \Lambda^{ij}\frac{\partial g_{ij}}{\partial g_{\alpha\beta}}\right]h_{\alpha\beta}$$

Factorization of the coefficients of $\Lambda^{ij,kl}$, $\Lambda^{ij,k}$ and Λ^{ij} gives $\overline{F} = (a) + (b) + (c)$ with

$$(a) = \Lambda^{ij,kl}\left[\frac{\partial g_{ij,kl}}{\partial g_{\alpha\beta,\gamma\lambda}}h_{\alpha\beta,\gamma\lambda} + \frac{\partial g_{ij,kl}}{\partial g_{\alpha\beta,\gamma}}h_{\alpha\beta,\gamma} + \frac{\partial g_{ij,kl}}{\partial g_{\alpha\beta}}h_{\alpha\beta}\right]$$

$$(b) = \Lambda^{ij,k}\left[\frac{\partial g_{ij,k}}{\partial g_{\alpha\beta,\gamma}}h_{\alpha\beta,\gamma} + \frac{\partial g_{ij,k}}{\partial g_{\alpha\beta}}h_{\alpha\beta}\right]$$

$$(c) = \Lambda^{ij}\left[\frac{\partial g_{ij}}{\partial g_{\alpha\beta}}h_{\alpha\beta}\right]$$

According to the relations (3.19), the quantities in square brackets are simplified, (a) $= \Lambda^{ij,kl}[h_{ij,kl}]$, (b) $= \Lambda^{ij,k}[h_{ij,k}]$, (c) $= \Lambda^{ij}[h_{ij}]$ and thus $\overline{F} = F$. □

Lemma 3.4. *($F - \Lambda^{ij,kl}h_{ij|k|l}$) is a scalar.*

Proof. Let us notice $G := \Lambda^{ij,kl}h_{ij|k|l}$ and $\overline{G} := \Lambda^{\alpha\beta,\gamma\lambda}h_{\alpha\beta|\gamma|\lambda}$. By (3.22) we have

$$\Lambda^{\alpha\beta,\gamma\lambda}h_{\alpha\beta|\gamma|\lambda} = \Lambda^{ij,kl}J_i^\alpha J_j^\beta J_k^\gamma J_l^\lambda h_{\alpha\beta|\gamma|\lambda} \tag{3.25}$$

h being a type $(0,2)$ tensor, the second covariant derivative $h_{\alpha\beta|\gamma|\lambda}$ forms the components of a type $(0,4)$ tensor. Consequently $J_i^\alpha J_j^\beta J_k^\gamma J_l^\lambda h_{\alpha\beta|\gamma|\lambda} = h_{ij|k|l}$, thus $\overline{G} = G$. By previous lemma, we have $\overline{F} - \overline{G} = F - G$. □

Remark 3.4. A direct and simple proof may be obtained by observing that Λ^{ijkl} is in fact a type $(4,0)$ tensor and $h_{ij|k|l}$ is a type $(0,4)$ tensor, then their contraction is a scalar.

Consequently, $\Pi^{ij,k}h_{ij|k} + \Pi^{ij}h_{ij}$ is a scalar for an arbitrary type $(0,2)$ tensor **h**. Using the quotient theorem 3.3, $\Pi_{(S)}^{ij,k}$ and $\Pi_{(S)}^{ij}$ are the components of type $(3,0)$ and $(2,0)$ tensor respectively, and these tensors are also symmetric.

3.3.3 *Theorem*

Expressions of $\Pi_{(S)}^{ij,k}$ in (3.24) hold in an arbitrary coordinate system. In a normal coordinate system, the Christoffel symbols vanish, e.g. [5], and we have, from (3.24), $\Pi_{(S)}^{ij,k} = \Lambda^{ij,k}$. However, from (3.21) $\Lambda^{ij,k} = 0$ in normal coordinate systems, thus $\Pi_{(S)}^{ij,k} = 0$ in normal coordinate systems and this is also true for any other coordinate system, because $\Pi^{ij,k}_{(S)}$ are components of tensor. The first equation in (3.24) is simplified in any coordinate system

$$0 = \Lambda^{ij,k} + 2\Gamma^{i}_{al}\Lambda^{aj,kl} + 2\Gamma^{j}_{al}\Lambda^{ia,kl} + \Gamma^{k}_{bl}\Lambda^{ij,bl}. \tag{3.26}$$

We can establish the following theorem:

Theorem 3.4. *Let a scalar field* $\mathscr{L} = \mathscr{L}(g_{ij}, g_{ij,k}, g_{ij,kl})$ *be defined on a Riemannian manifold.*

$$\text{If } \frac{\partial \mathscr{L}}{\partial g_{ij,kl}} = 0 \quad \text{then} \quad \frac{\partial \mathscr{L}}{\partial g_{ij,k}} = 0$$

Proof. $\forall\, i, j, k, l$, the condition $\Lambda^{ij,kl} = 0$ is introduced into (3.26). □

Consequently, from the second equality of (3.24), we deduce $\frac{\partial \mathscr{L}}{\partial g_{ij}} = \Pi^{ij}_{(S)}$. An equivalent formulation of Theorem 3.4 may be found in [110].

Theorem 3.5. *On a Riemannian manifold* $\mathscr{B}$, *there does not exist a scalar density such that* $\mathscr{L} = \mathscr{L}(g_{ij}, \partial_k g_{ij})$ *that only depends explicitly on the metric components* g_{ij} *and their first partial derivatives* $g_{ij,k}$.

In any coordinate system, we have the following decomposition:

$$\begin{cases} \Pi^{ij} = (1/2)(\Pi^{ij} + \Pi^{ji}) + (1/2)(\Pi^{ij} - \Pi^{ji}) \\ \Pi^{ij,k} = (1/2)(\Pi^{ij,k} + \Pi^{ji,k}) + (1/2)(\Pi^{ij,k} - \Pi^{ji,k}) \end{cases} \tag{3.27}$$

In [110], the quantities Π^{ij} and $\Pi^{ij,k}$ are assumed to be symmetric with respect to the indices i and j: $\Pi^{ij} = \Pi^{ji}$, $\Pi^{ij,k} = \Pi^{ji,k}$ and it is proven that the quantities $\Pi^{ij,k}$ are always zero. In the present study it has been proven that only the symmetric part $\Pi^{ij,k}_{(S)}$ is null: the quantities $\Pi^{ij,k}$ are skew-symmetric $\Pi^{ij,k} = -\Pi^{ji,k}$. The present study is slightly more general than that presented in [110].

Remark 3.5. To obtain sound mathematical models and then to study the concept of fields in physics, the arguments have to be tensors, and such is the case for $\mathscr{L}(\mathbf{g})$ where the Lagrangian is only dependent on the metric tensor. Let us consider for a moment a model where the Lagrangian depends on the metric and its derivatives. To avoid the coordinate dependence, we should first introduce covariant derivatives. Then, to extend the list of arguments for $\mathscr{L}$, defined on a Riemannian manifold endowed with an affine connection ∇, the new form is then $\mathscr{L}(\mathbf{g}, \nabla\mathbf{g}, \nabla^2\mathbf{g})$ where all the arguments are tensors being the first and second covariant derivatives of the metric. The corresponding form in the coordinate system (x^α) is

$\mathscr{L}(g_{\alpha\beta}, \nabla_\gamma g_{\alpha\beta}, \nabla_\lambda \nabla_\gamma g_{\alpha\beta})$. Now, by assuming that the connection is metric compatible, we have $\nabla \mathbf{g} \equiv 0$. Indeed, according to the Lemma of Ricci e.g. [5], the covariant derivative of $\mathbf{g}$, by using Levi-Civita connection $\nabla \to \overline{\nabla}$, is identically equal to zero. In such a case, the Lagrangian is shown to necessarily and sufficiently take the form of $\mathscr{L}(\mathbf{g}, \overline{\mathfrak{R}})$ [110], depending only on the metric and the Riemann curvature. As a collateral consequence, a torsion-free Riemannian manifold cannot provide an appropriately geometric model of continuum with a translational dislocation field. This is a simple example of a warning against the use of *a priori* background geometry of continuous media which could contradict the desire to take into account this or that physical phenomena. This warning actually constitutes a serious indication concerning the choice of the initial hypothesis of the geometric structure of spacetime that we will use hereafter in this book, taking into account both gravitation and electromagnetism.

3.4 Invariance with Respect to the Connection

For Riemann-Cartan manifold, the metric is required to be covariantly constant, $\nabla \mathbf{g} \equiv 0$, to be compatible with the metric. This means that $\nabla \mathbf{g}$ cannot be an explicit argument of the Lagrangian $\mathscr{L}$. This is why we should consider ∇ as an argument rather than $\nabla \mathbf{g}$. We labeled the biconnection $\nabla^2 = \nabla \circ \nabla$. In order to extend the list of arguments of $\mathscr{L}$, we will consider the following forms: $\mathscr{L}(\mathbf{g}, \nabla)$, $\mathscr{L}(\mathbf{g}, \nabla^2)$ and $\mathscr{L}(\mathbf{g}, \nabla, \nabla^2)$ e.g. [169]. However, connection is not tensorial, thus we aim to obtain tensorial arguments built upon the connection and/or the bi-connection. The introduction of ∇, as advocated by Palatini, as Lagrangian function arguments is also related to the affine variational method in relativistic gravitation. It is used to obviate non-covariance properties which arise when the Lagrangian depends upon metric derivatives.[5] Beyond the compatibility of ∇ with the metric, we also should consider

[5] In the scope of relativistic gravitation, the underlying idea of Palatini e.g. [176] is to consider the connection $\Gamma^\gamma_{\alpha\beta}$ necessary to link two microcosms, independent from the spacetime metric $g_{\alpha\beta}$ as an argument for the Lagrangian density function. There is no reason to limit the connection to Levi-Civita connection deduced from the metric. Some studies on relativistic gravitation introduce the concept of bi-metric theory by considering two slightly different metrics such as $g_{\alpha\beta}$ and $\tilde{g}_{\alpha\beta} := f[\mathcal{R}(\Gamma^\gamma_{\alpha\beta})]\ g_{\alpha\beta}$. Nevertheless we limit to a rather general purpose by considering the metric, the connection and bi-connection as independent arguments of $\mathscr{L}$.

the compatibility of the volume-form with the connection e.g. [132]. For Riemann-Cartan manifolds, we consider the Lagrangian putting apart the volume-form ω_n which should be compatible with the connection as we will see later.

Let us now come back to the link between local frame and affine connection. Let $\mathscr{B}$ be a continuum — a metric-affine manifold — endowed with a metric $\mathbf{g}$ and an affine connection ∇. Every neighbor of a point $\mathbf{x}$ can be considered as a microcosm which has its own local Galilean group $\{\mathbf{Q}(x^i), \mathbf{v}_0(x^i)\}$ defined by the transformation (a local Newton-Cartan transformation): $\mathbf{OM}'(t) = \mathbf{Q}(x^i)[\mathbf{OM}(t)] + \mathbf{v}_0(x^i)t$ where $x^0 := t$, and where Latin indices are used for space coordinates $i = (1, 2, 3)$. Let two microcosms be $\mathbf{x} := (x^0, x^i)$ and $\mathbf{x} + d\mathbf{x} := (x^0 + dx^0, x^i + dx^i)$ with the line element $ds^2 = g_{\alpha\beta}dx^\alpha dx^\beta$; Greek indices hold $\alpha, \beta = 0, 1, 3$, say $x^\mu(s)$. The change of the local Galilean group from one microcosm to another is then obtained with parallel transport (affine connection) (Fig. 3.2):

$$\begin{aligned} d\mathbf{v}_0 = \nabla_{d\mathbf{x}}\mathbf{v}_0 &= \left(\partial_\beta v_0^\alpha + \Gamma^\alpha_{\beta\gamma} v_0^\gamma\right) dx^\beta \mathbf{e}_\alpha \\ d\mathbf{Q} = \nabla_{d\mathbf{x}}\mathbf{Q} &:= \left(\partial_\gamma Q^\alpha_\beta + \Gamma^\alpha_{\gamma\lambda} Q^\lambda_\beta - \Gamma^\lambda_{\gamma\beta} Q^\alpha_\lambda\right) dx^\gamma \mathbf{e}_\alpha \otimes \mathbf{e}^\beta \end{aligned}$$

with coefficients $\Gamma^\gamma_{\alpha\beta} = \mathbf{e}^\gamma\left(\nabla_{\mathbf{e}_\alpha}\mathbf{e}_\beta\right)$. It is then obvious that the affine connection defines a change of local Galilean reference frames e.g. [113]. The method may be extended to local Minkowski spacetime and even to generalized changes of frames of reference. In the following, we consider invariance with respect to connection.

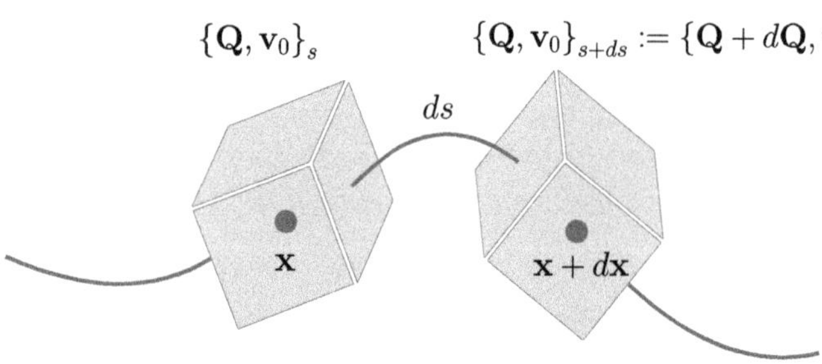

Fig. 3.2: Local Galilean reference frames, characterized by orthogonal transformation $\mathbf{Q}$ and translation velocity of frame $\mathbf{v}_0$, at two microcosms $\mathbf{x}$ and $\mathbf{x} + d\mathbf{x}$ depending on line element $ds := \sqrt{g_{\alpha\beta}dx^\alpha dx^\beta}$. For relative gravitation, the equivalence principle applied to microcosms $\mathbf{x}$ and $\mathbf{x} + d\mathbf{x}$ allows us to reconstruct the four-dimensional global structure of spacetime with gravitation by means of ∇.

3.4.1 *Preliminary*

We recall the covariance of a scalar field $\mathscr{L}(X,Y) = \mathscr{L}(X',Y')$ with both formal arguments X and Y defined in any two coordinate systems (with and without ').

Lemma 3.5. *Let us consider arbitrary constants K_1, K_2, C_1, C_2, C_3 and the variables x, x', y, y', p, p', q, q' which follow the transformations*

$$\begin{cases} x' = K_1\ x + K_2 \\ y' = K_1\ y \\ p' = C_1\ p + C_2\ (x+y) + C_3 \\ q' = C_1\ q \end{cases} \tag{3.28}$$

Now let us consider a scalar function $\mathscr{L}$ which satisfies the covariance:

(1) $\mathscr{L}(x,y) = \mathscr{L}(x',y')$
(2) $\mathscr{L}(p,q) = \mathscr{L}(p',q')$
(3) $\mathscr{L}(x,y,p,q) = \mathscr{L}(x',y',p',q')$.

If $(\partial K_1/\partial K_2) = 0$, $(\partial K_1/\partial C_3) = 0$, $(\partial K_2/\partial C_3) = 0$, $(\partial C_1/\partial C_3) = 0$ then, from equation (1) we have $\mathscr{L}(y) = \mathscr{L}(y')$, from equation (2) we have $\mathscr{L}(q) = \mathscr{L}(q')$, and from equation (3) we have $\mathscr{L}(y,q) = \mathscr{L}(y',q')$.

Proof. Equation (1). According to (3.28) we have $\mathscr{L}(x,y) = \mathscr{L}(K_1\ x + K_2, K_1\ y)$. We differentiate this equation with respect to K_2, to find

$$0 = \frac{\partial \mathscr{L}}{\partial x'}\frac{\partial x'}{\partial K_2} + \frac{\partial \mathscr{L}}{\partial y'}\frac{\partial y'}{\partial K_2}$$

$$0 = \frac{\partial \mathscr{L}}{\partial x'}\left[\frac{\partial K_1}{\partial K_2}\ x + 1\right] + \frac{\partial \mathscr{L}}{\partial y'}\frac{\partial K_1}{\partial K_2}\ y$$

which implies that $\partial \mathscr{L}/\partial x' = 0$ if $\partial K_1/\partial K_2 = 0$. Then, according to $(\partial \mathscr{L}/\partial x) = (\partial \mathscr{L}/\partial x')(\partial x'/\partial x)$, we prove that $\partial \mathscr{L}/\partial x = 0$. **Equation (2).** According to (3.28) we have $\mathscr{L}(p,q) = \mathscr{L}(C_1\ p + C_2\ (x+y) + C_3, C_1\ q)$. We differentiate this equation with respect to C_3, to find

$$0 = \frac{\partial \mathscr{L}}{\partial p'}\frac{\partial p'}{\partial C_3} + \frac{\partial \mathscr{L}}{\partial q'}\frac{\partial q'}{\partial C_3}$$

$$0 = \frac{\partial \mathscr{L}}{\partial p'}\left[\frac{\partial C_1}{\partial C_3}\ p + \frac{\partial C_2}{\partial C_3}\ (x+y) + 1\right] + \frac{\partial \mathscr{L}}{\partial q'}\frac{\partial C_1}{\partial C_3}\ q$$

which implies that $\partial\mathscr{L}/\partial p' = 0$ if $\partial C_1/\partial C_3 = 0$, the term in square brackets not vanishing. Then, according to $(\partial\mathscr{L}/\partial p) = (\partial\mathscr{L}/\partial p')(\partial p'/\partial p)$, we prove that $\partial\mathscr{L}/\partial p = 0$. Finally we obtain $\mathscr{L}(q) = \mathscr{L}(q')$. **Equation (3).** According to (3.28) we have $\mathscr{L}(x, y, p, q) = \mathscr{L}(K_1\ x + K_2, K_1\ y, C_1\ p + C_2\ (x + y) + C_3, C_1\ q)$. We differentiate this equation with respect to C_3, to find

$$0 = \frac{\partial\mathscr{L}}{\partial x'}\frac{\partial x'}{\partial C_3} + \frac{\partial\mathscr{L}}{\partial y'}\frac{\partial y'}{\partial C_3} + \frac{\partial\mathscr{L}}{\partial p'}\frac{\partial p'}{\partial C_3} + \frac{\partial\mathscr{L}}{\partial q'}\frac{\partial q'}{\partial C_3}$$

$$0 = \frac{\partial\mathscr{L}}{\partial x'}\left[\frac{\partial K_1}{\partial C_3}\ x + \frac{\partial K_2}{\partial C_3}\right] + \frac{\partial\mathscr{L}}{\partial y'}\frac{\partial K_1}{\partial C_3}\ y$$

$$+ \frac{\partial\mathscr{L}}{\partial p'}\left[\frac{\partial C_1}{\partial C_3}\ p + \frac{\partial C_2}{\partial C_3}\ (x + y) + 1\right] + \frac{\partial\mathscr{L}}{\partial q'}\frac{\partial C_1}{\partial C_3}\ q$$

which implies that $\partial\mathscr{L}/\partial p' = 0$ if $\partial K_1/\partial C_3 = 0$, $\partial K_2/\partial C_3 = 0$, $\partial C_1/\partial C_3 = 0$. Then we have too $\partial\mathscr{L}/\partial p = 0$. According to $(\partial\mathscr{L}/\partial x) = (\partial\mathscr{L}/\partial p')(\partial p'/\partial x)$, we prove that $\partial\mathscr{L}/\partial x = 0$. Finally, according to $(\partial\mathscr{L}/\partial x) = (\partial\mathscr{L}/\partial x')(\partial x'/\partial x) = (\partial\mathscr{L}/\partial x')\ K_1$, we show that $\partial\mathscr{L}/\partial x' = 0$. Then we obtain $\mathscr{L}(y, q) = \mathscr{L}(y', q')$. □

Remark 3.6. The previous proof is based on the principle of field invariance introduced by [110]. It is equivalent to form invariance [169].

In an arbitrary coordinate system (y^i), the metric components, the connection and the bi-connection coefficients are respectively g_{ij}, Γ^k_{ij} and $\Gamma^k_{ij,l} + \Gamma^m_{ij}\Gamma^k_{lm}$. The three expressions $\mathscr{L}(g, \nabla)$, $\mathscr{L}(g, \nabla^2)$ and $\mathscr{L}(g, \nabla, \nabla^2)$ are then explicitly written as $\mathscr{L}(g_{ij}, \Gamma^k_{ij})$, $\mathscr{L}(g_{ij}, \Gamma^k_{ij,l} + \Gamma^m_{ij}\Gamma^k_{lm})$ and $\mathscr{L}(g_{ij}, \Gamma^k_{ij}, \Gamma^k_{ij,l} + \Gamma^m_{ij}\Gamma^k_{lm})$, respectively. Let (x^α) be another coordinate system, and let us assume the diffeomorphism invariance of the scalar field $\mathscr{L}$ (three cases):

$$\begin{cases} \mathscr{L}(\Gamma^k_{ij}) = \mathscr{L}(\Gamma^\gamma_{\alpha\beta}) \\ \mathscr{L}(\Gamma^k_{ij,l} + \Gamma^m_{ij}\Gamma^k_{lm}) = \mathscr{L}(\Gamma^\gamma_{\alpha\beta,\lambda} + \Gamma^\mu_{\alpha\beta}\Gamma^\gamma_{\lambda\mu}) \\ \mathscr{L}(\Gamma^k_{ij}, \Gamma^k_{ij,l} + \Gamma^m_{ij}\Gamma^k_{lm}) = \mathscr{L}(\Gamma^\gamma_{\alpha\beta}, \Gamma^\gamma_{\alpha\beta,\lambda} + \Gamma^\mu_{\alpha\beta}\Gamma^\gamma_{\lambda\mu}) \end{cases} \tag{3.29}$$

where the metric components will be omitted for the sake of simplicity. For further applications, let us introduce the following components:

$$\begin{cases} \mathbb{T}^k_{ij} = (1/2)(\Gamma^k_{ij} - \Gamma^k_{ji}) \\ \mathbb{S}^k_{ij} = (1/2)(\Gamma^k_{ij} + \Gamma^k_{ji}) \\ \mathbb{B}^k_{lij} = (1/2)(\Gamma^k_{ij,l} + \Gamma^m_{ij}\Gamma^k_{lm} - \Gamma^k_{lj,i} - \Gamma^m_{lj}\Gamma^k_{im}) \\ \mathbb{A}^k_{lij} = (1/2)(\Gamma^k_{ij,l} + \Gamma^m_{ij}\Gamma^k_{lm} + \Gamma^k_{lj,i} + \Gamma^m_{lj}\Gamma^k_{im}) \end{cases} \tag{3.30}$$

According to (3.30), permutation between i and j allows the decomposition $\Gamma^k_{ij} = \mathbb{S}^k_{ij} + \mathbb{T}^k_{ij}$. A permutation between i and l allows the decomposition $\Gamma^k_{ij,l} + \Gamma^m_{ij}\Gamma^k_{lm} = \mathbb{A}^k_{lij} + \mathbb{B}^k_{lij}$. A first permutation between i and j, then another successive permutation between i and l, allow the simultaneous decompositions $\Gamma^k_{ij} = \mathbb{S}^k_{ij} + \mathbb{T}^k_{ij}$ and $\Gamma^k_{ij,l} + \Gamma^m_{ij}\Gamma^k_{lm} = \mathbb{A}^k_{lij} + \mathbb{B}^k_{lij}$. Details of the calculus are given in the annex. Thanks to these decompositions, covariance of $\mathscr{L}$ (3.29) becomes

$$\begin{cases} \mathscr{L}(\mathbb{S}^k_{ij}, \mathbb{T}^k_{ij}) = \mathscr{L}(\mathbb{S}^\gamma_{\alpha\beta}, \mathbb{T}^\gamma_{\alpha\beta}) \\ \mathscr{L}(\mathbb{A}^k_{lij}, \mathbb{B}^k_{lij}) = \mathscr{L}(\mathbb{A}^\gamma_{\lambda\alpha\beta}, \mathbb{B}^\gamma_{\lambda\alpha\beta}) \\ \mathscr{L}(\mathbb{S}^k_{ij}, \mathbb{T}^k_{ij}, \mathbb{A}^k_{lij}, \mathbb{B}^k_{lij}) = \mathscr{L}(\mathbb{S}^\gamma_{\alpha\beta}, \mathbb{T}^\gamma_{\alpha\beta}, \mathbb{A}^\gamma_{\lambda\alpha\beta}, \mathbb{B}^\gamma_{\lambda\alpha\beta}) \end{cases} \tag{3.31}$$

3.4.2 *Application: Covariance of* $\mathscr{L}$

On the one hand, the fact that the gravitational field is associated with the curvature of spacetime, and that the electromagnetic field (we will see this in the following chapters) is associated with the torsion field of spacetime, and on the other hand, that the geometric framework of general relativity does not consider *a priori* a flat spacetime as a reference, makes understanding and mastering the role and nature of a coordinate system in the covariant formulation of a Lagrangian with the metric, the connection and the derivative of the connection as arguments essential. Let us consider the following identification of variables x, x', y, y', p, p', q, q':

$$x = \mathbb{S}^k_{ij}, \quad y = \mathbb{T}^k_{ij}, \quad p = \mathbb{A}^k_{lij}, \quad q = \mathbb{B}^k_{lij}, \tag{3.32}$$

$$x' = \mathbb{S}^\gamma_{\alpha\beta}, \quad y' = \mathbb{T}^\gamma_{\alpha\beta}, \quad p' = \mathbb{A}^\gamma_{\lambda\alpha\beta}, \quad q' = \mathbb{B}^\gamma_{\lambda\alpha\beta}. \tag{3.33}$$

The transformation laws between the above variables take the form of (3.28) with (see annex)

$$\begin{cases} K_1 = J^i_\alpha J^j_\beta A^\gamma_k \\ K_2 = J^j_{\alpha\beta} A^\gamma_j \end{cases}, \quad \begin{cases} C_1 = J^i_\alpha J^j_\beta J^l_\lambda A^\gamma_k \\ C_2 = J^i_{\alpha\lambda} J^j_\beta A^\gamma_k + J^i_\lambda J^j_{\alpha\beta} A^\gamma_k + J^i_\alpha J^j_{\beta\lambda} A^\gamma_k \\ C_3 = J^i_\mu J^l_\lambda J^j_{\alpha\beta} A^\gamma_j A^\mu_{il} + J^i_{\mu\lambda} J^j_{\alpha\beta} A^\mu_i A^\gamma_j \end{cases} \tag{3.34}$$

We have $\partial K_1/\partial K_2 = 0$, $\partial K_1/\partial C_3 = 0$, $\partial K_2/\partial C_3 = 0$, $\partial C_1/\partial C_3 = 0$. According to Lemma 3.5, the covariance of $\mathscr{L}$ (3.31) means

$$\begin{cases} \mathscr{L}(\mathbb{T}^k_{ij}) = \mathscr{L}(\mathbb{T}^\gamma_{\alpha\beta}) \\ \mathscr{L}(\mathbb{B}^k_{lij}) = \mathscr{L}(\mathbb{B}^\gamma_{\lambda\alpha\beta}) \\ \mathscr{L}(\mathbb{T}^k_{ij}, \mathbb{B}^k_{lij}) = \mathscr{L}(\mathbb{T}^\gamma_{\alpha\beta}, \mathbb{B}^\gamma_{\lambda\alpha\beta}) \end{cases} \Longrightarrow \begin{cases} \mathscr{L}(\aleph^k_{ij}) = \mathscr{L}(\aleph^\gamma_{\alpha\beta}) \\ \mathscr{L}(\mathfrak{R}^k_{lij}) = \mathscr{L}(\mathfrak{R}^\gamma_{\lambda\alpha\beta}) \\ \mathscr{L}(\aleph^k_{ij}, \mathfrak{R}^k_{lij}) = \mathscr{L}(\aleph^\gamma_{\alpha\beta}, \mathfrak{R}^\gamma_{\lambda\alpha\beta}) \end{cases} \tag{3.35}$$

since we can identify the torsion by $\aleph_{ij}^k = 2\mathbb{T}_{ij}^k$ and the curvature by $\mathfrak{R}_{lij}^k = 2\mathbb{B}_{lij}^k$.

3.4.3 *Summary for Lagrangian Covariance*

We have temporarily omitted the argument g_{ij}. Adding this argument does not change the proof. The overall result then includes metric, connection, and bi-connection as arguments of the Lagrangian function $\mathscr{L}$. We then have:

$$\begin{cases} \mathscr{L}(g_{ij}, \aleph_{ij}^k) = \mathscr{L}(g_{\alpha\beta}, \aleph_{\alpha\beta}^\gamma) \\ \mathscr{L}(g_{ij}, \mathfrak{R}_{lij}^k) = \mathscr{L}(g_{\alpha\beta}, \mathfrak{R}_{\lambda\alpha\beta}^\gamma) \\ \mathscr{L}(g_{ij}, \aleph_{ij}^k, \mathfrak{R}_{lij}^k) = \mathscr{L}(g_{\alpha\beta}, \aleph_{\alpha\beta}^\gamma, \mathfrak{R}_{\lambda\alpha\beta}^\gamma) \end{cases} \tag{3.36}$$

All the arguments of $\mathscr{L}$ are components of tensors, invariant under the action of the diffeomorphism (Fig. 3.3) (in the sense that they transform covariantly according to usual tensor transformations depending on their type).

Remark 3.7. In general relativity, the application of the Minimal Coupling Procedure (MCP) would be a conversion of all partial derivatives $\partial_\gamma g_{\alpha\beta}$ and $\partial_\lambda \partial_\gamma g_{\alpha\beta}$ in the Minkowskian flat spacetime/continuum $\mathscr{M}$ into covariant derivatives $\nabla_\gamma g_{\alpha\beta}$ and $\nabla_\lambda \nabla_\gamma g_{\alpha\beta}$, and also choosing an appropriate volume-form ω_n in the underlying curved spacetime/continuum, possibly

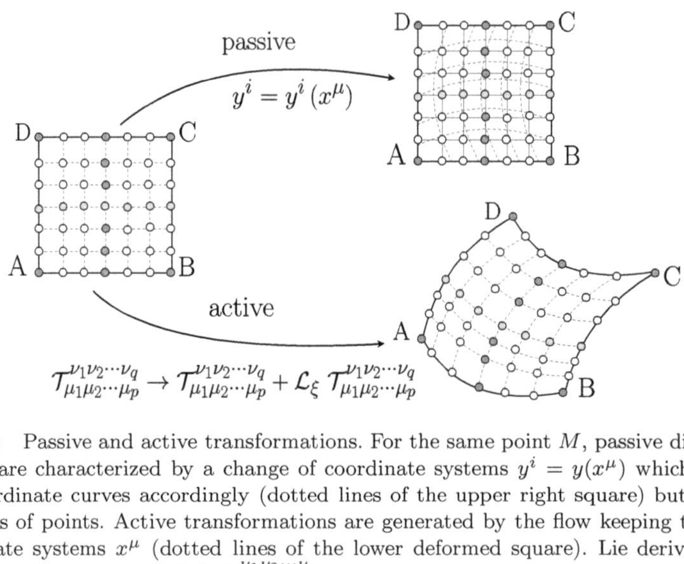

Fig. 3.3: Passive and active transformations. For the same point M, passive diffeomorphisms are characterized by a change of coordinate systems $y^i = y(x^\mu)$ which modify the coordinate curves accordingly (dotted lines of the upper right square) but not the positions of points. Active transformations are generated by the flow keeping the same coordinate systems x^μ (dotted lines of the lower deformed square). Lie derivatives of tensor variables are applied $\mathcal{L}_\xi\, \mathcal{T}^{\nu_1\nu_2\cdots\nu_q}_{\mu_1\mu_2\cdots\mu_p}$.

with torsion. For a metric-compatible connection, the standard MCP procedure would lead to $\nabla_\gamma g_{\alpha\beta} \equiv 0$ and $\nabla_\lambda \nabla_\gamma g_{\alpha\beta} \equiv 0$. The present approach is slightly different and can be considered as an extension of the MCP to Riemann-Cartan spacetime/continuum, since we assume the connection and the bi-connection to be independent arguments.

3.4.3.1 *Main Results*

Covariance does not impose any restrictions on the spacetime theories (Newtonian for classical mechanics, Minkowskian for special relativity, Riemannian or Riemann-Cartan for relative gravitation). The results are summarized in the following theorem:

Theorem 3.6. *Let a metric-affine manifold* $(\mathscr{B}, \mathbf{g}, \nabla)$ *where the affine connection is compatible with the metric* ($\nabla \mathbf{g} = 0$). *To the connection are associated the torsion tensor* $\aleph$ *and the curvature tensor* $\mathfrak{R}$. *For any scalar function* $\mathscr{L}$ *defined on* $\mathscr{B}$, *depending on the metric, the connection and the bi-connection, the covariance induces:*

$$\mathscr{L}(\mathbf{g}, \nabla) = \mathscr{L}(\mathbf{g}, \aleph), \tag{3.37}$$

$$\mathscr{L}(\mathbf{g}, \nabla^2) = \mathscr{L}(\mathbf{g}, \mathfrak{R}), \tag{3.38}$$

$$\mathscr{L}(\mathbf{g}, \nabla, \nabla^2) = \mathscr{L}(\mathbf{g}, \aleph, \mathfrak{R}) \tag{3.39}$$

The equations (3.37), (3.38), *and* (3.39) *can be read in three ways:*

(1) $\mathscr{L} = \mathscr{L}(\mathbf{g}, \nabla)$ *is covariant if and only if* $\mathscr{L} = \mathscr{L}(\mathbf{g}, \aleph)$
(2) $\mathscr{L} = \mathscr{L}(\mathbf{g}, \nabla^2)$ *is covariant if and only if* $\mathscr{L} = \mathscr{L}(\mathbf{g}, \mathfrak{R})$
(3) $\mathscr{L} = \mathscr{L}(\mathbf{g}, \nabla, \nabla^2)$ *is covariant if and only if* $\mathscr{L} = \mathscr{L}(\mathbf{g}, \aleph, \mathfrak{R})$

Covariance results conform to gravitational Utiyama theorem showing that the invariance under both the spacetime diffeomorphisms and the local Lorentz transformations impose the Lagrangian density to depend upon the tetrads, the connection and their derivatives only through the metric, torsion, and curvature e.g. [177, 178]. In the last chapter, we investigate the consequences of the gauge invariance (Principle of General Covariance). It should be recalled that the Lagrangian cannot explicitly depend on event coordinates [118, 177].

3.4.3.2 *General Remarks on Covariance*

Covariance is considered as extension of the Principle of Material Indifference, usually assumed for the constitutive laws in continuum mechanics e.g. [161]. In essence, the covariance of the Lagrangian is the generalization of the Principle of Material Indifference no longer by limiting itself to the group of Leibniz transformations but by considering the group of diffeomorphisms, knowing that the latter are passive, not active. Once again, it should be stressed that in no case is the theorem linked to any physical principle other than the independence of the formulation of phenomena modeling with respect to a diffeomorphism.

Material Frame Indifference and Covariance. Coming back to the analogy with classical continuum mechanics, we therefore retain the extension of the "Zaremba-Jaumann" Material Frame Indifference type which stipulates rather the invariance with respect to the observer, in this chapter naturally extended to the invariance with respect to a coordinate system in a four-dimensional spacetime (passive transformations).

It is more appropriate rather than to consider it as an invariance of constitutive laws with respect to a rigid motion of the continuum, formulation of the Material Frame Indifference à la "Hooke-Poisson-Cauchy", whose generalization would have been an invariance with respect to general motion with deformation of the considered continuum (active transformations). This would have been nonsense. On the other hand, the consequence of these active transformations will be treated in the last chapter, using the Lie derivative along a non-uniform vector field to derive the covariant conservation laws.

Chapter 4

General Covariance: Conservation Laws on Generalized Continua

We consider a Lagrangian of generalized continuum which depends only on metric, torsion and curvature without any other physical sources. Conservation laws are derived by their invariance under active diffeomorphisms which are generated by Lie derivatives of both the metric and the connection. The present chapter is only devoted to a generic shape of conservation laws in generalized continuum and do not explicitly include physics of either an electromagnetic field or a fluid with spin. The method uses the *Principle of General Covariance*, based on the Lie derivative of action along a non-uniform vector field ξ which is an extension of the classical Principle of Virtual Power in continuum mechanics. It is shown by deduction the necessity of introducing hyper-stress and polar stress analogous to hypermomenta in relative gravitation.

4.1 Introduction

First of all, it is useful to re-specify the vocabularies that we use in this book concerning the notions of invariance and covariance. In Chapter 3, we have clearly specified that the word *covariance* is a direct generalization of what some authors call in continuum mechanics *material frame indifference*. Other names are *background independence*, *passive general covariance*, *passive diff invariance* or *formal invariance* (see [163] for more detailed citations). In short, this is the invariance form of mathematical equations during a change of coordinate system e.g. [169]. The counterpart of this notion is what we call here *General Covariance* to express the invariance during an active diffeomorphism, which some authors have also called *active diff invariance*, *substance invariance*, and others. The controversy over the notion of passive and active diffeomorphisms appeared as early as 1917 because the spacetime of relative gravitation is at the same time a dynamical variable (via the metric) and also constitutes the background where physics happen e.g. [160]. This is a special case.

Whereas in Chapter 3, covariance is due to passive diffeomorphisms, the present chapter involves active diffeomorphisms to derive the conservation laws of generalized continuum, and by extension of Einstein-Cartan spacetime vacuum for which the geometric structure is defined by the metric $g_{\mu\nu}$, torsion $\aleph^{\gamma}_{\mu\nu}$, and curvature $\Re^{\gamma}_{\mu\nu\lambda}$. Introduction of torsion is required since it is linked to the electromagnetic field (photon) in the vacuum but can also be due to the presence of high-density spin fluid (fermion) occupying spacetime e.g. [88]. It is noticed that considering the region of the spacetime with neither physical sources nor matter as fluid does not exclude the presence of torsion without further precision, in addition to metric and curvature.

Einstein-Cartan spacetime. The geometry background of a Generalized Continuum Model (Riemann-Cartan manifold) is defined by:

(1) a metric field $g_{\alpha\beta}$, which represents the continuous deformation;
(2) a connection ∇ with torsion $\aleph^{\gamma}_{\alpha\beta}$ and curvature $\Re^{\gamma}_{\alpha\beta\lambda}$, which model defect distribution.

As a reminder, generalized continuum models that we consider in this work may be classified as follows:

(1) Euclidean continuum ($g_{\alpha\beta} \neq 0, \aleph^{\gamma}_{\alpha\beta} \equiv 0, \Re^{\gamma}_{\alpha\beta\lambda} \equiv 0$) for Newton and Minkowski spacetimes;
(2) Weitzenböck continuum ($g_{\alpha\beta} \neq 0, \aleph^{\gamma}_{\alpha\beta} \neq 0, \Re^{\gamma}_{\alpha\beta\lambda} \equiv 0$)
(3) Riemann continuum ($g_{\alpha\beta} \neq 0, \aleph^{\gamma}_{\alpha\beta} \equiv 0, \Re^{\gamma}_{\alpha\beta\lambda} \neq 0$) for Einstein spacetime;
(4) Riemann-Cartan continuum ($g_{\alpha\beta} \neq 0, \aleph^{\gamma}_{\alpha\beta} \neq 0, \Re^{\gamma}_{\alpha\beta\lambda} \neq 0$) for Einstein-Cartan spacetime.

4.1.1 *Symmetries and Conservation Laws*

Invariance by a symmetry group plays an important role for deriving laws of physics. At the beginning of the 20th century, the role of symmetry invariance which was thought as the basic foundation of natural laws was accepted for most domains, if not all, of modern physics. The link between symmetries and conservation laws gained in power with the famous theorems of E. Noether [186], where continuous symmetry was used. Basically, the application of Noether's theorems, under the impulsion of Hilbert, Klein and Einstein namely to better understand the problem of energy conservation in relative gravitation, was the *keystone* for the development of the symmetry method for theoretical physics. Newton's laws of mechanics embodied the invariance with respect to the Galilean group of transformations. Special relativity is based on Lorentz's group of transformations which reconcile classical mechanics with Maxwell's theory on electromagnetism. Minkowski spacetime includes mechanics and electromagnetism in a unique framework. The principle of local symmetry applied on the Hilbert-Einstein action under active diffeomorphims induced the gravitation equations. By extension, relative gravitation was seen as a gauge theory based on the local Poincaré's transformations group since the works of e.g. [118, 179–181]. What should be retained is however the difference between field equations, which are the Euler-Lagrange equations of a Lagrangian, and conservation laws. The former are obtained by using Euler variations whereas the latter are by active diffeomorphisms. In the same way, Poincaré's Gauge Theory is a well-known method for constructing field equations and conservation laws of general relativity e.g. [151, 176, 179, 182–184], and by extension spacetime with defects which are thought to be linked to the source of the Universe expansion by means of the cosmological constant Λ e.g. [185].

The basic problem motivating the writing of this chapter is summarized in Figure 4.1: We wonder if we could use the famous $\text{div}\ \sigma = 0$ of Cauchy in standard continuum mechanics for generalized continuum.

In standard continuum mechanics, the following steps are used:

(1) On a fictitious Clapeyron cutting, assume a vector stress $\mathbf{p}_n$ depending only on the unit normal $\mathbf{n}$ (it is valid also for the external normal on the boundary $\partial\mathscr{B}$);
(2) Write the linear momentum balance on the continuum $\int_{\mathscr{B}} \mathbf{p}_n \, da = 0$;
(3) The first step allows us to deduce the fundamental Cauchy theorem as the existence of a tensor σ such that $\mathbf{p}_n = \sigma(\mathbf{n})$;
(4) Apply the divergence theorem $\int_{\mathscr{B}} \sigma(\mathbf{n}) \, da = \int_{\mathscr{B}} \text{div}\ \sigma \, dv = 0$ and by using the Haar Lemma deduce the conservation laws $\text{div}\ \sigma = 0$.

The Cauchy approach is called the Tetrahedron method. The motivation of the present chapter is then to analyze if such a method is still valable and, if not, to derive the appropriate conservation laws for generalized continuum.

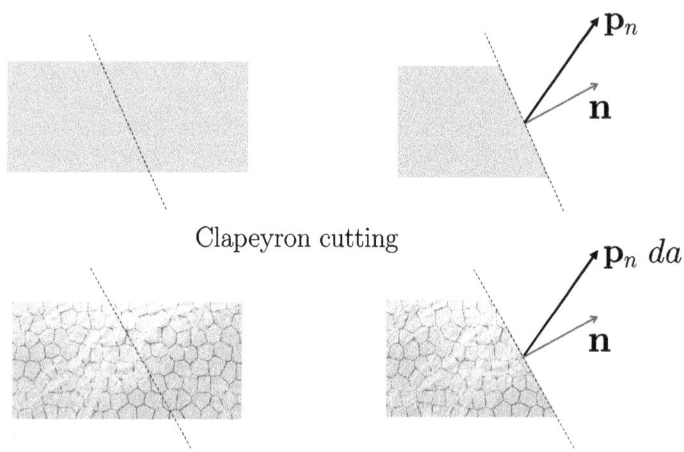

Fig. 4.1: (Top) Cauchy's approach for conservation laws of continuum mechanics: a stress vector is assumed to depend only on the normal $\mathbf{n}$ of the cutting. (Bottom) The same approach for deriving conservation laws for generalized continuum?

In this chapter, we rather directly use the Principle of General Covariance to derive conservation laws on a generalized continuum e.g. [129, 167]. This conforms to the so-called substantive covariance because we use active diffeomorphisms generated by the Lie derivative of the geometric variables along the flow. Generalized continuum, defined by the Lagrangian $\mathscr{L}(g_{\alpha\beta}, \aleph^{\gamma}_{\alpha\beta}, \Re^{\gamma}_{\alpha\beta\lambda})$ e.g. [6], is considered. Conservations laws are derived from the stationarity of the action. As an example of continuum with defect distribution, a sketch of constitutive laws is obtained by using the Principle of Maximum Dissipation which allows us to derive the nucleation and evolution, i.e., defect flow rule.

4.1.2 *Brief History of Generalized Continuum*

An interesting history of generalized continuum models was given by Maugin [9], in which the extension of standard continuum models, originally and essentially developed by Cauchy between 1822 and 1827, is based on abandoning three types of restrictive assumptions, namely:

(1) Cauchy's stress symmetry;
(2) Cauchy's postulate, which assumes that the stress vector depends exclusively on the normal (implicitly reducing the model to a first gradient model, thus using the metric tensor as an argument);

(3) the Euclidean nature of the manifold modeling a standard continuum, transforming it into a more general manifold with curvature and/or torsion.

The abandonment of Cauchy's stress symmetry had been introduced in 1909 by the Cosserat brothers to obtain media with a rigid microstructure. Deformable microstructure models, also called micromorphic models, were introduced later in the 1960s and 1970s, and a classification of these was provided, for example, by Eringen [187, 188]. The implicit or explicit abandonment of the Cauchy Postulate, which states that the stress vector acting on a local facet (following the Clapeyron cut-off hypothesis) depends only on the normal, but not on its derivatives, led to media of the second gradient or even higher order. This class also includes non-local continuous media for which the stress also depends on the deformation of the entire neighborhood or even of all other points in the continuum. Finally, the abandonment of the Euclidean nature of a Cauchy medium was proposed in the 1950s, particularly for modeling linear and surface defects such as dislocations and disclinations in crystalline media e.g. [189]. From a topological point of view, a *generalized continuum* was defined as connected, locally compact, separable metric space e.g. [150]. Conservation laws for generalized continuum are applicable not only for relative gravitation but also for continuum with size effects such as for metallic material by accounting for the influence of dislocation density, as in plasticity of materials e.g. [190]. The objective is to extend the linear momentum conservation law div $\sigma = 0$ obtained for simple material continuum (called also first gradient theory) in the sense of Noll [191] with neither external nor inertial force. In most parts of this chapter, the term Lagrangian function might be replaced by the term Helmholtz free energy e.g. [141]. The present chapter takes roots in previous works [6] and [192].

General framework. Inspired by works of Palatini and of Reichenbach, we define the generalized continuum by a covariant action e.g. [6]:

$$\mathscr{S} = \int_{\mathscr{B}} \mathscr{L}\left(\mathbf{g}, \aleph, \Re\right) \omega_n \tag{4.1}$$

where $\mathbf{g}$, $\aleph$, and $\Re$ are respectively the metric, torsion and curvature on a Riemann-Cartan manifold $\mathscr{B}$.

This is an extended action of not only the Einstein-Cartan gravitation if such a spacetime is considered e.g. [6], but also of an inelastic continuum model, such as generalization of the simple material of Noll e.g. [11]. In the relation (4.1), the volume-form is compatible with ∇ when it can be written as $\omega_n := e^{\vartheta(\mathbf{g},\aleph)}\,\overline{\omega}_n$ where $\overline{\omega}_n$ is the Riemann volume-form [134].

We consider here a connection compatible with the metric, say $\nabla\mathbf{g} \equiv 0$, for the sake of simplicity. It nevertheless should be recalled that the Weyl approach rather considers the non-metricity $\nabla\mathbf{g}$ as arguments even if such an approach is not as popular as the Einstein-Cartan approach e.g. [57] (for cosmology e.g. [4]).

Remark 4.1. It is not usual in continuum mechanics to construct a variational formulation deriving from a Lagrangian function, when dissipation occurs within the body. However, the presence of torsion and curvature implicitly engenders dissipation for defected materials e.g. [49, 193], since they describe the strain incompatibility as dislocations and disclinations (cf. [11]). Defects of spacetime are interestingly related to the electromagnetic fields within the spacetime.

4.2 Overview of the Variational Procedure

The Principle of Least Action is well-known for most field equations and conservations laws in discrete physics. In relative gravitation, the Principle of General Covariance is the extended version e.g. [130]. Continuous symmetries (Lie group) may be exploited to generate conservation laws (the most known method comes from Noether's Theorem) e.g. [194]. It is important to determine what are the consequences of a local translation group defining a Poincaré local transformation to primal variables: they define the gauge transformations.[1]

4.2.1 *Dynamical vs. Non-Dynamical Symmetries*

Before embarking on the variational procedure on general continuum, we consider the simple case of action depending on time, generalized coordinates at their time derivatives in a configurational space.

[1] In special relativity theory, starting from the causality principle, Lorentz transformations are not the only symmetries of the Minkowski spacetime. Translation in space and time also constitute symmetries. In the 1960s, [148] showed that the causality group is generated by the orthochronous Poincaré group together with dilations. Later, [149] demonstrated that the complete Poincaré group is the group that preserves time-like vectors of Minkowski spacetime and, even further, the Poincaré group follows from the principle of light speed invariance.

4.2.1.1 *Covariance of Action, Hamiltonian*

It seems necessary to give some additional details concerning the essential difference between covariance (*non-dynamical symmetries*) and general covariance (*dynamical symmetries*), starting from an elementary example of minimization of a path length in a configuration space. Let's consider a typical action borrowed from analytical mechanics:

$$\mathscr{S} := \int_{\tau_1}^{\tau_2} \mathscr{L}\left[\tau, q^\mu(\tau), \dot{q}^\mu(\tau)\right] d\tau \tag{4.2}$$

From the outset, covariance imposes that this action $\mathscr{S}$ must be a scalar and thus it remains invariant by a change of coordinate system; whether this action is extreme by a variational procedure or not, it must be covariant. We thus use a group of non-dynamical symmetries. Consider a re-parametrization of the curve $\mathcal{C}$ on which this action is calculated by introducing $\tau' := \tau + \epsilon(\tau)$ such that $\epsilon = 0$ at both ends τ_1 and τ_2. Assume that with this new parameterization τ' the action remains invariant for an infinitesimal variation ϵ, then we deduce that the quantity $\mathscr{H}$, generally called *Hamiltonian* in the context of analytical mechanics, vanishes:

$$\mathscr{H} := \frac{\partial \mathscr{L}}{\partial \dot{q}^\mu}\,\dot{q}^\mu - \mathscr{L} = 0 \tag{4.3}$$

The proof is straightforward by writing first the linearization (directional derivative):

$$\delta q^\mu := \lim_{\eta \to 0} \frac{1}{\eta}\left[q^\mu\left(\tau + \eta\epsilon\right) - q^\mu\left(\tau\right)\right] = \frac{dq}{d\tau}\,\epsilon$$

For small ϵ, we then have the two variations of q^μ and $\dot{q}^\mu$ respectively:

$$q^\mu(\tau + \epsilon) \simeq q^\mu(\tau) + \frac{dq^\mu}{d\tau}\,\epsilon, \quad \frac{dq^\mu}{d\tau}(\tau + \epsilon) \simeq \dot{q}^\mu + \frac{d}{d\tau}\left(\dot{q}^\mu\,\epsilon\right) \tag{4.4}$$

The variation of the action under this re-parametrization thus takes the form of:

$$\begin{aligned}\mathscr{S}[\mathbf{q}(\tau + \epsilon)] - \mathscr{S}[\mathbf{q}(\tau)] &= \int_{\tau_1}^{\tau_2} \left[\frac{\partial \mathscr{L}}{\partial \tau}\epsilon + \frac{\partial \mathscr{L}}{\partial q^\mu}\dot{q}^\mu\epsilon + \frac{\partial \mathscr{L}}{\partial \dot{q}^\mu}\frac{d}{d\tau}\left(\dot{q}^\mu\epsilon\right)\right] d\tau \\ &= \int_{\tau_1}^{\tau_2} \left[\frac{d\mathscr{L}}{d\tau}\,\epsilon + \left(\frac{\partial \mathscr{L}}{\partial \dot{q}^\mu}\dot{q}^\mu\right)\frac{d\epsilon}{d\tau}\right] d\tau\end{aligned}$$

Integrating by parts the first term of the last row, we get:

$$\mathscr{S}[\mathbf{q}(\tau + \epsilon)] - \mathscr{S}[\mathbf{q}(\tau)] = \left[\mathscr{L}\,\epsilon(\tau)\right]_{\tau_1}^{\tau_2} + \int_{\tau_1}^{\tau_2} \left(\frac{\partial \mathscr{L}}{\partial \dot{q}^\mu}\dot{q}^\mu - \mathscr{L}\right)\frac{d\epsilon}{d\tau}\,d\tau \tag{4.5}$$

By hypothesis, the quantity ϵ vanishes at both ends, the requirement that the action is invariant for the two parametrizations τ and τ' implies that the Hamiltonian (4.3) should equal zero $\mathscr{H} = 0$. The relation $\mathscr{H} = 0$ is a non-dynamical symmetry since nowhere did this calculus assume that the action $\mathscr{S}$ is extremal (no Hamiltonian Principle of Least Action is applied) nor that the generalized coordinates satisfy the corresponding Euler-Lagrange equations.

4.2.1.2 *Typical Examples*

We sketch the difference of dynamical and non-dynamical transformations with two examples of Lagrangians: arc length and kinetic energy. Now, we can look at it from another point of view by saying that we have an invariance by re-parametrization as long as the Hamiltonian is identically equal to zero. As a first example, this is obviously the case when the Lagrangian is identified to the arc length:

$$\mathscr{L}_1 := \sqrt{g_{\mu\nu}\ \dot{q}^\mu \dot{q}^\nu} \quad \Longrightarrow \quad \frac{\partial \mathscr{L}_1}{\partial \dot{q}^\mu}\dot{q}^\mu - \mathscr{L}_1 = \frac{2 g_{\mu\nu}\dot{q}^\nu}{2\sqrt{g_{\mu\nu}\ \dot{q}^\mu \dot{q}^\nu}}\ \dot{q}^\mu - \sqrt{g_{\mu\nu}\ \dot{q}^\mu \dot{q}^\nu} = 0$$

Care should be done at this stage: $\mathscr{H}_1 \equiv 0$ does not mean that there is no Hamiltonian. It expresses that the Lagrangian $\mathscr{L}_1$ has a non-dynamical degree of freedom that should be eliminated before a "real" Hamiltonian can be constructed. This only means that re-parametrization as $\tau \to \tau + \epsilon(\tau)$ constitutes a non-dynamical symmetry of the arc-length action $\mathscr{S}_1$. As a second example, let's consider another action based on the kinetic energy of a material test point of mass $m = 1$:

$$\mathscr{S}_2 := \int_{\tau_1}^{\tau_2} \mathscr{L}_2\ d\tau, \quad \mathscr{L}_2 := \frac{1}{2} g_{\mu\nu}\ \dot{q}^\mu \dot{q}^\nu \tag{4.6}$$

We easily obtain the Hamiltonian:

$$\mathscr{H}_2 = \frac{\partial \mathscr{L}_2}{\partial \dot{q}^\mu}\dot{q}^\mu - \mathscr{L}_2 = \frac{1}{2} g_{\mu\nu}\ \dot{q}^\mu \dot{q}^\nu \tag{4.7}$$

which is equal to the kinetic energy of a material point. In general when the point is in motion, this Hamiltonian is not equal to zero and then the kinetic-energy action is not invariant under the re-parametrization $\tau \to \tau + \epsilon(\tau)$, which is thus not a non-dynamical symmetry for this action $\mathscr{S}_2$.

4.2.1.3 *Case where $\partial_\tau \mathscr{L} = 0$*

For the sake of simplicity, consider the case where $\mathscr{L}$ does not depend explicitly on time τ; we remark that $\mathscr{L}_1$ and $\mathscr{L}_2$ constitute typical examples but they are not the only ones. For any Lagrangian, we have the following:

$$\frac{\partial \mathscr{L}}{\partial \tau} \equiv 0 \quad \Longrightarrow \quad \mathscr{S} = \int_{\tau_1}^{\tau_2} \mathscr{L}\left[q^\mu(\tau), \dot{q}^\mu(\tau)\right] d\tau \tag{4.8}$$

By applying the same procedure as previously, we directly obtain the variation:

$$\delta\mathscr{S} := \mathscr{S}\left[\mathbf{q}\left(\tau + \epsilon(\tau)\right)\right] - \mathscr{S}\left[\mathbf{q}\left(\tau\right)\right] = \int_{\tau_1}^{\tau_2} \left[\frac{\partial \mathscr{L}}{\partial q^\mu}\dot{q}^\mu \epsilon + \frac{\partial \mathscr{L}}{\partial \dot{q}^\mu}\frac{d}{d\tau}\left(\dot{q}^\mu \epsilon\right)\right] d\tau$$

After usual integration by parts, we arrive at the relation:

$$\delta\mathscr{S} = \int_{\tau_1}^{\tau_2} \frac{d}{d\tau}\left[\frac{\partial \mathscr{L}}{\partial \dot{q}^\mu}\dot{q}^\mu \epsilon\right] d\tau + \int_{\tau_1}^{\tau_2} \left[\frac{\partial \mathscr{L}}{\partial q^\mu} - \frac{d}{d\tau}\left(\frac{\partial \mathscr{L}}{\partial \dot{q}^\mu}\right)\right] \dot{q}^\mu \epsilon \, d\tau \tag{4.9}$$

We slightly modify the expression to obtain a well-known expression:

$$\delta\mathscr{S} = \left[p_\mu \delta q^\mu\right]_{\tau_1}^{\tau_2} + \int_{\tau_1}^{\tau_2} \left[\frac{\partial \mathscr{L}}{\partial q^\mu} - \frac{d}{d\tau}\left(\frac{\partial \mathscr{L}}{\partial \dot{q}^\mu}\right)\right] \delta q^\mu \, d\tau \tag{4.10}$$

by defining the generalized momentum associated with the generalized coordinate q^μ and the coordinate variation respectively:

$$p_\mu := \frac{\partial \mathscr{L}}{\partial \dot{q}^\mu}, \qquad \delta q^\mu := \dot{q}^\mu \epsilon \, d\tau \tag{4.11}$$

As long as the generalized coordinate variations are independent, and these variations vanish at both ends, we obtain the Euler-Lagrange equations if we impose the invariance of the action following this re-parametrization:

$$\frac{\partial \mathscr{L}}{\partial q^\mu} - \frac{d}{d\tau}\left(\frac{\partial \mathscr{L}}{\partial \dot{q}^\mu}\right) = 0, \quad \mu = 1, \dots n \tag{4.12}$$

which constitute the motion equations of the test particle in the configurational space. The condition that the Lagrangian does not depend explicitly on the time τ leads to the dynamical equations, assessing that the change of parameter $\tau \to \tau + \epsilon(\tau)$ becomes a *dynamical symmetry* for the action.

4.2.1.4 *Examples*

For the particular case of kinetic energy $\mathscr{L}_2$, the Euler-Lagrange equations lead to conservation of the momentum:

$$\frac{d}{d\tau}\left(\frac{\partial \mathscr{L}_2}{\partial \dot{q}^\mu}\right) = 0 \quad \Longrightarrow \quad p_\mu(\tau) := \frac{\partial \mathscr{L}_2}{\partial \dot{q}^\mu}(\tau) = p_\mu(\tau_0)$$

Given that the Lagrangian $\mathscr{L}_2$ is a symmetric and strictly definite positive quadratic form, we have:

$$p_\mu = \frac{\partial \mathscr{L}_2}{\partial \dot{q}^\mu} = g_{\mu\nu}\, \dot{q}^\nu \quad \Longrightarrow \quad \dot{q}^\nu = g^{\mu\nu}\, p_\mu$$

We accordingly obtain the particular form of the Hamiltonian:

$$\mathscr{H}_2(\tau) = g^{\mu\nu}\ p_\mu p_\nu(\tau) = \mathscr{H}_2(\tau_0) \tag{4.13}$$

which is also a conserved quantity. Keeping in mind these two examples, we observe that the mathematical identity $\mathscr{H}_1 = 0$ is not a conservation law as relation $\mathscr{H}_2 = \mathscr{H}_2(\tau_0)$, meaning that the Hamiltonian is constant. The first identity holds for any trajectory of the system with Lagrangian $\mathscr{L}_1$ and then constitutes a non-dynamical symmetry, whereas the conservation of Hamiltonian $\mathscr{H}_2$, and by extension the moment components $p_\mu = p_\mu(\tau_0)$, only holds for solutions of the Euler-Lagrange equations. They constitute a dynamical symmetry.

Remark 4.2. On the one hand, covariance embodied by non-dynamical symmetries reduces to changes of coordinate system that leave the action invariant (see chapter on covariance). On the other hand, general covariance embodied by dynamical symmetries acting on the physical variables leaves also the action invariant. Consequences of these second transformations will constitute the heart of this chapter since they are the basis for deriving the conservation laws of physics within generalized continuum.

4.2.2 *General Covariance and Active Diffeomorphism*

Covariance implies that Lagrangian $\mathscr{L}$ displays the same functional form in terms of transformed arguments as it does in terms of the original arguments[2]. Say $\mathscr{L}(g_{\alpha\beta}, \aleph^\gamma_{\alpha\beta}, \Re^\lambda_{\alpha\beta\mu})$. We now consider the dynamical symmetry group by taking into account local translations that generalize the uniform translations of Newtonian mechanics to (active) diffeomorphisms, which include *local translations* and *local rotations*. It constitutes in some sense Poincaré's group of transformations e.g. [127] for four-dimensional spacetime. In addition to the three rotation and three boost generators, Poincaré's group includes four translation generators. There are thus 10 generators in Poincaré's group. As for metric variables [195] and by analogy to the variations in [196], there are three kinds of variations[3]:

(1) the Lagrangian (comoving with particle) variations $(\Delta g_{\alpha\beta}, \Delta\aleph^\gamma_{\alpha\beta}, \Delta\Re^\lambda_{\alpha\beta\mu})$,

[2]The dependence on coordinates x^μ is dropped according to results in e.g. [118]. This is similar to the consideration of a Lagrangian satisfying $\partial_\tau \mathscr{L} = 0$.

[3]Lie derivative variations are related to the active diffeomorphisms. This allows us to formulate the transformed variables with respect to the coordinates x^μ before application of the transformations.

(2) the Eulerian (fixed point) variations $(\delta g_{\alpha\beta}, \delta \aleph^{\gamma}_{\alpha\beta}, \delta \Re^{\lambda}_{\alpha\beta\mu})$,
(3) the Lie derivative variations $(\mathcal{L}_\xi g_{\alpha\beta}, \mathcal{L}_\xi \aleph^{\gamma}_{\alpha\beta}, \mathcal{L}_\xi \Re^{\lambda}_{\alpha\beta\mu})$.

Lie derivatives along the arbitrary vector $\xi^\mu(x^\alpha)$ are a particular class of variations. The principle is then to express Lie derivatives along Lie derivative variations and to impose the stationarity to give conservation laws.

4.2.2.1 *Dynamical Symmetries and Lie Derivatives*

The Lie derivative variations may be regarded as active diffeomorphism in such a way that any point-event $M_0(x^\mu, \mu = 1,2,3) \in \mathscr{M}$ is changed into the point-event $M(y^\nu, \nu = 0,1,2,3) \in \mathscr{M}$ and accordingly all tensor fields $\mathcal{T}(M_0)$ according to the flow but the initial coordinates x^μ should be used to label either the point M or all tensor fields $\mathcal{T}$ on the spacetime manifold. The infinitesimal transformation generated by ξ may be written in coordinate terms:

$$y^\mu = x^\mu + \xi^\mu \, d\epsilon \quad \text{where } \xi \text{ is infinitesimal} \tag{4.14}$$

Coordinate $\tilde{x}^\mu$ corresponding to y^μ in terms of initial coordinates is obtained accordingly, given that transformations are infinitesimal, to first order in $d\epsilon$:

$$\tilde{x}^\mu := x^\mu - \xi^\mu \, d\epsilon \quad \Longrightarrow \quad x^\mu = \tilde{x}^\mu + \xi^\mu \, d\epsilon \tag{4.15}$$

so that $\tilde{x}^\mu(M) = x^\mu(M_0)$. The idea is to know how to formulate the values of any tensor field in terms of the initial coordinates after the dynamical symmetries. To start with, let's consider a 1-form $\omega = \omega_\mu dx^\mu$ decomposed into a coordinate basis. We have the transformation rule of 1-form:

$$\tilde{\omega}_\mu(x) = \frac{\partial x^\alpha}{\partial \tilde{x}^\mu}(x + \xi \, d\epsilon) \, \omega_\alpha (x + \xi \, d\epsilon) \tag{4.16}$$

Introducing the expression of the infinitesimal transformation, we have:

$$\tilde{\omega}_\mu(x) = \left[\delta^\alpha_\mu + \partial_\mu \xi^\alpha(x)\right] \left[\omega_\alpha(x) + \xi^\beta(x) \partial_\beta \omega_\alpha(x)\right] + \mathcal{O}(d\epsilon^2) \tag{4.17}$$

We obtain the relation with the coordinates x^μ:

$$\tilde{\omega}_\mu = \omega_\mu + \mathcal{L}_\xi \omega_\mu, \quad \text{with} \quad \mathcal{L}_\xi \omega_\mu := \xi^\alpha \partial_\alpha \omega_\mu + \omega_\alpha \partial_\mu \xi^\alpha \tag{4.18}$$

where we recognize the Lie derivative of a 1-form. In a similar manner, it is easy to express the dynamical symmetries of the metric:

$$\tilde{g}_{\mu\nu} = \frac{\partial x^\alpha}{\partial \tilde{x}^\mu}(x + \xi \, d\epsilon) \, \frac{\partial x^\beta}{\partial \tilde{x}^\nu}(x + \xi \, d\epsilon) \; g_{\alpha\beta}(x + \xi \, d\epsilon)$$

which leads to the formula, essential for the variational procedure:

$$\tilde{g}_{\alpha\beta} = g_{\alpha\beta} + \mathcal{L}_\xi g_{\alpha\beta} \quad \Longrightarrow \quad \Delta g_{\alpha\beta} := \tilde{g}_{\alpha\beta} - g_{\alpha\beta} = \mathcal{L}_\xi g_{\alpha\beta} \tag{4.19}$$

where we point out the notion of Lie derivative variation for dynamical symmetries. In the same way, after some straightforward calculus, the (dynamical) Lie derivative variations of torsion and curvature follow:

$$\Delta\aleph^{\gamma}_{\alpha\beta} = \mathcal{L}_{\xi}\aleph^{\gamma}_{\mu\nu}, \quad \Delta\Re^{\gamma}_{\alpha\beta\lambda} = \mathcal{L}_{\xi}\Re^{\gamma}_{\mu\nu\lambda} \tag{4.20}$$

Remark 4.3. Dynamical variations from the Lie derivatives (4.19) and (4.20) constitute the key point for elaborating the conservation laws (related to dynamical symmetries) for generalized continuum.

4.2.2.2 *Principle of General Covariance*

To begin with, the first step is to apply the Euler variation along metric $\delta g_{\alpha\beta}$, torsion $\delta\aleph^{\gamma}_{\alpha\beta}$, curvature $\delta\Re^{\gamma}_{\alpha\beta\lambda}$, and volume-form $\delta\omega_n$. Consider the action on the generalized continuum:

$$\mathscr{S} := \int_{\mathscr{B}} \mathscr{L}\left(g_{\alpha\beta}, \aleph^{\gamma}_{\alpha\beta}, \Re^{\gamma}_{\alpha\beta\lambda}\right)\omega_n \tag{4.21}$$

The Euler variation of this action takes the form of:

$$\delta\mathscr{S} := \int_{\mathscr{B}} \delta\mathscr{L}\left(g_{\alpha\beta}, \aleph^{\gamma}_{\alpha\beta}, \Re^{\gamma}_{\alpha\beta\lambda}\right)\omega_n + \int_{\mathscr{B}} \mathscr{L}\left(g_{\alpha\beta}, \aleph^{\gamma}_{\alpha\beta}, \Re^{\gamma}_{\alpha\beta\lambda}\right)\delta\omega_n$$

which we can rewrite in the following form, subject to the required compatibility between the volume-form ω_n and the connection ∇:

$$\delta\mathscr{S} = \int_{\mathscr{B}}\left(\frac{\partial\mathscr{L}}{\partial g_{\alpha\beta}}\,\delta g_{\alpha\beta} + \frac{\partial\mathscr{L}}{\partial\aleph^{\gamma}_{\alpha\beta}}\,\delta\aleph^{\gamma}_{\alpha\beta} + \frac{\partial\mathscr{L}}{\partial\Re^{\gamma}_{\alpha\beta\lambda}}\,\delta\Re^{\gamma}_{\alpha\beta\lambda}\right)\omega_n + \int_{\mathscr{B}}\mathscr{L}\,\delta\omega_n \tag{4.22}$$

Lie derivative variation of volume-form ω_n along vector $\xi(x^{\nu})$ gives:

$$\omega_n := e^{\vartheta}\,\overline{\omega}_n, \quad \mathcal{L}_{\xi}\omega_n \equiv \left(\nabla_{\nu}\xi^{\nu}\right)\omega_n \tag{4.23}$$

where ξ is arbitrary and volume-form ω_n compatible with ∇. The scalar function ϑ depends on the metric and torsion $\vartheta = \vartheta\left(\mathbf{g}, \aleph\right)$. Its Euler variation gives:

$$\delta\omega_n = \delta(e^{\vartheta})\,\overline{\omega}_n + e^{\vartheta}\,\delta\overline{\omega}_n$$

which induces:

$$\delta\omega_n = \left(\frac{\partial\vartheta}{\partial g_{\alpha\beta}} - \frac{g^{\alpha\beta}}{2}\right)\delta g_{\alpha\beta}\,\omega_n + \frac{\partial\vartheta}{\partial\aleph^{\gamma}_{\alpha\beta}}\,\delta\aleph^{\gamma}_{\alpha\beta}\,\omega_n \tag{4.24}$$

Remark 4.4. Another possibility is to express the volume-form in terms of the metric $\mathbf{g}$ and the contortion tensor $\mathfrak{T}$, which directly gives us the variation:

$$\delta\omega_n = \left(\frac{\partial\vartheta}{\partial g_{\alpha\beta}} - \frac{g^{\alpha\beta}}{2}\right)\delta g_{\alpha\beta}\,\omega_n + \frac{\partial\vartheta}{\partial\mathfrak{T}^{\gamma}_{\ \alpha\beta}}\,\delta\mathfrak{T}^{\gamma}_{\ \alpha\beta}\,\omega_n \tag{4.25}$$

which is rigorously equivalent. It should be nevertheless noticed that the variation $\delta\mathfrak{T}^{\gamma}_{\ \alpha\beta}$ is not skew-symmetric with respect to the lower indices.

Hereafter, we will always consider a divergence-free vector field ξ. The Euler variation of the action reduces to:

$$\delta\mathscr{S} = \int_{\mathscr{B}}\left(\sigma^{\alpha\beta}\,\delta g_{\alpha\beta} + \Sigma^{\alpha\beta}_{\gamma}\,\delta\aleph^{\gamma}_{\ \alpha\beta} + \Xi^{\alpha\beta\mu}_{\lambda}\,\delta\mathfrak{R}^{\gamma}_{\ \alpha\beta\lambda}\right)\omega_n \tag{4.26}$$

where the (dual) variables will be defined shortly. We are now applying the *gauge theory.* The Lagrangian is modified when the trajectory is shifted following an active diffeomorphism defined by translation $\xi(x^{\nu})$ on the manifold. For some shifting, the action may be left e.g. [197], and then these (virtual) translations define the symmetries of the problem. We need to find the trajectory that makes the action stationary (extension of the Principle of Least Action) and the resulting necessary (and hopefully sufficient) condition will be the conservation laws e.g. [176]. For that purpose, we focus on the change of the integral action under local translation ξ:

$$\delta_{\xi}\mathscr{S} := \mathscr{S}\left[x^{\mu}(\tau) + \xi^{\mu}[x^{\nu}(\tau)]\;d\lambda\right] - \mathscr{S}\left[x^{\mu}(\tau)\right] = 0 \tag{4.27}$$

where $d\lambda$ is a small parameter and τ the time parameter. Neglecting terms of higher order than $\mathscr{O}(d\lambda)$, this difference gives the Principle of General Covariance e.g. [130, 167]:

$$\delta_{\xi}\mathscr{S} = d\lambda\int_{B}\left(\sigma^{\alpha\beta}\mathcal{L}_{\xi}g_{\alpha\beta} + \Sigma^{\alpha\beta}_{\gamma}\mathcal{L}_{\xi}\aleph^{\gamma}_{\ \alpha\beta} + \Xi^{\alpha\beta\mu}_{\lambda}\mathcal{L}_{\xi}\mathfrak{R}^{\lambda}_{\ \alpha\beta\mu}\right)\omega_n = 0 \tag{4.28}$$

where we disregard the external environment for the sake of simplicity. Changing the Euler variations of the primal variables in (4.26) allows us to obtain directly the Lie derivative variation in (4.28).

4.2.2.3 *Constitutive Laws and Dual Variables*

Constitutive equations are not laws in the sense of fundamental laws of physics, but are mathematical relationships between kinetic and primal variables. The most usual cases in continuum physics are formulated in terms of elastic potential energy. In this section, we could extend the search for dual variables that has already been done, for example, for Einstein tensor in [198], or more recently e.g. [199].

Stress and Hypermomentum. For deriving the expression of the dual variables, we need to apply the Euler variation to the volume-form, given that the function ϑ depends on metric and torsion:

$$\delta\omega_n = \left(\frac{\partial\vartheta}{\partial g_{\alpha\beta}} - \frac{g^{\alpha\beta}}{2}\right)\delta g_{\alpha\beta}\ \omega_n + \frac{\partial\vartheta}{\partial \aleph^{\gamma}_{\alpha\beta}}\ \delta\aleph^{\gamma}_{\alpha\beta}\ \omega_n$$

Now, we define dual variables as being the derivatives of Lagrange $\mathscr{L}$ with respect to primal variables $g_{\alpha\beta}$, $\aleph^{\gamma}_{\alpha\beta}$ and $\Re^{\gamma}_{\alpha\beta\mu}$. As we do not consider external physical sources, we deduce from the internal power:

Hypermomenta. From Euler variations, dual variables are also named generalized Noether currents e.g. [151, 167, 200, 201]:

$$\begin{cases} \sigma^{\alpha\beta} & := \dfrac{\partial\mathscr{L}}{\partial g_{\alpha\beta}} + \mathscr{L}\left(\dfrac{\partial\vartheta}{\partial g_{\alpha\beta}} - \dfrac{g^{\alpha\beta}}{2}\right) \\ \Sigma^{\alpha\beta}_{\gamma} & := \dfrac{\partial\mathscr{L}}{\partial\aleph^{\gamma}_{\alpha\beta}} + \mathscr{L}\dfrac{\partial\vartheta}{\partial\aleph^{\gamma}_{\alpha\beta}} \\ \Xi^{\alpha\beta\mu}_{\gamma} & := \dfrac{\partial\mathscr{L}}{\partial\Re^{\gamma}_{\alpha\beta\mu}} \end{cases} \tag{4.29}$$

Relations (4.29) might not sometimes define effective constitutive laws, analogous to continuum mechanics, but by abuse of language, we can call them as such. In any case, they are called dual variables. In the absence of Lagrangian, the usual method is to postulate directly the three tensors $\sigma^{\alpha\beta}$, $\Sigma^{\alpha\beta}_{\gamma}$, and $\Xi^{\alpha\beta\mu}_{\gamma}$ as functions of the history of primal variables.

Remark 4.5. The variational procedure has three essential steps:

(1) the definition of the Lagrangian (the most subjective part since it directly relates to the physical phenomenon);
(2) the derivation of the hypermomenta, hyperstresses or currents;
(3) the construction of the covariant identities and the conservation laws by means of the stationarity of the action $\mathscr{S}$.

Remark 4.6. In relative gravitation, Hilbert defined the energy-momentum tensor $\sigma^{\alpha\beta}$ as a dual variable expressing a kind of "elastic property" of the spacetime and is similar to the stress tensor derived from hyperelasticity in continuum mechanics e.g. [200]. When applied to a local surface da_β, the stress tensor produces a local traction vector, considered as a local force $df^\alpha := \sigma^{\alpha\beta} da_\beta$. Similarly when the hypermomentum $\Sigma_\gamma^{\alpha\beta}$ is applied to a local line $d\ell^\gamma$, we deduce a "dipole force" as $df^{\alpha\beta} := \Sigma_\gamma^{\alpha\beta} d\ell^\gamma$. Physical interpretation of the polar hypermomentum $\Xi_\gamma^{\lambda\alpha\beta}$ is more complicated.

Some analogy with strain-gradient continuum. The equation (4.28) is analogous to internal power of the Principle of Virtual Power which includes the elastic, the continuum with defects and the gradient of defects e.g. [190]. The internal power is balanced by external power to deduce a somewhat classical procedure the traction condition of Cauchy, the local balance of linear momentum and the angular momentum. In continuum mechanics, the three terms are analogous respectively to stress, micro-stress and polar micro-stress e.g. [202, 203]. To give some details, first, they start with the multiplicative decomposition $\mathbf{F} = \mathbf{F}^e \mathbf{F}^p$ where $\mathbf{F}^p$ is applied on the initial configuration, and $\mathbf{F}^e$ on the intermediate configuration (it is not really a configuration). Then they define the kinematic variables:

$$\dot{\mathbf{E}}_e := \frac{1}{2}\left(\dot{\mathbf{F}}_e^{\mathrm{T}}\, \mathbf{F}_e + \mathbf{F}_e^{\mathrm{T}}\, \dot{\mathbf{F}}_e\right), \quad \mathbf{D}_p := \frac{1}{2}\left(\dot{\mathbf{F}}_p\, \mathbf{F}_p^{-1} + \mathbf{F}_p^{-\mathrm{T}}\, \dot{\mathbf{F}}_p^{\mathrm{T}}\right)$$

and the internal strain power:

$$\delta \mathscr{W}_{int} := \int_{\mathscr{B}_p} \left(\mathbf{T}_e : \dot{\mathbf{E}}_e + \mathbf{T}_p : \mathbf{D}_p + \mathbb{T}_p : \nabla \mathbf{D}_p\right) \omega_{np} \tag{4.30}$$

where $\mathbf{T}_e$, $\mathbf{T}_p$ and $\mathbb{T}_p$ are the dual of primal variables, respectively the elastic stress, the plastic micro-stress, and polar plastic micro-stress. Note that ∇ is here a gradient in the intermediate configuration. In fact, all these variables are defined on the intermediate configuration $\mathscr{B}_p$ whose volume-form is ω_{np}. Observe that the plastic strain rate $\mathbf{D}_p$ applies on the actual configuration by means of embedded base. The two equations (4.28) and (4.30) have some similarities; they have the main structure but differ for many aspects.

Remark 4.7. The analogy that we evoke is not subject to assumptions such as either the well-known multiplicative decomposition of the tetrad $\mathbf{F} = \mathbf{F}^e \mathbf{F}^p$ e.g. [190] — since the defect evolutions are directly defined from differential geometry variables, i.e., torsion and curvature — or additional assumptions such as irrotational plastic flow. Therefore, the present geometric approach is not subject to the problem of indeterminacy of the intermediate configuration. Moreover, the two equations are not applied on the same configuration.

4.2.3 *Lie Derivative of Metric, Torsion, and Curvature*

Without gravitation, spacetime reduces to Minkowski spacetime $\mathscr{M}$ with metric $\hat{g}_{\mu\nu} := \mathrm{diag}\{+1, -1, -1, -1\}$ uniform over all spacetime. The group of transformations: $x^{\mu} \to y^{\mu} = x^{\mu} + \Lambda^{\mu}_{\nu}\, x^{\nu} + \epsilon^{\mu}$, where $\Lambda^{\nu\mu}$ (such that $\Lambda^{\nu\mu} + \Lambda^{\mu\nu} = 0$), and ϵ^{μ} constitute 10 constant parameters and define the isometry group of the spacetime, called global Poincaré group of transformations (Lorentz rotations and translations) e.g. [127, 204]. In the following, we "localize" Poincaré's group in defining a non-uniform and arbitrary vector field $\xi^{\mu}(x^{\nu})$ on the manifold $\mathscr{B}$.

For deriving the field equations and the conservation laws, the Poincaré Gauge Theory is the most used procedure for writing action invariance. Instead of localizing the principle as in e.g. [182], the idea is to follow the spacetime evolution and introduce the notion of Lie derivative. Recall that the accounting of the total Lagrangian variation of any tensor, for instance $\Delta g_{\alpha\beta}$, includes the Eulerian variation and the variation due to the "Lie derivative variation" according to $\Delta g_{\alpha\beta} = \delta g_{\alpha\beta} + \mathcal{L}_{\xi} g_{\alpha\beta}$ [195, 196]. The same principle holds for torsion and curvature. The basic method is to generalize the translational invariance by considering both spatially varying shifts $\xi(\mathbf{x})$ e.g. [178] and coordinate transformations $\tilde{\mathbf{x}}(\mathbf{x})$ that leave the action $\mathscr{S}$ invariant e.g. [194]. For a 2-covariant tensor field on $\mathscr{M}$, we accordingly obtain (by neglecting second-order terms in $d\lambda$):

$$\begin{aligned} \tilde{g}_{\mu\nu}(\mathbf{x}) &= J^{\alpha}_{\mu}(\mathbf{x})\, J^{\beta}_{\nu}(\mathbf{x})\, g_{\alpha\beta}(\mathbf{x} + \xi\, d\lambda) \\ &= g_{\mu\nu}(\mathbf{x}) + [\xi^{\alpha}(\mathbf{x})\partial_{\alpha} g_{\mu\nu}(\mathbf{x}) + g_{\alpha\nu}(\mathbf{x})\partial_{\mu}\xi^{\alpha}(\mathbf{x}) \\ &\quad + g_{\mu\alpha}(\mathbf{x})\partial_{\nu}\xi^{\alpha}(\mathbf{x})]\, d\lambda \end{aligned} \tag{4.31}$$

to give:

$$\begin{aligned} \tilde{g}_{\mu\nu}(\mathbf{x}) = g_{\mu\nu}(\mathbf{x}) + [\xi^{\alpha}(\mathbf{x})\nabla_{\alpha} g_{\mu\nu}(\mathbf{x}) + g_{\alpha\nu}(\mathbf{x})\nabla_{\mu}\xi^{\alpha}(\mathbf{x}) + g_{\mu\alpha}(\mathbf{x})\nabla_{\nu}\xi^{\alpha}(\mathbf{x})]\, d\lambda \\ + \xi^{\alpha}\left(\aleph^{\gamma}_{\nu\alpha}\, g_{\mu\gamma} + \aleph^{\gamma}_{\mu\alpha}\, g_{\gamma\nu}\right) d\lambda = g_{\mu\nu}(\mathbf{x}) + \mathcal{L}_{\xi(\mathbf{x})} g_{\mu\nu}(\mathbf{x}) d\lambda \end{aligned} \tag{4.32}$$

where the Lie derivative of metric holds for compatible connections:

$$\mathcal{L}_{\xi} g_{\mu\nu} := g_{\alpha\nu}\nabla_{\mu}\xi^{\alpha} + g_{\mu\alpha}\nabla_{\nu}\xi^{\alpha} + \xi^{\alpha}\left(\aleph^{\gamma}_{\alpha\nu}\, g_{\mu\gamma} + \aleph^{\gamma}_{\alpha\mu}\, g_{\gamma\nu}\right) \tag{4.33}$$

It shows that the infinitesimal active transformations are generated by Lie derivatives, including the torsion effects e.g. [5]. It is pointed out that

$\mathcal{L}_{\xi(\mathbf{x})} g_{\mu\nu}(\mathbf{x}) \neq \xi^\alpha(\mathbf{x}) \nabla_\alpha g_{\mu\nu}(\mathbf{x}) = \nabla_{\xi(\mathbf{x})} g_{\mu\nu}(\mathbf{x})$. First, originally the Lie derivative needs no connection, second the vector $\xi(\mathbf{x})$ are not differentiated with respect to the covariant derivative, and third the torsion has influence apart for the Lie derivative. We recall that $\nabla_\gamma g_{\alpha\beta} \equiv 0$ on $\mathscr{B}$. For metric-compatible and torsion-free connection, the Lie derivative (4.33) reduces to e.g. [153]:

$$\mathcal{L}_{\xi(\mathbf{x})} g_{\mu\nu}(\mathbf{x}) = \nabla_\mu \xi_\nu(\mathbf{x}) + \nabla_\nu \xi_\mu(\mathbf{x}) \tag{4.34}$$

Active diffeomorphism invariance of the metric induces the vanishing of this tensor whose solutions ξ are the Killing vector fields e.g. [5].

Remark 4.8. For the two conditions of torsion absence and metric compatibility, the Lie derivative $\mathcal{L}_{\xi(\mathbf{x})} g_{\mu\nu}(\mathbf{x})$ is equivalent to the variation $\delta g_{\alpha\beta}$ along the vector field $\xi^\gamma(\mathbf{x})$. This is not the case when one of these two conditions is not satisfied.

If the previous transformation is considered as a coordinate system change, i.e., passive diffeomorphism $x^\mu = \tilde{x}^\mu + \xi^\mu(x^\alpha)\, d\lambda$ (re-labelling of coordinates), the transformation rule for 2-covariant tensor holds (by retaining first-order terms in $d\lambda$):

$$\begin{aligned}\tilde{g}_{\mu\nu}(\tilde{x}^\gamma) &= J_\mu^\alpha(x^\gamma)\, J_\nu^\beta(x^\gamma)\, g_{\alpha\beta}(x^\gamma)\\ &= g_{\mu\nu}(x^\gamma) + \left[g_{\alpha\nu}(x^\gamma)\, \partial_\mu \xi^\alpha(x^\gamma) + g_{\mu\alpha}(x^\gamma)\, \partial_\nu \xi^\alpha(x^\gamma)\right] d\lambda\end{aligned} \tag{4.35}$$

where $J_\mu^\alpha(x^\gamma) = \delta_\mu^\alpha + \partial_\mu \xi^\alpha(x^\gamma)\, d\lambda$ e.g. [205]. We observe that the right-hand-side of equations (4.31) and (4.35) look like each other for metric connection. With respect to a (local) normal coordinate system, the two equations are exactly the same, with however the difference that the first expresses the change of 2-covariant field by the active diffeomorphism whereas the second expresses the metric 2-covariant tensor in terms of new coordinates.

Remark 4.9. Lie derivative starts with transformations $J_\mu^\alpha(\mathbf{x}) := \delta_\mu^\alpha + \partial_\mu \xi^\alpha(\mathbf{x})\, d\lambda$, where λ is introduced to express smallness of amplitude. For non-flat manifolds, use covariant derivative instead of partial derivative:

$$J_\mu^\alpha(\mathbf{x}) := \delta_\mu^\alpha + \nabla_\mu \xi^\alpha(\mathbf{x}) d\lambda = \delta_\mu^\alpha + \partial_\mu \xi^\alpha(\mathbf{x}) d\lambda - \overline{\Gamma}_{\mu\gamma}^\alpha \xi^\gamma d\lambda - \mathcal{T}_{\mu\gamma}^\alpha \xi^\gamma d\lambda \tag{4.36}$$

where contortion $\mathcal{T}_{\mu\gamma}^\alpha$ is an additional gauge field. In [183], the skew-symmetric part of the contortion tensor is replaced by a rotation angle.

The set of all gauge parameters allows us to highlight what are called Hamiltonian gauge symmetries. However, it was shown that Poincaré (see hereafter) and Hamiltonian gauge symmetries are equivalent, modulo the trivial gauge transformations e.g. [183].

Lie derivative along a vector field is a way of differentiating tensors on a continuum e.g. [197]. It is an advantageous method because it is defined on $\mathscr{B}$ without any reference to metric $g_{\alpha\beta}$ or to extra structures such as a connection $\Gamma^{\lambda}_{\alpha\beta}$. Let's start with the definition of Lie derivative and apply to $(0,2)$ tensor $\mathbf{g}$, $(1,2)$ tensor $\aleph$, and $(1,3)$ tensor $\mathfrak{R}$ e.g. [110]:

$$\mathcal{L}_{\xi}\, g_{\alpha\beta} = \xi^{\gamma}\, \partial_{\gamma} g_{\alpha\beta} + g_{\gamma\beta}\, \partial_{\alpha}\xi^{\gamma} + g_{\alpha\gamma}\, \partial_{\beta}\xi^{\gamma} \tag{4.37}$$

$$\mathcal{L}_{\xi}\, \aleph^{\lambda}_{\alpha\beta} = \xi^{\gamma}\, \partial_{\gamma}\aleph^{\lambda}_{\alpha\beta} - \aleph^{\gamma}_{\alpha\beta}\, \partial_{\gamma}\xi^{\lambda} + \aleph^{\lambda}_{\gamma\beta}\, \partial_{\alpha}\xi^{\gamma} + \aleph^{\lambda}_{\alpha\gamma}\, \partial_{\beta}\xi^{\gamma} \tag{4.38}$$

$$\begin{aligned}\mathcal{L}_{\xi}\, \mathfrak{R}^{\lambda}_{\alpha\beta\mu} &= \xi^{\gamma}\partial_{\gamma}\, \mathfrak{R}^{\lambda}_{\alpha\beta\mu} - \mathfrak{R}^{\gamma}_{\alpha\beta\mu}\, \partial_{\gamma}\xi^{\lambda} + \mathfrak{R}^{\lambda}_{\gamma\beta\mu}\, \partial_{\alpha}\xi^{\gamma} \\ &\quad + \mathfrak{R}^{\lambda}_{\alpha\gamma\mu}\, \partial_{\beta}\xi^{\gamma} + \mathfrak{R}^{\lambda}_{\alpha\beta\gamma}\, \partial_{\mu}\xi^{\gamma}\end{aligned} \tag{4.39}$$

Field $\xi^{\alpha}(x^{\mu})$ embodies Poincaré's local group of transformations, and permits us to define the change following a tangent motion along a flow. For an affinely connected manifold, this permits us to consider the variation up to "second derivatives".

Lie derivative of metric, torsion, and curvature. We obtain the relations for the metric:

$$\begin{aligned}\mathcal{L}_{\xi}\, g_{\alpha\beta} &= \xi^{\gamma}\, \nabla_{\gamma} g_{\alpha\beta} + g_{\gamma\beta}\, \nabla_{\alpha}\xi^{\gamma} + g_{\alpha\gamma}\, \nabla_{\beta}\xi^{\gamma} \\ &\quad + \xi^{\gamma}\left(g_{\alpha\nu}\, \aleph^{\nu}_{\gamma\beta} + g_{\nu\beta}\, \aleph^{\nu}_{\gamma\alpha}\right),\end{aligned} \tag{4.40}$$

for torsion:

$$\begin{aligned}\mathcal{L}_{\xi}\, \aleph^{\lambda}_{\alpha\beta} &= \xi^{\gamma}\, \nabla_{\gamma}\aleph^{\lambda}_{\alpha\beta} - \aleph^{\gamma}_{\alpha\beta}\, \nabla_{\gamma}\xi^{\lambda} + \aleph^{\lambda}_{\gamma\beta}\, \nabla_{\alpha}\xi^{\gamma} + \aleph^{\lambda}_{\alpha\gamma}\, \nabla_{\beta}\xi^{\gamma} \\ &\quad - \xi^{\gamma}\, \aleph^{\lambda}_{\gamma\nu}\, \aleph^{\nu}_{\alpha\beta} + \xi^{\gamma}\, \aleph^{\nu}_{\gamma\alpha}\, \aleph^{\lambda}_{\nu\beta} + \xi^{\gamma}\, \aleph^{\nu}_{\gamma\beta}\, \aleph^{\lambda}_{\alpha\nu},\end{aligned} \tag{4.41}$$

and for curvature:

$$\begin{aligned}\mathcal{L}_{\xi}\, \mathfrak{R}^{\lambda}_{\alpha\beta\mu} &= \xi^{\gamma}\nabla_{\gamma}\mathfrak{R}^{\lambda}_{\alpha\beta\mu} - \mathfrak{R}^{\gamma}_{\alpha\beta\mu}\nabla_{\gamma}\xi^{\lambda} + \mathfrak{R}^{\lambda}_{\gamma\beta\mu}\nabla_{\alpha}\xi^{\gamma} + \mathfrak{R}^{\lambda}_{\alpha\gamma\mu}\nabla_{\beta}\xi^{\gamma} \\ &\quad + \mathfrak{R}^{\lambda}_{\alpha\beta\gamma}\nabla_{\mu}\xi^{\gamma} - \xi^{\gamma}\, \mathfrak{R}^{\nu}_{\alpha\beta\mu}\, \aleph^{\lambda}_{\gamma\nu} + \xi^{\gamma}\, \mathfrak{R}^{\lambda}_{\nu\beta\mu}\, \aleph^{\nu}_{\gamma\alpha} \\ &\quad + \xi^{\gamma}\, \mathfrak{R}^{\lambda}_{\alpha\nu\mu}\, \aleph^{\nu}_{\gamma\beta} + \xi^{\gamma}\, \mathfrak{R}^{\lambda}_{\alpha\beta\nu}\, \aleph^{\nu}_{\gamma\mu}\end{aligned} \tag{4.42}$$

Lie derivative of a tensor is also a tensor. For the proof, it is sufficient to rewrite it in terms of covariant derivatives (2.26) and torsion tensors.

Relations (4.40), (4.41), and (4.42) are the expressions of Lie derivatives for generalized continuum (Riemann-Cartan manifold). Connection ∇ appearing in these relationships includes torsion and cannot be easily utilized in practice to solve the conservation laws in continuum physics and theory of gravitation. Sometimes, these expressions should be modified to separate the Levi-Civita connection and the contortion tensor $\mathcal{T}^{\gamma}_{\alpha\beta}$, which becomes an unknown variable.

Remark 4.10. From relations (4.40), (4.41), and (4.42), it is confirmed that Lie derivatives differ from covariant derivatives for any tensor field. First, the original definition does not require connection ∇ since it is expressed merely with partial derivatives. Second, a Lie derivative involves derivatives of the vector field ξ while the covariant derivative does not. A Lie derivative expresses the metric, torsion and curvature changes as these tensors are pushed along the integral curves of vector field ξ. The directional derivatives tell us how these tensors are fixed at a point change as one moves through them in direction ξ.

4.3 Gauge Invariance and Conservation Laws

Major advancements of physics theory in the 20th century lie on the use of Noether's theorems. The first theorem associates global symmetries with conserved quantities such as linear momentum with translational symmetry, angular momentum with rotational symmetry, and energy with time translation. The second concerns local symmetries leading to the foundations of most gauge theories. Continuum mechanics as well as relative gravitation constitute classical fields of applications of gauge theory e.g. [206]. Generalizing formulation of conservation laws for gravitation with torsion is not so evident. See in e.g. [207] and particularly in presence of torsion without minimal coupling e.g. [208] which we will express by means of a variational procedure. We extend this way of reasoning to generalized continua. Our basic observation is that both spacetime of relative gravitation and generalized continua have the same geometrical roots: the Riemann-Cartan manifold. Indeed, it constitutes an appropriate manifold with compatible connection whose local symmetries are those of the local Poincaré group.

Remark 4.11. There is a difference between field equations and conservation laws. Field equations are similar to Euler-Lagrange equations of continuum deduced from an Eulerian variation, whereas conservation laws are obtained from Noether's theorems by using the Lie derivative variation.

4.3.1 *Lagrangian of the Form $\mathscr{L}(g_{\alpha\beta})$*

Classically, thermomechanics of macroscopic continuum $\mathscr{B}$ is described by a metric or by the triads/tetrads, $\mathbf{F}$, a stress tensor σ defined by means of Cauchy's theorem, Helmholtz free energy ψ, body force $\rho\mathbf{b}$ (ρ is the density), entropy s, and volume heating per unit mass r (see for example e.g. [209] for more details).

4.3.1.1 *Cauchy Approach for Conservation Laws*

In such a framework, most if not all previously published works classically used the Levi-Civita connection $\overline{\nabla}$ (and even an Euclidean connection) for calculating gradient, divergence and rotational operators of tensor fields on the continuum $\mathscr{B}$ e.g. [153], although some remarks on the use of affine connection with torsion and curvature were given in [141].

Euler-Cauchy conservation laws. By using the Levi-Civita and Newton's laws, as particularized by Euler — where the total forces acting on the continuum body are expressed as volume and surface integrals — the conservation of linear momentum takes the classical localized form e.g. [133, 141, 210]:

$$\overline{\nabla}_\alpha \sigma^\alpha_\gamma = 0 \tag{4.43}$$

We limit in this equation to statics without inertial forces and without external body force for the sake of simplicity.

This equation may be projected onto the concomitant basis to recover the continuum equations undergoing large deformations [11]. Some remarks should be made here. Stress tensor σ in (4.43) is defined by using Cauchy's theorem by first assuming (Cauchy's postulate) that a stress vector (or traction vector) $\mathbf{p}_n$ locally acting on a surface depends on the unit outward normal $\mathbf{n}$ of the surface but not for instance on the surface local curvature e.g. [210]. Second, it is shown that there exists a stress tensor σ linearly depending on normal $\mathbf{n}$ such as $\mathbf{p}_n = \sigma(\mathbf{n})$. This fundamental theorem, attributed to Cauchy, induces the linear momentum equation (4.43), and is commonly used and is nowadays applied to all models of continuum independent of material constitutive laws and independent of the amount and nature of deformations e.g. [133, 153]. This might be questionable.

4.3.1.2 *Conservation Laws*

Consider a continuum which may have internal defects but they remain unchanged with the deformation, acting merely as non-evolutive parameters [191]. The Lagrangian depends only on $g_{\alpha\beta}$ as explicit primal variables, and the action:

$$\mathscr{L} = \mathscr{L}\left(g_{\alpha\beta}\right), \quad \mathscr{S} := \int_{\mathscr{B}} \mathscr{L}\,\omega_n \tag{4.44}$$

where the metric is implicitly projected onto a base embedded in the continuum. However, for Riemann-Cartan manifold, it may model some elastic continua with a possible previous change of internal structure since the torsion $\aleph^{\gamma}_{\alpha\beta}$ is implicitly an argument as part of the connection $\Gamma^{\gamma}_{\alpha\beta}$. Application of the Euler variations on the action gives:

$$\delta\mathscr{S} = \delta\int_{\mathscr{B}} \mathscr{L}\,\omega_n = \int_{\mathscr{B}} \delta\left(\mathscr{L}\omega_n\right) = \int_{\mathscr{B}} \sigma^{\alpha\beta}\,\delta g_{\alpha\beta}\,\omega_n \tag{4.45}$$

where we define the stress/energy momentum:

$$\sigma^{\alpha\beta} := \frac{\partial\mathscr{L}}{\partial g_{\alpha\beta}} + \mathscr{L}\left(\frac{\partial\vartheta}{\partial g_{\alpha\beta}} - \frac{g^{\alpha\beta}}{2}\right) \tag{4.46}$$

This stress is obtained since the model assumes that $\delta\aleph^{\gamma}_{\alpha\beta} \equiv 0$ and $\delta\mathfrak{R}^{\gamma}_{\alpha\beta\lambda} \equiv 0$.

Theorem 4.1. *Let a generalized continuum* $(\mathscr{B}, \mathbf{g}, \nabla)$ *with a Lagrangian function* $\mathscr{L}(g_{\alpha\beta})$, *assumed to explicitly depend only on metric as a primal variable. We assume a metric-compatible connection. If the Lagrangian is gauge-invariant under the vector field* $\xi^{\mu}(x^{\alpha})$ *then the following conservation law holds:*

$$\boxed{\nabla_{\alpha}\sigma^{\alpha}_{\gamma} = 0} \tag{4.47}$$

where the stress σ *given by* (4.29) *defines the constitutive laws of the continuum.*

Remark 4.12. Torsion and curvature fields do not necessarily vanish on the continuum $(\mathscr{B}, \mathbf{g}, \nabla)$. They may be present but they do not influence explicitly the Lagrangian $\mathscr{L}$. The Lagrangian has in most cases the triads/tetrads as arguments $\mathscr{L}(\mathbf{f}_{\alpha}), \alpha = 1, 3$ or 4. By applying the Cauchy-Weyl theorem, it was shown that the Lagrangian is covariant if and only if there exists a form of the Lagrangian such that $\mathscr{L} = \tilde{\mathscr{L}}\left(\mathbf{f}_{\alpha}\cdot\mathbf{f}_{\beta}, \omega_n\left(\mathbf{f}_1, \ldots, \mathbf{f}_n\right)\right)$ e.g. [49].

Proof. The connection is metric-compatible. We proceed in two steps.

(1) The Lagrangian $\mathscr{L}(g_{\alpha\beta})$ depends only on metric. It is is gauge-invariant if for any vector field $\xi^{\mu}(x^{\nu})$, the action $\mathscr{S}$ remains unchanged. Then, we can define the constitutive laws (relating "stress" in function of "strain") and write:

$$\sigma^{\alpha\beta}\ \mathcal{L}_{\xi}g_{\alpha\beta} = \sigma^{\alpha\beta}\left[\xi^{\gamma}\nabla_{\gamma}g_{\alpha\beta} + g_{\gamma\beta}\nabla_{\alpha}\xi^{\gamma} + g_{\alpha\gamma}\nabla_{\beta}\xi^{\gamma}\right.$$
$$\left.+\ \xi^{\gamma}(g_{\alpha\nu}\aleph^{\nu}_{\gamma\beta} + g_{\nu\beta}\aleph^{\nu}_{\gamma\alpha})\right]$$

By construction, the stress is a symmetric tensor in the present case.

(2) We obtain a first group of terms including $\nabla\xi$:

$$\sigma^{\alpha\beta}\left(g_{\gamma\beta}\nabla_{\alpha}\xi^{\gamma} + g_{\alpha\gamma}\nabla_{\beta}\xi^{\gamma}\right) = 2\sigma^{\alpha}_{\gamma}\nabla_{\alpha}\xi^{\gamma} = 2\nabla_{\alpha}\left(\sigma^{\alpha}_{\gamma}\xi^{\gamma}\right) - \xi^{\gamma}\nabla_{\alpha}\sigma^{\alpha}_{\gamma}$$

and the second group proportional to ξ:

$$\xi^{\gamma}\ \sigma^{\alpha\beta}\left[\nabla_{\gamma}g_{\alpha\beta} + \left(g_{\alpha\nu}\ \aleph^{\nu}_{\gamma\beta} + g_{\nu\beta}\aleph^{\nu}_{\gamma\alpha}\right)\right] = 2\xi^{\gamma}\sigma^{\alpha}_{\nu}\aleph^{\nu}_{\gamma\alpha}$$

(3) Integrating over the continuum, we may write:

$$\int_{\mathscr{B}} 2\left(\nabla_{\alpha}\sigma^{\alpha}_{\gamma} + \aleph^{\nu}_{\alpha\gamma}\sigma^{\alpha}_{\nu}\right)\xi^{\gamma}\omega_{n} + \int_{\mathscr{B}} 2\nabla_{\alpha}\left(\sigma^{\alpha}_{\gamma}\xi^{\gamma}\right)\omega_{n} \equiv 0$$

(4) First, this variation equation should be verified for an arbitrary field ξ. By eliminating the divergence term which may be shifted at the boundary and choosing the gauge to vanish at the boundary $\partial\mathscr{B}$ and ξ uniform inside, we arrive at the relation:

$$\nabla_{\alpha}\sigma^{\alpha}_{\gamma} + \aleph^{\nu}_{\alpha\gamma}\sigma^{\alpha}_{\nu} = 0$$

(5) Second, we consider a uniform gauge field $\nabla\xi \equiv 0$. The second group of terms holds $\sigma^{\alpha}_{\nu}\ \aleph^{\nu}_{\gamma\alpha} \equiv 0$ for metric-compatible connections, which constitutes a particular case of Klein-Noether identities linking the stress and the density of dislocations. We deduce the conservation laws (4.47).

□

Conservation laws hold as div $\sigma = 0$ in coordinate-free form e.g. [153], and here in a covariant form. The symmetry of σ is *a priori* verified from its definition.

Remark 4.13. Apparently, conservation laws (4.43) and (4.47) have the same form. Their difference comes from the implicit presence of torsion (or accordingly the contortion) in the connection. The contortion tensor is a source of space fading of waves within the elastic continuum defined by the Lagrangian $\mathscr{L}(g_{\alpha\beta})$ (assumed to contain frozen dislocations and disclinations) e.g. [169, 211]. As a reminder, the general shape of connection coefficients may be decomposed as: $\Gamma^{\gamma}_{\alpha\beta} = \overline{\Gamma}^{\gamma}_{\alpha\beta} + D^{\gamma}_{\alpha\beta} + \Omega^{\gamma}_{\alpha\beta}$ with the Levi-Civita connection $\overline{\Gamma}^{\gamma}_{\alpha\beta}$, the skew-symmetric $\Omega^{\gamma}_{\alpha\beta} := (1/2)\aleph^{\gamma}_{\alpha\beta}$, and its symmetric part $D^{\gamma}_{\alpha\beta} := g^{\gamma\lambda} g_{\alpha\mu} \Omega^{\mu}_{\lambda\beta} + g^{\gamma\lambda} g_{\mu\beta} \Omega^{\mu}_{\lambda\alpha}$. They define the contortion tensor $\mathcal{T}^{\gamma}_{\alpha\beta} := \Omega^{\gamma}_{\alpha\beta} + D^{\gamma}_{\alpha\beta}$. Continuum $\mathscr{B}$ is a Riemann-Cartan manifold but the Lagrangian is assumed here to depend explicitly only on metric. The contortion tensor becomes a source of space fading of waves within the strain gradient elastic matter e.g. [169, 211]. Although the action involves only the metric tensor, we notice the implicit presence of torsion "force" in the conservation laws. Besides conservation laws, some conserved quantities or differential identities may exist according to the choice of the gauge field $\xi(\mathbf{x})$ e.g. [167], called Klein-Noether system of identities.

4.3.1.3 *Extended Navier Equation in Generalized Continuum*

We consider a generalized continuum $\mathscr{B}$ with the Helmholtz free energy $\mathscr{F}(g_{\alpha\beta})$ where the strain tensor $\varepsilon_{\alpha\beta}$ is calculated as the symmetrized part of $\nabla_{\alpha} u_{\beta}$ for material strain. We assume Lagrangian explicitly depends neither on the curvature nor on the torsion. For writing the conservation laws (4.47) the divergence of the stress tensor holds:

$$\nabla_{\beta} \sigma^{\alpha\beta} = (1/2) \nabla_{\beta} \left[\mathbb{E}^{\alpha\beta\mu\nu} \left(\nabla_{\mu} u_{\nu} + \nabla_{\nu} u_{\mu} \right) \right] = \mathbb{E}^{\alpha\beta\mu\nu} \nabla_{\beta} \nabla_{\mu} u_{\nu} \tag{4.48}$$

given the symmetry $\mathbb{E}^{\alpha\beta\mu\nu} = \mathbb{E}^{\alpha\beta\nu\mu}$ and the metric compatibility of the connection. Let us recall some properties of the torsion tensor (2.33) and the curvature (2.36) tensor in the framework of Riemann-Cartan geometry, for any scalar field Φ and vector field $\mathbf{u}$ on the manifold $\mathscr{B}$:

$$\begin{cases} -\aleph^{\gamma}_{\beta\mu} \nabla_{\gamma} \Phi = \nabla_{\beta} \nabla_{\mu} \Phi - \nabla_{\mu} \nabla_{\beta} \Phi \\ \Re^{\nu}_{\beta\mu\gamma} u^{\gamma} = \nabla_{\beta} \nabla_{\mu} u^{\nu} - \nabla_{\mu} \nabla_{\beta} u^{\nu} + \aleph^{\gamma}_{\beta\mu} \nabla_{\gamma} u^{\nu} \end{cases} \tag{4.49}$$

As example, we consider the case of linear isotropic elastic continuous medium for which the tangent tensor takes the form of:

$$\mathbb{E}^{\alpha\beta\mu\nu} := \lambda \, \hat{g}^{\alpha\beta} \hat{g}^{\mu\nu} + \mu \left(\hat{g}^{\alpha\mu} \hat{g}^{\beta\nu} + \hat{g}^{\alpha\nu} \hat{g}^{\beta\mu} \right) \tag{4.50}$$

Introduction of isotropic tangent tensor in (4.48) and accounting for (4.49) allows us to extend the Navier equation to give [211]:

$$\boxed{(\lambda + \mu) \nabla^{\alpha} \nabla_{\beta} u^{\beta} + \mu \nabla^{\beta} \nabla_{\beta} u^{\alpha} + \mu \hat{g}^{\alpha\mu} \left(\Re_{\mu\nu} \, u^{\gamma} - \aleph^{\gamma}_{\beta\mu} \nabla_{\gamma} u^{\beta} \right) = \rho \partial_t^2 u^{\alpha}} \tag{4.51}$$

where $\mathfrak{R}_{\mu\nu} := \mathfrak{R}^{\gamma}_{\gamma\mu\nu}$ is the Ricci curvature which entirely defines the curvature tensor in a three-dimensional space. Ricci curvature is related to the local change of volume due to the incompatible deformation.

Remark 4.14. Equation (4.51) includes the influence of torsion and curvature. Some aspects of the influence of torsion and curvature on the linear momentum conservation has been investigated in e.g. [212] in a rather general context. The form of the equation suggested similarity with the gravitation-spin theory.

We recover the elastic wave equation for non-curved and zero torsion continuum. The first two terms in (4.51) are analogous to the classical Navier equation. The third term is the contribution of torsion and curvature and they are only related to shear modulus μ although their interaction appears for all three displacement components. In the last two terms of the left hand side, the curvature part $\mu g^{\alpha\mu}\mathfrak{R}_{\mu\nu}\, u^{\gamma}$ induces an auto-oscillation of the continuum whereas the torsion part $\mu g^{\alpha\mu}\aleph^{\gamma}_{\beta\mu}\nabla_{\gamma}u^{\beta}$ engenders a space fading of wave.

Remark 4.15. Conversely to wave equations, it is difficult to separate the dilatational wave Φ and the shear wave $\mathbf{H}$ since they are highly coupled (4.51). The role of Ricci curvature in the Navier-Stokes equation in Riemann compact manifolds was already suggested in fluids e.g. [214]. In the conservation laws (4.51), torsion is involved for either continuum elastoplasticity or fluids with vortex e.g. [156].

4.3.2 *Einstein-Cartan Spacetime*

We design a class of spacetimes with evolving distribution of torsion and curvature fields. We do not consider point defects that need the introduction of connection non-metricity, say $\nabla\mathbf{g} \equiv 0$. Such a spacetime model is assumed to be sourceless of any physical action (vacuum spacetime).

4.3.2.1 *Constitutive Laws*

Consider the Lagrangian modeling a class of rate-independent generalized continuum. We then suggest a continuum $\mathscr{B}$ whose Lagrangian is defined by:

$$\mathscr{L} := \mathscr{L}(g_{\alpha\beta}, \aleph^{\gamma}_{\alpha\beta}, \mathfrak{R}^{\gamma}_{\alpha\beta\lambda}) \tag{4.52}$$

where metric, torsion, and curvature of the continuum are projected onto an embedded base. The action is defined accordingly:

$$\mathscr{S} := \int_{\mathscr{B}} \mathscr{L}(g_{\alpha\beta}, \aleph^{\gamma}_{\alpha\beta}, \Re^{\gamma}_{\alpha\beta\lambda})\ \omega_n \tag{4.53}$$

where the volume-form is that of a Riemann-Cartan manifold $\omega_n := e^{\vartheta(\mathbf{g},\aleph)}\ \overline{\omega}_n$. As a reminder, the action and its variation are given by:

$$\mathscr{S} = \int_{\mathscr{B}} \mathscr{L}\ \omega_n \quad \delta\mathscr{S} = \int_{\mathscr{B}} (\delta\mathscr{L})\,\omega_n + \int_{\mathscr{B}} \mathscr{L}\ \delta\omega_n \tag{4.54}$$

As per previously, Euler variation $\delta\mathscr{S}$ allows us to derive dual hypermomenta corresponding to each primal variable e.g. [200] or more precisely from (4.29) since we have to impose the compatibility of ω_n with $\mathbf{g}$:

$$\begin{cases} \sigma^{\alpha\beta} & := \dfrac{\partial\mathscr{L}}{\partial g_{\alpha\beta}} + \mathscr{L}\left(\dfrac{\partial\vartheta}{\partial g_{\alpha\beta}} - \dfrac{g^{\alpha\beta}}{2}\right) \\ \Sigma^{\alpha\beta}_{\gamma} & := \dfrac{\partial\mathscr{L}}{\partial\aleph^{\gamma}_{\alpha\beta}} + \mathscr{L}\dfrac{\partial\vartheta}{\partial\aleph^{\gamma}_{\alpha\beta}} \\ \Xi^{\alpha\beta\mu}_{\gamma} & := \dfrac{\partial\mathscr{L}}{\partial\Re^{\gamma}_{\alpha\beta\mu}} \end{cases}$$

In elastoplastic continuum theory, these quantities are called respectively stress (symmetric), micro-stress and polar micro-stress (the latter two are skew-symmetric with respect to indices (α, β)).

4.3.2.2 *Conservation Laws*

For deriving the conservation laws, in the spirit of Klein-Noether methods, we again consider the change of the action integral following Lie variations of the primal variables along ξ e.g. [129, 130]:

$$\delta_\xi\mathscr{S} = d\lambda \int_{x^0=\tau_i}^{x^0=\tau_f} \int_B \left(\sigma^{\alpha\beta}\mathcal{L}_\xi g_{\alpha\beta} + \Sigma^{\alpha\beta}_{\gamma}\mathcal{L}_\xi\aleph^{\gamma}_{\alpha\beta} + \Xi^{\alpha\beta\mu}_{\lambda}\mathcal{L}_\xi\Re^{\lambda}_{\alpha\beta\mu}\right)\omega_n \tag{4.55}$$

in which we have the relations respectively for metric:

$$\mathcal{L}_\xi\ g_{\alpha\beta} = \xi^\gamma\ \nabla_\gamma g_{\alpha\beta} + g_{\gamma\beta}\ \nabla_\alpha\xi^\gamma + g_{\alpha\gamma}\ \nabla_\beta\xi^\gamma + \xi^\gamma\left(g_{\alpha\nu}\ \aleph^{\nu}_{\gamma\beta} + g_{\nu\beta}\ \aleph^{\nu}_{\gamma\alpha}\right),$$

for torsion:

$$\begin{aligned}\mathcal{L}_\xi\ \aleph^{\lambda}_{\alpha\beta} = {} & \xi^\gamma\ \nabla_\gamma\aleph^{\lambda}_{\alpha\beta} \\ & - \aleph^{\gamma}_{\alpha\beta}\ \nabla_\gamma\xi^\lambda + \aleph^{\lambda}_{\gamma\beta}\ \nabla_\alpha\xi^\gamma + \aleph^{\lambda}_{\alpha\gamma}\ \nabla_\beta\xi^\gamma \\ & - \xi^\gamma\ \aleph^{\lambda}_{\gamma\nu}\ \aleph^{\nu}_{\alpha\beta} + \xi^\gamma\ \aleph^{\nu}_{\gamma\alpha}\ \aleph^{\lambda}_{\nu\beta} + \xi^\gamma\ \aleph^{\nu}_{\gamma\beta}\ \aleph^{\lambda}_{\alpha\nu},\end{aligned}$$

and for curvature:

$$\begin{aligned}\mathcal{L}_\xi \, \Re^\lambda_{\alpha\beta\mu} &= \xi^\gamma \nabla_\gamma \Re^\lambda_{\alpha\beta\mu} \\ &\quad - \Re^\gamma_{\alpha\beta\mu} \nabla_\gamma \xi^\lambda + \Re^\lambda_{\gamma\beta\mu} \nabla_\alpha \xi^\gamma + \Re^\lambda_{\alpha\gamma\mu} \nabla_\beta \xi^\gamma + \Re^\lambda_{\alpha\beta\gamma} \nabla_\mu \xi^\gamma \\ &\quad - \xi^\gamma \, \Re^\nu_{\alpha\beta\mu} \, \aleph^\lambda_{\gamma\nu} \\ &\quad + \xi^\gamma \, \Re^\lambda_{\nu\beta\mu} \, \aleph^\nu_{\gamma\alpha} + \xi^\gamma \, \Re^\lambda_{\alpha\nu\mu} \, \aleph^\nu_{\gamma\beta} + \xi^\gamma \, \Re^\lambda_{\alpha\beta\nu} \, \aleph^\nu_{\gamma\mu}\end{aligned}$$

Remark 4.16. The presence of both torsion and curvature in (4.55) could be considered as a similar action variation in elastoplastic models including the plastic spin; see e.g. [190] for a recent review on different models of strain gradient plasticity. Needless to say, the concept of generalized continuum model that we construct here can be applied both to vacuum spacetime and to any materials with translational and rotational dislocation defects.

Theorem 4.2. *Let a generalized continuum* $(\mathscr{B}, g_{\alpha\beta}, \Gamma^\gamma_\alpha)$ *be in motion with respect to an inertial reference with a Lagrangian* $\mathscr{L}(g_{\alpha\beta}, \aleph^\gamma_{\alpha\beta}, \Re^\gamma_{\alpha\beta\mu})$. *Then, the conservation law holds*:

$$\nabla_\alpha \tilde{\sigma}^\alpha_\gamma = 0 \tag{4.56}$$

with the ***generalized stress*** $\tilde{\sigma}^\alpha_\gamma$*:*

$$\begin{aligned}2\tilde{\sigma}^\alpha_\gamma &:= 2 \, \sigma^\alpha_\gamma + 2 \, \Sigma^{\alpha\beta}_\lambda \aleph^\lambda_{\gamma\beta} - \Sigma^{\lambda\beta}_\gamma \aleph^\alpha_{\lambda\beta} + 2 \, \Xi^{\alpha\beta\mu}_\lambda \Re^\lambda_{\gamma\beta\mu} \\ &\quad + \Xi^{\mu\beta\alpha}_\lambda \Re^\lambda_{\mu\beta\gamma} - \Xi^{\lambda\beta\mu}_\gamma \Re^\alpha_{\lambda\beta\mu}\end{aligned} \tag{4.57}$$

where hypermomenta are given by constitutive relations (4.29).

Proof. Consider the variation (4.55) and let's separate the terms associated with ξ and its covariant derivative $\nabla\xi$.

(1) For terms proportional to ξ^γ, we have:

$$\begin{aligned}0 &= \sigma^{\alpha\beta} \left(g_{\alpha\nu} \, \aleph^\nu_{\gamma\beta} + g_{\nu\beta} \, \aleph^\nu_{\gamma\alpha} \right) \\ &\quad + \Sigma^{\alpha\beta}_\lambda \left(\nabla_\gamma \aleph^\lambda_{\alpha\beta} - \aleph^\lambda_{\gamma\nu} \, \aleph^\nu_{\alpha\beta} + \aleph^\nu_{\gamma\alpha} \, \aleph^\lambda_{\nu\beta} + \aleph^\nu_{\gamma\beta} \, \aleph^\lambda_{\alpha\nu} \right) \\ &\quad + \Xi^{\alpha\beta\mu}_\lambda \left(\nabla_\gamma \Re^\lambda_{\alpha\beta\mu} - \Re^\nu_{\alpha\beta\mu} \, \aleph^\lambda_{\gamma\nu} + \Re^\lambda_{\nu\beta\mu} \, \aleph^\nu_{\gamma\alpha} + \Re^\lambda_{\alpha\nu\mu} \, \aleph^\nu_{\gamma\beta} \Re^\lambda_{\alpha\beta\nu} \, \aleph^\nu_{\gamma\mu} \right)\end{aligned}$$

These relations constitute the Klein-Noether identities for this continuum. They also allow us to prove more easily the theorem as for the previous case.

(2) For terms proportional to the gradient $\nabla\xi$, we write:

$$
\begin{aligned}
0 &= \sigma^{\alpha\beta}\left(g_{\gamma\beta}\ \nabla_\alpha\xi^\gamma + g_{\alpha\gamma}\ \nabla_\beta\xi^\gamma\right) \\
&\quad + \Sigma_\lambda^{\alpha\beta}\left(-\aleph^\gamma_{\alpha\beta}\ \nabla_\gamma\xi^\lambda + \aleph^\lambda_{\gamma\beta}\ \nabla_\alpha\xi^\gamma + \aleph^\lambda_{\alpha\gamma}\ \nabla_\beta\xi^\gamma\right) \\
&\quad + \Xi_\lambda^{\alpha\beta\mu}\left(-\Re^\gamma_{\alpha\beta\mu}\nabla_\gamma\xi^\lambda + \Re^\lambda_{\gamma\beta\mu}\nabla_\alpha\xi^\gamma + \Re^\lambda_{\alpha\gamma\mu}\nabla_\beta\xi^\gamma + \Re^\lambda_{\alpha\beta\gamma}\nabla_\mu\xi^\gamma\right)
\end{aligned}
$$

Factorization of the gradient and integration by parts lead to:

$$
\begin{aligned}
0 = \int &\Big[\nabla_\beta\left(2\ \sigma^\beta_\gamma\right)\xi^\gamma \\
&+ \nabla_\beta\left(2\ \Sigma_\lambda^{\alpha\beta}\aleph^\lambda_{\alpha\gamma} - \Sigma_\gamma^{\alpha\lambda}\aleph^\beta_{\alpha\lambda}\right)\xi^\gamma \\
&+ \nabla_\beta\left(2\ \Xi_\lambda^{\alpha\beta\mu}\Re^\lambda_{\alpha\gamma\mu} + \Xi_\lambda^{\alpha\mu\beta}\Re^\lambda_{\alpha\mu\gamma} - \Xi_\gamma^{\alpha\mu\lambda}\Re^\beta_{\alpha\mu\lambda}\right)\xi^\gamma\Big]\ \omega_n + \text{B.T.}
\end{aligned}
$$

From this integral, by using Lemma of Haar, we deduce the theorem, by arranging the dummy indices in the summation (Einstein convention).

□

Remark 4.17. Considering defect density as a primal variable and not internal variable as in plasticity theory may nevertheless imply analogous formulation of conservation laws deduced from the *Principle of Virtual Power* in e.g. [190, 202, 203] whenever micro-stress and polar stress are accounted for. Indeed, the physical phenomenon behind plastic deformation may be modeled by torsion and curvature as the variables capturing the internal changes such as dislocation density, grain transformations and relative motions. This may be helpful for experimental investigations e.g. [215].

Remark 4.18. Conservation laws (4.56) are similar to those derived in Chapter 18 on covariant conservation laws of the book [216]. Laws similar to (4.56) with generalized stress (4.57) were obtained in e.g. [217] by considering contortion tensor $\mathfrak{T}^\gamma_{\alpha\beta}$ instead of the torsion. Conservation laws are slightly different from (4.56) if contortion and its derivatives are used instead of torsion and curvature.

Conservation laws (4.56) are more general than (4.47), with the latter recovered by introducing zero hypermomenta within the expression of the generalized stress (4.57). Conservation laws (4.56) and constitutive laws (4.29) involve torsion and curvature fields on the manifold $\mathscr{B}$ immersed within a Galilean spacetime. At least theoretically, it is necessary to propose a "gedanken experiment" background to measure the torsion field e.g. [156, 218]. This is a main motivation to develop the model of generalized continuum moving within a non-zero torsion curved spacetime. Some attempts have been also made to suggest theories supporting the existence, and even the necessity, to adopt a Riemann-Cartan manifold to mimic the rotational flows of fluids with vorticity e.g. [49, 156].

Remark 4.19. Generalized stress (4.57) shows that it includes a kind of "reversible" stress, as for hyperelastic solid, and two other groups of stresses depending on torsion (and its dual) and curvature (and its dual) which respectively capture the density of dislocations and disclinations within the continuum. Additional stresses are "irreversible" stress due to defects and incompatibilities.

Remark 4.20. An alternative approach would be to consider the contortions as primal variables. It is worth splitting the connection with unique decomposition $\Gamma^{\gamma}_{\alpha\beta} = \overline{\Gamma}^{\gamma}_{\alpha\beta} + \mathcal{T}^{\gamma}_{\alpha\beta}$ where $\mathcal{T}^{\gamma}_{\alpha\beta}$ is the contortion tensor. Contortion tensor is sometimes called *disformation tensor*. The idea behind the use of contortion is to separate the influence of metric change and the torsion modification. We would like to derive the conservation laws thanks to Poincaré's invariance e.g. [6, 182]. This approach considers the following Lagrangians: $\mathscr{L}(g_{\alpha\beta}, \mathcal{T}^{\gamma}_{\alpha\beta})$ and $\mathscr{L}(\varepsilon_{\alpha\beta}, \mathcal{T}^{\gamma}_{\alpha\beta}, \overline{\nabla}_{\lambda}\mathcal{T}^{\gamma}_{\alpha\beta})$. An analogous method, although yet to be developed, could be used to derive the Klein-Noether identities and the associated conservation laws.

Remark 4.21. First, contortion $\mathcal{T}^{\gamma}_{\alpha\beta}$ is a tensor and therefore its intrinsic properties do not depend on the coordinate system x^{μ}. The conservation laws from this approach are expected to be covariant. Implicitly, the metric tensor $g_{\alpha\beta}$ captures a holonomic deformation whereas contortion $\mathcal{T}^{\gamma}_{\alpha\beta}$ captures the non-holonomic part of the deformation. In principle, additional conservation laws should be added due to the increase of unknowns.

(1) Similar equations were obtained for Newtonian gravitation in e.g. [129], for relative gravitation with spin e.g. [217], and for relative gravitation e.g. [167].
(2) The generalized stress tensor $\tilde{\sigma}^{\alpha}_{\gamma}$ extends the classical Cauchy symmetric stress σ^{α}_{γ}. It includes three contributions: the "standard" stress σ^{α}_{γ} due to strain, the torsion-stress $2\ \Sigma^{\alpha\beta}_{\lambda}\aleph^{\lambda}_{\gamma\beta} - \Sigma^{\lambda\beta}_{\gamma}\aleph^{\alpha}_{\lambda\beta}$ due to change of grain structure and dislocation density due to plastic deformation, and the micro-stress provoked by the density of disclinations

$2\ \Xi^{\alpha\beta\mu}_{\lambda}\Re^{\lambda}_{\gamma\beta\mu} + \Xi^{\mu\beta\alpha}_{\lambda}\Re^{\lambda}_{\mu\beta\gamma} - \Xi^{\lambda\beta\mu}_{\gamma}\Re^{\alpha}_{\lambda\beta\mu}$. All of them are obtained by deduction. The second is composed of the product of the micro-stress and the torsion whereas the third is given by the product of the polar micro-stress and the curvature. Generalized stress is not necessarily symmetric. The occurring of terms including torsion and its dual is not new since these contributions also exist for Einstein-Cartan gravitation and play a role in cosmology to avoid model singularities e.g. [4].

4.3.3 *Three-Dimensional Generalized Continuum*

First, symmetries of Rieman-Cartan curvature make the number of independent components in a n-dimensional space $n^2\left(n^2-1\right)/12$. Riemann-Cartan curvature on a 3-manifold is completely determined by its Ricci curvature $\mathcal{R}_{\alpha\beta} := \Re^{\lambda}_{\lambda\alpha\beta}$. Second, skew symmetry of torsion with respect to the lower indices allows us to separate the torsion into three 2-forms (Cartan's forms):

$$\aleph^{\gamma}_{\alpha\beta} \quad \rightarrow \quad \left\{\overset{(1)}{\omega}_{\alpha\beta}, \overset{(2)}{\omega}_{\alpha\beta}, \overset{(3)}{\omega}_{\alpha\beta}\right\} \tag{4.58}$$

Consider local deformed base $(\mathbf{f}_1, \mathbf{f}_2, \mathbf{f}_3))$.

Definition 4.1. A three-dimensional generalized continuum $\mathscr{B}$ is defined by the Lagrangian $\mathscr{L} := \mathscr{L}(g_{\alpha\beta}, \overset{(n)}{\omega}_{\alpha\beta}, \mathcal{R}_{\alpha\beta})$, $n = 1, 2, 3$, where $g_{\alpha\beta}$, $\overset{(n)}{\omega}_{\alpha\beta}$, and $\Re_{\alpha\beta}$ are respectively the metric, Cartan's forms, and the Ricci curvature.

The pounding factor of the volume-form compatible with the connection is suggested to take the form of $\vartheta = \vartheta(g_{\alpha\beta}, \overset{(n)}{\omega}_{\alpha\beta})$. We can write:

Theorem 4.3. *Let a three-dimensional continuum $(\mathscr{B}, g_{\alpha\beta}, \Gamma^{\gamma}_{\alpha})$ be in motion within an inertial reference with the Lagrangian $\mathscr{L}(g_{\alpha\beta}, \overset{(n)}{\omega}_{\alpha\beta}, \mathcal{R}_{\alpha\beta})$, $n = 1, 2, 3$. Then, the conservation law holds:*

$$\boxed{\nabla_{\alpha}\tilde{\sigma}^{\alpha}_{\gamma} = 0} \tag{4.59}$$

where we define the ***generalized stress*** $\tilde{\sigma}^{\alpha}_{\gamma}$:

$$\boxed{2\tilde{\sigma}^{\alpha}_{\gamma} := 2\ \sigma^{\alpha}_{\gamma} + 2\ \Sigma^{\alpha\beta}_{(n)}\overset{(n)}{\omega}_{\gamma\beta} + 2\ \Xi^{\alpha\beta}\mathcal{R}_{\gamma\beta}} \tag{4.60}$$

where hypermomenta are given by constitutive relations $n = 1, 2, 3$:

$$\begin{cases} \sigma^{\alpha\beta} := \dfrac{\partial \mathscr{L}}{\partial g_{\alpha\beta}} + \mathscr{L}\left(\dfrac{\partial \vartheta}{\partial g_{\alpha\beta}} - \dfrac{g^{\alpha\beta}}{2}\right) \\ \Sigma^{\alpha\beta}_{(n)} := \dfrac{\partial \mathscr{L}}{\partial \omega^{(n)}_{\alpha\beta}} + \mathscr{L}\,\dfrac{\partial \vartheta}{\partial \omega^{(n)}_{\alpha\beta}} \\ \Xi^{\alpha\beta} := \dfrac{\partial \mathscr{L}}{\partial \mathcal{R}_{\alpha\beta}} \end{cases} \tag{4.61}$$

Remark 4.22. $\sigma^{\alpha\beta}$ is symmetric whereas $\Sigma^{\alpha\beta}_{(n)}$ is skew-symmetric. The Ricci variable $\Xi^{\alpha\beta}$ dual to the curvature is not necessarily symmetric.

Proof. We use the same method as previously by expressing the Lie derivative of primal variables:

$$\mathcal{L}_\xi g_{\alpha\beta} = \xi^\gamma\left(\nabla_\gamma g_{\alpha\beta} + g_{\alpha\nu}\ \aleph^\nu_{\gamma\beta} + g_{\nu\beta}\ \aleph^\nu_{\gamma\alpha}\right) + \left(g_{\gamma\beta}\ \nabla_\alpha \xi^\gamma + g_{\alpha\gamma}\ \nabla_\beta \xi^\gamma\right)$$

$$\mathcal{L}_\xi \omega^{(n)}_{\alpha\beta} = \xi^\gamma\left(\nabla_\gamma \omega^{(n)}_{\alpha\beta} + \omega^{(n)}_{\nu\beta}\ \aleph^\nu_{\gamma\alpha} + \omega^{(n)}_{\alpha\nu}\ \aleph^\nu_{\gamma\beta}\right) + \left(\omega^{(n)}_{\gamma\beta}\nabla_\alpha \xi^\gamma + \omega^{(n)}_{\alpha\gamma}\nabla_\beta \xi^\gamma\right)$$

$$\mathcal{L}_\xi \mathcal{R}_{\alpha\beta} = \xi^\gamma\left(\nabla_\gamma \mathcal{R}_{\alpha\beta} + \mathcal{R}_{\nu\beta}\ \aleph^\nu_{\gamma\alpha} + \mathcal{R}_{\alpha\nu}\ \aleph^\nu_{\gamma\beta}\right) + \left(\mathcal{R}_{\gamma\beta}\nabla_\alpha \xi^\gamma + \mathcal{R}_{\alpha\gamma}\nabla_\beta \xi^\gamma\right)$$

(1) The three Klein-Noether identities take the following form for $\gamma = 1, 2, 3$:

$$\begin{aligned} 0 = {} & \sigma^{\alpha\beta}\left(\nabla_\gamma g_{\alpha\beta} + g_{\alpha\nu}\ \aleph^\nu_{\gamma\beta} + g_{\nu\beta}\ \aleph^\nu_{\gamma\alpha}\right) \\ & + \sum_{n=1}^{n=3}\left[\Sigma^{\alpha\beta}_{(n)}\left(\nabla_\gamma \omega^{(n)}_{\alpha\beta} + \omega^{(n)}_{\nu\beta}\ \aleph^\nu_{\gamma\alpha} + \omega^{(n)}_{\alpha\nu}\ \aleph^\nu_{\gamma\beta}\right)\right] \\ & + \Xi^{\alpha\beta}\left(\nabla_\gamma \mathcal{R}_{\alpha\beta} + \mathcal{R}_{\nu\beta}\ \aleph^\nu_{\gamma\alpha} + \mathcal{R}_{\alpha\nu}\ \aleph^\nu_{\gamma\beta}\right) \end{aligned}$$

(2) By accounting for these identities, we deduce the following equation by integrating by parts and choosing the gauge $\xi \equiv 0$ at the boundary $\partial\mathscr{B}$:

$$\int_{\mathscr{B}} \xi^\gamma \nabla_\alpha\left[2\ \sigma^\alpha_\gamma + 2\ \Sigma^{\alpha\beta}_{(n)}\omega^{(n)}_{\gamma\beta} + \left(\Xi^{\alpha\beta}\mathcal{R}_{\gamma\beta} + \Xi^{\beta\alpha}\mathcal{R}_{\beta\gamma}\right)\right]\omega_n = 0$$

By using the Lemma of Haar, we obtain the conservation laws (4.59).

□

Remark 4.23. Under different names, similar forms of generalized stress (4.60) were initially obtained in relative gravitation and in the case where spin occurs e.g. [217], where they are called hypermomentum, or total energy stress e.g. [129]. They are also increasingly considered also for strain gradient plasticity and are called micro-stress and polar micro-stress e.g. [202, 203].

Remark 4.24. As expected, the generalized elastic-plastic stress $2\tilde{\sigma}^{\alpha}_{\gamma}$ includes an "reversible" part σ^{α}_{γ} obtained as the derivative of the Lagrangian with respect to the metric strain, the contribution of the skew-symmetric forms $\omega^{(n)}_{\alpha\beta}$ (corresponding to a plastic spin, due to density of translational dislocations) and its dual $\Sigma^{\alpha\beta}_{(n)}$, and the contribution of the Ricci curvature and its dual $\Xi^{\alpha\beta}$ (due to the strain incompatibilities, associated with rotational dislocations/disclinations). Once again, the equation div $\sigma = 0$ holds, provided the stress tensor σ is replaced by its extended version accounting for incompatible deformations $\tilde{\sigma}$. Connection has a non-vanishing torsion, which makes its application more difficult in practice.

Remark 4.25. As for the general approach, we may write the total pressure:

$$\tilde{p} := \frac{1}{3}\,\mathrm{Tr}(\tilde{\sigma}) = \frac{1}{3}\,\mathrm{Tr}(\sigma) + \frac{1}{3}\sum_{n=1}^{n=3}\langle \Sigma_{(n)}, \omega^{(n)}\rangle + \frac{1}{3}\,\langle \Xi, \mathcal{R}\rangle \tag{4.62}$$

where the last two terms are like stored residual pressure due to the internal defects at the scale level of strain gradient plasticity. Such a finding might be helpful when studying pressure wave within plastically deformed material.

Remark 4.26. Similar models where metric, torsion, and curvature were considered in the past as arguments of the Lagrangian were developed by Noll in e.g. [191], where they were considered as "frozen" incompatibilities. Indeed, differential evolution laws of these variables were missing. Previous studies have proposed the use of the internal variables method to derive evolution laws of $\aleph^{\lambda}_{\alpha\beta}$ and $\Re^{\lambda}_{\alpha\beta\gamma}$. For this purpose, we refer the readers to a previous book [11]. Although it is not the purpose of the present work, the relation between Lagrangian and dissipation potential is not so evident. Considering a more general situation, the possibility of using Lagrange formalism to model continuum with dissipation remains a field of interest e.g. [219].

4.4 Generalized Continuum: Alternative Approach

When directly using the covariant derivative with ∇, equations depend implicitly on torsion and are not easily tractable in practice. We have introduced in the chapter devoted to the geometric bases the change of the connection ∇ during the deformation of the continuum $\mathscr{B}$ by means of the contortion tensor $\mathfrak{T}$.

4.4.1 *Gauge Invariance and Conservation Laws*

Assuming an initial state free from any defect, deformation includes the change of metric components $g_{\alpha\beta}$ and connection coefficients $\Gamma^{\gamma}_{\alpha\beta}$ on the embedded base $\mathbf{u}_{\alpha}, \alpha = 1, \ldots n$ deforming with the continuum $\mathscr{B}$. It may then be judicious to model the connection modification by the two contributions: that due to metric $g_{\alpha\beta}$ and that due to contortion $\mathfrak{T}^{\gamma}_{\alpha\beta}$. The contribution of the metric includes two aspects:

(1) the change of components $g_{\alpha\beta}$ (simple material framework);
(2) and eventually the implication on the Levi-Civita connection $\overline{\Gamma}^{\gamma}_{\alpha\beta}[\mathbf{g}, \partial\mathbf{g}]$ which can change the Riemann curvature $\overline{\mathfrak{R}}^{\gamma}_{\alpha\beta\lambda}$ (second gradient framework, but not first gradient).

The contortion also has two types of contributions vanishing when $\aleph = 0$:

(1) the symmetric part $D^{\gamma}_{\alpha\beta}$, corresponding to the first gradient plastic strain;
(2) the skew-symmetric part $\Omega^{\gamma}_{\alpha\beta}$, which is a first gradient spin.

We obtain a modified shape of the Lagrangian. We consider a Lagrangian function with a list of arguments: $\mathscr{L} := \mathscr{L}(g_{\alpha\beta}, \mathfrak{T}^{\gamma}_{\alpha\beta}, \overline{\nabla}_{\lambda}\mathfrak{T}^{\gamma}_{\alpha\beta})$ where:

$$\begin{cases} \mathfrak{T}^{\gamma}_{\alpha\beta} := \Gamma^{\gamma}_{\alpha\beta} - \overline{\Gamma}^{\gamma}_{\alpha\beta} \\ \mathfrak{K}^{\gamma}_{\alpha\beta\lambda} := \overline{\nabla}_{\alpha}\mathfrak{T}^{\gamma}_{\lambda\beta} - \overline{\nabla}_{\beta}\mathfrak{T}^{\gamma}_{\lambda\alpha} + \left(\mathfrak{T}^{\gamma}_{\beta\mu}\mathfrak{T}^{\mu}_{\alpha\lambda} - \mathfrak{T}^{\gamma}_{\alpha\mu}\mathfrak{T}^{\mu}_{\beta\lambda}\right) \end{cases}$$

Remark 4.27. This avoids the use of ∇ by using only $\overline{\nabla}$.

Application of Euler variations in the variational procedure gives exactly the same shape as previously. The Principle of General Covariance also takes the same form as previously. For any local translation $\xi(\mathbf{x})$, we should have:

$$\delta(\mathscr{L}\omega_n) = \left[\sigma^{\alpha\beta}\,\mathcal{L}_{\xi}g_{\alpha\beta} + \Sigma^{\alpha\beta}_{\gamma}\,\mathcal{L}_{\xi}\mathfrak{T}^{\gamma}_{\alpha\beta} + \Xi^{\lambda\alpha\beta}_{\gamma}\,\mathcal{L}_{\xi}\overline{\nabla}_{\lambda}\mathfrak{T}^{\gamma}_{\alpha\beta}\right]\omega_n = 0$$

Indeed, we have already mentioned before the volume-form variation as a function of metric and contortion:

$$\delta\omega_n = \left(\frac{\partial\vartheta}{\partial g_{\alpha\beta}} - \frac{g^{\alpha\beta}}{2}\right)\delta g_{\alpha\beta}\,\omega_n + \frac{\partial\vartheta}{\partial\mathfrak{T}^{\gamma}_{\alpha\beta}}\,\delta\mathfrak{T}^{\gamma}_{\alpha\beta}\,\omega_n \tag{4.63}$$

In the same way, we calculate the hypermomenta/stress, micro-stress, and polar stress from directional derivatives giving the dual variables:

$$\left\{\begin{aligned} \sigma^{\alpha\beta} &:= \frac{\partial \mathscr{L}}{\partial g_{\alpha\beta}} + \mathscr{L}\left(\frac{\partial \vartheta}{\partial g_{\alpha\beta}} - \frac{g^{\alpha\beta}}{2}\right) \\ \Sigma_{\gamma}^{\alpha\beta} &:= \frac{\partial \mathscr{L}}{\partial \mathfrak{T}_{\alpha\beta}^{\gamma}} + \mathscr{L}\,\frac{\partial \vartheta}{\partial \mathfrak{T}_{\alpha\beta}^{\gamma}} \\ \Xi_{\gamma}^{\lambda\alpha\beta} &:= \frac{\partial \mathscr{L}}{\partial \overline{\nabla}_{\lambda} \mathfrak{T}_{\alpha\beta}^{\gamma}} \end{aligned}\right. \tag{4.64}$$

To avoid a plethora of notations, we keep the same symbol for the polar hypermomentum. It seems interesting to highlight a correspondence with classical plasticity theory. The internal virtual power is given in e.g. [203], $\mathbf{F} := \mathbf{F}^e \mathbf{F}^p$:

$$\delta \mathscr{W}_i := \int_{\mathscr{B}} \mathbf{S}^e : \dot{\mathbf{F}}^e + \mathbf{T}^p : \mathbf{D}^p + \mathbb{T}^p : \nabla \mathbf{D}^p, \quad \mathbf{D}^p := \dot{\mathbf{F}}^p\, \mathbf{F}^{p-1}$$

with respectively the stress, the micro-stress, and the polar micro-stress with the conditions: $\mathbf{D}^p = \mathbf{D}^{p\mathrm{T}}$, $\mathrm{Tr}\mathbf{D}^p \equiv 0$.

Remark 4.28. To be covariant the two variables $\mathbf{D}^p$ and $\nabla \mathbf{D}^p$ should be merged into one variable $\dot{\mathbf{G}}$ where $\mathbf{G} := (\mathrm{Det}\mathbf{F}^p)^{-1}\mathbf{F}^p\ \overline{\mathrm{Rot}}\mathbf{F}^p$ in a study devoted to the Geometrically Necessary Dislocations in a continuum framework [220].

4.4.2 *Weitzenböck Continuum*

Let $\mathscr{B}$ a Weitzenböck continuum where the primal variables are metric and torsion, or contortion. Let us consider a continuum defined by its Lagrangian:

$$\mathscr{L} := \mathscr{L}(g_{\alpha\beta}, \mathfrak{T}_{\alpha\beta}^{\gamma}), \quad \mathscr{S} := \int_{\mathscr{M}} \mathscr{L}\, \omega_n \tag{4.65}$$

where the contortion tensor is independent of the metric. Application of the Euler variations on the Lagrangian (4.65) gives:

$$\delta\left(\mathscr{L}\, \omega_n\right) = \left(\sigma^{\alpha\beta}\ \delta g_{\alpha\beta} + \Sigma_{\gamma}^{\alpha\beta}\ \delta \mathfrak{T}_{\alpha\beta}^{\gamma}\right) \omega_n \tag{4.66}$$

In this relation, the constitutive laws give the two variables respectively dual to the metric and to the contortion:

$$\sigma^{\alpha\beta} := \frac{\partial \mathscr{L}}{\partial g_{\alpha\beta}} + \mathscr{L}\left(\frac{\partial \vartheta}{\partial g_{\alpha\beta}} - \frac{g^{\alpha\beta}}{2}\right), \quad \Sigma_{\gamma}^{\alpha\beta} := \frac{\partial \mathscr{L}}{\partial \mathfrak{T}_{\alpha\beta}^{\gamma}} + \mathscr{L}\,\frac{\partial \vartheta}{\partial \mathfrak{T}_{\alpha\beta}^{\gamma}} \tag{4.67}$$

The pounding factor of the volume-form depends on the metric and the contortion tensor $\vartheta = \vartheta(g_{\alpha\beta}, \mathcal{T}^{\gamma}_{\alpha\beta})$. In relative gravitation with spin and torsion, these dual variables $\sigma^{\alpha\beta}$ and $\Sigma^{\alpha\beta}_{\gamma}$ are respectively called energy-momentum and hypermomentum e.g. [217].

4.4.2.1 *Conservation Laws*

Application of the Principle of General Covariance which deals with the Lie derivative variations (active diffeomorphism) provides the conservation laws.

Theorem 4.4. *Consider a generalized continuum* $(\mathscr{B}, \mathbf{g}, \nabla)$ *with a Lagrangian* $\mathscr{L}(g_{\alpha\beta}, \mathcal{T}^{\gamma}_{\alpha\beta})$*. Then:*

$$\boxed{\overline{\nabla}_{\beta}\tilde{\sigma}^{\beta}_{\rho} + \mathcal{T}^{\alpha}_{\alpha\beta}\tilde{\sigma}^{\beta}_{\rho} = 0} \tag{4.68}$$

in which we define the generalized stress $\tilde{\sigma}^{\beta}_{\rho}$ *(4.67):*

$$\boxed{\tilde{\sigma}^{\beta}_{\rho} := \sigma^{\beta}_{\rho} + \frac{1}{2}\left(-\Sigma^{\alpha\gamma}_{\rho}\, \mathcal{T}^{\beta}_{\alpha\gamma} + \Sigma^{\beta\alpha}_{\gamma}\, \mathcal{T}^{\gamma}_{\rho\alpha} + \Sigma^{\alpha\beta}_{\gamma}\, \mathcal{T}^{\gamma}_{\alpha\rho}\right)} \tag{4.69}$$

Proof. 1. *Invariance of Poincaré.*

(1) Express the gauge invariance as a Lie derivative variation in (4.66):

$$\delta_{\xi}\left(\mathscr{L}\omega_n\right) = \left(\sigma^{\alpha\beta}\, \mathcal{L}_{\xi}g_{\alpha\beta} + \Sigma^{\alpha\beta}_{\gamma}\mathcal{L}_{\xi}\mathcal{T}^{\gamma}_{\alpha\beta}\right)\omega_n$$

as active diffeomorphism. The integrand should vanish.

(2) Compute the Lie derivatives of metric and contortion (tedious!)

$$\begin{cases} \mathcal{L}_{\xi}g_{\alpha\beta} = \overline{\nabla}_{\alpha}\xi_{\beta} + \overline{\nabla}_{\beta}\xi_{\alpha} + g_{\alpha\gamma}\mathcal{T}^{\gamma}_{\rho\beta}\xi^{\rho} + g_{\gamma\beta}\mathcal{T}^{\gamma}_{\rho\alpha}\xi^{\rho} \\ \mathcal{L}_{\xi}\mathcal{T}^{\gamma}_{\alpha\beta} = \xi^{\rho}\overline{\nabla}_{\rho}\mathcal{T}^{\gamma}_{\alpha\beta} - \mathcal{T}^{\rho}_{\alpha\beta}\overline{\nabla}_{\rho}\xi^{\gamma} + \mathcal{T}^{\gamma}_{\rho\beta}\overline{\nabla}_{\alpha}\xi^{\rho} + \mathcal{T}^{\gamma}_{\alpha\rho}\overline{\nabla}_{\beta}\xi^{\rho} \end{cases}$$

(3) We write the integrand (terms in brackets) $\delta_{\xi}\left(\mathscr{L}\omega_n\right) = 0$ and obtain:

$$\begin{aligned} &\sigma^{\alpha\beta}\left(\overline{\nabla}_{\alpha}\xi_{\beta} + \overline{\nabla}_{\beta}\xi_{\alpha}\right) + \sigma^{\alpha\beta}\left(g_{\alpha\gamma}\mathcal{T}^{\gamma}_{\rho\beta} + g_{\gamma\beta}\mathcal{T}^{\gamma}_{\rho\alpha}\right)\xi^{\rho} \\ &+ \left(\Sigma^{\alpha\beta}_{\gamma}\overline{\nabla}_{\rho}\mathcal{T}^{\gamma}_{\alpha\beta}\right)\xi^{\rho} + \Sigma^{\alpha\beta}_{\gamma}\left(-\mathcal{T}^{\rho}_{\alpha\beta}\overline{\nabla}_{\rho}\xi^{\gamma} + \mathcal{T}^{\gamma}_{\rho\beta}\overline{\nabla}_{\alpha}\xi^{\rho} + \mathcal{T}^{\gamma}_{\alpha\rho}\overline{\nabla}_{\beta}\xi^{\rho}\right) = 0 \end{aligned}$$

It should vanish for all possible diffeomorphisms (as for the Principle of Virtual Power when dealing with usual continuum mechanics e.g. [133]).

2. *Global Invariance.* By considering uniform field of gauge, we easily deduce an **identity**:

$$\sigma^{\alpha\beta}\left(g_{\alpha\nu}\mathcal{T}^{\nu}_{\rho\beta}+g_{\nu\beta\nu}\mathcal{T}^{\nu}_{\rho\alpha}\right)+\Sigma^{\alpha\beta}_{\gamma}\overline{\nabla}_{\rho}\mathcal{T}^{\gamma}_{\alpha\beta}=0$$

3. *Local Invariance.* We consider vector $\xi(\mathbf{x})$, factorize by $\overline{\nabla}_{\beta}\xi^{\rho}$, integrate by parts, drop all terms in divergence and obtain:

$$\int_{\mathcal{M}}\left(\tilde{\sigma}^{\beta}_{\rho}\nabla_{\beta}\xi^{\rho}\right)\omega_n\int_{\mathcal{M}}\left[\overline{\nabla}_{\beta}\left(\tilde{\sigma}^{\beta}_{\rho}\,\xi^{\rho}\right)-\xi^{\rho}\overline{\nabla}_{\beta}\tilde{\sigma}^{\beta}_{\rho}\right]\omega_n=0$$

where we have defined the generalized stress:

$$2\tilde{\sigma}^{\beta}_{\rho}:=2\,\sigma^{\beta}_{\rho}-\Sigma^{\alpha\gamma}_{\rho}\,\mathcal{T}^{\beta}_{\alpha\gamma}+\Sigma^{\beta\alpha}_{\gamma}\,\mathcal{T}^{\gamma}_{\rho\alpha}+\Sigma^{\alpha\beta}_{\gamma}\,\mathcal{T}^{\gamma}_{\alpha\rho}\tag{4.70}$$

Divergence theorem cannot be directly applied since the Levi-Civita connection is not compatible with the volume-form. It is worth introducing the contortion tensor $\nabla:=\overline{\nabla}+\mathcal{T}$ in such a way that the connection ∇ is compatible with the volume-form ω_n. We thus obtain:

$$\int_{\mathcal{M}}\left[\nabla_{\beta}\left(\tilde{\sigma}^{\beta}_{\rho}\xi^{\rho}\right)-\left(\overline{\nabla}_{\beta}\tilde{\sigma}^{\beta}_{\rho}+\mathcal{T}^{\alpha}_{\alpha\beta}\tilde{\sigma}^{\beta}_{\rho}\right)\xi^{\rho}\right]\omega_n=0\tag{4.71}$$

□

Remark 4.29. Other forms of conservation laws and identities may exist depending on the choice of gauge field $\xi(\mathbf{x})$ e.g. [167] (uniform, linear, nonlinear). Other remarks:

(1) Similar equations were obtained in the past for relative gravitation e.g. [4, 130, 167], and in Newton-Cartan Gravitation e.g. [129].
(2) As for the approach with torsion and curvature, the generalized stress tensor $\tilde{\sigma}^{\beta}_{\rho}$ extends the classical Cauchy symmetric stress σ^{β}_{ρ}.
(3) The generalized stress includes two contributions: "macro" σ^{α}_{ρ} and "micro" $\Sigma^{\alpha\beta}_{\gamma}\mathcal{T}^{\gamma}_{\rho\beta}$ due to change of microcosm structure and dislocation density from plastic deformation. Generalized stress is not necessarily symmetric.

Remark 4.30. The hypermomentum/hyperstress $\Sigma^{\alpha\beta}_{\gamma}$ may be decomposed into three contributions [200]:

$$\Sigma^{\alpha\beta}_{\gamma}=\Sigma^{[\alpha,\beta]}_{\gamma}+\frac{\delta^{\alpha}_{\gamma}}{n}\,\overline{\Sigma}^{\beta}+\overline{\Sigma}^{(\alpha,\beta)}_{\gamma}$$

(1) Spin-angular momentum/rotatory hyperstress

$$\Sigma^{[\alpha,\beta]}_{\gamma}\quad\text{skew-symmetric part}$$

(2) Proper hypermomentum/proper hyperstress

$$\Sigma_\gamma^{(\alpha,\beta)} \quad \text{symmetric part} \quad \text{and} \quad \Sigma_\alpha^{(\alpha,\beta)} = \overline{\Sigma}^\beta \quad \text{dilatational}$$

(3) Traceless proper hypermomentum/proper hyperstress

$$\overline{\Sigma}_\gamma^{(\alpha,\beta)} := \Sigma_\gamma^{(\alpha,\beta)} - \frac{\delta_\gamma^\alpha}{n}\,\overline{\Sigma}^\beta$$

Remark 4.31. The contortion tensor can be uniquely decomposed into a symmetric part and a skew-symmetric part as:

$$\mathcal{T}_{\alpha\beta}^\gamma = D_{\alpha\beta}^\gamma + \Omega_{\alpha\beta}^\gamma$$

where $D_{\alpha\beta}^\gamma$ is linear and homogeneous in terms of $\aleph$ and $\Omega_{\alpha\beta}^\gamma$ is half of the torsion tensor. An alternative approach would consider the decomposition:

$$\Sigma_\gamma^{[\alpha\beta]} := \frac{\partial \mathscr{L}}{\partial \Omega_{\alpha\beta}^\gamma}, \quad \Sigma_\gamma^{(\alpha\beta)} := \frac{\partial \mathscr{L}}{\partial D_{\alpha\beta}^\gamma}$$

but we retain the approach (4.67) for the sake of simplicity.

Remark 4.32. Stress σ is due to change of length and of angle between two material fibers, and is conventionally considered as elastic stress (it is not the usual convention). Hyperstresses Σ and Ξ express the modifications at the level of microstructures and contribute to the creation of heterogeneity of local stresses by re-arrangements of microcosms.

4.4.2.2 *Principle of Maximum Dissipation: Defect Evolution Rule*

Even if we work essentially on spacetime models of gravitation and electromagnetism, it is not uninteresting to make the analogy of spacetime made up of microcosms with an elastoplastic medium which is also composed of pieces of grains of matter that can move relative to each other (by translation and rotational dislocations). In this subsection, the analogy concerns the nucleation and development of relative motions of grains beyond some yield values of interactions such as in irreversible deformations e.g. [190]. Then, in the expressions of energy momenta $\sigma^{\alpha\beta}$ and $\Sigma_\gamma^{\alpha\beta}$, we have to introduce the values of the metric $g_{\alpha\beta}$ and the contortion $\mathcal{T}_{\alpha\beta}^\gamma$, which are primal variables. The expression of the contortion is obtained by integrating the rate $\zeta_\mathcal{T}$ of $\mathcal{T}$. It is again reminded that the use of an embedded base is the most sought after to work with the components $\dot{\mathcal{T}}_{\alpha\beta}^\gamma$ without concern about the problem of objectivity of time derivative and the influence of the rigid body motion of the continuum. It is not possible to calculate directly the contortion, so we use the same method as in material plasticity with the *Principle of Maximum Dissipation* recasted as follows:

$$\text{Max } \{\mathscr{D}(\zeta_{\mathfrak{T}}) \quad \text{such that} \quad \mathscr{D} - \mathbf{J}_{\mathfrak{T}} : \zeta_{\mathfrak{T}} \equiv 0\}$$

The maximization method consists of introducing a Lagrangian e.g. [228]:

$$\mathscr{L} := \mathscr{D} + \lambda\,(\mathscr{D} - \mathbf{J}_{\mathfrak{T}} : \zeta_{\mathfrak{T}}) \tag{4.72}$$

where λ is as usual a Lagrangian multiplier. Derivatives with $\zeta_{\mathfrak{T}}$ and Λ respectively give two equations:

$$\begin{cases} (1+\lambda)\dfrac{\partial \mathscr{D}}{\partial \zeta_{\mathfrak{T}}} - \lambda\,\mathbf{J}_{\mathfrak{T}} = 0 \\[2ex] \mathscr{D} - \mathbf{J}_{\mathfrak{T}} : \zeta_{\mathfrak{T}} = 0 \end{cases}$$

Multiplying the first equation by the rate $\zeta_{\mathfrak{T}}$ and defining a new Lagrange multiplier $\Lambda := (\lambda + 1)/\lambda$, we deduce the relation for a homogeneous function of degree 1:

$$\mathbf{J}_{\mathfrak{T}} = \frac{\mathscr{D}}{\dfrac{\partial \mathscr{D}}{\partial \zeta_{\mathfrak{T}}} : \zeta_{\mathfrak{T}}}\,\frac{\partial \mathscr{D}}{\partial \zeta_{\mathfrak{T}}} = \frac{\partial \mathscr{D}}{\partial \zeta_{\mathfrak{T}}} \tag{4.73}$$

For rate-independent behavior, the dissipation function is homogeneous of degree 1 and for a dissipation potential which is not differentiable at the origin, we obtain the plastic constitutive laws and its inverse:

$$\mathbf{J}_{\mathfrak{T}} \in \frac{\partial \mathscr{D}}{\partial \zeta_{\mathfrak{T}}} \qquad \zeta_{\mathfrak{T}} \in \frac{\partial \mathscr{D}^*}{\partial \mathbf{J}_{\mathfrak{T}}} \tag{4.74}$$

These relations reassemble to the elastoplastic flow rule with the existence of elastic zone and the normality rule. Integrating the rate of contortion:

$$\zeta^{\gamma}_{\mathfrak{T}\alpha\beta} := \dot{\mathfrak{T}}^{\gamma}_{\alpha\beta}, \quad \mathfrak{T}^{\gamma}_{\alpha\beta} = \int_0^t \dot{\mathfrak{T}}^{\gamma}_{\alpha\beta}\, dt$$

where the formula is simple when working in an embedded vector base. The description reduces to material description. It follows the expression of stress and micro-stress (hyperstress or hypermomentum) to obtain the stress (4.69):

$$\sigma^{\alpha\beta}\left(g_{\mu\nu}, \mathfrak{T}^{\lambda}_{\mu\nu}\right), \quad \Sigma^{\alpha\beta}_{\gamma}\left(g_{\mu\nu}, \mathfrak{T}^{\lambda}_{\mu\nu}\right) \quad \Longrightarrow \quad \tilde{\sigma}^{\alpha\beta}\left(g_{\mu\nu}, \mathfrak{T}^{\lambda}_{\mu\nu}\right)$$

For solving, $\tilde{\sigma}^{\alpha\beta}$ is introduced in the previous conservation laws (4.68).

Remark 4.33. In general relativity, the plastic continuum thus defined could be called "Weitzenböck" plasticity by analogy to Weitzenböck spacetime e.g. [216]. The plastic deformation is also described by the contortion, and by extension the torsion, but not curvature. In the continuum approach, it is usual to consider the incompatibility tensor of Einstein as the source of plastic deformation e.g. [221], when torsion is not accounted for. Dislocation density is accounted for in addition to the metric deformation.

Remark 4.34. This subsection and following developments do not aim to present the plastic deformation theory but to give an outline of how the problem of evolution of defects might be approached by a generalized continuum e.g. [11].

To anticipate next chapters, this paragraph is not a formal modeling of the spontaneous generation of electromagnetic waves in spacetime, but we could, on the one hand, note that Maxwell's electromagnetic wave equations are not at the same level as Newton's equations of material point since, unlike the latter, they are already analogous to Schrödinger's equations (after the first quantization) on the path to quantization e.g. [75]. On the other hand, we have been able to see throughout this book, as in other papers, that electromagnetism, at least some of its components, are directly linked to torsion fields e.g. [39, 50, 222]. From then on, there is a step, which we refrain from taking since we remain in a model of continuous media physics (spacetime), which seems to suggest that quantum modeling is somewhere linked to the consideration of torsion for the spacetime geometry. Now, by recalling in this subsection the modes of creation of irreversible plastic transformations, we have some maybe naive but mathematical elements to project towards a future that could outline the theoretical basis of modeling torsional nucleation, and thus, by transition, possibly electromagnetism in spacetime. Everything remains speculative, however.

4.4.3 *Generalized Continuum: Alternative Formulation*

It is known that accounting for defects in continua has been one of the main motivations for constructing continuum models that generalize Noll's simple materials [191]. The strain gradient approach, well known in elastoplasticity, allows us to compare the results that we can obtain with the present alternative approach.

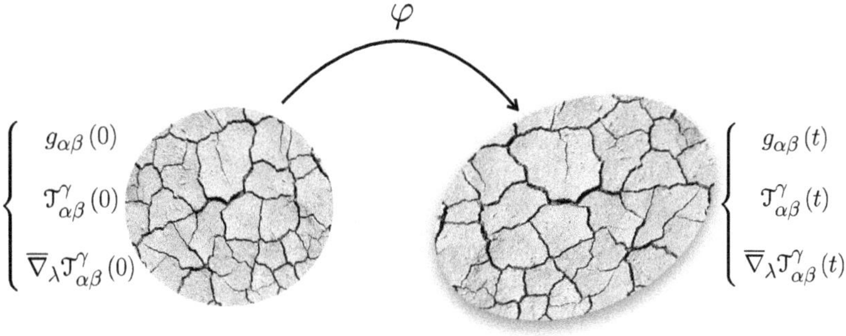

Fig. 4.2: (Left) Initial state of the continuum; (Right) Actual state of the continuum at instant t. The non-uniform representation of the continuum is intentional to mark the presence of discontinuous fields in the medium.

4.4.3.1 *Einstein-Cartan Spacetime as Generalized Continuum*

In the theory of continuum plasticity, the development of the so-called *strain gradient plasticity* models has been very significant over the past 50 years. Theoretical and numerical work has grown significantly over the past two decades (see the review by Voyijadis & Song e.g. [190] and references therein). Of course, plasticity is strongly related to the presence of dislocations and disclinations (Fig. 4.2). The main reason for using these two equivalent dislocation variables is to avoid the presence of the torsion implicitly present in the connection ∇. Different approaches might be observed according to the choice of primal variables at the base of the models; most theories are based on multiplicative decomposition. In this section, we start with a very different path by considering the contortion and its Levi-Civita gradient as additional primal variables:

$$(g_{\alpha\beta}, \aleph^{\gamma}_{\alpha\beta}, \Re^{\gamma}_{\alpha\beta\lambda}) \quad \longrightarrow \quad (g_{\alpha\beta}, \mathcal{T}^{\gamma}_{\alpha\beta}, \overline{\nabla}_{\lambda}\mathcal{T}^{\gamma}_{\alpha\beta}) \tag{4.75}$$

The present approach, geometrically based, is more general than following the standard approach of multiplicative decomposition in elastoplasticity theory.

4.4.3.2 *Constitutive Laws*

We do not build our model on the classical multiplicative decomposition even though it is very popular. We rely instead on the geometrical approach of Riemann-Cartan manifolds by considering the connection as a variable measuring irreversibility.

Einstein-Cartan spacetime as gradient continuum. Consider a generalized continuum whose Lagrangian depends on metric, contortion and contortion gradient:

$$\mathscr{L} := \mathscr{L}(g_{\alpha\beta}, \mathfrak{T}^{\gamma}_{\alpha\beta}, \overline{\nabla}_{\lambda}\mathfrak{T}^{\gamma}_{\alpha\beta}) \tag{4.76}$$

where we note that the gradient is done with the Levi-Civita connection $\overline{\nabla}$.

Let's again write the Euler variations on the associated action:

$$\delta\mathscr{S} = \int_{\mathscr{B}} \left[\sigma^{\alpha\beta}\delta g_{\alpha\beta} + \Sigma^{\alpha\beta}_{\gamma}\ \delta\mathfrak{T}^{\gamma}_{\alpha\beta} + \Xi^{\alpha\beta\lambda}_{\gamma}\ \delta\overline{\nabla}_{\lambda}\mathfrak{T}^{\gamma}_{\alpha\beta}\right]\omega_n \tag{4.77}$$

where we have applied an integration by parts and assumed the vanishing of divergence, or equivalently the vanishing of the flux at the boundary $\partial\mathscr{B}$. Let us recall constitutive laws giving stress, micro-stress and "polar micro-stress" as follows:

$$\begin{cases} \sigma^{\alpha\beta} := \dfrac{\partial\mathscr{L}}{\partial g_{\alpha\beta}} + \mathscr{L}\left(\dfrac{\partial\vartheta}{\partial g_{\alpha\beta}} - \dfrac{g^{\alpha\beta}}{2}\right) \\ \Sigma^{\alpha\beta}_{\gamma} := \dfrac{\partial\mathscr{L}}{\partial\mathfrak{T}^{\gamma}_{\alpha\beta}} + \mathscr{L}\,\dfrac{\partial\vartheta}{\partial\mathfrak{T}^{\gamma}_{\alpha\beta}} \\ \Xi^{\alpha\beta\lambda}_{\gamma} := \dfrac{\partial\mathscr{L}}{\partial\overline{\nabla}_{\lambda}\mathfrak{T}^{\gamma}_{\alpha\beta}} \end{cases} \tag{4.78}$$

As usual, dual variables measure the reaction forces following deformation and changes in the density of dislocations and disclinations. The same notation $\Xi^{\alpha\beta\lambda}_{\gamma}$ for polar micro-stress was kept to avoid using too many symbols.

4.4.3.3 *Gauge Invariance and Conservation Laws*

Generalized continuum models can be achieved through at least two different approaches. The first is the *Postulation à la Cauchy*, which generalizes the Tetrahedron Principle e.g. [223], a complex generalized constraint that takes into account the various internal interactions. The second is obtained through *Postulation à la D'Alembert*, which consists of defining a virtual work function of internal interactions. In certain situations, these two approaches can be equivalent. Nevertheless, we choose the second approach as advocated by Germain in the 1970s [224]: the Principle of General Covariance by writing the stationarity equation:

$$\delta_\xi \mathscr{S} = \int_{\mathscr{B}} \left(\sigma^{\alpha\beta}\, \mathcal{L}_\xi g_{\alpha\beta} + \Sigma_\gamma^{\alpha\beta}\, \mathcal{L}_\xi \Upsilon^\gamma_{\alpha\beta} + \Xi_\gamma^{\alpha\beta\lambda}\, \mathcal{L}_\xi \partial \overline{\nabla}_\lambda \Upsilon^\gamma_{\alpha\beta} \right) \omega_n = 0 \quad (4.79)$$

where we consider active diffeomorphism generated by Lie derivatives e.g. [130].

Theorem 4.5. *Consider a generalized continuum* $(\mathscr{B}, \mathbf{g}, \nabla)$ *with a Lagrangian* $\mathscr{L} := \mathscr{L}(g_{\alpha\beta}, \Upsilon^\gamma_{\alpha\beta}, \overline{\nabla}_\lambda \Upsilon^\gamma_{\alpha\beta})$. *Then we deduce the conservation laws:*

$$\boxed{\overline{\nabla}_\beta \tilde{\sigma}^\beta_\rho + \Upsilon^\alpha_{\alpha\beta} \tilde{\sigma}^\beta_\rho = 0} \quad (4.80)$$

with the ***generalized stress*** $\tilde{\sigma}^\beta_\rho$ *extending* (4.78)

$$\begin{aligned} \tilde{\sigma}^\beta_\rho &:= \sigma^\beta_\rho + \frac{1}{2}\left(-\Sigma_\rho^{\alpha\gamma}\, \Upsilon^\beta_{\alpha\gamma} + \Sigma_\gamma^{\beta\alpha}\, \Upsilon^\gamma_{\rho\alpha} + \Sigma_\gamma^{\alpha\beta}\, \Upsilon^\gamma_{\alpha\rho}\right) \quad (4.81) \\ &+ \frac{1}{2}\left(-\Xi_\rho^{\lambda\alpha\gamma} \overline{\nabla}_\lambda \Upsilon^\beta_{\alpha\gamma} + \Xi_\gamma^{\beta\alpha\lambda} \overline{\nabla}_\rho \Upsilon^\gamma_{\alpha\lambda} + \Xi_\gamma^{\lambda\beta\alpha} \overline{\nabla}_\lambda \Upsilon^\gamma_{\rho\alpha} + \Xi_\gamma^{\lambda\alpha\beta} \overline{\nabla}_\lambda \Upsilon^\gamma_{\alpha\rho}\right) \end{aligned}$$

Remark 4.35. Before deriving the proof, we observe two points:

(1) The generalized stress includes contributions σ^β_ρ, $\Sigma_\gamma^{\alpha\beta}\Upsilon^\gamma_{\rho\beta}$ and $\Xi_\gamma^{\lambda\alpha\beta}\overline{\nabla}_\lambda \Upsilon^\gamma_{\rho\beta}$, where $\Xi_\gamma^{\lambda\alpha\beta}$ corresponds to (mass) quadrupole moment in relative gravitation e.g. [130, 225].
(2) The same form of conservation laws is obtained; only the definition of the generalized stress $\tilde{\sigma}^\beta_\rho$ differs.

Proof. Let's consider the steps for the proof:

(1) Express the active variation of Lagrangian:

$$\delta\left(\mathscr{L}\omega_n\right) = \left(\sigma^{\alpha\beta}\mathcal{L}_\xi g_{\alpha\beta} + \Sigma_\gamma^{\alpha\beta}\mathcal{L}_\xi \Upsilon^\gamma_{\alpha\beta} + \Xi_\gamma^{\lambda\alpha\beta}\mathcal{L}_\xi \overline{\nabla}_\lambda \Upsilon^\gamma_{\alpha\beta}\right)\omega_n$$

with the Lie derivatives $\mathcal{L}_\xi g_{\alpha\beta}$ and $\mathcal{L}_\xi \Upsilon^\gamma_{\alpha\beta}$.
(2) Compute the Lie derivative of the covariant derivative:

$$\begin{aligned} \mathcal{L}_\xi \overline{\nabla}_\lambda \Upsilon^\gamma_{\alpha\beta} &= \xi^\rho \overline{\nabla}_\rho \overline{\nabla}_\lambda \Upsilon^\gamma_{\alpha\beta} - \overline{\nabla}_\lambda \Upsilon^\rho_{\alpha\beta} \overline{\nabla}_\rho \xi^\gamma \\ &+ \overline{\nabla}_\rho \Upsilon^\gamma_{\alpha\beta} \overline{\nabla}_\lambda \xi^\rho + \overline{\nabla}_\lambda \Upsilon^\gamma_{\rho\beta} \overline{\nabla}_\alpha \xi^\rho + \overline{\nabla}_\lambda \Upsilon^\gamma_{\alpha\rho} \overline{\nabla}_\beta \xi^\rho \end{aligned}$$

(3) Consider a uniform gauge field, then we obtain the *identity*:

$$\sigma^{\alpha\beta}\left(g_{\alpha\gamma}\mathfrak{T}^{\gamma}_{\rho\beta}+g_{\gamma\beta}\mathfrak{T}^{\gamma}_{\rho\alpha}\right)+\Sigma^{\alpha\beta}_{\gamma}\overline{\nabla}_{\rho}\mathfrak{T}^{\gamma}_{\alpha\beta}+\Xi^{\lambda\alpha\beta}_{\gamma}\overline{\nabla}_{\rho}\overline{\nabla}_{\lambda}\mathfrak{T}^{\gamma}_{\alpha\beta}=0$$

(4) Factorize the remaining term by $\overline{\nabla}_{\beta}\xi^{\rho}$, and obtain:

$$\begin{aligned}\delta_{\xi}\left(\mathcal{L}\omega_n\right)=&\left[\sigma^{\alpha\beta}\left(\overline{\nabla}_{\alpha}\xi_{\beta}+\overline{\nabla}_{\beta}\xi_{\alpha}\right)\right.\\&+\Sigma^{\alpha\beta}_{\gamma}\left(-\mathfrak{T}^{\rho}_{\alpha\beta}\overline{\nabla}_{\rho}\xi^{\gamma}+\mathfrak{T}^{\gamma}_{\rho\beta}\overline{\nabla}_{\alpha}\xi^{\rho}+\mathfrak{T}^{\gamma}_{\alpha\rho}\overline{\nabla}_{\beta}\xi^{\rho}\right)\\&-\Xi^{\lambda\alpha\gamma}_{\rho}\,\overline{\nabla}_{\lambda}\mathfrak{T}^{\beta}_{\alpha\gamma}\,\overline{\nabla}_{\beta}\xi^{\rho}+\Xi^{\beta\alpha\lambda}_{\gamma}\,\overline{\nabla}_{\rho}\mathfrak{T}^{\gamma}_{\alpha\lambda}\,\overline{\nabla}_{\beta}\xi^{\rho}\\&\left.+\Xi^{\lambda\beta\alpha}_{\gamma}\,\overline{\nabla}_{\lambda}\mathfrak{T}^{\gamma}_{\rho\alpha}\,\overline{\nabla}_{\beta}\xi^{\rho}+\Xi^{\lambda\alpha\beta}_{\gamma}\,\overline{\nabla}_{\lambda}\mathfrak{T}^{\gamma}_{\alpha\rho}\,\overline{\nabla}_{\beta}\xi^{\rho}\right]\omega_n\end{aligned}$$

This expression can be easily written in a condensed form:

$$\left(\tilde{\sigma}^{\beta}_{\rho}\overline{\nabla}_{\beta}\xi^{\rho}\right)\omega_n=0\quad\Longrightarrow\quad\left[\overline{\nabla}_{\beta}(\tilde{\sigma}^{\beta}_{\rho}\,\xi^{\rho})-\xi^{\rho}\overline{\nabla}_{\beta}\tilde{\sigma}^{\beta}_{\rho}\right]\omega_n=0\tag{4.82}$$

By introducing the contortion tensor $\mathfrak{T}_{\alpha\beta\gamma}:=\Gamma^{\gamma}_{\alpha\beta}-\overline{\Gamma}^{\gamma}_{\alpha\beta}$ in the second equation (only the first term), we have:

$$\left[\nabla_{\beta}(\tilde{\sigma}^{\beta}_{\rho}\,\xi^{\rho})-\left(\overline{\nabla}_{\beta}\tilde{\sigma}^{\beta}_{\rho}+\mathfrak{T}^{\alpha}_{\alpha\beta}\tilde{\sigma}^{\beta}_{\rho}\right)\xi^{\rho}\right]\omega_n=0\tag{4.83}$$

□

Remark 4.36. The using of Riemann-Cartan volume-form ω_n is mandatory to derive conservation laws. The presence of contortion in (4.80) is essentially related to volume-form choice. The use of contortion together with the Levi-Civita connection $\overline{\nabla}$ rather than ∇ is dictated by the search for more convenient formulation. However, an additional term appears and breaks the minimal coupling principle.

Remark 4.37. There are a few comments to make on generalized stress. Roughly speaking, the generalized stress includes three contributions:

(1) the analogy of classical Cauchy stress/energy-momentum $\sigma^{\alpha\beta}$ deforms the underlying continuum spacetime;
(2) the hyperstress/hypermomentum current $\Sigma^{\alpha\beta}_{\gamma}\mathfrak{T}^{\gamma}_{\rho\beta}$ (and $\Sigma^{\alpha\beta}_{\gamma}\mathfrak{T}^{\gamma}_{\alpha\rho}$) induces change of density of dislocations/spacetime defects at the continuum point (discontinuity of scalar field);
(3) the polar hyperstress or also (mass) quadrupole moment: $\Xi^{\lambda\alpha\beta}_{\gamma}\overline{\nabla}_{\lambda}\mathfrak{T}^{\gamma}_{\rho\beta}$ (and $\Xi^{\lambda\alpha\beta}_{\gamma}\overline{\nabla}_{\lambda}\mathfrak{T}^{\gamma}_{\alpha\rho}$) drives the relative motions of grains/spacetime microcosms (discontinuity of vector field).

4.4.3.4 *Generalized Continuum: Example of Constitutive Laws*

For this example, the two defect variables $\mathcal{T}^{\gamma}_{\alpha\beta}$ and $\overline{\nabla}_{\lambda}\mathcal{T}^{\gamma}_{\alpha\beta}$ capturing density of dislocations are considered as internal variables. A *thermodynamical process* of a generalized continuum is assumed to be defined by the set [226]:

(1) spatial position of each point $M \in \mathscr{B}$: $\mathbf{x} = \varphi(\mathbf{X}, t)$;
(2) temperature $\theta(M, t)$;
(3) stress tensor $\sigma(M, t)$ (micro-stress $\Sigma(M, t)$ and polar stress $\Xi(M, t)$) and external body force $\rho\mathbf{b}(M, t)$;
(4) entropy $s(M, t)$;
(5) Helmholtz free energy $\phi(M, t)$;
(6) heat flux $\mathbf{q}(M, t)$;
(7) heat source $r(M, t)$;
(8) extended set of internal variables: contortion $\mathcal{T}$ and its Levi-Civita covariant derivative $\overline{\nabla}\mathcal{T}$ e.g. [193].

Definition 4.2 ($\mathscr{B}$-derivative). Let $\mathbf{A}$ be a tensor of type (p, q) on $\mathscr{B}$. The time derivative of $\mathbf{A}$ with respect to $\mathscr{B}$ is a tensor of the same type as $\mathbf{A}$, which satisfies for any p-uplet of vectors $(\mathbf{f}_1, \ldots, \mathbf{f}_p)$ and for any q-uplet of 1-forms $(\omega^1, \ldots, \omega^q)$, macroscopically embedded in $\mathscr{B}$,:

$$\frac{d^{\mathscr{B}}\mathbf{A}}{dt}(\mathbf{f}_1, \ldots, \mathbf{f}_p, \omega^1, \ldots, \omega^q) \equiv \frac{d}{dt}[\mathbf{A}(\mathbf{f}_1, \ldots, \mathbf{f}_p, \omega^1, \ldots, \omega^q)]$$

The $\mathscr{B}$-derivatives of the primal/internal variables are given by:

$$\zeta_g := \frac{1}{2}\frac{d^{\mathscr{B}}\mathbf{g}}{dt}, \quad \zeta_{\mathcal{T}} := \frac{d^{\mathscr{B}}\mathcal{T}}{dt}, \quad \zeta_{\overline{\nabla}\mathcal{T}} := \frac{d^{\mathscr{B}}\overline{\nabla}\mathcal{T}}{dt}$$

This definition includes the so-called objective rate of tensor of classical continuum mechanics (and Lie derivatives on Riemann-Cartan manifolds) [133].

Theorem 4.6. *A generalized continuum of the rate type $\mathscr{B}$ is defined by constitutive tensor functions $\mathfrak{F} = \{\sigma, \Sigma, \Xi, \phi, s\}$:*

$$\mathfrak{F} = \tilde{\mathfrak{F}}\left(\mathbf{g}, \mathcal{T}, \overline{\nabla}\mathcal{T}, \zeta_g, \zeta_{\mathcal{T}}, \zeta_{\overline{\nabla}\mathcal{T}}\right)$$

Then the free energy ϕ takes necessarily the form $\phi = \tilde{\phi}\left(\mathbf{g}, \mathcal{T}, \overline{\nabla}\mathcal{T}\right)$ and the entropy inequality as:

$$\mathbf{J}_g : \zeta_g + \mathbf{J}_{\mathcal{T}} : \zeta_{\mathcal{T}} + \mathbf{J}_{\overline{\nabla}\mathcal{T}} : \zeta_{\overline{\nabla}\mathcal{T}} \geq 0$$

with dual dissipation variables $(\mathbf{J}_g, \mathbf{J}_{\mathcal{T}}, \mathbf{J}_{\overline{\nabla}\mathcal{T}})$.

Proof. This is an extension of Coleman & Noll's theorem e.g. [193]. □

To ensure the *entropy inequality*, introduce a *Dissipation Potential* ψ which is positive, convex and zero when the rates are equal to zero [227]. Now, we consider the Principle of Maximum Dissipation (v. Mises, 1928), from which we derive the evolution rates of the defects e.g. [228].

(1) The Principle of Maximum Dissipation associated with a constrained optimization holds:

$$\text{Max}\,\{\psi\,(\zeta_{\mathcal{T}},\zeta_{\overline{\nabla}\mathcal{T}}) \quad \text{such that} \quad \psi-(\mathbf{J}_{\mathcal{T}}:\zeta_{\mathcal{T}}+\mathbf{J}_{\overline{\nabla}\mathcal{T}}:\zeta_{\overline{\nabla}\mathcal{T}})\equiv 0\}$$

(2) The classical approach of optimization is to introduce a extended dissipation function to maximize:

$$\psi_{\text{ext}} := \psi+\lambda\,[\psi-(\mathbf{J}_{\mathcal{T}}:\zeta_{\mathcal{T}}+\mathbf{J}_{\overline{\nabla}\mathcal{T}}:\zeta_{\overline{\nabla}\mathcal{T}})]$$

(3) Necessary stationarity conditions from derivatives give three (in)-equations:

$$\begin{cases} (1+\lambda)\,\dfrac{\partial\psi}{\partial\zeta_{\mathcal{T}}}-\lambda\,\mathbf{J}_{\mathcal{T}}=0 \quad [\ \ni 0\ ] \\[2ex] (1+\lambda)\,\dfrac{\partial\psi}{\partial\zeta_{\overline{\nabla}\mathcal{T}}}-\lambda\,\mathbf{J}_{\overline{\nabla}\mathcal{T}}=0 \quad [\ \ni 0\ ] \\[2ex] \psi-(\mathbf{J}_{\mathcal{T}}:\zeta_{\mathcal{T}}+\mathbf{J}_{\overline{\nabla}\mathcal{T}}:\zeta_{\overline{\nabla}\mathcal{T}})=0 \end{cases}$$

This approach allows us to derive the (non)-smooth evolution rule as follows.

(1) As first step, we define a new Lagrange multiplier $\Lambda=(\lambda+1)/\lambda$, and assume a homogeneous function ψ of degree 1 to deduce:

$$\begin{cases} \mathbf{J}_{\mathcal{T}}=\Lambda\dfrac{\partial\psi}{\partial\zeta_{\mathcal{T}}} \\[2ex] \mathbf{J}_{\overline{\nabla}\mathcal{T}}=\Lambda\dfrac{\partial\psi}{\partial\zeta_{\overline{\nabla}\mathcal{T}}} \end{cases} \quad \text{extended to} \quad \begin{cases} \mathbf{J}_{\mathcal{T}}\in\partial\psi_{\zeta_{\mathcal{T}}}\,(\zeta_{\mathcal{T}},\zeta_{\overline{\nabla}\mathcal{T}}) \\[2ex] \mathbf{J}_{\overline{\nabla}\mathcal{T}}\in\partial\psi_{\zeta_{\overline{\nabla}\mathcal{T}}}\,(\zeta_{\mathcal{T}},\zeta_{\overline{\nabla}\mathcal{T}}) \end{cases}$$

(2) Then, we invert by using Legendre-Fenchel transform e.g. [227]:

$$\psi^*\,(\mathbf{J}_{\mathcal{T}},\mathbf{J}_{\overline{\nabla}\mathcal{T}}) := \text{Sup}_{(\zeta_{\mathcal{T}},\zeta_{\overline{\nabla}\mathcal{T}})}\,[\mathbf{J}_{\mathcal{T}}:\zeta_{\mathcal{T}}+\mathbf{J}_{\overline{\nabla}\mathcal{T}}:\zeta_{\overline{\nabla}\mathcal{T}}-\psi\,(\zeta_{\mathcal{T}},\zeta_{\overline{\nabla}\mathcal{T}})]$$

(3) Finally, we use conjugate function ψ^* to obtain the *defect evolution rule*:

$$\begin{cases} \zeta_{\mathcal{T}}=\dfrac{\partial\psi^*}{\partial\mathbf{J}_{\mathcal{T}}} \\[2ex] \zeta_{\overline{\nabla}\mathcal{T}}=\dfrac{\partial\psi^*}{\partial\mathbf{J}_{\zeta_{\overline{\nabla}\mathcal{T}}}} \end{cases} \quad \text{extended to} \quad \boxed{\begin{cases} \zeta_{\mathcal{T}}\in\partial\psi^*_{\mathbf{J}_{\mathcal{T}}}\,(\mathbf{J}_{\mathcal{T}},\mathbf{J}_{\overline{\nabla}\mathcal{T}}) \\[2ex] \zeta_{\overline{\nabla}\mathcal{T}}\in\partial\psi^*_{\mathbf{J}_{\overline{\nabla}\mathcal{T}}}\,(\mathbf{J}_{\mathcal{T}},\mathbf{J}_{\overline{\nabla}\mathcal{T}}) \end{cases}}$$

In the framework of Normal Dissipative Material e.g. [49, 219], we can introduce the particular Dissipation Potential as follows.

(1) *Indicator function* ($\simeq$ rate-independent plasticity): A choice of convex, homogeneous of degree one Dissipation Potential $\psi = \hat{\psi}\left(\zeta_{\mathcal{T}}, \zeta_{\overline{\nabla}\mathcal{T}}\right)$ that vanishes when rates are zero:

$$\hat{\psi}^{*}\left(\mathbf{J}_{\mathcal{T}}, \mathbf{J}_{\overline{\nabla}\mathcal{T}}\right) := \mathrm{Sup}_{\zeta_{\mathcal{T}}, \zeta_{\overline{\nabla}\mathcal{T}}}\left[\mathbf{J}_{\mathcal{T}} : \zeta_{\mathcal{T}} + \mathbf{J}_{\overline{\nabla}\mathcal{T}} : \zeta_{\overline{\nabla}\mathcal{T}} - \hat{\psi}\left(\zeta_{\mathcal{T}}, \zeta_{\overline{\nabla}\mathcal{T}}\right)\right]$$

$$= \begin{cases} 0 \text{ if } \left(\mathbf{J}_{\mathcal{T}}, \mathbf{J}_{\overline{\nabla}\mathcal{T}}\right) \in \mathscr{C} \\ \infty \text{ if } \left(\mathbf{J}_{\mathcal{T}}, \mathbf{J}_{\overline{\nabla}\mathcal{T}}\right) \notin \mathscr{C} \end{cases}$$

allows us to satisfy the entropy inequality!

(2) The Helmholtz free energy ϕ takes the form of:

$$\phi = \hat{\phi}\left(\mathbf{g}, \mathcal{T}, \overline{\nabla}\mathcal{T}\right)$$

Remark 4.38. $\zeta_{\mathcal{T}}$ and $\zeta_{\overline{\nabla}\mathcal{T}}$ are the *covariant rate of defect evolution* analogous to plastic deformation rates $\mathbf{D}^p := \dot{\mathbf{F}}^p\ \mathbf{F}^{p-1}$ and $\overline{\nabla}\mathbf{D}^p$ used in the framework of Geometrically Necessary Dislocations e.g. [220].

Remark 4.39. The flow rule of defects is thus derived by means of the Principle of Maximum Dissipation e.g. [212]. Since we have exactly the same steps for the derivation of flow rule by considering the principle, we obtain the flow rule without going into details of the calculus. For homogeneous degree 1 dissipation function, not differentiable at the origin, we again obtain the irreversible constitutive laws and its inverse:

$$\mathbf{J}_{\mathcal{T}} \in \frac{\partial \mathscr{D}}{\partial \zeta_{\mathcal{T}}} \quad \zeta_{\mathcal{T}} \in \frac{\partial \mathscr{D}^{*}}{\partial \mathbf{J}_{\mathcal{T}}} \tag{4.84}$$

These relations reassemble to the defect flow rule with the existence of reversible zone with the normality rule. Integrating the rate of contortion:

$$\zeta_{\mathcal{T}\alpha\beta}^{\gamma} := \dot{\mathcal{T}}_{\alpha\beta}^{\gamma}, \quad \mathcal{T}_{\alpha\beta}^{\gamma} = \int_0^t \dot{\mathcal{T}}_{\alpha\beta}^{\gamma}\, dt$$

It follows the expression of stress, micro-stress (hyperstress or hypermomentum), and polar micro-stress to obtain the generalized stress (4.81):

$$\sigma^{\alpha\beta}\left(g_{\mu\nu}, \mathcal{T}_{\mu\nu}^{\lambda}\right), \quad \Sigma_{\gamma}^{\alpha\beta}\left(g_{\mu\nu}, \mathcal{T}_{\mu\nu}^{\lambda}\right), \quad \Xi_{\gamma}^{\alpha\beta\lambda}\left(g_{\mu\nu}, \mathcal{T}_{\mu\nu}^{\lambda}\right) \implies \tilde{\sigma}^{\alpha\beta}\left(g_{\mu\nu}, \mathcal{T}_{\mu\nu}^{\lambda}\right)$$

For problem solving, the generalized stress $\tilde{\sigma}^{\alpha\beta}$ is introduced in the conservation laws (4.80). The main difference with the previous model is the presence of the polar micro-stress and the ease of using the metric connection $\overline{\nabla}$.

Remark 4.40. A model admitting metric, contortion and gradient of contortion is more or less similar to the strain gradient plasticity e.g. [203]. Nevertheless, the main difference is the definition of elasticity and the accounting for the plastic spin.

4.4.3.5 *Some Analogy with the Strain Gradient Continuum*

In the hopes of a future investigation for models of spontaneous creation of electromagnetism in vacuum spacetime, let us recall that all the spacetime models seen in this book remain in the continuous domain belonging to the class of generalized continuum models (Riemann-Cartan manifolds). Several currents are observed for strain gradient continuum (particularly strain gradient plastic continuum) according to the choice of primal variables, but the ones which comes closest to our approach in this section is that of a plastic deformation and its gradient e.g. [202, 203] and which also considers plastic flow with a "rotational plastic flow". Most if not all these theories are based on the multiplicative decomposition $\mathbf{F} = \mathbf{F}_e\, \mathbf{F}_p$ where mathematically $\mathbf{F}_e$ and $\mathbf{F}_p$ are Pfaffian forms rather than deformation gradients. We do not build our model upon this decomposition although it is very well-accepted. To make the analogy with Einstein-Cartan spacetimes, models based on $\mathbf{F}^p$ as a plastic variable are related to continua with the Lagrangian $\mathscr{L}(g_{\alpha\beta}, \mathfrak{T}^{\gamma}_{\alpha\beta})$, where the Pfaffian $\mathbf{F}^p$ has no dimension. A representative class of strain gradient plasticity is defined by the expression of its internal strain power:

$$\mathscr{W}_{int} := \int_{\mathscr{B}} \left(\mathbf{T}^{(e)} : \dot{\mathbf{E}}^{(e)} + \mathbf{T}^{(p)} : \mathbf{D}^{(p)} + \mathbb{K}^{(p)} : \overline{\nabla}\mathbf{D}^{(p)} \right) dV \tag{4.85}$$

in which are defined e.g. [203]: (a) the elastic stress $\mathbf{T}^{(e)}$ conjugate to the elastic strain rate $\mathbf{E}^{(e)} := (1/2)(\mathbf{F}_e^{\mathrm{T}}\mathbf{F}_e - \mathbf{I})$; (b) the plastic stress $\mathbf{T}^{(p)}$ associated with the plastic strain rate $\mathbf{D}^{(p)}$ (symmetric part of $\dot{\mathbf{F}}_p\mathbf{F}_p^{-1}$); and (c) the so-called third-order polar micro-stress $\mathbb{K}^{(p)}$ associated with the gradient of the plastic strain rate $\overline{\nabla}\mathbf{D}^{(p)}$. Given that this approach is built upon multiplicative decomposition, the necessity to fix the intermediate configuration (which is not unique) remains. For small strain, this is not really a problem. As usual with this assumption, the plastic strain rate $\mathbf{D}_p$ may be calculated by integrating the flow rule and then deducing the plastic triads $\mathbf{F}_p$. Anyway, the strain power (4.85) has obviously the same structure as the Euler variation of the action obtained from (4.76). But this is only an appearance because each term of the integrand has its own meaning.

We have the correspondence:

$$\begin{cases} \sigma : \dfrac{d^{\mathscr{B}}}{dt}\mathbf{g} & \longleftrightarrow & \mathbf{T}^{(e)} : \dot{\mathbf{E}}^{(e)}, \\ \Sigma : \dfrac{d^{\mathscr{B}}}{dt}\mathfrak{T} & \longleftrightarrow & \mathbf{T}^{(p)} : \mathbf{D}^{(p)}, \\ \Xi : \dfrac{d^{\mathscr{B}}}{dt}\overline{\nabla}\mathfrak{T} & \longleftrightarrow & \mathbb{K}^{(p)} : \overline{\nabla}\mathbf{D}^{(p)} \end{cases}$$

where the contortion power clearly includes power due to plastic strain and plastic spin. By extension, this decomposition in strain and spin has also consequences on the gradient of contortion $\overline{\nabla}\mathfrak{T}$. In the previous models which consider the contortion $\mathfrak{T}$ and its gradient $\overline{\nabla}\mathfrak{T}$ as primal variables of plasticity, we note that the curvature $\mathfrak{R}$ is not entirely represented. Indeed, the gradient of the contortion models is only part of the curvature. A class of elastoplastic continuum could be defined by a Lagrangian whose arguments are the metric tensor and the Riemann curvature:

$$\mathscr{L} = \mathscr{L}\left(g_{\alpha\beta}, \overline{\mathfrak{R}}^{\gamma}_{\alpha\beta\lambda}\right) \tag{4.86}$$

in which torsion plays no role and is assumed to vanish $\aleph^{\gamma}_{\alpha\beta} \equiv 0$. Therefore this class of continuum is based on a Riemann manifold and no longer on a Riemman-Cartan manifold. As we have shown in the chapter on the geometry of plastic media and in several books and textbooks of continuum mechanics, the curvature tensor measures the incompatibility of the deformation e.g. [133, 212, 221, 229]. If in the initial configuration the continuum has zero curvature, an elastic transformation does not create curvature and it remains zero. Therefore Lagrangian (4.86) expresses the presence of incompatible deformations considered as changes of disclination density, which is part of irreversible plastic deformations in a medium. From the Riemann theorem, metric connection without torsion is reduced to the Levi-Civita connection and the curvature is entirely formulated with metric derivatives as e.g. [6]:

$$\begin{aligned} \overline{\mathfrak{R}}^{\lambda}_{\alpha\beta\mu} &= (1/2)g^{\lambda\sigma}\left(\partial_\mu\partial_\alpha g_{\sigma\beta} - \partial_\mu\partial_\beta g_{\sigma\alpha} + \partial_\sigma\partial_\beta g_{\mu\alpha} - \partial_\sigma\partial_\alpha g_{\mu\beta}\right) \\ &\quad + (1/4)g^{\lambda\sigma}\left(\partial_\alpha g_{\sigma\gamma} + \partial_\gamma g_{\alpha\sigma} - \partial_\sigma g_{\alpha\gamma}\right)g^{\gamma\kappa}\left(\partial_\beta g_{\kappa\mu} + \partial_\mu g_{\beta\kappa} - \partial_\kappa g_{\beta\mu}\right) \\ &\quad - (1/4)g^{\lambda\sigma}\left(\partial_\beta g_{\sigma\gamma} + \partial_\gamma g_{\beta\sigma} - \partial_\sigma g_{\beta\gamma}\right)g^{\gamma\kappa}\left(\partial_\alpha g_{\kappa\mu} + \partial_\mu g_{\alpha\kappa} - \partial_\kappa g_{\alpha\mu}\right) \end{aligned}$$

This expression states that the Lagrangian defines an inelastic continuum due to the presence of curvature but not an elastic of the second gradient continuum whenever the rate does not vanish $\dot{\overline{\mathfrak{R}}}^{\lambda}_{\alpha\beta\mu} \neq 0$. More generally, continuum models where torsion and curvature are considered as primal variables but do not depend on time (cannot vary) constitute a theory with constrained dislocations and disclinations e.g. [191]. They cannot model continuum plasticity whenever torsion and curvature are *frozen*. From the theorem on covariance [169] it is assessed that the Lagrangian is covariant with its arguments. Furthermore, any attempt to build a continuum model whose Lagrangian depend on $g_{\alpha\beta}$, $\partial_\gamma g_{\alpha\beta}$, and $\partial_\lambda \partial_\gamma g_{\alpha\beta}$ is covariant under passive diffeomorphisms if and only if it takes the form of (4.86). For a more in-depth study of the continua defined by (4.86), we can draw inspiration from relative gravitation theory whose primal variables are precisely the metric and the Riemann curvature.

4.5 Summary

A class of generalized continua was suggested to model different media as elastoplastic deformation by using metric, torsion, and curvature as primal variables. This model is also designed to define the Einstein-Cartan space-time in the framework of extended relative gravitation, however limited to line and surface defects excluding point defects. The model is specified by a Lagrangian function $\mathscr{L}$ whose arguments are these primal variables.

Although Poincaré's Gauge Theory is the most used variational procedure for deriving the field equations and conservation laws in special relativity and relative gravitation, we adopt in this book a slightly different version by considering combined Euler variations and Lie derivative variations. Field equations are obtained by using Euler variations, whereas conservation laws are derived from Lie derivative variations, which are generators of active diffeomorphisms, thanks to Noether's theorems e.g. [230]. Indeed, equations of relative gravitation constitute a good illustration of this distinction between them e.g. [6].

First, starting from a Lagrangian $\mathscr{L}(\mathbf{g}, \aleph, \mathfrak{R})$, we apply the *Principle of General Covariance* as the variational procedure which includes the Euler variations to define stress σ associated with metric, hyper-stress Σ associated with torsion (dislocation density), and polar hyper-stress Ξ associated with curvature (disclination density). At a second step, we consider stationarity of the action along active diffeomorphisms to express the conservation laws (4.56), and (4.59) and with the Klein-Noether differential identities.

$$\nabla_\alpha \tilde{\sigma}^\alpha_\gamma = 0, \quad \text{where} \quad \tilde{\sigma} = \sigma + \tilde{\mathfrak{F}}[\Sigma, \aleph, \Xi, \mathfrak{R}]$$

It is deduced that the total stress $\tilde{\sigma}$ involved in conservation laws is constituted by an equivalent object as "reversible" stress σ in standard continuum mechanics, and an additional stress $\tilde{\mathfrak{F}}$ function of two hyper-stresses Σ and Ξ associated with torsion and curvature respectively (4.57) and (4.60). They are similar to irreversible contributions for the total stress e.g. [11]. In relative gravitation, the total stress that we define here is similar to the generalized energy momentum (cf. [4, 231]) in the theory of metric-affine gravity. Thirdly, we analyze the role of contortion for generalized continuum $\mathscr{B}$ by considering a Lagrangian $\mathscr{L}\left(\mathbf{g}, \mathfrak{T}, \overline{\nabla}\mathfrak{T}\right)$ function of contortion tensor and its Levi-Civita gradient. Again, application of active diffeomorphisms allows us to obtain conservation laws (4.80) and constitutive laws (4.81) of the form:

$$\overline{\nabla}_\alpha \tilde{\sigma}^\alpha_\gamma = 0, \quad \text{where} \quad \tilde{\sigma} = \sigma + \tilde{\mathfrak{F}}\left[\Sigma, \mathfrak{T}, \Xi, \overline{\nabla}\mathfrak{T}\right]$$

Again, each of the stress $\tilde{\sigma}$ includes a reversible part and an irreversible contribution depending on the densities of dislocations and disclinations. This is an alternative to the approach based on the search of higher-order contact interactions from Tetrahedron Principle à la Cauchy developed by using the concept of distributions in a more analytical framework e.g. [223]. The present work is rather a geometric approach.

As an example of non-smooth constitutive laws, tensors $\tilde{\mathfrak{F}}$ are obtained using the Principle of Maximal Dissipation, or by the Method of Pseudo-Potential of Dissipation. Beyond the analogies with materials with irreversible deformations such as inelastic deformations, we must not lose sight of the fact that this chapter is first and foremost an outline of the derivation of the generic shape of conservation laws of a vacuum spacetime more complex than that of Einstein relative gravitation.

Generalized Continuum. Indeed, we must recall that the construction of a model of generalized continuous media involves, broadly speaking, three relaxations of the fundamental assumptions that Cauchy proposed in the 1820s, namely [9]: (1) abandoning the symmetry of the constraint σ (Cauchy constraint), (2) abandoning the theory of the first gradient and localization, and finally (3) abandoning the Euclidean nature of the continuous medium. This last assumption, the strongest, forms the basis of our work by considering a Generalized Continuous Medium (which has no *a priori* connection with relativity [49]) that is a connected and compact manifold. This chapter is simply an attempt to write the conservation laws associated with this generalized medium, which take a different form than simple divergence like that of Euler-Cauchy. This model will be the starting point for the second part of this book.

PART 2

Applications of Generalized Continuum to Spacetime Models: Relative Gravitation and Cosmology

Chapter 5

Einstein Spacetime: Relative Gravitation

This chapter gradually introduces the relation between acceleration, affine connection and geodesic curves. Expression of acceleration in Newtonian mechanics and general relativity, respectively in three and four dimensions, is derived from the velocity field and metric connection. This chapter deals with Einstein gravitation for which is curved but not twisted. Classical spacetimes are mentioned: Schwarzschild, (anti)-de Sitter, Reissner-Nordström, and FLRW. Geodesics and geodesic deviation equations are revisited for various cases.

5.1 Introduction

Geodesic curves represent the trajectories of test points under the gravitation force alone. In this chapter, we consider only gravitation forces either for Newton or Einstein spacetime. The definition of geodesics remains the same, but spacetime does not have the same meaning when working in Newtonian mechanics versus Einstein gravitation.

In a Euclidean flat space without physical sources, geodesic curves reduce to straight lines. Such is also the trajectory of material points without any interaction with other parts of the Universe. Under gravity, geodesics–representing the trajectories of material objects such as planets (e.g., the Earth)—take the form of elliptical orbits around massive bodies such as stars (e.g., the Sun), or hyperbolic paths in the case of unbound satellites. In Newtonian gravitation, motions of all bodies are not only relative to each other but also assumed to be relative to an *a priori* absolute spacetime. The key point is that the measure of any motion is, at the end, conceptually depending on the light speed (or more technically depending on some signal which is recorded on an experimental apparatus). However, famous experiments conducted by Michelson (1881) and Morley (1887) established that the speed of the Earth has no effect on the light speed (which is revealed to be finite). More precisely, the light emitted by two bodies, with different velocities, will travel with the same speed and cannot reveal any absolute velocity. These experimental results undermined the notion of absolute spacetime, while confirming the invariance of the speed of light.

Einstein then proposed the two principles of special relativity that completely changed the spacetime concept in 1905, ending the supremacy of Newton's theories on gravitation and light: Finiteness (and constancy in vacuum) of light speed, and existence of inertial frames (in relation to which all laws have the same form). Special relativity remains nevertheless a kinematic theory. Assuming these two principles make it possible to derive the Lorentz transformations, which allow one to compare the velocities of test particles when changing observers, that is, to define the relations between inertial frames. It was Minkowski (1908) who rigorously developed the mathematical shape of Lorentz transformations (extending Galilean transformations), merging time and space into a four-dimensional spacetime. Nevertheless, it was Poincare, in 1898, who first merged space and time into one entity. In 1900, he also recognized that Lorentz local time (which appeared coupled with space) is actually what is indicated by moving clocks when assuming constant light speed. Adding gravitation, it will take almost ten years for Einstein (1915) to formulate a coherent theory from Newton gravitation and Minkowski four-dimensional spacetime by abandoning flat Euclidean geometry and incorporating gravitation by means of curvature into spacetime geometry.

5.2 Notions of Geodesics in Newton Gravitation

Modeling of space presupposes premises which are self-evident and do not need demonstration. Any deductive approach to define space is based on axioms that cannot be demonstrated but are *a priori* assumed to be true.

5.2.1 *History of Action-At-A-Distance*

The concept of gravitational field had taken several centuries to mature (see e.g. [3] and references herein). In the Middle Ages, actions-at-a-distance were not popular because forces were not accepted as being able to act at a distance. Despite a few exceptions such as the observation of magnetic forces, necessarily at a distance, by Pierre de Maricourt (Petrus Peregrinus) in 1269 or the early suggestion of Abu Ma'shar in 850 concerning the causality between the attraction of Moon and Earth's tides several centuries before, action-at-a-distance, which is closely linked to the notion of field, had illustrious opponents such as Descartes and Galileo. It was Kepler who highlighted in 1609 the influence of planets and stars as sources of attractive forces, and which generated circular or elliptical trajectories, such as that of Mars for example. Kepler refined his theory for more than 25 years before arriving at arguments that were accepted, between 1618 and 1634. Nevertheless, opposition to the existence of actions-at-a-distance, which are ultimately gravitational fields, remained strong at that time and Newton, to get his ideas across, mentioned the notion of gravitational forces as only a mathematical but not physical explanation in his *Principia* in 1687. Thanks to Kepler, gravitation became an integral part of what is called a field. Boscovich and later Kant elevated field to the rank of action-at-a-distance. In 1777, Priestley mentioned that empty space is no longer empty but contains a substance called "a field", which would later be associated with the energy of the vacuum (which is not indeed empty).

Matter, space and time are the three entities which constitute the foundations of physics observed by human beings. Intuitively and gradually, it was necessary to pool space and time on the one hand and to put matter on the other. In Western civilization, Aristotle's four books on physics provided the foundation of natural science for thousands of years. In Aristotle's paradigm, there is an absolute time and an absolute independent space with an absolute reference which is constituted by the Earth. In 1687, in Newton's *Principia*, time remains absolute independent of absolute space (*in fine* asymptotic concept) and which is the same for any observer (human or chronometer). On the other hand, there is not just an absolute inertia frame of reference but a class of inertia frames characterized by Galileo's transformations. The *Principia* served as a reference for nearly 150 years, until the advent of Maxwell's theory of electromagnetism (1861), whose equations are not covariant with respect to Galileo's transformations.

To reconcile point mechanics and electromagnetism, Einstein abolished the existence of absolute time and merged space and time into a single four-dimensional spacetime entity thus forming Minkowski spacetime as the underlying geometry of the theory of special relativity (1905). In this theory the mass remains pointlike and has no influence on spacetime, which partly preserves Aristotle's schema of space and time: spacetime remains a domain where physical phenomena take place without modifying it.

Inspired by Riemann geometry where non-vanishing curvatures allow us to say that all spaces are not necessarily Euclidean, Einstein integrated progressively from 1907 to 1915 this non-flat geometry to include gravitation and merged space and time into a curved spacetime whose curvature models gravitational actions. In this chapter, however, it must be emphasized that the notion of geodesic curves in Newtonian mechanics is not that of spacetime empty of all physical phenomena including gravitation, but that of spacetime in which gravitational forces act. This differs fundamentally from the geodesics of relative gravitation from a conceptual point of view.

5.2.2 *Laws of Newtonian Mechanics*

Newtonian gravitation may be represented geometrically by sections of the bundle of Euclidean frames, or in more common words represented by the motion of a body whose configuration is immersed within a three-dimensional space, and such that the movement is parameterized by one-dimensional time $t \in \mathbb{R}$ uncoupled from the space e.g. [1]. Time is assumed to be absolute without any relationship with the external environment: it is a Euclidean line provided with a norm. Newtonian mechanics then assumes the existence of an absolute three-dimensional space and without any relationship with the external environment and decoupled from time: it is a Euclidean space equipped with a metric, and may be represented without loss of generality by $\mathbb{R}^3$ e.g. [111]. Spacetime $\mathscr{B}$ is the product of a chronometer $\mathfrak{T}$ and a three-dimensional space $\mathscr{E}$.

Remark 5.1. Time t in Newtonian mechanics is reduced to a parameter for measuring the distance between instants but remains totally decoupled from space.

It is then absolute but it is not unique because any spacetime obtained by a Galilean transformation is also absolute. Newtonian spacetime is thus an immobile asymptotic abstraction and ultimately an idealization. The four coordinates of a material point is given by the quadruplet (t, x^1, x^2, x^3) with respect to the coordinate system of space $\mathscr{E}$. The position of a material point M at time t with respect to the reference frame is then: $\mathbf{OM} = \mathbf{OM}(t, \mathbf{x}) \in \mathscr{E}$ with O the origin of the affine space. Since the space is Euclidean, there is at least one reference frame, called Cartesian frame, with base $(\mathbf{e}_1, \cdots \mathbf{e}_n)$ over all space such that: $\mathbf{OM} = x^1(t)\ \mathbf{e}_1 + \cdots + x^n(t)\ \mathbf{e}_n$

(here $n = 3$). The distance of two points (A, B) is calculated:

$$\|\mathbf{AB}\|^2 = \left(x_B^1 - x_A^1\right)^2 + \left(x_B^2 - x_A^2\right)^2 + \left(x_B^3 - x_A^3\right)^2 = \delta_{ij}(x_B^i - x_A^i)(x_B^j - x_A^j)$$

where the Krönecker symbols δ_{ij} represent the space metric projected onto the orthonormal base $(\mathbf{e}_1, \mathbf{e}_2, \mathbf{e}_3)$. The traveled distance of a particle M, between instants t and $t + dt$, in Galilean spacetime is then given by:

$$\|d\mathbf{M}\|^2 = \delta_{ij}\ dx^i dx^j, \quad \text{where} \quad d\mathbf{M} = dx^i\ \mathbf{e}_i, \quad n = 1, 3$$

Considering curvilinear coordinate system of space $\mathscr{E}$ is possible. However, referential system should not be confused with coordinate system. Consider a referential system including a reference body with an origin O and a base $(\mathbf{e}_1, \mathbf{e}_2, \mathbf{e}_3)$, and a chronometer to measure lapse of time. Vector **OM** defines the particle position with its velocity with respect to the frame $(O, \mathbf{e}_1, \mathbf{e}_2, \mathbf{e}_3)$:

$$\mathbf{v} := \frac{d\mathbf{OM}}{dt} \tag{5.1}$$

This gives us a local referential frame $(M, \mathbf{e}_1, \mathbf{e}_2, \mathbf{e}_3)$ showing that the velocity is a tangent vector at point M.

Newton's laws of classical mechanics. Without going into details that we do not need in this work, we recall Newton's three laws of motion (*Principia*):

(1) (Principle of Inertia) A free particle, isolated from any external action in the space, moves with a constant velocity.

(2) (Fundamental Principle of Dynamics) The total force applied on a particle moving within an inertial frame (absolute spacetime, that is Galilean spacetime) is equal to the time derivative of the linear momentum $m\mathbf{v}$, where m is the velocity with respect to the spacetime.

(3) (Principle of Action-Reaction) When two particles (1) and (2) are in interaction, the force $\mathbf{F}_{12}$ exerted by (1) on (2) is opposite (equal in magnitude) to the force $\mathbf{F}_{21}$ exerted by (2) on (1).

The third law assumes that the two forces are on the same line and are not only opposite vectors. This third law was inspired by the fact that there is a force of attraction (gravitation) between the two masses (1) and (2).

Given an inertial frame $\mathscr{R}_x$ with coordinates $(x^0 = ct, x^1, x^2, x^3)$, we can deduce an infinite number of others that are in rectilinear and uniform motion with respect to $\mathscr{R}_x$. Assuming that the motion of such a frame $\mathscr{R}_y$ with coordinates $(y^0 = ct', y^1, y^2, y^3)$ is along the $Ox^3 \parallel O'y^3$ axis, the transformations that connect the inertial frames are those of Galileo:

$$\begin{cases} y^0 = x^0 + u^0 \\ y^i = R^i_j \quad x^j(x^0) + \beta^i x^0 + u^i \end{cases} \quad \text{with} \quad \beta^i := \frac{v^i}{c} \tag{5.2}$$

where v^i is the velocity of the origin O' of $\mathscr{R}_y$ and c the light speed. Tranformations (5.2) are adopted to compare the Galilean transformations with those of Lorentz.

Remark 5.2. (Inertial frames and Galileo's Principle) Galileo's Principle of Relativity states that the fundamental laws of classical mechanics are valid in all frames of reference used to describe motion. Frames of reference that satisfy this principle are called inertial frames in Newtonian mechanics.

Remark 5.3. In a more philosophical consideration, Newton's approach to spacetime belongs to the class of substantialists (space is an absolute substance even if it is invisible) while the other approach would be that of Leibniz which is based on relationism or functionalism (spacetimes are not substances but exist in relation to their movement to each other). Historical reviews on the spacetime concept point out that Galileo's spacetime connects these two points of view e.g. [139].

5.2.3 *Geodesics in Newtonian Gravitation*

Once again, it must be remembered that the geodesics we are considering are those of spacetime in which physical forces, notably gravitational forces, act. For the sake of clarity and convenience, we use Lagrangian formulations for Newtonian mechanics. Moreover, the so-called Euler-Lagrange equations are covariant, which is not the case for Newton's equations of motion.

5.2.3.1 *Configuration Space and Euler-Lagrange Equations*

Let's consider the motion of a particle with respect to an inertial frame. For this particle belonging to a system $M \in (\Sigma)$, the coordinates are assumed to be uniquely defined by a certain set of parameters:

$$\mathbf{x} = \mathbf{x}\left(t, q^1, \cdots, q^n\right), \quad \forall\, t \in \mathbb{R} \tag{5.3}$$

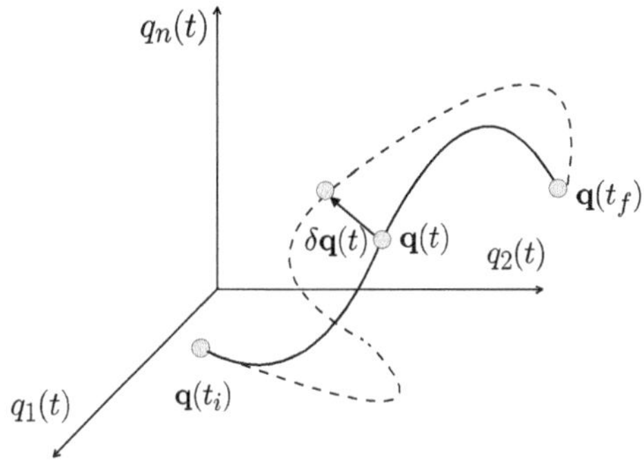

Fig. 5.1: Configurational space for Lagrangian mechanics. Generalized coordinates $(q_1, \cdots q_n)$ and Euler variation $\delta\mathbf{q}(t)$.

where n (degree of freedom of (Σ)) is the necessary number of variables to uniquely define the point position in the space $\mathscr{E}$. Due to the independence of t with respect to coordinates (x^i), it becomes a special variable which makes it possible to construct the trajectory of the particle in the space $\mathscr{E}$, as stated by a parameterized curve. Generalized coordinates $(q^\alpha(t), \alpha = 1, \cdots n$ are a function of time.

Definition 5.1. Configuration space is the n-dimensional manifold $\mathscr{C}$ engendered by coordinates $(q^1, \cdots, q^n)$. The path of the particle in the configurational space is given by the curve $(q^\alpha(t))$ (Fig. 5.1).

Consider an arbitrary coordinate system defining the position of a point M of $\mathscr{E}$ where we denote $\mathbf{q} = (q^1, \cdots, q^n)$. The configuration space $\mathscr{C}$ is the space generated by the coordinates $\mathbf{q}$ of dimension n. We assume no constraints between coordinates. It is an abstract space which does not correspond to physical space in which the particle M evolves. For this purpose, we adopt the Lagrangian approach by considering the action where $\mathbf{OM}\,(t, q^1, \cdots, q^n)$:

$$\mathscr{S}\,[q^\alpha] := \int_{t_i}^{t_f} \mathscr{L}\,[t, q^\alpha(t), \dot{q}^\alpha(t)]\, dt, \quad \mathscr{L} := \mathscr{T}\,[t, q^\alpha(t), \dot{q}^\alpha(t)] - \mathscr{U}\,[t, q^\alpha(t)] \tag{5.4}$$

where q^α and $\dot{q}^\alpha$ are respectively the generalized coordinates and its time derivative, in which $\alpha = 1, n$. $\mathscr{L}$, defined as the difference between the kinetic energy $\mathscr{T}$ and the potential $\mathscr{U}$, is the Lagrangian which specifies the point motion.

Axiom 5.1. The Hamilton Principle of Least Action states that the true path taken by the particle is an extremum of the action $\mathscr{S}$, say $\delta\mathscr{S} = 0$.

In Newtonian mechanics of a particle, the kinetic energy of the mass m and the potential generated by a mass M are defined respectively as:

$$\mathscr{T} := \frac{m}{2}\,\|\mathbf{v}\|^2, \quad \text{and} \quad \mathscr{U} := -GM\,\frac{\mathbf{OM}}{\|\mathbf{OM}\|^2} \tag{5.5}$$

where G is the constant of gravitation and O is the center of the mass M. The kinetic energy is formed by making use of the metric of the Euclidean space $\mathscr{T} := (m/2)\,\delta_{ij}\,v^i v^j$.

The first high-precision experimental measurement of the gravitational constant $G = 6,74 \times 10^{-11}\,[m^3 kg^{-1} s^{-2}]$ (only about 1% of the modern accepted value) had been carried out by Cavendish in 1798, for the first time since Newton had written his law of gravitation a century earlier. He used a simple but ingenious apparatus consisting of a torsion pendulum, with two small lead balls at the ends of a beam, each ball being subjected respectively to the attraction of a large ball of 158 $[kg]$ near them. The value of the constant G was then deduced from the oscillation period of the pendulum thus constituted. Since then, sophisticated and modern devices have always been keeping Cavendish's measurements as a reliable reference.

Remark 5.4. According to the Newtonian approach, all movement takes place in spacetime which is considered absolute, and such that the class of inertial frames of reference is defined by Galilean transformations. In Newtonian mechanics, the potential $\mathscr{U}$ is added as an *ad hoc* source of gravitation.

Stationarity of the action gives the *Euler-Lagrange equations*. The mass positivity implies that this extremum is a minimum. To derive the Euler-Lagrange equations, let's consider a different path slightly closed to the true (unknown) path in the configurational space of generalized coordinates $q^\alpha(t) \longrightarrow q^\alpha(t) + \delta q^\alpha(t)$, where $\delta q^\alpha(t)$ is an Euler variation of the generalized coordinates. By keeping the initial and final endpoints, say $\delta q^\alpha(t_i) = 0$ and $\delta q^\alpha(t_f) = 0$, the stationarity of the action gives:

$$\begin{aligned}
\delta\mathscr{S} &= \delta \int_{t_i}^{t_f} \mathscr{L}\left[t, q^\alpha(t), \dot{q}^\alpha(t)\right] dt \\
&= \int_{t_i}^{t_f} \left[\frac{\partial \mathscr{L}}{\partial q^\alpha} - \frac{d}{dt}\left(\frac{\partial \mathscr{L}}{\partial \dot{q}^\alpha}\right)\right] \delta q^\alpha\, dt + \left[\frac{\partial \mathscr{L}}{\partial \dot{q}^\alpha}\,\delta q^\alpha\right]_{t_i}^{t_f} = 0
\end{aligned}$$

where we have used Leibniz's rule for derivatives and integration by parts to obtain the second term. For fixed endpoints and arbitrariness of the Euler variations, we deduce the Euler-Lagrangian equations:

$$\frac{\partial \mathscr{L}}{\partial q^\alpha} - \frac{d}{dt}\left(\frac{\partial \mathscr{L}}{\partial \dot{q}^\alpha}\right) = 0, \quad \alpha = 1, \cdots n \tag{5.6}$$

The shape of these equations do not depend on the coordinate choice: they are covariant. On the contrary, Newton's laws are not covariant. To illustrate this statement, we consider a mass m in a plane subjected to a force derived from a potential $\mathscr{U}$. Newton's equations $\mathbf{F} = -\nabla\mathscr{U} = m\mathbf{a}$ in two different coordinate systems give:

(1) in polar coordinates (r, θ) where $\mathbf{OM}(r, \theta, t) = t(t)\ \mathbf{e}_r(\theta)$

$$m(\ddot{r} - r\dot{\theta}^2) = -\frac{\partial \mathscr{U}}{\partial r}, \qquad m(r\ddot{\theta} + 2\dot{r}\dot{\theta}) = -\frac{1}{r}\frac{\partial \mathscr{U}}{\partial \theta} \tag{5.7}$$

(2) in Cartesian coordinate (x, y) where $\mathbf{OM}(x, y, t) = x\ \mathbf{e}_x + y\ \mathbf{e}_y$

$$m\ddot{x} = -\frac{\partial \mathscr{U}}{\partial x}, \quad m\ddot{y} = -\frac{\partial \mathscr{U}}{\partial y} \tag{5.8}$$

Their shapes obviously depend on the coordinate choice although they describe the same motion of a material point with respect to one referential frame. On the contrary, (5.6) keeps the same shape either in polar or in Cartesian coordinates. The shape of (5.6) remains the same for any coordinate transformations considered as passive diffeomorphisms $Q^i = Q^i(q^\alpha), i = 1, n; \alpha = 1, n$.

5.2.3.2 *Geodesics of Free Particles in Physical Space*

Originally, the geodesic is obtained by optimizing the length of curves between two arbitrary points. Consider two points of the affine space $\mathscr{E}$. At any point, the metric tensor can be written $g_{ij}(\mathbf{x})$ where $\mathbf{x}(t)$ allows us to calculate the line element at this point. The length of the covered distance by a material point between two instants t_1 and t_2 along its trajectory is:

$$ds := \sqrt{g_{ij}(\mathbf{x})\dot{x}^i\dot{x}^j}\ dt \quad \Longrightarrow \quad \mathscr{D}_{12} := \int_{t_1}^{t_2} \sqrt{g_{ij}(\mathbf{x})\dot{x}^i\dot{x}^j}\ dt \tag{5.9}$$

The mass of the material test point does not play any role here. The geodesic is defined as the curve with the shortest covered distance. The appropriate

Lagrangian is defined here as $\mathscr{L} := \sqrt{g_{ij}(\mathbf{x})\dot{x}^i\dot{x}^j}$. Within a Euclidean space, the Euler-Lagrange equations obviously reduce to:

$$\delta_{ij}\ \dot{x}^j = p_{i0}\ \sqrt{\delta_{ij}\ \dot{x}^i\dot{x}^j} \quad \Longrightarrow \quad p_{i0}\ \dot{x}^i = \sqrt{\delta_{ij}\ \dot{x}^i\dot{x}^j}$$

We recover the line integral which is known as the *Maupertius form*:

$$\mathscr{I}_M := \int_1^2 p_i\ dx^i \tag{5.10}$$

Now we recall the relation between geodesics and Euler-Lagrange equations for a free material point. Using a Cartesian frame, for a point of mass m moving within a Galilean spacetime without external potential, the Lagrangian is given by:

$$\mathscr{L} = \frac{m}{2}\|\mathbf{v}\|^2 = \frac{m}{2}\left(\dot{x}_1^2 + \dot{x}_2^2 + \dot{x}_3^2\right)$$

where (t, x_1, x_2, x_3) are Cartesian coordinates of the point with respect to Galilean reference. The Euler-Lagrange equations are then:

$$m\ \frac{d^2}{dt^2}x_i = 0, \quad i = 1, 2, 3 \tag{5.11}$$

The particle moves in a straight line within $\mathscr{E}$ with a constant velocity. Integrating (5.11) and expressing it in component-free form, *Newton's first law of inertia* is written:

$$\mathbf{p} := m\ \mathbf{v}(t) = m\ \mathbf{v}(0), \quad \forall t \in \mathbb{R} \tag{5.12}$$

The same motion of the material point is described by laws of the same form with respect to any Galilean frame. The transition from one Galilean frame of reference to another does not modify Newton's law of gravitation. From (5.11), it is obvious that kinetic energy $\mathscr{T} = \mathscr{L}$ along the geodesic remains constant.

5.2.3.3 *Geodesics in the Configuration Space*

For extending the straight line path, let's reformulate the Lagrangian within a curved space, called configurational space, but staying in the framework of Newtonian mechanics. The Lagrangian for a particle without external applied force reduces to the kinetic energy:

$$\boxed{\mathscr{L} := \frac{m}{2}\ g_{\alpha\beta}\dot{q}^\alpha\dot{q}^\beta} \tag{5.13}$$

in which the Cartesian coordinates x_i are merely replaced by the generalized Lagrangian coordinates q^α. The Lagrangian is reduced here to the kinetic energy. We have the two derivatives:

$$\frac{\partial \mathscr{L}}{\partial \dot{q}^\alpha} = m g_{\alpha\gamma}\, \dot{q}^\gamma \quad \Longrightarrow \quad \frac{d}{dt}\left(\frac{\partial \mathscr{L}}{\partial \dot{q}_\alpha}\right) = m \frac{\partial g_{\alpha\gamma}}{\partial q^\beta}\, \dot{q}^\beta \dot{q}^\gamma + m g_{\alpha\gamma}\, \ddot{q}^\gamma$$

and

$$\frac{\partial \mathscr{L}}{\partial q^\alpha} = \frac{m}{2} \frac{\partial g_{\beta\gamma}}{\partial q^\alpha}\, \dot{q}^\beta \dot{q}^\gamma$$

We deduce from these two relations the expression of the Euler-Lagrange equations, divided by mass m:

$$g_{\alpha\gamma}\, \ddot{q}^\gamma + \left(\frac{\partial g_{\alpha\gamma}}{\partial q^\beta} - \frac{1}{2}\frac{\partial g_{\beta\gamma}}{\partial q^\alpha}\right) \dot{q}^\beta \dot{q}^\gamma = 0$$

Because the first term in brackets is contracted with $\dot{q}^\beta \dot{q}^\gamma$, we can use the symmetry to obtain:

$$g_{\alpha\gamma}\, \ddot{q}^\gamma + \frac{1}{2}\left(\frac{\partial g_{\alpha\gamma}}{\partial q^\beta} + \frac{\partial g_{\alpha\beta}}{\partial q^\gamma} - \frac{\partial g_{\beta\gamma}}{\partial q^\alpha}\right) \dot{q}^\beta \dot{q}^\gamma = 0$$

Inverting the equation by the tensor $g^{\alpha\beta}$, we finally obtain a convenient form of the Euler-Lagrange equations, called also geodesic equations:

$$\boxed{\ddot{q}^\alpha + \overline{\Gamma}^\alpha_{\beta\gamma}\, \dot{q}^\beta \dot{q}^\gamma = 0, \qquad \overline{\Gamma}^\alpha_{\beta\gamma} := \frac{1}{2} g^{\alpha\lambda}\left(\frac{\partial g_{\lambda\gamma}}{\partial q^\beta} + \frac{\partial g_{\lambda\beta}}{\partial q^\gamma} - \frac{\partial g_{\beta\gamma}}{\partial q^\lambda}\right)} \tag{5.14}$$

where we notice the Levi-Civita connection. $\overline{\Gamma}^\alpha_{\beta\gamma}$ are Christoffel symbols e.g. [5].

Definition 5.2. Equation (5.14) is called the geodesic equation in the configurational space (space of generalized coordinates) and solutions are called geodesics.

Remark 5.5. Non-vanishing Christoffel symbols $\overline{\Gamma}^\alpha_{\beta\gamma}$ do not necessarily mean that the spacetime is curved. This situation may arise with curvilinear coordinates. For instance, from Cartesian coordinates (x^1, x^2, x^3) with zero coefficients to spherical coordinates (r, θ, ϕ) with non-zero coefficients of connection is a very pedagogical example. The connection coefficients $\overline{\Gamma}^\alpha_{\beta\gamma}$ are not components of a tensor and cannot be used as primal variables for continuum mechanics e.g. [169]. The acceleration of the particle is then defined from the formula (5.14):

$$a^\alpha := \ddot{q}^\alpha + \overline{\Gamma}^\alpha_{\beta\gamma}\, \dot{q}^\beta \dot{q}^\gamma \tag{5.15}$$

in a generalized coordinate system. It is nevertheless worth searching for an alternative form, in order to extend it to four-dimensional formulation.

Remark 5.6. Lagrangian (5.4) and the related geodesic equation (5.14) are easily extended and is only mandatory to the existence of a metric tensor, and therefore constitutes a geodesic equation within a Riemann (configurational) manifold.

5.2.4 *Geodesic Deviation and Newtonian Gravitation*

We consider a Galilean spacetime endowed with a 1-form τ (time variable) and a metric $g_{\alpha\beta}$ (space variable). Gravitation is generated by a potential $\mathscr{U}$. We derive Jacobi's equation which describes the gap movement ξ of two neighboring point masses under gravitation.

5.2.4.1 *Field Equation and Conservation Laws for Gravitation*

Two basic equations govern Newton's gravitation. The first is Poisson's field equation which expresses the potential $\mathscr{U}(t, \mathbf{x}) := \mathscr{U}(x^\mu)$:

$$\Delta \mathscr{U} = 4\pi\rho G \tag{5.16}$$

where $\rho(t, \mathbf{x})$ is the density of mass source of a body engendering the field $\mathscr{U} i$, and G a constant of proportionality. The second equation is Newton's second law for a particle test under gravitation:

$$\mathbf{a} := \ddot{\mathbf{x}} = -\nabla \mathscr{U} \tag{5.17}$$

with respect to a Galilean spacetime where t is absolute and the space is uncoupled from the time. For the potential value outside the body source of gravitation, i.e. where the density $\rho = 0$, Poisson's equation admits the solution:

$$\mathscr{U} = -\frac{GM}{r} \tag{5.18}$$

where M denotes the total of the body mass source of gravitation, obtained by integration of the density, and r is the distance of the particle to the center of the gravitating mass.

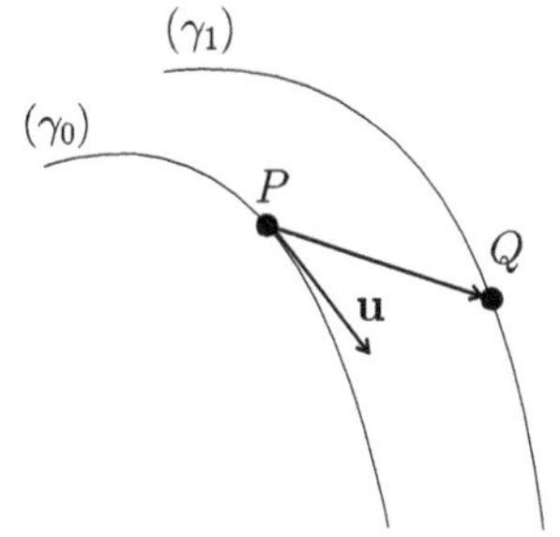

Fig. 5.2: Curves (γ_0) and (γ_1) are geodesics. At each time t of τ, points P and Q are separated by the vector $\xi := \mathbf{PQ}$ depending on the time. The vector $\mathbf{u}$ is a unit vector tangent to the geodesic lines satisfying $\nabla_{\mathbf{u}}\mathbf{u} = 0$. The two vectors satisfy also $\nabla_{\xi}\mathbf{u} = \nabla_{\mathbf{u}}\xi$. t is the parameter along the trajectories of particles, and τ is a parameter along the geodesic.

Remark 5.7. Equations (5.15) and (5.17) cannot be identified. The second leads to Newton-Cartan gravitation with a non-symmetric connection whereas the first is the expression of the acceleration but does not include gravitation, for instance the symmetry of $\overline{\Gamma}^{\alpha}_{\beta\gamma}$ with respect to the lower indices is limitative. To compare the two expressions, the Lagrangian should include the potential $\mathscr{L} := \mathscr{T} - \mathscr{U}$.

5.2.4.2 *Geodesic Deviation*

Whether in Newton or in Einstein gravitation, characterization of spacetime may be conducted by the analytical and relative study of two neighboring test particles (Fig. 5.2) which move without external effort other than gravitation e.g. [35]. The present subsection recalls the basics for geodesic equations and deviation in Newtonian mechanics with gravitation due to a potential $\mathscr{U}(x^{\mu})$.

For Newtonian mechanics, a geodesic is a curve along which a particle moves as a free falling particle. Search of geodesic deviation is based on the comparison of two geodesic curves in the spacetime (t, x^a) for Newtonian spacetime and (x^{α}) for Einstein or Einstein-Cartan spacetime. The vector gap between the two particles is denoted $\xi(t)$. The equations of motions are respectively for the two particles:

$$\ddot{x}^a(t) = -\nabla^a\mathscr{U}(P), \qquad \ddot{z}^a(t) = -\nabla^a\mathscr{U}(Q) = -\nabla^a\mathscr{U}(P) + \ddot{\xi}(t) \quad (5.19)$$

Expanding the gradient of gravitation potential about the point P gives:

$$-\mathscr{U}(Q) = -\mathscr{U}(P) - \nabla_b\mathscr{U}(P)\,\xi^b - \mathscr{O}(\xi)$$

This gives the expression of the acceleration of the geodesic deviation:

$$\boxed{\frac{d^2\xi^a}{dt^2} = -\nabla^a\nabla_b\mathscr{U}\ \xi^b} \tag{5.20}$$

This gives the expression of the distance acceleration of two particles falling in a non-uniform gravitational field $\mathscr{U}(t, x^a)$.

> **Action-at-a-distance and geodesics.** Under this quite innocuous expression (5.20), the hypothesis of action-at-a-distance is strongly linked to the hypothesis of the existence of the gravitational field $\mathscr{U}$. Acceptance of such a field is equivalent to assuming that actions can be at a distance without contact, that is an unmediated force.

5.2.4.3 *Newtonian Gravitation: Spherical Body Example*

Consider a spherical body. Formulation and resolution of Newtonian gravitation with a spherical symmetry may be found in [43] and more recently in e.g. [232]. Related solutions were useful for studying planets and satellite orbits and their eventual anomalies. In this section we give as much as possible details which will allow us to understand most of the resolution steps. From the outset, gravitational potential $\mathscr{U}$ is a solution of the field equation e.g. [125] $\Delta\mathscr{U} = 4\pi G\rho$ in the presence of mass density $\rho(x^\mu)$. Relations between Cartesian and spherical coordinates are:

$$x_1 = r\ \sin\theta\cos\varphi, \quad x_2 = r\ \sin\theta\sin\varphi, \quad x_3 = r\ \cos\varphi \tag{5.21}$$

as illustrated in Figure 5.3. Using orthonormal non-coordinate base $(\mathbf{e}_r, \mathbf{e}_\theta, \mathbf{e}_\varphi)$, the gradient of the potential is:

$$\nabla\mathscr{U} = \left(\frac{\partial\mathscr{U}}{\partial r}, \frac{1}{r}\frac{\partial\mathscr{U}}{\partial\theta}, \frac{1}{r\sin\theta}\frac{\partial\mathscr{U}}{\partial\varphi}\right) \tag{5.22}$$

Time derivative of orthonormal spherical base $(\mathbf{e}_r, \mathbf{e}_\theta, \mathbf{e}_\varphi)$ (which is not a coordinate base) is given by e.g. [133]:

$$\begin{cases} \dot{\mathbf{e}}_r &= \dot{\theta}\ \mathbf{e}_\theta + \dot{\varphi}\ \sin\theta\ \mathbf{e}_\varphi \\ \dot{\mathbf{e}}_\theta &= -\dot{\theta}\ \mathbf{e}_r + \dot{\varphi}\ \cos\theta\ \mathbf{e}_\varphi \\ \dot{\mathbf{e}}_\varphi &= -\dot{\varphi}\ \sin\theta\ \mathbf{e}_r + \dot{\varphi}\ \cos\theta\ \mathbf{e}_\theta \end{cases} \tag{5.23}$$

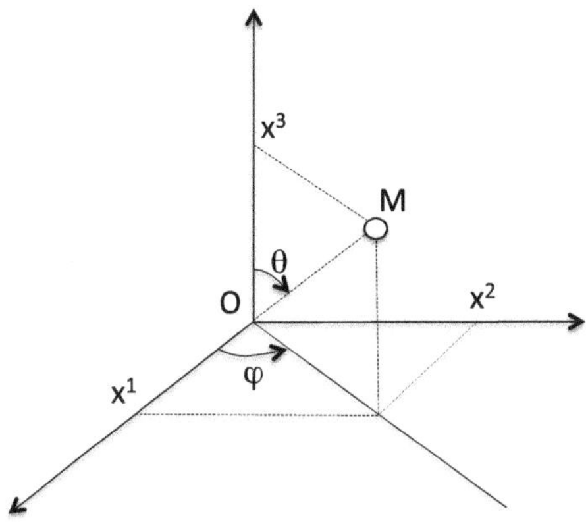

Fig. 5.3: Spherical coordinates (r, θ, φ) with orthonormal base $(\mathbf{e}_r, \mathbf{e}_\theta, \mathbf{e}_\varphi)$, deduced from the differential $d\mathbf{M} = dr\ \mathbf{e}_r + rd\theta\ \mathbf{e}_\theta + r\sin\theta d\varphi\ \mathbf{e}_\varphi$. Cartan constants of structure are not equal to zero for this orthonormal base $[\mathbf{e}_\alpha, \mathbf{e}_\beta] \neq 0$.

The gradient of a vector field $\mathbf{v}(t, r, \theta, \varphi)$ in $(\mathbf{e}_r, \mathbf{e}_\theta, \mathbf{e}_\varphi)$ is given by:

$$\nabla\mathbf{v} = \begin{pmatrix} \dfrac{\partial v_r}{\partial r} & \dfrac{1}{r}\left(\dfrac{\partial v_r}{\partial \theta} - v_\theta\right) & \dfrac{1}{r\sin\theta}\dfrac{\partial v_r}{\partial \varphi} - \dfrac{v_\varphi}{r} \\ \dfrac{\partial v_\theta}{\partial r} & \dfrac{1}{r}\left(\dfrac{\partial v_\theta}{\partial \theta} + v_r\right) & \dfrac{1}{r\sin\theta}\dfrac{\partial v_\theta}{\partial \varphi} - \dfrac{\cos\theta}{\sin\theta}\dfrac{v_\varphi}{r} \\ \dfrac{\partial v_\varphi}{\partial r} & \dfrac{1}{r}\dfrac{\partial v_\varphi}{\partial \theta} & \dfrac{1}{r\sin\theta}\dfrac{\partial v_\varphi}{\partial \varphi} + \dfrac{\cos\theta}{\sin\theta}\dfrac{v_\theta}{r} + \dfrac{v_r}{r} \end{pmatrix} \tag{5.24}$$

where the use of an orthonormal base allows us to identify a 1-form and a vector,

$$\left(v_r = \frac{\partial \mathscr{U}}{\partial r}, v_\theta = \frac{1}{r}\frac{\partial \mathscr{U}}{\partial \theta}, v_\varphi = \frac{1}{r\sin\theta}\frac{\partial \mathscr{U}}{\partial \varphi}\right)$$

Outside a sphere object of mass M, the solution is the potential $\mathscr{U} = -GM/r$ and:

$$\nabla\mathbf{v} = \frac{\partial^2 \mathscr{U}}{\partial r^2}\ \mathbf{e}_r \otimes \mathbf{e}_r + \frac{1}{r}\frac{\partial \mathscr{U}}{\partial r}\mathbf{e}_\theta \otimes \mathbf{e}_\theta + \frac{1}{r}\frac{\partial \mathscr{U}}{\partial r}\mathbf{e}_\varphi \otimes \mathbf{e}_\varphi \tag{5.25}$$

Consider now the gap between two particles to determine the geodesic deviation. Let's calculate the left-hand side of (5.20). The gap vector is defined by the small transformation of spherical coordinates:

$$(r, \theta, \varphi) \quad \Longrightarrow \quad (r + \xi_r, \theta + \xi_\theta, \varphi + \xi_\varphi) \tag{5.26}$$

where the triplet ($\xi_r = dr, \xi_\theta = d\theta, \xi_\varphi = d\varphi$) is variation (r, θ, φ). The associated physical gap vector is deduced accordingly:

$$\eta := \xi_r \, \mathbf{e}_r + r\xi_\theta \, \mathbf{e}_\theta + r \sin\theta \, \xi_\varphi \, \mathbf{e}_\varphi \tag{5.27}$$

To avoid too lengthy equations, let's focus on the particular geodesic curves with $\theta = \pi/2$. In such a case, the components of the vector are:

$$\eta_r := \xi_r, \quad \eta_\theta := r \, \xi_\theta, \quad \eta_\varphi := r \sin\theta \, \xi_\varphi = r \, \xi_\varphi \tag{5.28}$$

The second time derivative of the small variation is directly given by:

$$\begin{aligned}\ddot{\eta} &= (\ddot{\eta}_r - \dot{\varphi} \, \dot{\eta}_\varphi - \ddot{\varphi} \, \eta_\varphi) \, \mathbf{e}_r + \ddot{\eta}_\theta \, \mathbf{e}_\theta + (\dot{\eta}_r \, \dot{\varphi} + \eta_r \, \ddot{\varphi} + \ddot{\eta}_\varphi) \, \mathbf{e}_\varphi \\ &\quad + \left(\dot{\varphi} \, \dot{\eta}_r - \dot{\varphi}^2 \, \eta_\varphi\right) \mathbf{e}_\varphi - (\dot{\varphi}^2 \, \eta_r + \dot{\eta}_\varphi \, \dot{\varphi}) \mathbf{e}_r\end{aligned}$$

Using the geodesic deviation equation (5.20), we obtain a system of equations:

$$\begin{cases} \ddot{\eta}_r - \dot{\varphi} \, \dot{\eta}_\varphi - \ddot{\varphi} \, \eta_\varphi - \left(\dot{\varphi}^2 \, \eta_r + \dot{\eta}_\varphi \, \dot{\varphi}\right) = 2\dfrac{GM}{r^3} \, \eta_r \\ \ddot{\eta}_\theta = -\dfrac{GM}{r^3} \, \eta_\theta \\ \dot{\eta}_r \, \dot{\varphi} + \eta_r \, \ddot{\varphi} + \ddot{\eta}_\varphi + \dot{\varphi} \, \dot{\eta}_r - \dot{\varphi}^2 \, \eta_\varphi = -\dfrac{GM}{r^3} \, \eta_\varphi \end{cases}$$

where the right-hand terms represent the second gradient $\nabla\nabla\mathscr{U}$. We denoted $m := GM/c^2$ and $\omega_\theta := m/r^3$. For uniform motion, the term $\ddot{\varphi}$ vanishes.

5.2.4.4 *Greenberg Solution*

As an illustration of problem (5.20), Greenberg in the 1970s found the geodesic deviation equation for a spherical body (earth model) as [43]:

$$\begin{cases} \dfrac{d^2\eta_r}{d\tau^2} - 2\omega_\theta \dfrac{d\eta_\varphi}{d\tau} - 3\omega_\theta^2 \, \eta_r = 0 \\ \dfrac{d^2\eta_\theta}{d\tau^2} + \omega_\theta^2 \, \eta_\theta = 0 \\ \dfrac{d^2\eta_\varphi}{d\tau^2} + 2\omega_\theta^2 \, \dfrac{d\eta_r}{d\tau} = 0 \end{cases} \tag{5.29}$$

in a spherical coordinate (r, θ, φ) where he has chosen a circular orbit in the equatorial plane $r_0 = R, \theta_0 = \pi/2, \varphi_0(\sqrt{m/r^3})\ x^0$. Determination of this particular orbit is obvious. We temporarily use the notation $x^0 := \tau = ct$ to avoid odd notation as $d(x^0)^2$. The solution of the geodesic deviation equation (5.29) holds [45]:

$$\begin{cases} \eta_r = C_1 + C_2\ \sin(\omega_\theta \tau) + C_3\ \cos(\omega_\theta \tau) \\ \eta_\theta = C_4\ \sin(\omega_\theta \tau) + C_5\ \cos(\omega_\theta \tau) \\ \eta_\varphi = -\dfrac{3}{2} C_1 \omega_\theta\ t + 2\left[C_2 \sin(\omega_\theta \tau) - C_3\ \cos(\omega_\theta \tau)\right] + C_6 \end{cases}$$

where constants C_i correspond to various possibilities to perturb the geodesic under study. It was found that the effect of earth oblateness on deviation $\xi(t)$ dominates by far any effect of relative gravitation [43]. This supports the use of Newtonian mechanics for most past space research and satellite launches. Nevertheless, future satellite missions in faraway space, similar to Voyager 2, would require knowledge of relativistic effects of gravitation, particularly on effects of tidal forces.

5.2.5 *Four-Dimensional Extension of Newtonian Gravitation*

Newtonian mechanics is often referred to as vector mechanics, with the concepts of position, velocity, and acceleration all defined in a single vector space. Generalizing to a four-dimensional formulation becomes complicated theoretically without introducing the notion of connection. The most controversial (though often overlooked) point is the calculus of the difference in velocities knowing that these vectors do not belong to the same (tangent) vector space. This naturally highlights the interweaving of connection and acceleration.

5.2.5.1 *Introduction to Newton-Cartan Gravitation*

To develop a theory of gravitation without resorting to Einstein gravitation, Cartan (1923) introduced a way to combine the four-dimensional formulation in the presence of gravitation while remaining in Newtonian mechanics. Geometrization of gravitation in Newtonian mechanics includes two steps.

(1) The first step is to assume that time is uncoupled from space. The spacetime remains a flat Newtonian spacetime initiated by Cartan

which considers a four-dimensional manifold endowed with a metric composed of two degenerate metrics $g_{\alpha\beta}$ related to space and $\tau_\alpha\tau_\beta$ related to time that can be written $g_{\alpha\beta} := \mathrm{diag}[0,-1,-1,-1]$ and $\tau_\alpha\tau_\beta = \mathrm{diag}[c^2,0,0,0]$ by using a particular Cartesian coordinate system $(x^0 = \tau = ct, x^1, x^2, x^3)$. We introduce the light speed c and t the usual time of classical mechanics. The two metrics satisfy the orthogonality condition $g^{\alpha\beta}\tau_\alpha\tau_\beta = 0$. Moreover, their sum satisfies the Hilbert causality $\tau_\alpha\tau_\beta + g_{\alpha\beta}$. Havas [123] defined separate time and space by starting from the total spacetime metric as follows:

$$g_{\alpha\beta} = \begin{bmatrix} c^2 & 0 & 0 & 0 \\ 0 & g_{11} & g_{12} & g_{13} \\ 0 & g_{21} & g_{22} & g_{23} \\ 0 & g_{31} & g_{32} & g_{33} \end{bmatrix} \quad \text{and} \quad g^{\alpha\beta} = \begin{bmatrix} c^{-2} & 0 & 0 & 0 \\ 0 & g^{11} & g^{12} & g^{13} \\ 0 & g^{21} & g^{22} & g^{23} \\ 0 & g^{31} & g^{32} & g^{33} \end{bmatrix} \tag{5.30}$$

Given that $c >> 1$ and then $c^2\ g^{ij} >> 1$ we obtain the limit:

$$\lim_{c\to\infty} g^{\alpha\beta}(c)\cdot c^2 \simeq c^2 \begin{bmatrix} 0 & 0 & 0 & 0 \\ 0 & g^{11} & g^{12} & g^{13} \\ 0 & g^{21} & g^{22} & g^{23} \\ 0 & g^{31} & g^{32} & g^{33} \end{bmatrix} \tag{5.31}$$

This limit tends to the space metric. Consider now the limit:

$$\lim_{c\to\infty} \frac{g_{\alpha\beta}(c)}{c^2} \simeq \begin{bmatrix} 1 & 0 & 0 & 0 \\ 0 & 0 & 0 & 0 \\ 0 & 0 & 0 & 0 \\ 0 & 0 & 0 & 0 \end{bmatrix} \tag{5.32}$$

which is the tensor product $\tau \otimes \tau$. Again the extremely large value of c^2 means that $g_{ij}/c^2 \to 0$. This limit tends to the time metric.

A recent and detailed review of Newton-Cartan gravitation was made in [142] and references therein in which it was mentioned that the introduction of light speed and separation of metric into two distinct parts (time and space) is due to Friedrichs (1927) to better estimate the influence of relativity in Newton-Cartan theory.

Trautman (1966) was the first to axiomatize Newton-Cartan mechanics, while the most significant recent developments are related to

Bargmann algebra. We are not trying to provide an exhaustive history here; this very brief reminder recalls that Newton-Cartan theory is located between Newton flat spacetime enriched *ad hoc* by gravitational forces and the curved spacetime of relative gravitation.

(2) The second step is the search of an extra connection ∇, with coefficients $\Gamma^{\gamma}_{\alpha\beta}$, to model gravitation. ∇ will be determined using the gravitational field. For the sake of simplicity, assume again a Cartesian coordinate system $(x^0 = \tau = ct, x^1, x^2, x^3)$ and rewrite (5.17):

$$\delta_{ij} \frac{d^2 x^j}{d\tau^2} + \frac{\partial \mathscr{U}}{\partial x^i} = 0$$

The idea is to extract the geometric structure behind the gravitational potential. This three-dimensional viewpoint of trajectories is interpreted within Cartan geometry as the geodesics in four-dimensional curved spacetime:

$$\frac{d^2 x^\gamma}{d\tau^2} + \Gamma^{\gamma}_{\alpha\beta} \frac{dx^\alpha}{d\tau} \frac{dx^\beta}{d\tau} = 0 \tag{5.33}$$

Comparing the two equations suggests us to choose the *ansatz*:

$$\delta_{ij}\ \Gamma^{j}_{00} = \frac{\partial \mathscr{U}}{\partial x^i}, \quad \text{all other coefficients} \quad \Gamma^{\gamma}_{\alpha\beta} = 0 \tag{5.34}$$

where Latin indices stand for $i, j = (1, 2, 3)$. Rewriting $\Gamma^{i}_{00} = \delta^{ij}\ \partial_j \mathscr{U}$, we deduce that only non-vanishing terms of the Riemann-Cartan curvature matter:

$$\mathfrak{R}^{i}_{j00} = \delta^{ik} \partial_j \partial_k \mathscr{U}, \quad \text{all other components} \quad \mathfrak{R}^{\gamma}_{\alpha\beta\lambda} = 0 \tag{5.35}$$

Therefore, the Poisson field equation governing the potential becomes:

$$\Delta \mathscr{U} = \delta^{ij} \partial_i \partial_j \mathscr{U} = \mathcal{R}_{00} = 4\pi\rho G \tag{5.36}$$

showing that the potential is equal to the (00) component of Ricci curvature. Geometrization of gravitation is thus effective.

Newton-Cartan spacetime is thus defined by a compact and connected $\mathscr{B}$ manifold $(\mathscr{B}, \mathbf{g} + \tau \otimes \tau, \nabla)$. In sum, it is a covariant reformulation of Newton gravitation in a language of differential geometry that brings it closer to relative gravitation than to classical mechanics.

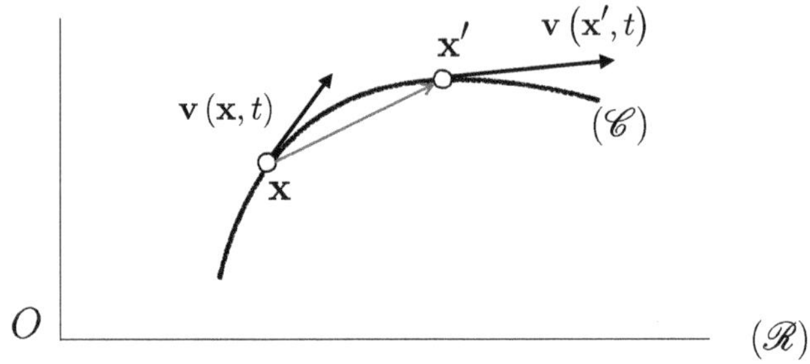

Fig. 5.4: Vectors $\mathbf{v}(\mathbf{x}, t)$ and $\mathbf{v}(\mathbf{x}', t)$ do not belong to the same vector space. Their difference needs the "connection" of their tangent spaces. Connection is the tool for (parallel) transport of velocity at $\mathbf{x}'$ to tangent space at $\mathbf{x}$.

5.2.5.2 *Connection and Acceleration*

We consider it judicious at this stage to make the link between classical mechanics and relative gravitation, and how Newton relates, explicitly or implicitly, the acceleration of a point with connection (Fig. 5.4).

Assuming an absolute space and time, it is straightforward to write the equation:

$$\mathbf{v}(\mathbf{x}', t') = \mathbf{v}(\mathbf{x}', t') - \mathbf{v}(\mathbf{x}', t) + \mathbf{v}(\mathbf{x}', t)$$

where we define $\mathbf{x}' := \mathbf{x} + d\mathbf{x}$ and $\mathbf{x}$ as the spatial position of the material point M at times t and $t' := t + dt$ respectively. Assuming that value of dt and norm of $d\mathbf{x}$ are infinitesimally small, we may write:

$$\mathbf{v}(\mathbf{x}', t') = \frac{\partial \mathbf{v}}{\partial t}(\mathbf{x}', t)\ dt + \mathbf{v}(\mathbf{x}', t)$$

Now, we consider a parallel transport of the vector $\mathbf{v}$ from $\mathbf{x}$ to $\mathbf{x}'$, and write explicitly:

$$\mathbf{v}(\mathbf{x}', t) = \mathbf{v}(\mathbf{x}, t) + \nabla_{d\mathbf{x}}\mathbf{v}(\mathbf{x}, t) \tag{5.37}$$

where the "small" displacement $d\mathbf{x}$ is generated by velocity $\mathbf{v}(\mathbf{x}, t)$ during a short lapse of time dt. This allows us to write, thanks to $d\mathbf{x} = \mathbf{v}(\mathbf{x}, t)dt$:

$$\mathbf{v}(\mathbf{x}', t') = \mathbf{v}(\mathbf{x}, t) + \frac{\partial \mathbf{v}}{\partial t}(\mathbf{x}', t)\ dt + \nabla_{\mathbf{v}(\mathbf{x},t)}\mathbf{v}(\mathbf{x}, t)\ dt \tag{5.38}$$

From this equation, neglecting terms of higher order than one, we deduce the acceleration where the connection appears:

$$\boxed{\mathbf{a}(\mathbf{x}, t) := \frac{\partial \mathbf{v}}{\partial t}(\mathbf{x}, t) + \nabla_{\mathbf{v}(\mathbf{x},t)}\mathbf{v}(\mathbf{x}, t)} \tag{5.39}$$

Remark 5.8. Introduction of connection ∇ is not insignificant in the conceptualization of Newtonian space and time. The expression (5.39) is classical for Euler's description of fluid mechanics, but it is not covariant. Another point of view was also proposed in e.g. [49] by relating material and spatial description with connection and the objective derivative with respect to a continuum.

For modeling gravitation, hereafter, we move from three-dimensional formulation of Newtonian space to four-dimensional spacetime of Einstein gravitation, and more particularly we stress the role of connection and frame of reference.

5.2.5.3 *Premises for the Generalization to Four Dimensions*

Spacetime of Newtonian gravitation is based on the principle of *Absolute Simultaneity* which partitions the four-dimensional spacetime $\mathscr{M}$ of events $((x^0, x^1, x^2, x^3) \in \mathscr{M})$ into classes of three-dimensional Euclidean spaces (called instantaneous spaces because the light speed is assumed to be infinite). For convenience, we note x^0 as the time coordinate.

In Newtonian gravitation, a Euclidean spacetime, including absolute time and absolute space, is an affine fibre bundle $(\mathscr{M}, \mathfrak{T}, \mathscr{E}, \tau)$ (e.g. [143], [233]), where $\mathscr{M}$ is a four-dimensional manifold, set of event $\mathbf{x}$, time $\mathfrak{T}$ is a four-dimensional ordered Euclidean space, and τ is a projection map $\tau : \mathscr{M} \to \mathfrak{T}$ which defines the absolute time $\tau(\mathbf{x}) = x^0$ of the event $\mathbf{x} \in \mathscr{M}$. $\mathscr{E}$ is a three-dimensional Euclidean space for which the structure group of the bundle is the group of isometries of $\mathscr{E}$. Three-dimensional *instantaneous spaces* are the fibres $\mathscr{M}_0 := \{\mathbf{x} \in \mathscr{M} : \tau(\mathbf{x}) = x^0 \in \mathfrak{T}\}$, subsets of the manifold $\mathscr{M}$. The evolution of any event $\gamma : x^0 \in \mathfrak{T} \to \mathbf{x} \in \mathscr{M}$ defines the worldline in the manifold $\mathscr{M}$. The worldline represents the motion of a particle by its absolute position all along the time $\mathbf{x}(x^0)$ and the tangent of the worldline is the particle's absolute velocity.

Newtonian spacetime. Without gravity, *Newtonian vacuum spacetime* is defined by a metric $g_{\alpha\beta}$, a linear form τ^γ and a compatible symmetric connection $\Gamma^\gamma_{\alpha\beta}$. These three fields satisfy the conditions e.g. [117]:

$$g_{\alpha\beta}\,\tau^\beta = 0, \quad \nabla^\gamma g_{\alpha\beta} = 0 \quad \text{and} \quad \nabla_\alpha \tau^\beta = 0, \quad \Re^\gamma_{\alpha\beta\lambda} = 0 \qquad (5.40)$$

for any indices $(\alpha, \beta, \gamma, \lambda)$, where ∇^γ and $\Re^\gamma_{\alpha\beta\lambda}$ are respectively the covariant derivative and the curvature.

In addition, the metric should verify the semi-definite negativity and be of rank 3. In such a case, there exist a family of coordinate systems in which we have:

$$g_{\alpha\beta} = \text{diag}\,[0,-1,-1,-1] \quad \text{and} \quad \tau^\gamma = (c^{-1},0,0,0) \quad \text{and} \quad \Gamma^\gamma_{\alpha\beta} = 0 \qquad (5.41)$$

Each coordinate system of this family is related to one another by means of Galilean transformations. Assumption of infinite light speed implies the obligation to introduce a 1-form τ in addition to metric $\mathbf{g}$. This is a relic of the three-dimensional formulation where time and space are decoupled. In the more general situation where we have a four-dimensional spacetime with a metric $g_{\alpha\beta}$ (non-degenerate) and a connection ∇^γ, then it is convenient, at least in a thought exercise, to generalize the acceleration formula by (5.39) in Einstein gravitation.

5.2.5.4 *Context for Galilean Transformations*

From a conceptual point of view, Leibniz had already raised this point by suggesting that if Newtonian absolute space exists, then the Universe would be filled with objects with absolute speeds. Since no experiment, even intellectual, can measure absolute speed, then Newtonian mechanics manipulates phenomena that cannot be measured in absolute terms. This was Leibniz's main argument for proposing the approach of a *relationist* rather than a *substantivalist* spacetime. *These opposing points of view were resolved by the introduction of Galilean transformations, thus arriving at Galilean spacetime.* It is at this level that instead of identifying material point trajectories across time, spacetime is endowed with an affine structure that is able to distinguish straight trajectories as a function of time compared to curved trajectories. Affine structure is materialized by affine connection. At the same time, affine connection allows us to construct an absolute acceleration without having to first define an absolute speed e.g. [139].

5.3 Geodesics for Curved Spacetime

For centuries, Newtonian gravitation remained a successful theory, mainly used by astronomers to analyze the motions and paths of planets. Inertial frames are flat.

5.3.1 *Role of Connection*

We have expressed the acceleration or geodesic curves in at least three ways for Newtonian mechanics:

(1) *(Newton)* In a configuration space described by a Lagrangian with generalized coordinates $(x^1, \cdots, x^n)$, and where the configuration point moves on a Riemann manifold endowed with metric $g_{\alpha\beta}$, covariant expression of the acceleration is written as:

$$a^\gamma := \frac{d^2x^\gamma}{dt^2} + \overline{\Gamma}^\gamma_{\alpha\beta} \frac{dx^\alpha}{dt}\frac{dx^\beta}{dt} = 0$$

in the absence of gravitation. Spacetime being flat, there exists an inertial frame with an appropriate coordinate system such that the Levi-Civita connection has zero coefficients and this equation reduces to $a^\gamma := \ddot{x}^\gamma = 0$. Geodesics of Newtonian absolute spacetime are straight lines.

(2) *(Newton-Cartan)* When the material test point is subject to a gravitational potential $\mathscr{U}$, geodesic curves are described by:

$$a^\gamma := \frac{d^2x^\gamma}{dt^2} + \Gamma^\gamma_{\alpha\beta} \frac{dx^\alpha}{dt}\frac{dx^\beta}{dt} = 0$$

where connection coefficients are calculated with the potential $\mathscr{U}$. Despite the resemblance, this equation does not have the same meaning as the previous equation since connections contain external forces. Coefficients $\overline{\Gamma}^\gamma_{\alpha\beta}$ result from change of coordinate systems whereas $\Gamma^\gamma_{\alpha\beta}$ are due to external gravitation. But their similarity was the main inspiration to suggest the Equivalence Principle.

(3) *(Newton)* For a material point in the presence or absence of gravitation, acceleration takes the following form (non-indexed):

$$\mathbf{a} = \frac{\partial \mathbf{v}}{\partial t} + \nabla_{\mathbf{v}}\mathbf{v}$$

where the connection is not *a priori* defined. In the absence of external gravitation, the connection is Euclidean without curvature.

Hereafter we generalize these equations to relative gravitation. To go further, it is worth giving a more precise framework for the notion of spacetime e.g. [234].

Definition 5.3. A spacetime is a four-dimensional continuum (compacted and connected), endowed with a pseudo-Riemann hyperbolic metric $\mathbf{g}$ (verifying the Hilbert causality condition) of signature $(1, 3)$ and with a connection ∇.

Spacetime is curved when the curvature tensor does not vanish everywhere (Einstein spacetime), and is curved and twisted when curvature and torsion are not equal to zero. We deal with a particular case for which the connection is compatible with the metric, say $\nabla\mathbf{g} = 0$ (Einstein-Cartan spacetime).

5.3.2 *Particle Test Within Curved Spacetime*

Special relativity, originally proposed by Einstein in his 1905 paper, has profound consequences on our view of space and time. Roughly speaking, Einstein's special relativity reconciles Newtonian mechanics and Maxwell's electromagnetism. In special relativity, the material test point has no gravitational consequences and therefore does not influence the gravitational field of the spacetime.

5.3.2.1 *Minkowski Spacetime (Notion of Special Relativity)*

The essential difference between classical Newtonian mechanics and Einstein's special relativity is that the former admits instantaneous propagation of action-at-a-distance between two bodies. The second admits the possibility of action-at-a-distance but the speed of propagation remains below that of light.

Postulates of special relativity. The two postulates of special relativity are:

(1) There are inertial reference frames. All physical phenomena proceed in an identical manner with respect to inertial reference frames if the initial conditions are the same.
(2) The light speed is finite in a vacuum and is constant for all inertial frames of reference, in all directions. It does not depend on the speed of the source nor that of the observer. This speed is therefore invariant in any inertial frame of reference.

Remark 5.9. The first postulate makes it possible to reconcile electromagnetism with mechanics because Maxwell's equations are not covariant for Galilean transformation. This is Galileo's principle of relativity from classical mechanics extended to electromagnetic phenomena.

Indeed, the appearance of electromagnetism theory and the high magnitude of its wave propagation speed was the first earthquake that shook the established classical mechanics theory, since classical mechanics and electromagnetism theories were not compatible: the first is based on Galilean transformations whereas Lorentz transformations are required for the covariance for the second. The light speed is a barrier and cannot be surpassed by the velocity of any material point. Minkowski spacetime remains flat.

Minkowski spacetime: action and distance. To start, let's consider a particle of mass m in a Minkowski spacetime. The four coordinates of the mass test is denoted $x^\alpha = (x^0 := ct, x^1, x^2, x^3)$ with respect to a Cartesian reference. The metric components are given by:

$$\eta_{\alpha\beta} := \mathrm{diag}[1, -1, -1, -1] \tag{5.42}$$

The distance between two events is calculated as:

$$ds^2 = \eta_{\alpha\beta}\, dx^\alpha dx^\beta, \qquad ds^2 = (dx^0)^2 - (dx^1)^2 - (dx^2)^2 - (dx^3)^2 \tag{5.43}$$

By using metric η, we have three situations for a four-vector $\mathbf{u}$:

(1) $\mathbf{u}$ is a timelike vector if $\eta(\mathbf{u}, \mathbf{u}) > 0$
(2) $\mathbf{u}$ is a spacelike vector if $\eta(\mathbf{u}, \mathbf{u}) < 0$
(3) $\mathbf{u}$ is a lightlike vector if $\eta(\mathbf{u}, \mathbf{u}) = 0$.

As a quick reminder, the *Lorentz transformation* which connects the coordinate system $x^\alpha, \alpha = 0, 1, 2, 3$ with a new coordinate system $y^i, i = 1, 2, 3$, following a uniform translation of velocity v of the new frame $\mathscr{R}_y$ with respect to the frame $\mathscr{R}_x$ is given by:

$$\begin{cases} y^0 = \gamma\left(x^0 - \beta x^3\right) \\ y^1 = x^1 \\ y^2 = x^2 \\ y^3 = -\gamma(\beta x^0 - x^3) \end{cases} \quad \text{with} \quad \beta := \frac{v}{c}, \quad \gamma := (1 - \beta^2)^{-1/2} \tag{5.44}$$

given that the uniform translation is along the direction 3. The *Lorentz invariance* of the metric is checked accordingly by straightforward calculus:

$$\begin{aligned}\eta_{ij}\, y^i y^j &= (y^0)^2 - (y^1)^2 - (y^2)^2 - (y^3)^2 \\ &= \gamma^2 \left(x^0 - \beta x^3\right)^2 - (x^1)^2 - (x^2)^2 - \gamma^2 \left(\beta x^0 - x^3\right)^2 \\ &= (x^0)^2 - (x^1)^2 - (x^2)^2 - (x^3)^2 = \eta_{\alpha\beta}\, x^\alpha x^\beta\end{aligned}$$

The translation velocity v being constant, the Lorentz invariance also applies after differentiating the coordinates leading to:

$$ds^2 = \eta_{ij}\, dy^i dy^j = \eta_{\alpha\beta}\, dx^\alpha dx^\beta$$

ensuring that any Lagrangian depending on ds is invariant under Lorentz transformations. In other words, the set of Lorentz transformations is the group of invariance of special relativity. Consider the path of such a particle within a Minkowski spacetime (i.e., without gravitation). As for Newtonian spacetime formulation, we can use an anonymous labeling to describe the motion by using a parameter σ. The events at initial time and final time are denoted respectively:

$$x_i^\alpha := x^\alpha(\sigma_i), \quad x_f^\alpha := x^\alpha(\sigma_f) \tag{5.45}$$

Physically, the Principle of Least Action for a particle within Minkowski spacetime is designed to extremize the distance between starting and end points.

Definition 5.4. The action of a material point of mass m evolving within Minkowski spacetime with metric η is defined by:

$$\mathscr{S} := mc \int_{\sigma_i}^{\sigma_f} \sqrt{ds^2} = mc \int_{\sigma_i}^{\sigma_f} \sqrt{\eta_{\alpha\beta}\, dx^\alpha dx^\beta} \tag{5.46}$$

where the causality of Hilbert ensures that the term under the $\sqrt{}$ is positive.

m is called rest mass. The particle follows a timelike path and the term c in the definition is introduced to have a dimension of "energy × time".

Remark 5.10. The description should not depend on the parametrization σ. Consider another arbitrary parameter $\overline{\sigma}(\sigma)$. The expression of the action is then:

$$\overline{\mathscr{S}} := mc \int_{\overline{\sigma}_i}^{\overline{\sigma}_f} \sqrt{\overline{ds}^2} = mc \int_{\overline{\sigma}_i}^{\overline{\sigma}_f} \sqrt{\overline{\eta}_{\alpha\beta}\, dx^\alpha(\overline{\sigma}) dx^\beta(\overline{\sigma})}$$

that can be written as follows:

$$\overline{\mathscr{S}} = mc \int_{\overline{\sigma}_i}^{\overline{\sigma}_f} \sqrt{\overline{\eta}_{\alpha\beta} \frac{dx^\alpha}{d\overline{\sigma}} \frac{dx^\beta}{d\overline{\sigma}}}\, d\overline{\sigma} = mc \int_{\overline{\sigma}_i}^{\overline{\sigma}_f} \sqrt{\eta_{\alpha\beta} \frac{dx^\alpha}{d\sigma} \frac{dx^\beta}{d\sigma}} \left(\frac{d\overline{\sigma}}{d\sigma}\right) d\sigma = \mathscr{S}$$

by accounting for the transformation rules for metric η and also for coordinates. The action does not depend on the parametrization of the *worldline* (the set of events when the parameter spans the interval $(0, \sigma)$, and its value between two events 0 and σ has an intrinsic meaning).

Definition 5.5. The infinitesimal proper time $d\tau$ and the proper time τ as its integral are respectively defined by:

$$\boxed{d\tau := \frac{1}{c}\sqrt{\eta_{\alpha\beta}\, dx^\alpha dx^\beta}}, \quad \tau = \frac{1}{c}\int_0^\sigma \sqrt{\eta_{\alpha\beta} \frac{dx^\alpha}{d\sigma'} \frac{dx^\alpha}{d\sigma'}}\, d\sigma' \qquad (5.47)$$

where $\sigma' = 0$ at the starting event and $\sigma' = \sigma$ at the actual event.

It is therefore worth using the proper time as the parameter for describing the path (worldline) of a particle within Minkowski spacetime $x^\alpha = x^\alpha(\tau)$. It is also convenient to notice the relation $ds = cd\tau$ from (5.47). In a particle's own rest frame — in which $dx^1 = dx^2 = dx^3 = 0$ — the proper time coincides with the "time" t. In other words, the proper time is the time measured by a clock in the rest frame (proper clock). Assuming a global Cartesian frame, the metric has the components:

$$\eta_{\alpha\beta} = \mathrm{diag}\,[+1, -1, -1, -1] \qquad (5.48)$$

Definition 5.6. Let $x^\alpha(\tau)$ be the coordinates of a particle event in a Minkowski spacetime; the four-velocity and four-acceleration are respectively defined by:

$$u^\alpha := \frac{dx^\alpha}{d\tau}, \qquad a^\alpha := \frac{du^\alpha}{d\tau} \qquad (5.49)$$

where τ is the proper time. The action of a particle with coordinates $x^\alpha(\tau)$ is rewritten as:

$$\boxed{\mathscr{S} = mc \int_0^\tau \sqrt{\eta_{\alpha\beta}\, u^\alpha u^\beta}\, d\tau} \qquad (5.50)$$

The action has four dynamical degrees of freedom $x^\alpha(\tau), \alpha = 0,1,2,3$ instead of three in classical mechanics of a particle.

"Inertial" motion of a test particle. By choosing the time t used by an observer other than the particle, we have $d\tau = cdt$, and we recover the more usual formulation in special relativity (sign "$-$" is used to conform to Newtonian mechanics for velocity $\|\mathbf{v}\| << c$):

$$\mathscr{S} = -mc^2 \int_0^t \sqrt{1 - \delta_{\alpha\beta}\frac{\dot{x}^\alpha}{c}\frac{\dot{x}^\beta}{c}}dt, \qquad \dot{x}^\alpha := \frac{dx^\alpha}{dt}, \quad \dot{x}^\beta := \frac{dx^\beta}{dt}$$

Let's consider a test particle of mass m_0 within Minkowski spacetime. Classical form of the action may be written as follows when projected onto a Cartesian frame:

$$\mathscr{S} = -m_0c^2 \int_0^t \sqrt{1 - \frac{\|\mathbf{v}\|^2}{c^2}}dt, \quad \text{with} \quad \|\mathbf{v}\|^2 := \sum_{i=1}^{i=3} \left(v^i\right)^2$$

where $\mathbf{v}$ is the particle velocity in the Lorentz frame, given that the mass is assumed to not influence the structure of spacetime. Expression of the Euler-Lagrange deduced from this action is obvious:

$$\frac{d}{dt}\left(\frac{\partial \mathscr{L}}{\partial v^i}\right) - \frac{\partial \mathscr{L}}{\partial x^i} = 0 \quad \Longrightarrow \quad = \frac{d}{dt}\left(\frac{m_0 v_i}{\sqrt{1 - \|\mathbf{v}\|^2/c^2}}\right) = 0 \tag{5.51}$$

since the light speed c is constant. The coordinates (x^i) are cyclic and the momenta are constant (term in brackets). By defining the mass of the particle as:

$$m := \frac{m_0}{\sqrt{1 - \|\mathbf{v}\|^2/c^2}}, \qquad p_i := \frac{\partial \mathscr{L}}{\partial v^i} = m\, v_i \tag{5.52}$$

we recover the analogy of the inertial motion of the particle's mass at rest m_0 as $\mathbf{p} := m\ \mathbf{v} =$ Cste in special relativity. With respect to a referential frame such that $\mathbf{v} = 0$ we have the identity $m = m_0$, which is the reason we qualify m_0 as the mass at rest.

Remark 5.11. *(Photon)* Definition of the mass (5.52) shows that it is possible to have particles with a rest mass $m_0 \to 0$ in special relativity; such particles called *photons* move with the light speed $\|\mathbf{v}\| \to c$ with the indeterminacy:

$$m := \lim_{m_0 \to 0,\ \|\mathbf{v}\| \to c} \frac{m_0}{\sqrt{1 - \|\mathbf{v}\|^2/c^2}} \to m_{photon} \text{ (finite mass)} \tag{5.53}$$

Such a particle cannot exist in Newtonian mechanics. In special relativity relationships (5.52) do not contradict the existence of particles moving at the speed of light but only in the case where the rest mass of these particles is zero.

5.3.2.2 *Curved Spacetime*

The second step is to combine gravitation with special relativity. The paradigm totally changes and the forces are replaced by geometric characteristics of the spacetime which couples both time and space. Gravitational potential $\mathscr{U}$ is extended to enrich the flat spacetime. The Principle of Equivalence assumes that the local spacetime of relative gravitation is equivalent to a Minkowski spacetime. The geometric variable becomes metric $\mathbf{g}$ depending on time and space. Einstein spacetime is a Riemann manifold.

Extension of the action from special relativity (5.55) to relative gravitation is straightforward. Einstein spacetime of relative gravitation $\mathscr{G}$ is a manifold endowed with a metric $g_{\alpha\beta}(x^\mu)$, satisfying the Hilbert causality [6]. This means that it is a four-dimensional manifold endowed with an hyperbolic pseudo-Riemann structure of signature $(1,3)$. There is at least a vector base of the tangent space $\mathcal{T}_M\mathscr{G}$ such that the metric takes the form of:

$$\mathbf{g}(\mathbf{u},\mathbf{v}) = u^0\, v^0 - \sum_{i=1}^{i=3} u^i\, v^i$$

Otherwise, we can write the general formula $\mathbf{g}(\mathbf{u},\mathbf{v}) = g_{\alpha\beta}\, u^\alpha v^\beta$. Similarly to special relativity, the distance between two events within Einstein spacetime is calculated as:

$$ds^2 = g_{\alpha\beta}\, dx^\alpha dx^\beta, \qquad ds^2 = (dx^0)^2 - (dx^1)^2 - (dx^2)^2 - (dx^3)^2 \quad (5.54)$$

where the second expression is valid only at a local point contrary to special relativity. By using metric $\mathbf{g}$, we have three situations:

(1) $\mathbf{u}$ is a timelike vector if $\mathbf{g}(\mathbf{u},\mathbf{u}) > 0$
(2) $\mathbf{u}$ is a spacelike vector if $\mathbf{g}(\mathbf{u},\mathbf{u}) < 0$
(3) $\mathbf{u}$ is a lightlike vector if $\mathbf{g}(\mathbf{u},\mathbf{u}) = 0$.

Definition 5.7. Let there be a particle of mass m in an Einstein spacetime with metric $g_{\alpha\beta}(x^\mu)$. The action of this particle with coordinates $x^\alpha(\tau)$ is defined as:

$$\boxed{\mathscr{S} = mc \int_0^{\mathcal{T}} \sqrt{g_{\alpha\beta}\ u^\alpha u^\beta}\ d\tau}, \qquad u^\alpha := \frac{dx^\alpha}{d\tau} \tag{5.55}$$

where $d\tau$ is the proper time element of the particle e.g. [128]:

$$\boxed{d\tau := \frac{1}{c}\sqrt{g_{\alpha\beta}\ dx^\alpha dx^\beta}} \tag{5.56}$$

in which the metric is event-dependent $g_{\alpha\beta}(x^\mu)$.

Remark 5.12. In relative gravitation, trajectories of material test points curve and Einstein spacetime replaces Minkowski spacetime which is flat.

5.3.2.3 *Light Cone*

Light cone is a concept which allows us, from an event point (x^μ) in spacetime $\mathscr{B}$, to distinguish past events, future events (in the gray area) and events inaccessible from (x^μ) (Figure 5.5). In special relativity, points of spacetime other than (x^μ) are divided into three categories: absolute past and absolute future of (x^μ) on the one hand — these events occurring inside the gray zone of the light cone — and on the other hand, other events. Thanks to the Hilbert causality principle e.g. [6], events inside the cone can be causally linked to (x^μ). Events located outside the cone are

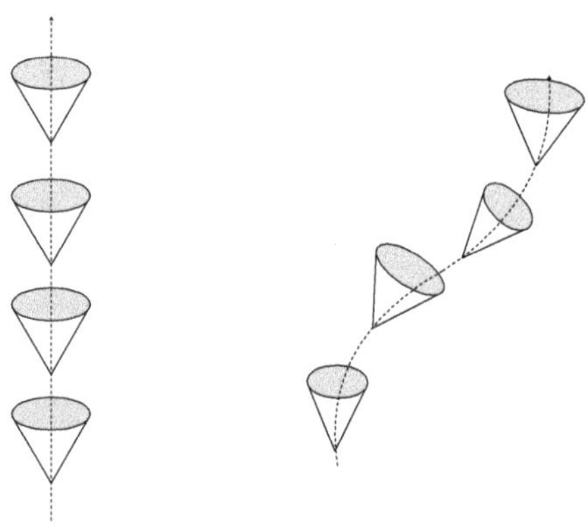

Fig. 5.5: (Left) Path of a particle within a Minkowski spacetime; (Right) Path of a particle within curved spacetime with non-homogeneous metric tensor field $g_{\alpha\beta}(x^\mu)$. Orientation and shape of the cone depend on the point-event.

causally disconnected from (x^μ) and cannot influence it or be influenced by it.

In relative gravitation, to each point $(x^\mu) \in \mathscr{B}$ is attached a local infinitesimal light cone, which concerns infinitely close events (in the sense of Einstein metrics). While in special relativity the light cones of all events (in a given frame of reference) are parallel to each other (Figure 5.5, left), this is not the case in relative gravitation (Figure 5.5, right), due to non-homogeneity of metrics and especially the spacetime curvature $\Re^{\gamma}_{\alpha\beta\lambda} \neq 0$. Definition (5.56) extends the proper time used in special relativity (5.47). Locally, metrics should verify the Hilbert causality e.g. [6]. To put it simply, Hilbert causality could be assimilated, without losing the generality of its purposes, as the condition that a metric projected onto an appropriate basis admits as components diag $[1, -1, -1, -1]$.

5.3.3 *Inertial Frame and Geodesics*

Geodesics are closely related to inertial systems. Roughly speaking, an inertial system is a frame of reference (time and space) that is not undergoing any acceleration. On the other hand, an acceleration is defined with respect to a frame of reference. Although it is an asymptotic view in Newtonian mechanics where there are an infinite number of inertial references, we have to give a more precise definition in relative gravitation.

5.3.3.1 *Inertial Frame*

In special relativity, an inertial frame of reference is defined as a system of reference (chronometer and three-dimensional rules) where a free test point, that is, a particle (without mass *in fine*), is subject to no external forces, whether gravitational or electromagnetic. Moreover, the test point does not generate any physical fields in its surroundings and moves with a constant velocity vector. Its trajectory is therefore a straight line. The concept of inertial frames of reference is more subtle in the presence of gravitation and especially in relative gravitation theory, because gravitation is a force-at-a-distance with an infinite action radius. The only way to define an inertial frame of reference is to consider locally each point-event and to approximately assume it as being far away from any gravitating matter or radiating body. Therefore, by extrapolation, in the absence of matter we can define inertial frames of reference where the geometry of spacetime is pseudo-Euclidean, globally for special relativity and locally for relative gravitation. The following definition holds.

Definition 5.8. Let $\mathscr{G}$ be an Einstein spacetime and $\mathbf{x}_0 = (x_0^0, x_0^1, x_0^2, x_0^3)$ be an event of $\mathscr{G}$. An inertial system at $\mathbf{x}_0$ is a coordinate chart $(\mathscr{U}, \varphi)$ of the manifold satisfying the two conditions:

(1) the metric $g_{\alpha\beta}(\mathbf{x}_0)$ at point $\mathbf{x}_0$ takes the shape of a Minkowski metric:

$$g_{00}(\mathbf{x}_0) = 1, \quad g_{ij}(\mathbf{x}_0) = -\delta_{ij}, \quad i, j = 1, 2, 3 \tag{5.57}$$

(2) the partial derivatives of the metric vanish at point $\mathbf{x}_0$ for all indices (α, β, γ):

$$\frac{\partial}{\partial x^\gamma} g_{\alpha\beta}(\mathbf{x}_0) = 0 \tag{5.58}$$

Remark 5.13. Mathematically, it means that locally the metric is Minkowski-type. The second condition implies that $\overline{\Gamma}^\gamma_{\alpha\beta}(\mathbf{x}_0) = 0$. It can be shown that for any $\mathbf{x}_0$ of Einstein spacetime, there are inertial frames, by using normal coordinates on a manifold e.g. [128]. This definition conforms to the aspects of invariance where covariance means an indifference with respect to a coordinate system, which is in fact an inertial frame, locally.

Theorem 5.1. *Let $(\mathscr{U}, \varphi)$ and $(\tilde{\mathscr{U}}, \tilde{\varphi})$ be two coordinate charts of the manifold $\mathscr{G}$ (not necessarily Einstein spacetime). Coordinates with respect to the first coordinate chart are denoted (x^α) and to the second coordinate chart $\tilde{x}^i$. Say $\Gamma^\gamma_{\alpha\beta}$ and $\tilde{\Gamma}^k_{ij}$ are the coefficients of any connection ∇ onto these two coordinate charts respectively. Then:*

$$\tilde{\Gamma}^k_{ij} = \frac{\partial x^\alpha}{\partial \tilde{x}^i} \frac{\partial x^\beta}{\partial \tilde{x}^j} \frac{\partial \tilde{x}^k}{\partial x^\gamma} \Gamma^\gamma_{\alpha\beta} + \frac{\partial^2 x^\ell}{\partial \tilde{x}^i \partial \tilde{x}^j} \frac{\partial \tilde{x}^k}{\partial x^\ell} \tag{5.59}$$

Proof. It is straightforward by writing:

$$\nabla_{\tilde{\mathbf{e}}_i} \tilde{\mathbf{e}}_j = \tilde{\Gamma}^k_{ij}\ \tilde{\mathbf{e}}_k = \nabla_{\left(\frac{\partial x^\alpha}{\partial \tilde{x}_i}\ \mathbf{e}_\alpha\right)} \left(\frac{\partial x^\beta}{\partial \tilde{x}_j}\ \mathbf{e}_\beta\right)$$

Without going into details, after developing the calculus thanks to the properties of an affine connection, we arrive easily at the result, given that $\nabla_{\mathbf{e}_\alpha} \mathbf{e}_\beta = \Gamma^\gamma_{\alpha\beta}\ \mathbf{e}_\gamma$:

$$\nabla_{\tilde{\mathbf{e}}_i} \tilde{\mathbf{e}}_j = \left(\frac{\partial x^\alpha}{\partial \tilde{x}^i} \frac{\partial x^\beta}{\partial \tilde{x}^j} \frac{\partial \tilde{x}^k}{\partial x^\gamma} \Gamma^\gamma_{\alpha\beta} + \frac{\partial^2 x^\ell}{\partial \tilde{x}^i \partial \tilde{x}^j} \frac{\partial \tilde{x}^k}{\partial x^\ell} \right) \tilde{\mathbf{e}}_k$$

□

Lemma 5.1. *For an inertial frame, coordinate chart with coordinates $(\tilde{x}^1, \cdots, \tilde{x}^1)$, for any coordinate chart $(\mathscr{U}, \varphi)$ with coordinates $(x^1, \cdots, x^n)$, we deduce the connection coefficients:*

$$\Gamma^{\gamma}_{\alpha\beta}(\mathbf{x}_0) = \frac{\partial^2 \tilde{x}^{\ell}}{\partial x^{\alpha} \partial x^{\beta}} \frac{\partial x^{\gamma}}{\partial \tilde{x}^{\ell}} \tag{5.60}$$

Proof. It is a direct consequence of connection transformation rule (5.59) (by inverting the inertial connection (which vanishes) and the considered connection in the relation). □

Remark 5.14. Definition of inertial frame is directly related to Einstein's principle of equivalence for relative gravitation.

Theorem 5.2. *Whenever a tensor field at a point $\mathbf{x}_0 \in \mathscr{G}$ is written as follows in an inertial frame:*

$$\frac{\partial}{\partial \tilde{x}^{\alpha}} \tilde{\mathscr{T}}^{\beta_1 \cdots \beta_q}_{\alpha_1 \cdots \alpha_p}(\mathbf{x}_0) \tag{5.61}$$

where $\mathscr{T}$ is a (p, q)-type tensor, then in any arbitrary coordinate chart $(x^1, \cdots, x^n)$, it is written as:

$$\nabla_{\alpha} \mathscr{T}^{\beta_1 \cdots \beta_q}_{\alpha_1 \cdots \alpha_p}(\mathbf{x}_0) \tag{5.62}$$

meaning that partial derivative ∂_{α} is replaced by a covariant derivative ∇_{α}.

Proof. The case of tensor $(0, 0)$-type is trivial. For the proof, it suffices to consider a tensor $(1, 0)$-type or $(0, 1)$-type.

(1) For a tensor $(1, 0)$-type, we have at any point $\mathbf{x}_0 \in \mathscr{G}$ in the inertial frame and in any other atlas, by using the previous relation (5.59):

$$\mathscr{S}^{\lambda}_{\mu} = \frac{\partial x^{\lambda}}{\partial \tilde{x}^{\alpha}} \frac{\partial \tilde{x}^{\beta}}{\partial x^{\mu}} \frac{\partial \tilde{\mathscr{T}}^{\alpha}}{\partial \tilde{x}^{\beta}}, \qquad \mathscr{S}^{\lambda}_{\mu} := \left(\frac{\partial x^{\lambda}}{\partial \tilde{x}^{\alpha}} \frac{\partial \tilde{x}^{\beta}}{\partial x^{\mu}} \right) \frac{\partial}{\partial x^{\rho}} \left(\frac{\partial \tilde{x}^{\alpha}}{\partial x^{\nu}} \mathscr{T}^{\nu} \right) \frac{\partial x^{\rho}}{\partial \tilde{x}^{\beta}}$$

in which the first relation expresses the change of coordinates for the tensor (5.61) whereas the second relation highlights the change of coordinates for tensor $\mathscr{T}$ together with the rule of derivative composition. The term $\mathscr{S}^{\lambda}_{\mu}$ then takes the form:

$$\mathscr{S}^{\lambda}_{\mu} = \left(\frac{\partial x^{\lambda}}{\partial \tilde{x}^{\alpha}} \frac{\partial \tilde{x}^{\beta}}{\partial x^{\mu}} \right) \frac{\partial \tilde{x}^{\alpha}}{\partial x^{\nu}} \frac{\partial \mathscr{T}^{\nu}}{\partial x^{\rho}} \frac{\partial x^{\rho}}{\partial \tilde{x}^{\beta}} + \left(\frac{\partial x^{\lambda}}{\partial \tilde{x}^{\alpha}} \frac{\partial \tilde{x}^{\beta}}{\partial x^{\mu}} \right) \frac{\partial}{\partial x^{\rho}} \left(\frac{\partial \tilde{x}^{\alpha}}{\partial x^{\nu}} \right) \frac{\partial x^{\rho}}{\partial \tilde{x}^{\beta}} \mathscr{T}^{\nu}$$

which can be simplified as:

$$\mathscr{S}_{\mu}^{\lambda} = \delta_{\mu}^{\rho}\, \delta_{\nu}^{\lambda}\, \frac{\partial \mathscr{T}^{\nu}}{\partial x^{\rho}} + \frac{\partial x^{\lambda}}{\partial \tilde{x}^{\alpha}}\, \delta_{\mu}^{\rho}\, \frac{\partial}{\partial x^{\rho}} \left(\frac{\partial \tilde{x}^{\alpha}}{\partial x^{\nu}} \right) \mathscr{T}^{\nu}$$

Thanks to Lemma (5.1), we deduce:

$$\mathscr{S}_{\mu}^{\lambda} = \frac{\partial \mathscr{T}^{\lambda}}{\partial x^{\mu}} + \Gamma_{\mu\nu}^{\lambda}\, \mathscr{T}^{\nu} = \nabla_{\mu} \mathscr{T}^{\nu}$$

(2) Let's now consider a tensor $(0,1)$-type. In an inertial frame, we can write the tensor:

$$\frac{\partial \tilde{\mathscr{T}}_{\alpha}}{\partial \tilde{x}^{\beta}}(\mathbf{x}_0)$$

Elevation of the index α by means of the components of the metric projected onto the inertial frame gives:

$$\tilde{g}^{\alpha\gamma}(\mathbf{x}_0)\, \frac{\partial \tilde{\mathscr{T}}_{\gamma}}{\partial \tilde{x}^{\beta}}(\mathbf{x}_0) = \frac{\partial}{\partial \tilde{x}^{\beta}} \left(\tilde{g}^{\alpha\gamma}\, \tilde{\mathscr{T}}_{\gamma} \right) = \frac{\partial \tilde{\mathscr{T}}^{\alpha}}{\partial \tilde{x}^{\beta}}$$

given the definition of the inertial frame (the partial derivative of the metric in this frame vanishes). According to step (1), this term is a tensor $\nabla_{\beta} \mathscr{T}^{\alpha}$. Lowering the index by a compatible metric gives $\nabla_{\beta} \mathscr{T}_{\alpha}$ which shows that it is also a covariant derivative then a tensor.

□

Remark 5.15. This theorem is tightly connected to the Principle of Minimal Coupling. Indeed, application of the Minimal Coupling Principle amounts to replacing a partial derivation ∂ and a metric tensor η_{ij} of Minkowski spacetime, in a flat spacetime, by a covariant derivation ∇ and the metric $g_{\alpha\beta}$ respectively, in an Einstein curved spacetime. This principle presupposes the absence of torsion and in the case of a connection $\overline{\nabla}$ compatible with the metric. Problems appear in Einstein-Cartan spacetime. This aspect is all the more crucial when we are interested in the notion of volume-form, which has drastic consequences for the rigorous formulation of the divergence theorem e.g. [131].

5.3.3.2 *Connection and Inertial Frames in a Flat Spacetime*

Basically, special relativity involves two referential frames in uniform translation to each other. In four-dimensional spacetime, we then have the classic Lorentz transformations. More generally, however, the choice of coordinate system for each reference frame may be arbitrary, due to the requirement of form invariance (covariance) of the equations. What concerns us in this subsection is to determine the general criterion for deciding which connection should be attributed to flat spacetime viewed as a continuum. For the metric, the situation is straightforward. By contrast, the connection is not a tensor: its coefficients depend on the choice of coordinates used to describe the motion of a set of material points, and therefore this criterion cannot be tensorial.

Proposition 5.1. *(Krause, 1976) In the framework of special relativity, let* $\left(\mathscr{M}, g_{\mu\nu}, \Gamma^{\lambda}_{\mu\nu}\right)$ *be a Minkowski flat spacetime and say a referential frame on* $\mathscr{M}$ *where an arbitrary coordinate system* $(x^{\mu}) := (x^0,\ x^i,\ i = 1,2,3)$ *is chosen. This referential frame is locally inertial if and only if* $\Gamma^{i}_{0\nu}(\mathbf{x}) = 0$.

Proof. Details of the proof may be found in [116], but for completeness, we report some parts of it. We have to only consider internal transformations for coordinate systems and their inverse:

$$\begin{cases} \tilde{x}^0 = \tilde{\varphi}^0\left(x^0, x^j\right) \\ \tilde{x}^i = \tilde{\varphi}^i\left(x^j\right) \end{cases} \quad \text{and} \quad \begin{cases} x^0 = \varphi^0\left(\tilde{x}^0, \tilde{x}^j\right) \\ x^i = \varphi^i\left(\tilde{x}^j\right) \end{cases} \tag{5.63}$$

where x^{μ} and $\tilde{x}^{\mu}$ are coordinate systems respectively on referential frames $\mathscr{S}$ and $\tilde{\mathscr{S}}$ (Fig. 5.6). Assume first that the frame $\mathscr{M}$ is (locally) inertial,

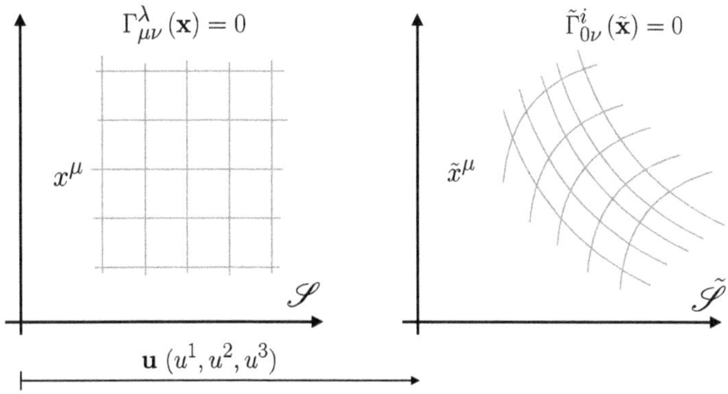

Fig. 5.6: The two inertial frames $\mathscr{S}$ and $\tilde{\mathscr{S}}$ are in relative uniform translation. On each flat spacetime, coordinate systems x^{μ} and $\tilde{x}^{\mu}$ may be defined. For a Cartesian choice, the connection coefficients are identically equal to zero, whereas for a curvilnear coordinate, the coefficient should satisfy conditions.

then $\Gamma^{\lambda}_{\mu\nu}(\mathbf{x}) = 0$. By applying the transformation rules for connection coefficients (5.59), we can write:

$$\tilde{\Gamma}^{\lambda}_{\mu\nu}(\tilde{\mathbf{x}}) = \frac{\partial \tilde{x}^{\lambda}}{\partial x^{\gamma}} \frac{\partial^2 x^{\gamma}}{\partial \tilde{x}^{\mu} \partial \tilde{x}^{\nu}} \tag{5.64}$$

since the set x^{μ} is inertial (Galilean). On the one hand, considering internal transformations, we obtain more explicitly:

$$\begin{cases} \tilde{\Gamma}^{0}_{\mu\nu}(\tilde{\mathbf{x}}) = \dfrac{\partial \tilde{x}^{0}}{\partial x^{\gamma}} \dfrac{\partial^2 x^{\gamma}}{\partial \tilde{x}^{\mu} \partial \tilde{x}^{\nu}} \\ \tilde{\Gamma}^{i}_{0\nu}(\tilde{\mathbf{x}}) = \dfrac{\partial \tilde{x}^{i}}{\partial x^{\gamma}} \dfrac{\partial^2 x^{\gamma}}{\partial \tilde{x}^{0} \partial \tilde{x}^{\nu}} = 0 \\ \tilde{\Gamma}^{k}_{i\nu}(\tilde{\mathbf{x}}) = \dfrac{\partial \tilde{x}^{k}}{\partial x^{\gamma}} \dfrac{\partial^2 x^{\gamma}}{\partial \tilde{x}^{i} \partial \tilde{x}^{j}} \end{cases}$$

On the other hand, given the transformation rules of the connection coefficients, we have particularly for further internal transformations to a new coordinate $\tilde{\tilde{x}}^{\mu}$:

$$\begin{cases} \tilde{\tilde{x}}^{0} = \tilde{\tilde{\varphi}}^{0}\left(\tilde{x}^{0}, \tilde{x}^{j}\right) \\ \tilde{\tilde{x}}^{i} = \tilde{\tilde{\varphi}}^{i}\left(\tilde{x}^{j}\right) \end{cases}$$

thus we obtain the relations:

$$\tilde{\tilde{\Gamma}}^{i}_{0\nu}(\tilde{\tilde{\mathbf{x}}}) = \left(\frac{\partial \tilde{\tilde{x}}^{i}}{\partial \tilde{x}^{j}}\right)\left(\frac{\partial \tilde{x}^{0}}{\partial \tilde{\tilde{x}}^{0}}\right)\left(\frac{\partial \tilde{x}^{\alpha}}{\partial \tilde{\tilde{x}}^{\nu}}\right)\tilde{\Gamma}^{j}_{0\alpha}(\mathbf{x})$$

To ensure that $\tilde{\mathscr{S}}$ is an inertial frame of reference, it is necessary to satisfy the condition $\tilde{\Gamma}^{j}_{0\alpha}(\tilde{\mathbf{x}}) = 0$ and consequently $\tilde{\tilde{\Gamma}}^{i}_{0\nu}(\tilde{\tilde{\mathbf{x}}}) = 0$, which is an invariant property for arbitrary internal transformations.

It is sufficient to assume $\Gamma^{i}_{0\nu}(\mathbf{x}) = 0$. Let's separate the transformations into a "time-preserving" part and a "space-preserving" part respectively as follows:

$$\tilde{x}^{\nu}\begin{pmatrix} \tilde{x}^{0} = x^{0} \\ \tilde{x}^{i} = \tilde{\varphi}^{i}(x^{0}, x^{j}) \end{pmatrix} \quad \text{and its inverse} \quad x^{\nu}\begin{pmatrix} x^{0} = \tilde{x}^{0} \\ x^{i} = \varphi^{i}(\tilde{x}^{0}, \tilde{x}^{j}) \end{pmatrix} \tag{5.65}$$

and

$$\tilde{\tilde{x}}^{\nu}\begin{pmatrix} \tilde{\tilde{x}}^{0} = \tilde{\tilde{\varphi}}^{0}(\tilde{x}^{0}, \tilde{x}^{j}) \\ \tilde{\tilde{x}}^{i} = \tilde{x}^{i} \end{pmatrix} \tag{5.66}$$

On the one hand, assuming that the coordinate system x^μ is inertial (Galilean), say $\Gamma^\gamma_{\alpha\beta} = 0$, the transformation rules during the whole transformation give:

$$\tilde{\tilde{\Gamma}}^i_{0\nu} = \left(\frac{\partial \tilde{\tilde{x}}^i}{\partial x^\alpha}\right)\left(\frac{\partial^2 x^\alpha}{\partial \tilde{\tilde{x}}^0 \partial \tilde{\tilde{x}}^\nu}\right)$$

$$= \left(\frac{\partial \tilde{\tilde{x}}^i}{\partial \tilde{x}^0}\frac{\partial \tilde{x}^0}{\partial x^\alpha} + \frac{\partial \tilde{\tilde{x}}^i}{\partial \tilde{x}^j}\frac{\partial \tilde{x}^j}{\partial x^\alpha}\right)\left(\frac{\partial^2 x^\alpha}{\partial \tilde{\tilde{x}}^0 \partial \tilde{\tilde{x}}^\nu}\right) = \left(\frac{\partial \tilde{\tilde{x}}^i}{\partial \tilde{x}^j}\right)\left(\frac{\partial^2 x^i}{\partial \tilde{\tilde{x}}^0 \partial \tilde{\tilde{x}}^\nu}\right)$$

where we observe that the first term of the second row vanishes. On the other hand, transformation rules for the space-preserving part give by hypothesis on Galilean property of the frame:

$$\tilde{\tilde{\Gamma}}^i_{0\nu} = \left(\frac{\partial \tilde{\tilde{x}}^i}{\partial \tilde{x}^0}\right)\left(\frac{\partial \tilde{\tilde{x}}^0}{\partial \tilde{x}^0}\right)\left(\frac{\partial \tilde{x}^\alpha}{\partial \tilde{\tilde{x}}^\nu}\right)\tilde{\Gamma}^j_{0\alpha} + 0 = 0$$

Combining the above two relations, we necessarily obtain:

$$\frac{\partial^2 x^i}{\partial \tilde{\tilde{x}}^0 \partial \tilde{\tilde{x}}^\nu} = 0 \quad \Longrightarrow \quad \begin{cases} x^0 = \tilde{\tilde{x}}^0 \\ x^i = u^i\, \tilde{\tilde{x}}^0 + f^i(\tilde{\tilde{x}}^j) \end{cases}, \qquad \frac{dx^i}{dx^0} = u^i \quad (\tilde{\tilde{x}}^j \text{ fixed})$$

where u^i are three arbitrary constants (assimilated to components of a spatial vector) and f^i are three arbitrary time-independent functions. This shows that coordinates $\tilde{x}^\mu$ and $\tilde{\tilde{x}}^\mu$ are attached to inertial frames of reference. □

Remark 5.16. Non-vanishing connection $\Gamma^i_{0\nu} \neq 0$ means inertial forces, and this will be observed for Newton-Cartan gravitation e.g. [144]. Another point remarked by [116] is that Lorentz transformations are not the only transformations connecting inertial frames in special relativity.

5.3.3.3 *Acceleration and Geodesics in Curved Spacetime*

We have derived the expression of acceleration using generalized coordinates and we have obtained the expression (5.15). The idea is now to extend this concept to relative gravitation.

Theorem 5.3. *The path of a moving material point which is only subject to gravitational fields is on a geodesic curve of the spacetime.*

Proof. Assume that the point occupies the event-point $\mathbf{x}_0 \in \mathscr{G}$ of Einstein spacetime. Let's consider an inertial frame, and recall the four-dimensional velocity and the acceleration:

$$u^\alpha := \frac{dx^\alpha}{d\tau}, \qquad a^\alpha := \frac{d^2 x^\alpha}{d\tau^2} \tag{5.67}$$

where τ is the proper time. We can rewrite this acceleration as the derivative of the four-velocity:

$$a^\alpha = \frac{du^\alpha}{d\tau}$$

With respect to an inertial frame (think Principle of Equivalence for physical interpretation), the equation of motion at point $\mathbf{x}_0$ is:

$$\frac{d\tilde{u}^\alpha}{d\tau} = 0 \tag{5.68}$$

where the mass behaves as in a local microcosm where no gravitation is felt. Only gravitational field is considered. We then obtain by chain rule:

$$\frac{d\tilde{u}^\alpha}{d\tau}(\mathbf{x}_0) = \frac{\partial \tilde{u}^\alpha}{\partial \tilde{x}^\beta}(\mathbf{x}_0)\ \frac{d\tilde{x}^\beta}{d\tau}(\mathbf{x}_0) = \frac{\partial \tilde{u}^\alpha}{\partial \tilde{x}^\beta}(\mathbf{x}_0)\ \tilde{u}^\beta(\mathbf{x}_0) = 0$$

Thanks to the proposition concerning the relation (5.62), we can apply for an arbitrary coordinate chart to give:

$$\frac{\partial \tilde{u}^\alpha}{\partial \tilde{x}^\beta}(\mathbf{x}_0)\ \tilde{u}^\beta(\mathbf{x}_0) = 0 \quad \Longrightarrow \quad u^\beta(\mathbf{x}_0)\ \nabla_\beta u^\alpha(\mathbf{x}_0) = 0 \tag{5.69}$$

This induces the conclusion that the material point follows the geodesic and it is expressed in coordinate-free formulation:

$$\boxed{\mathbf{a} := \nabla_{\mathbf{u}} \mathbf{u} = 0} \tag{5.70}$$

The path is then a geodesic as expressed by the equation e.g. [128], [125]. □

Remark 5.17. This four-dimensional formulation of the acceleration is similar to the acceleration (5.39) in three dimensions. We could have immediately introduced a four-dimensional framework for the Newtonian approach but that would not have changed anything since time t is only a decoupled parameter in the classical approach.

To point out the relation between acceleration of Newtonian mechanics and that of relative gravitation, it is straightforward to write the four-velocity:

$$u^\alpha := \frac{dx^\alpha}{d\tau}, \qquad d\tau = \frac{1}{c}\sqrt{g_{\alpha\beta}dx^\alpha dx^\beta} = \frac{1}{c}\sqrt{c^2dt^2 - g_{ii}\left(dx^i\right)^2} \qquad (5.71)$$

where the expression of the proper time is simplified without losing the generality of our purpose. The four-acceleration is rewritten as:

$$u^\beta\left(x^\alpha + u^\alpha d\tau\right) - u^\beta\left(x^\alpha\right) = \left(u^\alpha\ \nabla_\alpha u^\beta\right)d\tau := a^\beta\ d\tau \qquad (5.72)$$

For the velocity, we may approximate:

$$u^0 = \frac{dx^0}{d\tau} = \frac{cdt}{\frac{1}{c}\sqrt{c^2dt^2 - g_{ii}\left(dx^i\right)^2}} \simeq \frac{c^2dt}{cdt} \simeq c \qquad (5.73)$$

when the light speed is much greater that any test point velocity. The other components of the four-velocity hold $i = 1, 2, 3$:

$$u^i = \frac{dx^i}{d\tau} = \frac{dx^i}{\frac{1}{c}\sqrt{c^2dt^2 - g_{ii}\left(dx^i\right)^2}} \simeq \frac{cdx^i}{cdt} = v^i \qquad (5.74)$$

where v^i are the components of the three-dimensional velocity field. We then deduce easily the acceleration:

$$a^0 = u^i\nabla_i u^0 \simeq 0, \quad a^i = u^0\nabla_0 u^i + u^j\nabla_j u^i$$

Given that $x^0 = ct$, $u^i \simeq v^i$, we recover the tri-dimensional form of the acceleration in Newtonian mechanics:

$$a^i = c\ \partial(ct)v^i + v^j\nabla_j v^i \quad \Longrightarrow \quad \mathbf{a} = \partial_t\mathbf{v} + \nabla_\mathbf{v}\mathbf{v} \qquad (5.75)$$

Remark 5.18. Conversely to (5.14), formula (5.70) is more convenient if one limits oneself to solve tangent vectors u^α without solving curves themselves. In (5.70) acceleration is a curved spacetime, where connection may include torsion. This relation is also valid for Einstein-Cartan gravitation.

5.3.3.4 *Geodesics and Autoparallels*

Definition 5.9. (Autoparallels) Consider a manifold $\mathscr{M}$ endowed with a metric $\mathbf{g}$ and a connection ∇ independent from the metric. A curve $\gamma(\sigma)$ such that $\sigma \in \mathbb{R} \to x^\alpha(\sigma) \in \mathscr{M}$ is defined to be autoparallel if the tangent vector $\mathbf{t}$ is transported along the curve by parallel transport, embodied by the connection ∇:

$$\boxed{\nabla_{\mathbf{t}}\mathbf{t} = 0} \tag{5.76}$$

Definition of self-parallel curves (5.76) had not been proposed *ex nihilo* because it simply expresses the nullity of acceleration either in three-dimensional formulation of Newton (5.39), or in four-dimensional formulation of Einstein (5.70), using connection. In terms of components, this coordinate-free expression may be projected onto a base:

$$\nabla_{\mathbf{t}}\mathbf{t} = 0 \quad \Longrightarrow \quad (t^{\nu}\nabla_{\nu}t^{\mu})\,\mathbf{e}_{\mu} = 0$$

Using covariant derivative of a vector leads to:

$$t^{\nu}\left(\frac{\partial t^{\mu}}{\partial x^{\nu}} + \Gamma^{\mu}_{\rho\nu}\, t^{\rho}\right) = 0$$

Introducing the four-velocity, which is tangent to the line-event, we have

$$t^{\mu} = u^{\mu} := \frac{dx^{\mu}}{d\tau} \quad \Longrightarrow \quad \frac{dx^{\nu}}{d\tau}\left[\frac{\partial}{\partial x^{\nu}}\left(\frac{dx^{\mu}}{d\tau}\right) + \Gamma^{\mu}_{\rho\nu}\frac{dx^{\rho}}{d\tau}\right] = 0$$

that is:

$$\boxed{\frac{d^2x^{\mu}}{d\tau^2} + \Gamma^{\mu}_{\rho\nu}\frac{dx^{\rho}}{d\tau}\frac{dx^{\nu}}{d\tau} = 0} \tag{5.77}$$

We have two expressions (5.76) and (5.77) of autoparallel curves. In (5.77), connection is not necessarily symmetric with respect to the two lower indices.

Remark 5.19. The two equations (5.76) and (5.77) do not give the same curves in presence of torsion. The set of curves obtained by (5.76) includes those of (5.77) since the last ones are limited to the symmetric part of the connection (they are not reduced to Christoffel's symbols in the general case).

5.3.3.5 *Geodesics as Shortest Distance Path*

The idea is to make similar developments as we have done for the Hamiltonian principle. To begin with, consider the distance between two points as action. Let's consider the distance as the Lagrangian for a particle without external applied force:

$$\mathscr{L} := \sqrt{g_{\alpha\beta}\, dx^\alpha dx^\beta} \quad \Longrightarrow \quad \mathscr{D}(P,Q) := \int_P^Q \sqrt{g_{\alpha\beta}\, dx^\alpha dx^\beta} \tag{5.78}$$

where $\mathscr{D}(P,Q)$ is the path length from event P to event Q.

Definition 5.10. (Geodesics) Consider a Riemann manifold endowed with a metric $\mathbf{g}$. A curve $\gamma(\sigma)$ such that $\sigma \in \mathbb{R} \to x^\alpha(\sigma) \in \mathscr{M}$ is defined to be a geodesic if the distance between any two points P and Q on $\mathscr{M}$ is the shortest by following γ.

Let's parametrize the curve by τ (arbitrary parameter, but the proper time can be chosen in relative gravitation). The distance is rewritten as:

$$\mathscr{D}(P,Q) := \int_{\tau_1}^{\tau_2} \sqrt{g_{\alpha\beta}\, \frac{dx^\alpha}{d\tau}\frac{dx^\beta}{d\tau}}\; d\tau = \int_{\tau_1}^{\tau_2} \mathscr{L}\left(x^\mu(\tau), x^\mu(\tau), \dot{x}^\mu(\tau), \tau\right) d\tau$$

Finding the shortest distance between these two points amounts to finding the minimum of this functional. It is effective by considering deformed curves "close" to the solution by applying Euler variation $x^\mu \to x^\mu + \delta x^\mu$. The Euler-Lagrange equation (see (5.14)) is deduced:

$$\frac{d}{dt}\left(\frac{1}{\mathscr{L}}\, g_{\beta\gamma}\, \dot{x}^\gamma\right) - \frac{1}{2\mathscr{L}}\, \frac{\partial g_{\lambda\gamma}}{\partial x^\beta}\dot{x}^\lambda\, \dot{x}^\gamma = 0$$

At this step, it is worth choosing the parameter τ as the proper time which is equivalent to imposing $\mathscr{L} \equiv 1$. This implies a simpler equation:

$$\frac{d}{dt}\left(g_{\beta\gamma}\, \dot{x}^\gamma\right) - \frac{1}{2}\, \frac{\partial g_{\lambda\gamma}}{\partial x^\beta}\, \dot{x}^\lambda \dot{x}^\gamma = 0 \quad \Longrightarrow \quad g_{\beta\gamma}\, \ddot{x}^\gamma + \frac{\partial g_{\beta\gamma}}{\partial x^\lambda}\, \dot{x}^\lambda \dot{x}^\gamma$$
$$- \frac{1}{2}\, \frac{\partial g_{\lambda\gamma}}{\partial x^\beta}\, \dot{x}^\lambda \dot{x}^\gamma = 0$$

Knowing that one can invert the indices in front of the term $\dot{x}^\lambda \dot{x}^\gamma$, and by multiplication with the inverse of the metric, we deduce by rearranging the indices:

$$\boxed{\frac{d^2x^\alpha}{d\tau^2} + \overline{\Gamma}^\alpha_{\beta\gamma}\, \frac{dx^\beta}{d\tau}\, \frac{dx^\gamma}{d\tau} = 0, \quad \overline{\Gamma}^\alpha_{\beta\gamma} := \frac{1}{2} g^{\alpha\lambda}\left(\frac{\partial g_{\lambda\gamma}}{\partial x^\beta} + \frac{\partial g_{\lambda\beta}}{\partial x^\gamma} - \frac{\partial g_{\beta\gamma}}{\partial x^\lambda}\right)} \tag{5.79}$$

The geodesic equation (shortest distance) (5.79) has exactly the same shape as the autoparallel equation (5.77). However, the connections involved in these equations are not the same: an affine connection is used for autoparallels, whereas the Levi-Civita connection is used for geodesics. Accordingly, autoparallel curves are used in Einstein-Cartan gravitation, while geodesic curves are used in Einstein gravitation.

Remark 5.20. The concept of geodesic (or autoparallel) curves is local because it is essentially based on the path that a test particle would take in the gravitational field generated by a certain mass distribution in spacetime. By hypothesis, on the one hand, it is assumed that the test particle does not disturb the initial gravitational field. On the other hand, it is assumed that dimensions and implicitly mass of the test particle have no influence on its movement along the geodesic curves. As an extension, a rigorous study by Ehlers and Geroch [235] has shown that within a certain limit of its size and mass, a body moving in the spacetime of general relativity follows a geodesic curve.

Obtention of the geodesic curve equation (5.79) requires some additional developments.

5.3.3.6 *Kinetic Energy, Shortest Distance and Geodesics*

Comparing the previous approach with the approach considering kinetic energy as a Lagrangian may bring new insights. Consider a test particle with mass $m = 1$ without losing the generality of our purpose. The element of distance can be expressed in terms of the kinetic energy:

$$\mathscr{L} = \sqrt{\mathscr{T}}, \quad \text{where} \quad \mathscr{T} := \frac{1}{2} g_{\beta\gamma} \frac{dx^\beta}{d\tau} \frac{dx^\gamma}{d\tau} \tag{5.80}$$

where τ is the time (not necessarily its proper time). The Euler-Lagrange associated with the shortest distance gives after straightforward simplification:

$$\frac{1}{\sqrt{\mathscr{T}}} \frac{\partial \mathscr{T}}{\partial x^\alpha} - \frac{d}{d\tau} \left(\frac{1}{\sqrt{\mathscr{T}}} \frac{\partial \mathscr{T}}{\partial x'^\alpha} \right) = 0$$

where $x'^\alpha := dx^\alpha / d\tau$. By introducing (5.80) and after some calculus (see details in e.g. [128]), we obtain the Euler-Lagrange equations:

$$\frac{d^2 x^\alpha}{d\tau^2} + \overline{\Gamma}^\alpha_{\beta\gamma} \frac{dx^\beta}{d\tau} \frac{dx^\gamma}{d\tau} - \frac{dx^\alpha}{d\tau} \frac{\dfrac{d^2 s}{d\tau^2}}{\dfrac{ds}{d\tau}} = 0 \tag{5.81}$$

in which the arc length is:

$$s(\sigma) := \int_0^{\sigma} \sqrt{g_{\beta\gamma} \frac{dx^\beta}{d\sigma} \frac{dx^\gamma}{d\sigma}} \, d\sigma$$

with σ any arbitrary parameter of the arc. The arc length definition does not depend on parametrization. However, the length of arc traveled depends on the lapses of time τ. We have shown previously that if we choose the kinetic energy as Lagrangian $\mathscr{L} = \mathscr{T}$ (where $m = 1$), the geodesic equation takes the form of:

$$\frac{d^2 x^\alpha}{d\tau^2} + \overline{\Gamma}^\alpha_{\beta\gamma} \frac{dx^\beta}{d\tau} \frac{dx^\gamma}{d\tau} = 0$$

It is observed that along a geodesic curve, the kinetic energy of a particle is constant. For that purpose, on the one hand, we lay down the Euler-Lagrange equation as:

$$\frac{\partial \mathscr{T}}{\partial x^\alpha} - \frac{\mathrm{d}}{d\tau}\left(\frac{\partial \mathscr{T}}{\partial x'^\alpha}\right) = 0$$

On the other hand, let's calculate the rate of the kinetic energy and factorize by introducing the previous equation:

$$\frac{d\mathscr{T}}{d\tau} = \frac{\partial \mathscr{T}}{\partial x^\alpha} \frac{dx^\alpha}{d\tau} + \frac{\partial \mathscr{T}}{\partial x'^\alpha} \frac{d^2 x^\alpha}{d\tau^2} = \frac{\mathrm{d}}{d\tau}\left(x'^\alpha \frac{\partial \mathscr{T}}{\partial x'^\alpha}\right)$$

We deduce:

$$\mathscr{T} - x'^\alpha \frac{\partial \mathscr{T}}{\partial x'^\alpha} = \text{Cste} \tag{5.82}$$

Since kinetic energy is a homogeneous function of degree 2, we have:

$$x'^\alpha \frac{\partial \mathscr{T}}{\partial x'^\alpha} = 2\,\mathscr{T} \qquad \Longrightarrow \qquad \mathscr{T} = \text{Cste} \tag{5.83}$$

Finally, it is assessed that kinetic energy remains constant all along geodesic curves. It is nothing more than the analogy of a uniform (constant) velocity along a straight line for a particle without external force. In sum, kinetic energy is constant instead of velocity. Now, coming back to traveled arc length along a geodesic:

$$s(\tau) := \int_0^{\tau} \sqrt{g_{\beta\gamma} \frac{dx^\beta}{d\tau} \frac{dx^\gamma}{d\tau}} \, d\tau = \int_0^{\tau} \sqrt{2\mathscr{T}} \, d\tau = \sqrt{2\mathscr{T}}\, \tau \tag{5.84}$$

The scalar velocity is constant $s'(\tau) = \sqrt{2\mathscr{T}}$ and the scalar acceleration $s''(\tau) = 0$ is equal to zero, that is to deduce that (5.81) is also the geodesic equation if the arc length is chosen as the parameter.

5.4 Einstein Relative Gravitation

Vacuum spacetime is supposed to be empty of any material substance. However, vacuum is not really an empty region because gravitational and electromagnetic fields can be present and propagate themselves, even without material support. Unlike flat spacetime, this spacetime can be curved or twisted meaning there is dependence of metric and connection on the point event $x^\mu = (x^0, x^1, x^2, x^3)$. We take first a curved spacetime where torsion $\aleph$ is zero. We then obtain a Riemann manifold with metric $\mathbf{g}$ and the associated Levi-Civita connection $\overline{\nabla}$ whose coefficients are denoted $\overline{\Gamma}^\gamma_{\alpha\beta}$ e.g. [5]:

$$\overline{\Gamma}^\gamma_{\alpha\beta} = (1/2) g^{\gamma\lambda} (\partial_\alpha g_{\lambda\beta} + \partial_\beta g_{\alpha\lambda} - \partial_\lambda g_{\alpha\beta}) \tag{5.85}$$

formally deduced from the assumption of metric compatibility $\nabla^\gamma g_{\alpha\beta} = 0$ and of symmetry $\Gamma^\gamma_{\alpha\beta} = \Gamma^\gamma_{\beta\alpha}$.

5.4.1 *Einstein Field Equation*

Definition 5.11. An Einstein spacetime is a four-dimensional manifold (compacted and connected), endowed with a pseudo-Riemann hyperbolic metric $\mathbf{g}$ (verifying the Hilbert causality condition) of signature $(1, 3)$ and with a Levi-Civita connection $\overline{\nabla}$.

5.4.1.1 *Variational Procedure*

The variational approach for Einstein gravitation begins with the definition of an action $\mathscr{S}$ functional of the metric of the spacetime:

$$\mathscr{S} := \int_{\mathscr{B}} \mathscr{L}\left(g_{\alpha\beta}, \partial^\gamma g_{\alpha\beta}, \partial_\lambda \partial^\gamma g_{\alpha\beta}\right) \overline{\omega}_n \tag{5.86}$$

where the volume-form $\overline{\omega}_n := \sqrt{\det g_{\alpha\beta}}\, dx^0 \wedge \cdots \wedge dx^3$, compatible with the Levi-Civita connection. The Lagrangian is covariant if and only if it takes the form of e.g. [110], [169]:

$$\mathscr{S} := \int_{\mathscr{B}} \mathscr{L}\left(g_{\alpha\beta}, \overline{\Re}^{\gamma}_{\alpha\beta\lambda}\right) \overline{\omega}_n \tag{5.87}$$

where $\overline{\Re}^{\gamma}_{\alpha\beta\lambda}$ is the curvature calculated with Levi-Civita connection. Derivation of field equations and then conservation law equations includes (variational procedure):

(1) the first step is the Euler variation:

$$\delta\mathscr{S} := \int_{\mathscr{B}} \delta\mathscr{L}\left(g_{\alpha\beta}, \overline{\Re}^{\gamma}_{\alpha\beta\lambda}\right) \overline{\omega}_n + \mathscr{L}\left(g_{\alpha\beta}, \overline{\Re}^{\gamma}_{\alpha\beta\lambda}\right) \delta\overline{\omega}_n = 0 \tag{5.88}$$

where the primal variable is the metric; its Euler variation holds $\delta g_{\alpha\beta}$;

(2) the second step is the application of the Lie derivative variation ($\delta g_{\alpha\beta} \to \mathcal{L}_\xi g_{\alpha\beta}$):

$$\delta_\xi\mathscr{S} := \int_{\mathscr{B}} \mathcal{L}_\xi\mathscr{L}\left(g_{\alpha\beta}, \overline{\Re}^{\gamma}_{\alpha\beta\lambda}\right) \overline{\omega}_n + \mathscr{L}\left(g_{\alpha\beta}, \overline{\Re}^{\gamma}_{\alpha\beta\lambda}\right) \mathcal{L}_\xi\overline{\omega}_n = 0 \tag{5.89}$$

where vector ξ is assumed to be divergence-free $\mathcal{L}_\xi\overline{\omega}_n = (\overline{\nabla}_\alpha\xi^\alpha)\overline{\omega}_n = 0$. Introduction of the Lie derivative of the metric induces the conservation laws.

Remark 5.21. Two compatibility conditions are used during the variational procedure: compatibility of metric and connection $\overline{\nabla}^{\gamma} g_{\alpha\beta} = 0$, and compatibility of connection and volume-form $\mathcal{L}_\xi\overline{\omega}_n = (\overline{\nabla}_\alpha\xi^\alpha)\overline{\omega}_n$. In addition, the choice of a divergence-free vector field for Lie derivative $\overline{\nabla}_\alpha\xi^\alpha = 0$ is another condition to obtain the Einstein field equations (sometimes overlooked in the literature).

5.4.1.2 *Einstein Field Equation for Relative Gravitation*

We deal with a vacuum spacetime, without matter. A vacuum spacetime is not exactly an empty spacetime since fields may occur. As far as we know, the only two physical interactions that do not require a material medium to exist and propagate at distance are gravitation and electromagnetism in classical physics. For the moment, we are limiting ourselves to gravitation.

Before arriving at the definition of the *Hilbert-Einstein action*, there were developments in the theory of relative gravitation, almost in parallel but sometimes in competition, between that of Einstein, well documented, and that of Hilbert, who established the theory on more mathematical bases, namely by also including electromagnetism. The result was the covariant formulation of field equations of relative gravitation in late 1915.

Hilbert-Einstein action. The Lagrangian associated with vacuum spacetime of relative gravitation is defined as:

$$\mathscr{S}_{HE} := \frac{1}{2\chi} \int_{\mathscr{B}} \overline{\mathcal{R}}\, \overline{\omega}_n \tag{5.90}$$

where $\overline{\mathcal{R}}$ is the scalar curvature associated with the Levi-Civita connection and $\overline{\omega}_n$ the Riemann volume-form.

The Hilbert-Einstein action should be perceived as a beautiful illustration of the outcome of complementary work by a physicist and a mathematician, rather than a competition e.g. [146]. Two paths may be used for establishing the conservation laws of vacuum spacetime in Einstein gravitation.

Einstein field equation: first approach. The first approach directly starts with $\overline{\omega}_n := \sqrt{|\mathrm{Det}\mathbf{g}|}\, dx^0 \wedge \cdots \wedge dx^3$, the variation with respect to the metric giving:

$$\delta\overline{\omega}_n = -\frac{1}{2} g_{\alpha\beta} \delta g^{\alpha\beta}\, \overline{\omega}_n, \quad \text{or} \quad \delta\overline{\omega}_n = -\frac{1}{2} g^{\alpha\beta} \delta g_{\alpha\beta}\, \overline{\omega}_n \tag{5.91}$$

By using the usual path of variational procedure on the action (5.90), the field equation of Einstein gravitation for vacuum spacetime is given by:

$$\boxed{\overline{G}_{\alpha\beta} := \overline{\mathfrak{R}}_{\alpha\beta} - \frac{\overline{\mathfrak{R}}}{2}\, g_{\alpha\beta} = 0, \quad \overline{\mathfrak{R}}_{\alpha\beta} := \overline{\mathfrak{R}}^{\gamma}_{\ \gamma\alpha\beta}} \tag{5.92}$$

Boundary terms are dropped. This tensor has a specific property which could be confused with the conservation laws sometimes. Using Lie derivative variation $\mathcal{L}_\xi g_{\alpha\beta} = \overline{\nabla}_\alpha \xi_\beta + \overline{\nabla}_\beta \xi_\alpha$ gives the well-known result stating that the covariant derivative of Einstein tensor $\overline{G}_{\alpha\beta}$ vanishes:

$$\overline{\nabla}_\beta \overline{G}^{\alpha\beta} = 0 \tag{5.93}$$

in spacetime regions where there is no source. This approach is appealing but hides important facts behind the theory. In fact, Einstein did not derive his field equation via this first approach.

Einstein field equation: second approach. A slightly different approach may be taken. Let's consider again with the Hilbert-Einstein action $\mathscr{S}_{HE}$ (5.90) of a vacuum spacetime and let us apply directly the Lie derivative (such approach is related to the so-called Principle of General Covariance e.g. [130]):

$$\delta_\xi \mathscr{S}_{HE} = \frac{1}{2\chi} \int_{\mathscr{B}} \left(\mathcal{L}_\xi \overline{\mathcal{R}}\right) \overline{\omega}_n + \overline{\mathcal{R}}\, \mathcal{L}_\xi \overline{\omega}_n$$

where $\overline{\mathcal{R}} := g^{\alpha\beta}\overline{\Re}_{\alpha\beta}$. Assuming that volume-form is compatible with connection, as well as metric, we have respectively:

$$\mathcal{L}_\xi \overline{\omega}_n = \overline{\nabla}_\alpha \xi^\alpha\, \overline{\omega}_n, \qquad \overline{\nabla}\mathbf{g} = 0$$

Given that the Lie derivative of Ricci curvature can be written as a divergence e.g. [49], by assuming that the vector field ξ is divergence-free, the boundary terms may be dropped and we thus recover the variational form equation:

$$\mathcal{L}_\xi \mathscr{S}_{HE} = \frac{1}{2\chi} \int_{\mathscr{B}} \overline{\Re}_{\alpha\beta}\, \mathcal{L}_\xi g^{\alpha\beta}\, \overline{\omega}_n + \text{boundary terms} \equiv 0$$

For an arbitrary vector field and by extension an arbitrary 2-contravariant tensor field $\mathcal{L}_\xi g^{\alpha\beta}$, we deduce the field equations for the vacuum spacetime e.g. [125]:

$$\boxed{\overline{\Re}_{\alpha\beta} = 0} \qquad \text{(ten equations)} \tag{5.94}$$

This is initially the Einstein field equations and seems different to the usual formulation (5.92). The key is in fact coming from the compatibility condition which is expressed by means of Bianchi's second identity for Riemann manifolds.

Remark 5.22. In the presence of matter, we might guess therefore a constitutive law between momentum energy tensor and Ricci curvature tensor $T^{\alpha\beta} = 2\chi\, \overline{\Re}^{\alpha\beta}$. This is not possible because the covariant derivative of momentum-energy vanishes whereas that of Ricci curvature does not. We recall (see (2.45)) that it is the Einstein tensor which has a zero derivative covariant. Therefore, the correct law may be suggested as $T^{\alpha\beta} = 2\chi\, \overline{G}^{\alpha\beta}$.

For the second approach, it is interesting to be aware that relation (5.93) can be obtained by using Bianchi's first identities e.g. [125] given the properties of Ricci curvature:

$$\overline{\nabla}_\beta \overline{G}^{\alpha\beta} = 0 \tag{5.95}$$

This is not a conservation law but an identity conservation deduced directly from Bianchi's identities resulting solely from Ricci curvature properties in the region where there is no source. In the next subsection, we derive the motion law of a particle of mass m subject to a gravitational field, where the mass is too small to modify the field. In the meantime, the true conservation law takes the form of:

$$\overline{\nabla}_\beta T^{\alpha\beta} = 0 \tag{5.96}$$

provided the correct constitutive laws and the Bianchi identities (5.95). An interesting historical comment on the introduction of Bianchi's identities in the derivation of Einstein field equations may be found in e.g. [237].

5.4.2 *Jacobi Equation for Einstein Gravitation*

Let's consider two geodesic curves γ_0 and γ_1 in Einstein spacetime $(\mathscr{B}, g_{\alpha\beta})$. At the same proper time τ, we define the separation ξ between these two curves in the spacetime $\mathscr{M}$ which connects a point-event $\mathbf{x}(\tau)$ on the geodesic γ_0 to a point-event $\mathbf{x}(\tau) + \xi(\tau)$. The separation is assumed small in such a way that any Taylor expansion of tensors in terms of ξ with respect to ξ can be truncated to the first-order terms.

5.4.2.1 *Geodesic Deviation Equation in Component Form*

It is reminded that the relativistic acceleration a^α of two particles is defined as the second derivative of the separation vector ξ^α as the two material points move along their respective geodesics defined by (5.70). Let's define the separation velocity and deduce the separation acceleration as follows:

$$v^\alpha := u^\beta \overline{\nabla}_\beta \xi^\alpha, \qquad a^\alpha := u^\beta \overline{\nabla}_\beta v^\alpha \tag{5.97}$$

where $u^\beta := dx^\beta/d\tau$ is the four-vector velocity (timelike vector). From (5.105), we write:

$$\begin{cases} v^\alpha = u^\beta \left(\partial_\beta \xi^\alpha + \overline{\Gamma}^\alpha_{\beta\gamma} \xi^\gamma\right) = \dfrac{d\xi^\alpha}{d\tau} + \overline{\Gamma}^\alpha_{\beta\gamma} u^\beta \xi^\gamma \\[2ex] a^\alpha = u^\beta \left(\partial_\beta v^\alpha + \overline{\Gamma}^\alpha_{\beta\varepsilon} v^\varepsilon\right) = \dfrac{dv^\alpha}{d\tau} + \overline{\Gamma}^\alpha_{\delta\varepsilon} u^\delta v^\varepsilon \end{cases} \tag{5.98}$$

where we have introduced the expressions:

$$\frac{d\xi^\alpha}{d\tau} := u^\beta \partial_\beta \xi^\alpha, \qquad \frac{dv^\alpha}{d\tau} := u^\beta \partial_\beta v^\alpha$$

after the general expression $d\varphi/d\tau := \partial f/\partial_\beta \; dx^\beta/d\tau$. First, substituting the separation velocity into the expression of the separation acceleration gives:

$$a^\alpha = \frac{d^2\xi^\alpha}{d\tau^2} + \frac{d}{d\tau}\left(\overline{\Gamma}^\alpha_{\beta\gamma} u^\beta \xi^\gamma\right) + \overline{\Gamma}^\alpha_{\delta\varepsilon}\left(\frac{d\xi^\varepsilon}{d\tau} + \overline{\Gamma}^\varepsilon_{\beta\gamma} u^\beta \xi^\gamma\right) u^\delta \tag{5.99}$$

Second, let's write the geodesic equations for the two curves γ_0 and γ_1:

$$\begin{cases} \dfrac{d^2x^\alpha}{d\tau^2} + \overline{\Gamma}^\alpha_{\beta\gamma}(x^\mu)\dfrac{dx^\beta}{d\tau}\dfrac{dx^\gamma}{d\tau} = 0 \\ \dfrac{d^2\left(x^\alpha + \xi^\alpha\right)}{d\tau^2} + \overline{\Gamma}^\alpha_{\beta\gamma}(x^\mu + \xi^\mu)\dfrac{d\left(x^\beta + \xi^\beta\right)}{d\tau}\dfrac{d\left(x^\gamma + \xi^\gamma\right)}{d\tau} = 0 \end{cases} \tag{5.100}$$

Since all components ξ^α are small, we retain only the first-order terms after expansion of the second equation. We thus obtain:

$$\frac{d^2\xi^\alpha}{d\tau^2} + \left(\overline{\Gamma}^\alpha_{\beta\gamma} + \overline{\Gamma}^\alpha_{\gamma\beta}\right) u^\beta \frac{d\xi^\gamma}{d\tau} + \frac{\partial \overline{\Gamma}^\alpha_{\beta\gamma}}{\partial x^\delta} u^\beta u^\gamma \xi^\delta = 0 \tag{5.101}$$

Introducing (5.101) in the separation acceleration leads to:

$$a^\alpha = \left(\overline{\Gamma}^\alpha_{\beta\gamma} - \overline{\Gamma}^\alpha_{\gamma\beta}\right) u^\beta \frac{d\xi^\gamma}{d\tau} + \left(\partial_\delta \overline{\Gamma}^\alpha_{\beta\gamma} - \partial_\gamma \overline{\Gamma}^\alpha_{\beta\delta} + \overline{\Gamma}^\alpha_{\delta\varepsilon}\overline{\Gamma}^\varepsilon_{\beta\gamma} - \overline{\Gamma}^\alpha_{\gamma\varepsilon}\overline{\Gamma}^\varepsilon_{\beta\delta}\right) u^\beta \xi^\gamma u^\delta \tag{5.102}$$

where we have used the orthogonality condition

$$\frac{du^\beta}{d\tau} = u^\varepsilon \partial_\varepsilon u^\beta = u^\varepsilon \overline{\nabla}_\varepsilon u^\beta - u^\varepsilon \overline{\Gamma}^\beta_{\varepsilon\delta} \xi^\delta \tag{5.103}$$

Conventionally, we then deduce that the separation acceleration is written as follows for Levi-Civita connection in a Riemann manifold e.g. [32], [33]:

$$\boxed{\frac{D^2\xi^\alpha}{D\tau^2} = \overline{\Re}^\alpha_{\delta\gamma\beta}\; u^\beta u^\delta \; \xi^\gamma} \tag{5.104}$$

The geodesic deviation equation due to Levi-Civita connection for Riemann manifold shows that the curvature produces acceleration of the separation between two neighboring geodesics γ_0 and γ_1. This provides a geometrical interpretation of the curvature tensor.

5.4.2.2 *Coordinate-Free Expression of Jacobi Equation*

First, let's consider two geodesic curves in the pseudo-Riemann manifold $(\mathscr{M}, \mathbf{g})$ denoted by γ_0 and γ_1 respectively. At the same proper time τ, we define the separation four-vector ξ of $\mathscr{M}$ which connects a point (event) $x(\tau)$ of the geodesic γ_0 to a point (event) $x(\tau) + \xi(\tau)$ of a nearby geodesic γ_1. The separation ξ is small in such a way that any Taylor expansion with respect to ξ can be truncated to only the first-order terms. The relativistic acceleration $\mathbf{a}$ of two material points is defined as the second derivative of ξ as the two material points move along their respective geodesics. Separation velocity and separation acceleration along the geodesic curve are:

$$\mathbf{v} := \nabla_{\mathbf{u}}\xi, \qquad \mathbf{a} := \nabla_{\mathbf{u}}\mathbf{v} \tag{5.105}$$

where components $\mathbf{u} := d\mathbf{x}/d\tau$ define the four-vector velocity (timelike vector).

Definition 5.12. The acceleration of the separation ξ is defined as:

$$\boxed{\frac{D^2\xi}{D\tau^2} := \nabla_{\mathbf{u}}\nabla_{\mathbf{u}}\xi} \tag{5.106}$$

which remains valid for either Einstein or Einstein-Cartan spacetime.

Nevertheless, it is well-known that the connection is not uniquely defined for a spacetime, so it is necessary to rewrite this definition (5.106) in a covariant form. For this purpose, let us recall some basic assumptions and relations:

(1) A geodesic on $\mathscr{M}$ endowed with a connection ∇ is defined as a curve $\gamma(t)$ such that parallel transport along the curve preserves the tangent vector to the curve. Say $\mathbf{u}$ a four-vector along a geodesic curve, then we have:

$$\nabla_{\mathbf{u}}\mathbf{u} = 0 \tag{5.107}$$

(2) The orthogonality condition implies the nullity of Lie derivative of vector ξ along the tangent to the geodesic curve: $\mathcal{L}_{\mathbf{u}}\xi := \nabla_{\mathbf{u}}\xi - \nabla_{\xi}\mathbf{u} - \aleph(\xi, \mathbf{u}) = 0$. This allows us to deduce two relations:

$$\begin{cases} \nabla_{\xi}\mathbf{u} = \nabla_{\mathbf{u}}\xi - \aleph(\xi, \mathbf{u}) \\ \nabla_{\mathbf{u}}\xi = \nabla_{\xi}\mathbf{u} + \aleph(\xi, \mathbf{u}) \end{cases} \tag{5.108}$$

(3) We also recall the relation for the curvature operator for any three vectors:

$$\nabla_{\mathbf{u}}\nabla_{\mathbf{v}}\mathbf{w} - \nabla_{\mathbf{v}}\nabla_{\mathbf{u}}\mathbf{w} - \nabla_{[\mathbf{u},\mathbf{v}]}\mathbf{w} = \Re\left(\mathbf{u},\mathbf{v},\mathbf{w}\right) \tag{5.109}$$

From (5.109), we have:

$$\nabla_{\mathbf{u}}\nabla_{\xi}\mathbf{u} - \nabla_{\xi}\nabla_{\mathbf{u}}\mathbf{u} - \nabla_{[\mathbf{u},\xi]}\mathbf{u} = \Re\left(\mathbf{u},\xi,\mathbf{u}\right) \tag{5.110}$$

where the second and third terms vanish because vector $\mathbf{u}$ is tangent to a geodesic, and the Lie derivative of ξ along the geodesic curve is zero.

Geodesic deviation equation. We deduce the intrinsic Jacobi equation called geodesic deviation equation, expressed by means of the Levi-Civita connection e.g. [32], [33]:

$$\frac{D^2\xi}{D\tau^2} = \overline{\Re}\left(\mathbf{u},\xi,\mathbf{u}\right), \qquad \frac{D^2\xi^\alpha}{D\tau^2} = \overline{\Re}^{\alpha}_{\delta\gamma\beta}\ u^\beta u^\delta\ \xi^\gamma \tag{5.111}$$

which is valid within an Einstein spacetime without torsion. $\xi \to \overline{\Re}\left(\mathbf{u},\xi,\mathbf{u}\right)$ is called Jacobi operator along the curve $\gamma(\tau)$ with the tangent $\mathbf{u}$.

The geodesic deviation equation (5.111) shows that the curvature accelerates the separation between two neighboring geodesics γ_0 and γ_1. In a flat spacetime, the separation will be linear. Equation (5.111) allows us to analyze numerous motions of particles in a gravitational field, such as the chaotic behavior of particle orbits, but they are not well-suited to studying spinning particles, either for quantum or macroscopic bodies with intrinsic spin e.g. [238].

Remark 5.23. For a non-uniform potential $\mathscr{U}(t,x^a)$ due to a massive body, we recall the expression of the acceleration of the geodesic deviation:

$$\frac{d^2\xi^a}{dt^2} = -\nabla^a\nabla_b\mathscr{U}\ \xi^b \tag{5.112}$$

In weak field condition (for Earth gravitation this means that $GM/(c^2R) << 1$; G is the constant of gravitation, M earth mass, R earth radius, and c light speed) and low speed motion, i.e., $v/c << 1$, we can relate Newtonian gravitation with Einstein gravitation e.g. [239]. Let's recall $d\tau = cdt$ where c is the light speed, and rewrite the geodesic equation of the Newtonian limit case as:

$$\boxed{\frac{d^2\xi^a}{d\tau^2} = -\Re^a_{b00}\ \xi^b} \tag{5.113}$$

where we introduce $\Re^a_{b00} = c^2\ \nabla^a\nabla_b\mathscr{U}$. This equation was used to analyse the oblateness of the earth e.g. [43].

Remark 5.24. Geodesic deviation equation can be used to characterize spacetime geometry and the right-hand term of (5.111) is called *Tidal Forces.*

5.4.2.3 *Sketch of Resolution*

Even if the form of the equation is linear to the second order in ξ, it turns out that the term in front of ξ in the second member is an unknown factor of the problem in the sense that it is necessary to solve the problem of the geodesics to extract curves whose tangent field is u^μ. The Jacobi equation is a non-trivial differential equation since tidal force is time-dependent and the form of this force is complex. The general form of the equation is written as:

$$\frac{D^2\xi^\alpha}{D\tau^2} + F^\alpha_\beta\,(\mathbf{u},\tau)\,\xi^\beta = 0 \tag{5.114}$$

where $F^\alpha_\beta\,(\mathbf{u},\tau)$ is quadratic in $\mathbf{u}$. Resolution of (5.111) includes two steps:

(1) resolution of geodesics to extract $\mathbf{u}$ (nonlinear first-order differential equations);
(2) resolution of the Jacobi differential equation in ξ (linear second-order differential equations).

In view of (5.114), the equations of geodesic deviation are tightly linked to those of geodesic motion (solve vector $\mathbf{u}$) within a given spacetime. Whenever possible, the Lagrange method or the Hamilton-Jacobi formulation may be used to find part of the solutions for geodesics. However, even if the geodesic Hamiltonian is a motion constant, due to energy conservation in the absence of dissipation, the linearized Hamiltonian for the geodesic deviation is explicitly time-dependent e.g. [240], in which authors have shown how to integrate the geodesic deviation equation for any spacetime admitting integrable geodesics.

5.4.2.4 *Ricci Curvature and Volume Change*

We are interested in Ricci curvature for its role in the evolution of a volume element of an inelastic continuum. To do this, we start from a volume V generated by the vector basis $(\mathbf{f}_1, \cdots, \mathbf{f}_n)$ which deforms with the continuum. The continuum $\mathscr{B}$ is endowed with a volume-form ω_n and a connection ∇:

$$V := \omega_n (\mathbf{f}_1, \cdots, \mathbf{f}_n) = \omega_n \left(f^{\alpha_1}_{(1)}\, \mathbf{f}_{10}, \cdots, f^{\alpha_n}_{(n)}\, \mathbf{f}_{n0} \right) \tag{5.115}$$

where the embedded base is projected onto the initial base, fixed with respect to the time. $\mathbf{f}_\mu = f^{\alpha_\mu}_{(\mu)}\, \mathbf{f}_{\mu 0}$, with $\mu = 1, \cdots, n$. The volume-form is a non-vanishing form everywhere, so we can write the relation:

$$V := \omega_n (\mathbf{f}_1, \cdots, \mathbf{f}_n) = \left(f^{\alpha_1}_{(1)} \cdots f^{\alpha_n}_{(n)} \right) \omega_n (\mathbf{f}_{10}, \cdots, \mathbf{f}_{n0}) \tag{5.116}$$

where the initial volume element does not depend on time, meaning that its time derivative is identically equal to zero. Let us now write the time derivative of the volume according to direction $\mathbf{u}$:

$$\frac{DV}{D\tau} := \nabla_{\mathbf{u}} V = \sum_{\mu=1}^{n} \nabla_{\mathbf{u}} f^{\alpha_\mu}_{(\mu)} \prod_{\nu=1, \nu\neq\mu}^{n} f^{\alpha_\nu}_{(\nu)} \omega_n (\mathbf{f}_{10}, \cdots, \mathbf{f}_{n0}) \tag{5.117}$$

The second derivative again along direction of time $\mathbf{u}$ gives:

$$\begin{aligned} \frac{D^2 V}{D\tau^2} &= \sum_{\mu=1}^{n} \nabla_{\mathbf{u}} \nabla_{\mathbf{u}} f^{\alpha_\mu}_{(\mu)} \prod_{\nu=1, \nu\neq\mu}^{n} f^{\alpha_\nu}_{(\nu)} \omega_n (\mathbf{f}_{10}, \cdots, \mathbf{f}_{n0}) \\ &+ \sum_{\mu,\nu=1, \nu\neq\mu}^{n} \nabla_{\mathbf{u}} f^{\alpha_\mu}_{(\mu)}\, \nabla_{\mathbf{u}} f^{\alpha_\nu} \prod_{\rho=1, \rho\neq\mu,\nu}^{n} f^{\alpha_\rho}_{(\rho)}\, \omega_n (\mathbf{f}_{10}, \cdots, \mathbf{f}_{n0}) \end{aligned} \tag{5.118}$$

By definition, for an assumed Riemann manifold, the first row of the right-hand side of the previous equation induces:

$$\nabla_{\mathbf{u}} \nabla_{\mathbf{u}} f^{\alpha_\mu}_{(\mu)} := \frac{D^2}{D\tau^2} f^{\alpha_\mu}_{(\mu)} = \overline{\Re}^{\alpha_\mu}_{\delta\alpha_\rho\beta}\, u^\beta\, u^\delta\, f^{\alpha_\rho}_{(\rho)} = -\overline{\Re}^{\alpha_\mu}_{\alpha_\rho\delta\beta}\, u^\beta\, u^\delta\, f^{\alpha_\rho}_{(\rho)} \tag{5.119}$$

In this relation, we can see that if ρ is equal to any value of ν except for $\rho = \mu$ then we have the determinant where two columns are equal, and therefore the term vanishes. The only possibility to have non-zero term is the case where $\rho = \mu$. We deduce the second derivative of the volume element:

$$\frac{D^2 V}{D\tau^2} = -\overline{\Re}^{\alpha_\rho}_{\alpha_\rho \delta\beta}\, u^\beta\, u^\delta \prod_{\nu=1}^{n} f^{\alpha_\nu}_{(\nu)} \omega_n\left(\mathbf{f}_{10}, \cdots, \mathbf{f}_{n0}\right)$$
$$+ \sum_{\mu,\nu=1,\nu\neq\mu}^{n} \nabla_{\mathbf{u}} f^{\alpha_\mu}_{(\mu)}\, \nabla_{\mathbf{u}} f^{\alpha_\nu} \prod_{\rho=1,\rho\neq\mu,\nu}^{n} f^{\alpha_\rho}_{(\rho)}\, \omega_n\left(\mathbf{f}_{10}, \cdots, \mathbf{f}_{n0}\right) \tag{5.120}$$

(5.120) shows the physical meaning of the Ricci curvature tensor which governs the evolution of a volume element propagated along a geodesic. Since the second row can be considered as the volume change in the absence of curvature, (5.120) can be interpreted as:

$$\boxed{\frac{D^2_{\text{inel}} V}{D\tau^2} = -\overline{\Re}^{\alpha_\rho}_{\alpha_\rho \delta\beta}\, u^\beta\, u^\delta\, V = -\overline{\mathcal{R}}_{\delta\beta}\, u^\beta\, u^\delta\, V} \tag{5.121}$$

Remark 5.25. The curvature tensor has several physical interpretations. In particular, it can be seen as a quantity that measures variations in the density of disclinations. In addition, an important component of the curvature—the Ricci tensor—also measures the evolution of a volume element during inelastic transformations.

Remark 5.26. In relative gravitation, knowing that the Ricci curvature is the main variable describing Einstein field equations, we thus retain its essential role in the description of the volume change of the Universe by following the geodesic lines (i.e., its expansion).

5.4.3 *Application to Schwarzschild Spacetime*

We are interested in the motion of small objects near a spherical massive body engendering gravitational fields in relative gravitation. Various solutions of the Einstein equations have been found and intensively analyzed e.g. [241] and references therein. The goal is not to study all of them but rather to sketch the method for solving geodesics, autoparallels and, above all, to draw a path towards Einstein-Cartan spacetime models, and its relation with electromagnetism.

We thus consider the geodesic deviation within a Schwarzschild spacetime with a spherically symmetric body, and particularly for sn exterior field. More detailed development on the applicability of (5.111) can be found in e.g. [232] both in Newtonian and Einstein gravitation for comparing the effects of relativity on the orbits of satellites. Various perturbations have been considered.

5.4.3.1 *Schwarzschild Metric*

Conforming to Synge *conformastat* vocabulary, a general static Universe is represented by the line element $ds^2 := f^2(x^i)dt^2 - g^2(x^i)dx^i dx^j$ where t and $x^i, i = 1, 2, 3$ are time and spatial coordinates respectively [242]. For these static spacetimes (meaning non-transient), metric components depend only on spatial coordinates. In the present example, we have to find a metric in a spherically symmetric spacetime. The classical method is to search for the spacetime metric defined by the line element e.g. [125]; proof may be also found in the Appendix (8.11):

$$\boxed{ds^2 := e^{2\nu}\ (dx^0)^2 - e^{2\mu}\ dr^2 - r^2\ d\theta^2 - r^2 \sin^2\theta d\varphi^2} \tag{5.122}$$

where functions $\nu(r)$ and $\mu(r)$ depend only on radius r. This line element constitutes the most general form for spherical symmetry. For the sake of simplicity, we directly consider here a non-transient case. To begin with, we only consider the case where torsion vanishes, which reduces the model to classical Schwarzschild metric e.g. [125]. The only non-vanishing coefficients of Levi-Civita connection $\overline{\nabla}$ are the symbols of Christoffel calculated by means of partial derivative of metric components:

$$\overline{\Gamma}^0_{01} = \Gamma^0_{10} = \nu' \tag{5.123}$$

$$\overline{\Gamma}^1_{00} = \nu'\ e^{2\nu-2\mu}, \quad \overline{\Gamma}^1_{11} = \mu', \quad \overline{\Gamma}^1_{22} = -r\ e^{-2\mu},$$
$$\overline{\Gamma}^1_{33} = -r\ e^{-2\mu} \sin^2\theta \tag{5.124}$$

$$\overline{\Gamma}^2_{12} = \Gamma^2_{21} = \frac{1}{r}, \quad \overline{\Gamma}^2_{33} = -\sin\theta\cos\theta \tag{5.125}$$

$$\overline{\Gamma}^3_{13} = \overline{\Gamma}^3_{31} = \frac{1}{r}, \quad \overline{\Gamma}^3_{23} = \overline{\Gamma}^3_{32} = \cot\theta \tag{5.126}$$

The *a priori* non-vanishing components of the Ricci curvature $\overline{\Re}_{\alpha\beta} := \overline{\Re}^{\gamma}_{\gamma\alpha\beta}$ hold:

$$\begin{cases} \overline{\Re}_{00} = e^{2\nu-2\mu}\left[\nu'' + (\nu')^2 - \nu'\mu' + (2\nu'/r)\right] \\ \overline{\Re}_{11} = -\left[\nu'' + (\nu')^2 - \nu'\mu' - (2\mu'/r)\right] \\ \overline{\Re}_{22} = e^{-2\mu}\left(-1 + r\ \mu' - r\ \nu'\right) + 1 \\ \overline{\Re}_{33} = \left[e^{-2\mu}\left(-1 + r\ \mu' - r\ \nu'\right) + 1\right]\sin^2\theta \end{cases} \tag{5.127}$$

The off-diagonal components all vanish. By writing the field equations (5.94), we have $\overline{\Re}_{\alpha\beta} = 0$ which reduce to three independent differential equations:

$$\begin{cases} \nu'' + (\nu')^2 - \nu'\mu' + (2\nu'/r) = 0 \\ -\left[\nu'' + (\nu')^2 - \nu'\mu' - (2\mu'/r)\right] = 0 \\ e^{-2\mu}\,(-1 + r\,\mu' - r\,\nu') + 1 = 0 \end{cases}$$

Adding the first two equations gives the condition $\nu' + \mu' = 0$ showing that $\nu + \mu = C$. The system can be rewritten as:

$$\begin{cases} \nu'' + 2(\nu')^2 + (2\nu'/r) = 0 \\ \nu' + \mu' = 0 \\ e^{-2\mu}\,(-1 + 2\,r\,\mu') + 1 = 0 \end{cases}$$

From the equation $\overline{\Re}_{22} = 0$ we make the change of variable $V(r) := 1 - e^{-2\mu}$:

$$V(r) + r\,V'(r) = 0 = 0 \quad \Longrightarrow \quad e^{-2\mu} = 1 - \frac{2m}{r}$$

where m is a constant of integration. The Schwarzschild metric is then expressed as e.g. [241], whenever we choose $C = 0$ if we admit an asymptotic Minkowski spacetime:

$$\boxed{\begin{cases} g_{00} = 1 - 2m/r \\ g_{rr} = -\,(1 - 2m/r)^{-1} \end{cases} \qquad \begin{cases} g_{\theta\theta} = -r^2 \\ g_{\varphi\varphi} = -r^2 \sin^2\theta \end{cases}} \tag{5.128}$$

Remark 5.27. Obtention of solution (5.128) is not so smooth as it seems. The solution in $\nu(r)$ is not so easy to find at first glance, since we have to satisfy the three equations. In the original paper of Schwarzschild, the following solution was proposed (historical aspect may be found in e.g. [243]):

$$ds^2 = (1 - \alpha/R)\,dt^2 - \frac{dR^2}{1 - \alpha/R} - R^2\left(d\vartheta^2 + \sin^2\vartheta\right), \text{with } R := \left(r^2 + \alpha^3\right)^{1/3}$$

This is quite different from (5.128), particularly for the interpretation of R which is not the radial coordinate. α plays the role of the constant $2m$. The details of the calculus in the original paper is quite complicated. During his search for a solution, Schwarzschild's original method wanted to respect Einstein's covariance condition which restricted transformations to only those with a determinant equal to 1, which gave him inspiration for special, non-standard spherical coordinates $R := \left(r^2 + \alpha^3\right)^{1/3}$. This method was more or less abandoned subsequently when covariance transformations were no longer restricted to that with the determinant 1. It was Droste in 1916 in his thesis in Dutch (a less widely used language than German at the time) who established the Schwarzschild metric in its currently recognized form.

There is another path to calculate the metric by means of the Einstein tensor. The scalar curvature is given by the relation $\overline{\mathcal{R}} := g^{\alpha\beta}\, \overline{\Re}_{\alpha\beta}$:

$$\overline{\mathcal{R}} = e^{-2\mu}\left[2\nu'' + 2(\nu')^2 - 2\nu'\mu' + 4\frac{\nu' - \mu'}{r} + \frac{2}{r^2}\right] - \frac{2}{r^2} \tag{5.129}$$

Einstein tensor $\overline{G}_{\alpha\beta} := \overline{\Re}_{\alpha\beta} - (\overline{\mathcal{R}}/2) g_{\alpha\beta}$ has also four *a priori* non-vanishing diagonal components after the relation (5.127):

$$\begin{cases} \overline{G}_{00} = e^{2\nu}\left[e^{-2\mu}\left(\dfrac{2\mu'}{r} - \dfrac{1}{r^2}\right) + \dfrac{1}{r^2}\right] \\ \overline{G}_{11} = e^{2\mu}\left[e^{-2\mu}\left(\dfrac{2\nu'}{r} + \dfrac{1}{r^2}\right) - \dfrac{1}{r^2}\right] \\ \overline{G}_{22} = r^2\, e^{-2\mu}\left[\nu'' + (\nu')^2 - \nu'\mu' + \dfrac{\nu' - \mu'}{r}\right] \\ \overline{G}_{33} = r^2\, e^{-2\mu}\left[\nu'' + (\nu')^2 - \nu'\mu' + \dfrac{\nu' - \mu'}{r}\right] \sin^2\theta \end{cases} \tag{5.130}$$

Remark 5.28. Components (5.130) are the projection of the Einstein tensor onto the dyadic: $\{dx^0 \otimes dx^0, dr \otimes dr, d\theta \otimes d\theta, d\varphi \otimes d\varphi\}$. It is sometimes worth defining the coframe for tensor decomposition:

$$\left\{\vartheta^0 := e^{\nu}\, dx^0, \vartheta^1 := e^{\mu}\, dr, \vartheta^2 := r\, d\theta, \vartheta^3 := r \sin\theta\, d\varphi\right\} \tag{5.131}$$

The expression of the metric tensor in this base is given by:

$$\mathbf{g} = \vartheta^0 \wedge \vartheta^0 - \vartheta^r \wedge \vartheta^r - \vartheta^\theta \wedge \vartheta^\theta - \vartheta^\varphi \wedge \vartheta^\varphi \tag{5.132}$$

By projecting Einstein tensor $\overline{\mathbf{G}}$ onto the base $\vartheta^\mu \wedge \vartheta^\nu$, we write:

$$\overline{\mathbf{G}} = \overline{G}_{(\mu\nu)}\, \vartheta^\mu \wedge \vartheta^\nu \tag{5.133}$$

We obtain the "physical components" of the tensor:

$$\begin{cases} \overline{G}_{(00)} = \left[e^{-2\mu} \left(\frac{2\mu'}{r} - \frac{1}{r^2} \right) + \frac{1}{r^2} \right] \\ \overline{G}_{(11)} = \left[e^{-2\mu} \left(\frac{2\nu'}{r} + \frac{1}{r^2} \right) - \frac{1}{r^2} \right] \\ \overline{G}_{(22)} = e^{-2\mu} \left[\nu'' + (\nu')^2 - \nu'\mu' + \frac{\nu' - \mu'}{r} \right] \\ \overline{G}_{(33)} = e^{-2\mu} \left[\nu'' + (\nu')^2 - \nu'\mu' + \frac{\nu' - \mu'}{r} \right] \end{cases} \tag{5.134}$$

It could be checked that for a vacuum spacetime without torsion, Einstein tensor vanishes and by means of the first two field equations (5.134), we deduce that $\nu' + \mu' = 0$. From the first component $\overline{G}_{(00)} = 0$ we exactly obtain the solution:

$$e^{-2\mu} = 1 - \frac{2m}{r}$$

without any approximation or additional assumption. The equation $\overline{G}_{(11)} = 0$ was implicitly replaced by $\nu' + \mu' = 0$ due to the addition operation. Either for the equation $\overline{\Re}_{00} = 0$ or $\overline{G}_{(22)} = \overline{G}_{(33)} = 0$, the same problem holds—we have to solve the equation $\nu'' + 2(\nu')^2 + (2\nu'/r) = 0$ which is not obvious.

By assuming that the spacetime tends to a Minkowski spacetime, a solution of the Einstein field equation for vacuum spacetime with spherical symmetry is then provided by the same metric as previously by also choosing $\nu + \mu = 0$, say the Schwarzschild metric (5.128):

$$g_{00} = 1 - \frac{2m}{r}, \quad g_{rr} = -\left(1 - \frac{2m}{r}\right)^{-1}, \quad g_{\theta\theta} = -r^2, \quad g_{\varphi\varphi} = -r^2 \sin^2\theta$$

projected onto coordinate spherical base $(\partial_{ct}, \partial_r, \partial_\theta, \partial_\varphi)$. m is an integration constant. The line element of Schwarzschild spacetime holds:

$$ds^2 = \left(1 - \frac{2m}{r}\right)(dx^0)^2 - \left(1 - \frac{2m}{r}\right)^{-1} dr^2 - r^2 d\theta^2 - r^2 \sin^2\theta d\varphi^2 \tag{5.135}$$

The same remark as for the original solution of Schwarzschild still holds.

Remark 5.29. It is consistent with Schwarzschild's remark on his solution's validity, that when we neglect the second-order term $(\nu')^2 << \nu'$ then the three equations are compatible and the solution found is valid.

Remark 5.30. The Schwarzschild solution occupies an important place in understanding the concept of spacetime in relative gravitation. Being the simplest solution of the Einstein field equation, the solution is so simple that Einstein himself was amazed. In fact, in Schwarzschild's original solution, the coordinate r which appears in the metric did not have the same interpretation as the radius vector in spherical coordinates, but a function of $R = f(r)$. This provoked numerous debates between Einstein and Painlevé, but also the scepticisms of Hilbert e.g. [243]. The object of the misunderstandings was found in the interpretation of singularity at the origin $r = 0$ and in the Schwarzschild sphere (whose radius is defined at the place of change of sign of g_{00}) linked to the behavior of black holes.

Remark 5.31. In most cases, the term $2m << r$. This seems to exclude the aspects of singularity in the line of Schwarzschild (5.136). Lanczos in 1922 suggested another radial coordinate $\overline{r} := r + m$ but keeping the definitions of the angle θ and φ. The line element is modified as follows:

$$ds^2 = \left(\frac{\overline{r} - m}{\overline{r} + m}\right)(dx^0)^2 - \left(\frac{\overline{r} + m}{\overline{r} - m}\right)d\overline{r}^2 - (\overline{r} + m)^2\left(d\theta^2 + \sin^2\theta d\varphi^2\right) \tag{5.136}$$

which is nothing more than another expression of the Schwarzschild line element. In fact, generally speaking, the choice of another valid coordinate $r := f(\overline{r})$ (namely requiring the inversibility of the function f) is always possible and does not constitute another solution, but would only make it possible to better carry out an analysis of the same physical phenomenon which is the gravitational field due to a mass body with spherical symmetry.

5.4.3.2 *Schwarzschild Geodesics*

Staying in the example of satellite launches and their relativistic motions, it is worth considering only timelike geodesics, which are the trajectories of massive particles at subliminal speed. Anyhow, in order to apply geodesic deviation equations, we should start searching for general forms of geodesic curves in Schwarzschild spacetime by means of geodesic equations at the initial time. The geodesic equation (5.70) then takes the following generic form for acceleration:

$$\frac{Du^\gamma}{D\tau} \equiv \frac{du^\gamma}{d\tau} + \overline{\Gamma}^\gamma_{\mu\nu} u^\mu u^\nu = 0 \tag{5.137}$$

For spherically symmetric Schwarzschild spacetime, we can derive the geodesic equations where $e^{2\nu} = 1 - 2m/r$:

$$\left\{ \begin{aligned} \frac{du^0}{d\tau} + 2\nu' \left(u^0 u^1\right) &= 0 \\ \frac{du^1}{d\tau} + \nu' \, e^{4\nu}(u^0)^2 - \nu'(u^1)^2 - re^{2\nu}(u^2)^2 - r\sin^2\theta e^{2\nu}(u^3)^2 &= 0 \\ \frac{du^2}{d\tau} + \frac{2}{r} \left(u^1 u^2\right) - \sin\theta\cos\theta \left(u^3\right)^2 &= 0 \\ \frac{du^3}{d\tau} + \frac{2}{r} \left(u^1 u^3\right) + 2\cot\theta (u^2 u^3) &= 0 \end{aligned} \right. \tag{5.138}$$

The coefficients of Levi-Civita connection for a Schwarzschild spacetime are directly calculated from the metric (5.128):

$$\overline{\Gamma}^1_{00} = \frac{m}{r^3}(r - 2m), \quad \overline{\Gamma}^0_{01} = \overline{\Gamma}^0_{10} = \frac{m}{r(r-2m)}, \quad \overline{\Gamma}^1_{11} = -\frac{m}{r(r-2m)} \tag{5.139}$$

$$\overline{\Gamma}^2_{12} = \overline{\Gamma}^2_{21} = \frac{1}{r}, \quad \overline{\Gamma}^1_{22} = -(r - 2m) \tag{5.140}$$

$$\overline{\Gamma}^3_{13} = \overline{\Gamma}^3_{31} = \frac{1}{r}, \quad \overline{\Gamma}^1_{33} = -(r - 2m)\sin^2\theta \tag{5.141}$$

$$\overline{\Gamma}^2_{33} = -\sin\theta\cos\theta, \quad \overline{\Gamma}^3_{23} = \overline{\Gamma}^3_{32} = \frac{\cos\theta}{\sin\theta}, \quad \text{others} = 0 \tag{5.142}$$

Remark 5.32. These are the connection coefficients of vacuum spacetime due to the presence of equivalent mass $m := GM : c^2$ where G is the universal gravitation constant, M the mass of the spherical body generating the gravitational field, and c the light speed. The considered region is outside the body of mass M.

Let's consider an example of equatorial motion. For our particular case, we obtain the Schwarzschild geodesic equation:

$$\begin{cases} \dfrac{du^0}{d\tau} + \dfrac{2m}{r(r-2m)} u^0 u^1 = 0 \\ \dfrac{du^1}{d\tau} + \dfrac{m}{r^3}(r-2m)(u^0)^2 - \dfrac{m}{r(r-2m)}(u^1)^2 - (r-2m)[(u^2)^2 \\ \quad + \sin^2\theta (u^3)^2] = 0 \\ \dfrac{du^2}{d\tau} + \dfrac{2}{r} u^1 u^2 - \sin\theta\cos\theta (u^3)^2 = 0 \\ \dfrac{du^3}{d\tau} + \dfrac{2}{r} u^1 u^3 + 2\dfrac{\cos\theta}{\sin\theta} u^2 u^3 = 0 \end{cases} \tag{5.143}$$

From metric (5.128), we also write the line element divided by $d\tau$:

$$\left(\frac{ds}{d\tau}\right)^2 = \left(1 - \frac{2m}{r}\right)(u^0)^2 - \frac{1}{1-2m/r}(u^1)^2 - r^2(u^2)^2 - r^2\sin^2\theta(u^3)^2 \tag{5.144}$$

where $u^\mu := dx^\mu/d\tau$ denotes the four-velocity ($x^0 := ct$). When we choose the time coordinate as $x^0 := t$, the timelike line element is equal to $ds^2 := c^2 d\tau^2$, whereas if we adopt $x^0 := ct$, then we have $ds^2 = d\tau^2$. The four-velocity is normalized, then the term $ds/d\tau = -1, 0, +1$ takes one of these three constants. Lightlike geodesic $ds/d\tau$ corresponds to massless particles. First, we consider radial geodesic spacetime curves, thus we can limit to the first two rows and consider $u^2 \equiv 0$ and $u^3 \equiv 0$ which are compatible. Given that $u^1 := dr/d\tau$, we can integrate the first equation of (5.143) (related to the conservation of energy):

$$\frac{du^0}{u^0} + \frac{2m}{r(r-2m)} dr = 0 \quad \Longrightarrow \quad u^0\left(1 - \frac{2m}{r}\right) = E$$

If there is no initial motion in the spatial directions, the line element (5.144) gives the initial four-velocity in the time direction:

$$u^0(0) := \frac{dx^0}{d\tau} = \frac{cdt}{d\tau} = \frac{1}{\sqrt{1-2m/r}}, \qquad E = \sqrt{1 - \frac{2m}{r_0}}$$

where r_0 is the initial radial position. For determining the radial component for timelike orbits $ds = d\tau$, we now introduce u^0 into (5.144) to find u^1:

$$1 \equiv \left(\frac{ds}{d\tau}\right)^2 = \left(1 - \frac{2m}{r}\right)(u^0)^2 - \frac{1}{1-2m/r}(u^1)^2 \quad \Longrightarrow$$

$$u^1 = \sqrt{\frac{2m}{r} - \frac{2m}{r_0}}$$

For azimuthal geodesic orbits, we come back to the general equations (5.143). The spherical symmetry allows us to consider orbits which remain in a plane by choosing $\theta \equiv \pi/2$ and $u^2(0) = 0$ so that u^2 remains zero. The previous system (5.143) reduces accordingly. The first equation leads to the same form as previous radial geodesics, and the fourth equation gives u^3 (recall that $u^1 \equiv dr/d\tau$): $u^0(1-2m/r) = E$, and $u^3 = L/r^2$. Solution in u^3 may be rewritten as follows: $r^2 u^3 := r^2 d\varphi/d\tau = L$, which states the conservation on angular momentum, where L is called specific angular momentum. We can compute the component u^1 from the line element by constraining the search to timelike orbits or the null geodesics. To begin from relation (5.128), we write the line element (5.144) with $\theta \equiv \pi/2$:

$$\left(\frac{ds}{d\tau}\right)^2 = \left(1 - \frac{2m}{r}\right)(u^0)^2 - \left(1 - \frac{2m}{r}\right)^{-1}(u^1)^2 - r^2(u^3)^2$$

For timelike orbits $ds = d\tau$ (timelike geodesics describe the motions of massive particles at subliminal speed) we deduce:

$$u^1 := \frac{dr}{d\tau} = \sqrt{E^2 - \left(1 + \frac{L^2}{r^2}\right)\left(1 - \frac{2m}{r}\right)}$$

For timelike geodesics, this equation can be integrated directly to give an elliptic integral. Substituting the expression of u^1 leads to the second derivative of r:

$$u^1 := \frac{dr}{d\tau} \quad \Longrightarrow \quad \frac{d^2r}{d\tau^2} + \frac{m}{r^2} + \frac{L^2}{r^3}\left(\frac{3m}{r} - 1\right) = 0$$

Finally, for circular orbits $r \equiv R$ (meaning that $dr/d\tau = 0$ and $d^2r/d\tau^2 = 0$) we obtain the radius in terms of angular momentum L and the Schwarzschild radius from the vanishing of second derivatives:

$$R = \frac{L^2}{2m}\left(1 + \sqrt{1 - \frac{12m^2}{L^2}}\right), \quad R = \frac{L^2}{2m}\left(1 - \sqrt{1 - \frac{12m^2}{L^2}}\right) \tag{5.145}$$

The other solution would be $R = \infty$. Coming back to the equation of geodesic deviation (5.111), we can analyze the deviation acceleration for the circular orbit of radius R (5.145) with the four-velocity:

$$\left\{u^0 = E\left(1-\frac{2m}{R}\right)^{-1}, \quad u^1 = \sqrt{E^2-\left(1+\frac{L^2}{R^2}\right)\left(1-\frac{2m}{R}\right)}, \right.$$
$$\left. u^2 \equiv 0, \quad u^3 = \frac{L}{R^2}\right\} \tag{5.146}$$

5.4.3.3 *Schwarzschild Geodesic Deviation*

In the reference, Fuchs derived the general solution of the equation of geodesic deviation for Schwarzschild static spherically symmetric spacetimes [244], namely for the particular case of circular geodesics [245].

Remark 5.33. From Schwarzschild metric (5.128), another way would be to consider the Lagrangian:

$$\mathscr{L} = m^*\left[\left(1-\frac{2m}{r}\right)(\dot{x}^0)^2 - \left(1-\frac{2m}{r}\right)^{-1}\dot{r}^2 - r^2\left(\dot{\theta}^2+\sin^2\dot{\varphi}^2\right)\right] \tag{5.147}$$

Since no explicit dependence on x^0 and on φ occurs, then the derivative of the Lagrangian with respect to $\dot{x}^0$ and $\dot{\varphi}$ are constant of motions:

$$\begin{cases} m^*\dfrac{dx^0}{d\tau}\left(1-\dfrac{2m}{r}\right) = E \\ m^*\dfrac{d\varphi}{d\tau}\,r^2\sin^2\theta = L \end{cases} \Longrightarrow \begin{cases} u^0 = \dfrac{E}{m^*}\left(1-\dfrac{2m}{r}\right)^{-1} \\ u^\varphi = \dfrac{L}{m^*}\dfrac{1}{r^2\sin^2\theta} \end{cases} \tag{5.148}$$

where E is the energy, and L the analogous angular momentum in relativistic gravitation. This elegant method, called the method of first integrals, is classic in Lagrangian analytical mechanics. These equations conform to the previous solutions.

Remark 5.34. The radial solution can be rewritten as follows:

$$\frac{1}{2}E^2 = \frac{1}{2}\left(\frac{dr}{d\tau}\right)^2 + V(r), \quad V(r) := \frac{1}{2} - \frac{m}{r} + \frac{L^2}{2r^2} - \frac{mL^2}{r^3} \tag{5.149}$$

where the first two terms are respectively the Newtonian potential and the contribution from angular momentum (same for Newtonian and relative gravitation). The third term is the proposed contribution of Einstein gravitation. The above equation describes an analogous motion of a particle (unit mass) moving in a one-dimensional r with potential $V(r)$. At a large distance from the center, Newtonian mechanics matches relative gravitation.

Although the following aspect does not constitute the main motivation of this work, it is judicious to give some snippets of information concerning geodesics and their deviations in relative gravitation. To obtain the geodesic deviation equation, we calculate the components of curvature. We then obtain the Riemann curvature components by using its definition:

$$\begin{aligned}
&\overline{\Re}^0_{101} = -\frac{2m}{r^2(r-2m)}, \quad \overline{\Re}^0_{202} = \frac{m}{r}, \quad \overline{\Re}^0_{303} = \frac{m}{r}\,\sin^2\theta,\\
&\overline{\Re}^1_{010} = -\frac{2m}{r^4}(r-2m), \quad \overline{\Re}^1_{212} = \frac{m}{r}, \quad \overline{\Re}^1_{313} = \frac{m}{r}\,\sin^2\theta,\\
&\overline{\Re}^2_{020} = \frac{m}{r^4}(r-2m), \quad \overline{\Re}^2_{121} = \frac{m}{r^2(r-2m)}, \quad \overline{\Re}^2_{323} = -\frac{2m}{r}\,\sin^2\theta\\
&\overline{\Re}^3_{030} = \frac{m}{r^4}(r-2m), \quad \overline{\Re}^3_{131} = \frac{m}{r^2(r-2m)}, \quad \overline{\Re}^3_{232} = -\frac{2m}{r}
\end{aligned}$$

These components are introduced in (5.111):

$$\frac{D^2\xi^\alpha}{D\tau^2} = \overline{\Re}^\alpha_{\delta\gamma\beta}\; u^\beta u^\delta\; \xi^\gamma$$

where the four-velocity u^α are replaced by the previous results on geodesics (5.146). Fuchs developed a general method based on the Killing vector field to solve the geodesic deviation equation and applied it to static Schwarzchild spacetime [244]. Exhaustive resolution and detailed results may be found in [44].

Remark 5.35. At this point, we have resumed the results concerning Hilbert-Einstein action on Einstein spacetime (5.90). The discovery at the end of the 1990s that Universe is in accelerated expansion has reoriented research in cosmology by introducing again by necessity an *ad hoc* cosmological constant Λ to account for dark energy, unknown but which could be at the origin of this acceleration. In the following, we sketch the development of a spacetime which models dark energy by shifting only the zero point of scalar curvature $\mathfrak{R}$ of the Universe. This will be detailed in the next chapter.

5.4.4 *De Sitter Spacetime*

The spacetime attributed to de Sitter is a manifold space used in relative gravitation, consistent with the observed expansion of the Universe (Hubble). De Sitter spacetime constitutes a solution of Einstein field equations with a negative cosmological constant Λ which corresponds to a repulsive vacuum energy density and negative pressure. It is assumed that the Universe itself is asymptotically de Sitter, for which the far future will be dominated by dark energy. The constant Λ was introduced originally by Einstein himself to avoid a model with an expanding Universe e.g. [24]. However, Hubble's observations showed that the Universe is expanding, beginning with the Big Bang. Since then, its was found that this constant is not equal to zero but positive and very small $|\Lambda| \simeq 10^{-52}m^2$ e.g. [44]. For short, $\Lambda < 0$ (de Sitter) corresponds to an expanding Universe as assessed by Hubble measurements, whereas $\Lambda > 0$ means a contraction of the Universe (anti-de Sitter spacetime). More recent investigations make some assumptions on the existence of both de Sitter and anti-de Sitter spacetimes (called bubbles) which interact each other within the same Universe e.g. [25]. This concept follows the hypothesis that Universe is inside a black hole and is subjected to fluctuations of anti-de Sitter bubbles generating de Sitter bubbles. More recent findings associate the nature of dark matter and dark energy to the cosmological constant e.g. [23], [38].

5.4.4.1 *Lagrangian for (anti)-de Sitter Spacetimes*

Recent advances of cosmology theory consider a Riemann Universe dominated by Cold Dark Matter (CDM) which is non-baryonic (stars, planets and so on) and by a cosmological constant Λ, associated with the inflation since the very early Universe. All measurements are more or less consistent with the concept of dark energy represented by Λ. The two merge into a model of cosmology which is called curved ΛCDM. Such a model remains a phenomenological model. By slightly extending the Hilbert-Einstein action, the de Sitter spacetimes are obtained from the action for vacuum spacetime:

$$\mathscr{S}_{AdS} := \frac{1}{2\chi} \int_{\mathscr{B}} \left(\overline{\mathfrak{R}} - 2\Lambda\right) \overline{\omega}_n \tag{5.150}$$

where Λ is introduced to account for the expansion/contraction of the spacetime. It should be noticed that a very small non-vanishing positive constant Λ is necessary to satisfy two independent observations in cosmology: the Hubble constant which explains expansion of the Universe, and the estimated age of the Universe. It is easy to obtain the Einstein-de Sitter field equation by means of the variational procedure:

$$\overline{G}_{\alpha\beta} + \Lambda\ g_{\alpha\beta} = 0, \qquad \overline{G}_{\alpha\beta} := \overline{\mathcal{R}}_{\alpha\beta} - \frac{\overline{\mathcal{R}}}{2}\ g_{\alpha\beta} \tag{5.151}$$

where the *cosmological constant* is added to the classical Einstein tensor $\overline{G}_{\alpha\beta}$. Despite its apparent simplicity, it should be emphasized that astronomical investigations and results over the years highlighted the paramount importance of the cosmological constant in the Einstein field equations (5.151).

Remark 5.36. Assumption of the existence of dark energy is the most common way to explain observations that the Universe is expanding at an accelerating rate. In the standard model of cosmology, dark energy currently accounts for nearly $\simeq 3/4$ of the total mass-energy of the Universe. And the two suggested forms of the dark energy are cosmological constant Λ and scalar fields such as the quintessence whose energies are dependent on the event-point (x^ν) of spacetime. Nevertheless, the exact nature and physics of this dark energy remains today a matter of speculation; some authors suggest new physics for that purpose.

5.4.4.2 *Schwarzschild-de Sitter Spacetime*

Again, according to the conformastat spacetimes in [242], the situation with a non-vanishing cosmological constant is not so easy in the general case since the resolution reduces to solving a nonlinear Laplace equation of the form $\Delta f(x^i) = 2\Lambda\ f^3$ e.g. [97] where function f is related to the component of metric $f^2 := g_{00}$, and we do not pursue in this way. Extension of the Schwarzschild solution is obtained by introducing the line element:

$$ds^2 = e^{2\nu}\ (dx^0)^2 - e^{2\mu}\ dr^2 - r^2 d\theta^2 - r^2 \sin^2\theta\ d\varphi^2$$

in a spherical coordinate system, and where λ and ν are functions of x^μ. The calculus is similar to that of Schwarzschild and, without going into details, we arrive easily at the slightly modified Einstein field equation, see equation (5.130):

$$\begin{cases} \overline{G}_{00} = e^{2\nu}\left[e^{-2\mu}\left(\dfrac{2\mu'}{r}-\dfrac{1}{r^2}\right)+\dfrac{1}{r^2}+\Lambda\right]=0 \\ \overline{G}_{11} = e^{2\mu}\left[e^{-2\mu}\left(\dfrac{2\nu'}{r}+\dfrac{1}{r^2}\right)-\dfrac{1}{r^2}-\Lambda\right]=0 \\ \overline{G}_{22} = r^2\, e^{-2\mu}\left[\nu''+(\nu')^2-\nu'\mu'+\dfrac{\nu'-\mu'}{r}-\Lambda\right]=0 \\ \overline{G}_{33} = r^2\, e^{-2\mu}\left[\nu''+(\nu')^2-\nu'\mu'+\dfrac{\nu'-\mu'}{r}-\Lambda\right]\sin^2\theta=0 \end{cases} \tag{5.152}$$

The first two equations of (5.152) induce the conditions $\nu' + \mu' = 0$. Solving the field equation (5.151) consists then in defining the function:

$$U(r) = e^{-2\mu} \quad \Longrightarrow \quad -\left(rU(r)\right)' + 1 + \Lambda r^2 = 0$$

This gives us the classical solution e.g. [241]:

$$e^{2\nu} = 1 - \frac{2m}{r} + \frac{\Lambda}{3}\, r^2 \tag{5.153}$$

where $2m$ is an integration constant, and voluntarily chosen to conform to the usual Schwarzschild solution.

Schwarzshild-(anti)-de Sitter metric. The metric for (anti)-de Sitter spacetime is given by:

$$\begin{cases} g_{00} = 1 - \dfrac{2m}{r} + \dfrac{\Lambda}{3}r^2 \\ g_{rr} = -\left(1 - \dfrac{2m}{r} + \dfrac{\Lambda}{3}r^2\right)^{-1} \end{cases}, \quad \begin{cases} g_{\theta\theta} = -r^2 \\ g_{\varphi\varphi} = -r^2\sin^2\theta \end{cases} \tag{5.154}$$

This constitutes a generalization of the Schwarzschild line element, and was discovered by Kottler (1918), Weyl (1919), and Trefftz (1922) according to e.g. [241]; this reference contains numerous exact solutions for spacetimes in Einstein gravitation. The line element compared to the Schwarzschild metric (5.128) is slightly different and accounts for Λ:

$$ds^2 = \left(1 - \frac{2m}{r} + \frac{\Lambda}{3}r^2\right)(dx^0)^2 - \left(1 - \frac{2m}{r} + \frac{\Lambda}{3}r^2\right)^{-1} \times dr^2 - r^2 d\theta^2 - r^2\sin^2\theta\, d\varphi^2 \tag{5.155}$$

The associated metric is called Schwarzschild-de Sitter metric. The spacetime is not Minkowskian (flat) asymptotically, and there is a pre-existing initial curvature, *in fine* associated to dark matter and particularly the dark energy concept. We have three cases according to the sign of the cosmological constant:

(1) when $\Lambda < 0$, we have a de Sitter spacetime;
(2) when $\Lambda = 0$, we recover the Schwarzschild spacetime;
(3) when $\Lambda > 0$, we have an anti-de Sitter spacetime.

Care should be taken because the names are interchanged according to the sign convention of the metric. Here we have chosen the signature $(+---)$.

Remark 5.37. In the year 1917, Einstein and de Sitter discovered the static cosmological model (5.154) incorporating the cosmological constant Λ. However, they faced the problem that such a model is not *Machian* in the sense that the cosmological constant is not associated to matter content. The spatial part is related to a fictitious radius $r_{\mathscr{A}d\mathscr{S}} := \sqrt{3|\Lambda|}$. Initially, Einstein discarded this spacetime considering it as physically irrelevant. Nevertheless, ten years later, the discovery of Hubble on the relation between radial distance and recession of galaxies made evident that the Universe is expanding [12]. Since then, Hubble's law was confirmed with more and more fine measurements and for larger and larger portions of the cosmic Universe.

Remark 5.38. The tight relation between (anti)-de Sitter spacetime and the expansion or the contraction of the Universe is not so obvious. Some studies suggested the electromagnetic field as possible origins of dark energy e.g. [98], [96] or at least of the cosmological constant by considering both Yang-Mills and Hilbert-Einstein actions e.g. [39].

5.4.4.3 *(Anti)-de Sitter Spacetime as Holograms*

In several respects, the spacetime of de Sitter (and also anti-de Sitter) has numerous interests both in terms of geometric interpretation of the Universe and in the long journey of the unification of relative gravitation (infinitely large domains for cosmology) and quantum field theory (infinitely small objects). In this subsection, we recall the classic approach of the hologram. We start from a higher dimensional spacetime (of dimension 5) equipped with the coordinate system $(y^0, \cdots, y^4)$. This five-dimensional spacetime is of Minkowski type with the line element, *in extenso*:

$$ds^2 = (dy^0)^2 - (dy^1)^2 - (dy^2)^2 - (dy^3)^2 - (dy^4)^2 \tag{5.156}$$

We consider the following function which defines immersed subspaces of dimension four:

$$\begin{cases} y^0 = \sqrt{a^2 - r^2}\ \sinh\left(x^0/a\right) \\ y^1 = \sqrt{a^2 - r^2}\ \cosh\left(x^0/a\right) \end{cases}, \qquad \begin{cases} y^2 = r\ \cos\theta \\ y^3 = r\ \sin\theta\ \cos\varphi \\ y^4 = r\ \sin\theta\ \sin\varphi \end{cases} \tag{5.157}$$

where $x^0 := ct$ is time and (r, θ, φ) are classical three-dimensional spherical coordinates. An easy straightforward calculus gives:

$$ds^2 = \left(1 - \frac{\Lambda}{3}\ r^2\right)(dx^0)^2 - \left(1 - \frac{\Lambda}{3}\ r^2\right)^{-1} dr^2 - r^2 d\theta^2 - r^2 \sin^2\theta\ d\varphi^2 \tag{5.158}$$

by defining $\Lambda := 3/a^2$. We recover exactly the de Sitter line element, obtained by using action with cosmological constant. For the sake of simplicity, we do not consider the Schwarzschild contribution here. Now, consider the Minkowski space equipped with the coordinates $(y^0, y^1, \cdots, y^4)$ and, by analogy, consider the sub-manifold defined by:

$$\begin{cases} y^0 = \sqrt{a^2 - r^2}\ \sin\left(x^0/a\right) \\ y^1 = \sqrt{a^2 - r^2}\ \cos\left(x^0/a\right) \end{cases}, \qquad \begin{cases} y^2 = r\ \cos\theta \\ y^3 = r\ \sin\theta\ \cos\varphi \\ y^4 = r\ \sin\theta\ \sin\varphi \end{cases} \tag{5.159}$$

where we again have $x^0 := ct$ as time and (r, θ, φ) as classical three-dimensional spherical coordinates. The line element of the five-dimensional Minkowski space is

$$ds^2 = (dy^0)^2 + (dy^1)^2 - (dy^2)^2 - (dy^3)^2 - (dy^4)^2 \tag{5.160}$$

(be cautious with the sign of the second term). A similar calculus without difficulty gives the following line element:

$$ds^2 = \left(1 + \frac{\Lambda}{3}\ r^2\right)(dx^0)^2 - \left(1 + \frac{\Lambda}{3}\ r^2\right)^{-1} dr^2 - r^2 d\theta^2 - r^2 \sin^2\theta\ d\varphi^2 \tag{5.161}$$

which is clearly the line element of the anti-de Sitter spacetime, also interpreted as a hologram of a five-dimensional manifold. Again, the Schwarzschild contribution is not considered for the sake of clarity.

Remark 5.39. As a reminder, line element (5.161) defines metric on a spherically symmetric de Sitter spacetime clearly obtained as a hologram of a five-dimensional Minkowski space. The main outcome of the hologram concept was discovered by Maldacena in the long path to unify relative gravitation and quantum physics theories by means of the ADS/CFT conjecture [246]. Of course, the interpretation of number a is important to better understand the process of immersion. Indirectly, the merging of the ADS/CFT conjecture of Maldacena and the hologram concept seems to support the idea that quantum effects contribute to the cosmological constant as proposed by the seminal paper [29].

5.4.4.4 *Geodesics for (anti)-de Sitter Spacetime*

As for the spacetimes of Schwarzschild and Reissner-Nordström, solutions for de Sitter spacetime are plethora in the literature of relative gravitation. Theoretical and practical bases can be found in the works of e.g. [41], [44], [247], to name but a few. We simply recall the Levi-Civita connection:

$$\overline{\Gamma}^0_{01} = \Gamma^0_{10} = \nu' \tag{5.162}$$

$$\overline{\Gamma}^1_{00} = \nu' \, e^{'\nu}, \quad \overline{\Gamma}^1_{11} = \mu', \quad \overline{\Gamma}^1_{22} = -r \, e^{2\nu}, \quad \overline{\Gamma}^1_{33} = -r \, e^{2\nu} \sin^2\theta \tag{5.163}$$

$$\overline{\Gamma}^2_{12} = \Gamma^2_{21} = \frac{1}{r}, \quad \overline{\Gamma}^2_{33} = -\sin\theta\cos\theta \tag{5.164}$$

$$\overline{\Gamma}^3_{13} = \overline{\Gamma}^3_{31} = \frac{1}{r}, \quad \overline{\Gamma}^3_{23} = \overline{\Gamma}^3_{32} = \cot\theta \tag{5.165}$$

These relations are valid for either positive or negative value of Λ. For spherically symmetric spacetime, we obtain the geodesics with $e^{2\nu} = 1 - 2m/r + \Lambda r^2/3$:

$$\boxed{\begin{cases} \dfrac{du^0}{d\tau} + 2\nu' \, (u^0 u^1) = 0 \\ \dfrac{du^1}{d\tau} + \nu' \, e^{4\nu}(u^0)^2 - \nu'(u^1)^2 - re^{2\nu}(u^2)^2 - r\sin^2\theta e^{2\nu}(u^3)^2 = 0 \\ \dfrac{du^2}{d\tau} + \dfrac{2}{r} \, (u^1 u^2) - \sin\theta\cos\theta \, (u^3)^2 = 0 \\ \dfrac{du^3}{d\tau} + \dfrac{2}{r} \, (u^1 u^3) + 2\cot\theta (u^2 u^3) = 0 \end{cases}} \tag{5.166}$$

Remark 5.40. We have exactly the same form of equations as (5.138) and (6.84); only the expression of function $e^{2\nu}$ differs.

5.4.4.5 *Lagrangian Method*

For illustration, consider the geodesic in the equatorial plane $\theta = \pi/2$. As for Schwarzschild and Reissner-Nordström spacetimes, geodesics can also be studied through the Lagrangian of a particle of mass m^*:

$$\mathscr{L} = m^* \left[\left(1 - \frac{2m}{r} + \frac{\Lambda}{3} r^2\right) (\dot{x}^0)^2 - \left(1 - \frac{2m}{r} + \frac{\Lambda}{3} r^2\right)^{-1} \dot{r}^2 \right.$$
$$\left. - r^2 \left(\dot{\theta}^2 + \sin^2\theta \, \dot{\varphi}^2\right)\right] \tag{5.167}$$

No explicit dependence on x^0 and φ occurs, so the derivative of the Lagrangian with respect to $\dot{x}^0$ and $\dot{\varphi}$ are constant of motions:

$$\begin{cases} m^* \dfrac{dx^0}{d\tau} \left(1 - \dfrac{2m}{r} + \dfrac{\Lambda}{3} r^2\right) = E \\ m^* \dfrac{d\varphi}{d\tau}\, r^2 \sin^2\theta = L \end{cases} \Longrightarrow \begin{cases} u^0 = \dfrac{E}{m^*} \left(1 - \dfrac{2m}{r} + \dfrac{\Lambda}{3} r^2\right)^{-1} \\ u^\varphi = \dfrac{L}{m^*} \dfrac{1}{r^2 \sin^2\theta} \end{cases} \tag{5.168}$$

where E is the energy and L the analogous angular momentum in relative gravitation. These equations also conform to previous Schwarzschild solutions. Complete solutions of Schwarzschild-de Sitter geodesics may be found in literature e.g. [41], [44], [244]. For example, Cruz *et al.* [41] investigated the geodesics for anti-de Sitter black holes to highlight all the possible motions of photons and test particles according to various energy levels. They described both radial and non-radials geodesics, which are similar to Schwarzschild geodesics.

Remark 5.41. So far, previous sections have been devoted to Einstein gravitation with zero torsion. If we consider a non-vanishing torsion, we advance towards the geometry of Riemann-Cartan, and from then on we have to generalize models towards Einstein-Cartan gravitation. Several reasons push us to go in this direction. Among them, we wish to highlight the role of electromagnetism coupled with gravitation e.g. [248]. In addition, Einstein-Cartan gravitation makes it possible to overcome some disadvantages such as the presence of singularity at the initial instant of the Universe as well as the impossibility of expansion during the initial instants when using torsion-free Einstein spacetime [4].

5.5 Summary and Remarks on Einstein Gravitation

(1) *Special relativity to relative gravitation, celerity of light and curved four-dimensional spacetime.*

The transition from special to general relativity is not of the same nature as the transition from Newtonian mechanics to special relativity. The latter is essentially linked to the finiteness of light speed and thus the search for a formulation compatible with classical point mechanics (satisfying the invariance by the Galileo transformation group) and the equations of electromagnetism (covariant with respect to the Lorentz transformations). These transformations are carried out in a four-dimensional spacetime because time and space are no longer considered absolute and decoupled. The first transition, on the other hand, dispenses with the systematic use of a flat Minkowski spacetime in special relativity and considers a curved spacetime which constitutes the geometric foundations of the gravitational field in the Universe.

(2) *Role of affine connections in Newtonian and relative gravitation.*

Affine connection is omnipresent in mechanics and more generally in physics. Connection draws its usefulness as soon as we consider the playground of physical phenomena as a differentiable manifold. In particular, in the various theories of gravitation, of either Newton (Newton-Cartan) or Einstein, connection represents the gravitational fields where connection has a non-zero curvature. But the affine connection is also omnipresent both in the definition of an inertial reference frame, at least locally, and in the calculation of acceleration of material points within spacetimes. The notion of free fall is primordial in the development of relative gravitation by means of the Principle of Equivalence. Despite their important role in the modeling of gravitation, covariance (i.e., form invariance of action and equations under passive diffeomorphism that are merely change of coordinate systems) implies action dependence on metric and curvature rather than on connection and its first and second derivatives.

(3) *Gravitation alone is not sufficient to describe vacuum spacetime energy.*

The main feature of relative gravitation is that curvature of manifold

$\mathscr{B}$ constitutes the gravitation energy of vacuum spacetime. The curvature is often attributed to non-uniform metrics. It must nevertheless be realized that curvature is by definition associated with the affine connection in the absence of other simplifying hypotheses. Strictly speaking, curvature is first associated with connection. Presence of spacetime curvature is the main source of gravitation. It can be considered in its own right as an energy in the vacuum which allows action-at-a-distance between massive bodies. In Einstein gravitation, Ricci curvature is the relevant component. In a Riemann manifold, the curvature also allows us to determine the deviation of geodesic curves in spacetime: it is the acceleration of the gap between two test particles in a gravitational field.

(4) *(Anti)-de Sitter spacetime and acceleration of the Universe expansion.*

The discovery of the accelerated expansion of the Universe has shaken many certainties in the understanding of cosmology. Scientists generally agree that there is a form of dark energy behind this acceleration and that it is no less than almost 70% of the total matter-energy of the Universe. This unknown but essential quantity has been approached by the introduction of a cosmological constant Λ. This has led us to consider de Sitter and anti-de Sitter spacetimes. Therefore, these spacetime models have not only a geometric interest, especially the introduction of Λ, but also a need to highlight their physical meanings.

However, it is worth recalling that the two *actions-at-a-distance* (among the four fundamental forces) that may be present in large-scale physics are gravitation and electromagnetism. Weak and strong nuclear forces are practically contact interactions at Planck length scale. For reasons that will be supported in the next chapter, but will not thus restrict to Einstein gravitation by studying only curvature of spacetime, we will extend the playground to Riemann-Cartan manifolds by accounting for other geometry characteristics such as torsion, to hopefully capture electromagnetism.

Chapter 6

Einstein-Cartan Spacetimes: Gravitation and Electromagnetism

This chapter extends relative gravitation to Einstein-Cartan spacetime. On the one hand, by means of Yang-Mills action, we add an electromagnetic field to derive the Maxwell equations and the coupled Einstein-Maxwell equations. On the other hand, the combination of relative gravitation with electromagnetism within an Einstein-Cartan spacetime makes it possible to show a direct link (formally an additional equation) between torsion tensor and electromagnetic field. Concerning the geometric structure of spacetime, we define, in addition to the analogous Einstein tensor $G_{\mu\nu}$, a distortion tensor $\mathscr{D}_{\lambda}^{\mu\nu}$ the value of which will depend on gravitation and the electromagnetic background. Optical helicity including both electric and magnetic helicity, and the spin orbital momentum (represented by the Cherns-Simons current) are physical sources of the Einstein-Cartan spacetime distortion. The geodesic deviation equation is modified accordingly to account for these two fields.

6.1 Introduction

Gravitation and electromagnetism are fields that can exist in a vacuum, and constitute the typical action-at-a-distance of physics. Their range of action is infinite unlike the other two fundamental forces (weak and strong nuclear forces). The previous chapter is a reminder of Einstein gravitation for which vacuum spacetime is filled with curvature fields (and therefore energy). The existence of background fields is possible in vacuum parts of the Universe as interstellar voids. In this chapter, we integrate the electromagnetic field, which can also exist in a material medium, but which propagates like waves in a vacuum. As we have already recalled, the coexistence of Newtonian classical mechanics and electromagnetic waves could not be modeled without the introduction of the theory of special relativity. In this chapter we consider the coupling of gravitation with the electromagnetic field. The geometric structure of spacetime containing these two fields simultaneously requires an extension of Riemann geometry towards the Riemann-Cartan manifold that we will gradually construct. It is commonly accepted that the classical Maxwell equations in an empty spacetime reveal additional facets from a physical point of view; the most important thing is to know that electromagnetism can be considered as part of quantum physics (first quantization) e.g. [75]. This point will be the subject of only a short paragraph in this chapter since we limit ourselves to classical physics without entering into quantum phenomena. As soon as electromagnetism is added in empty spacetime, it would be as if we were implicitly mixing gravitation with quantum physics since the Schrödinger equations of photons can be put in parallel with the Maxwell equations in vacuum by introducing the Weber vector, more or less carefully assimilated to a wave function of the photon, which we will see later but briefly; some detailed aspects may be found in e.g. [249].

Although we limit ourselves to classical physics, we cannot ignore the fact that the electromagnetic field in vacuum plays a primordial role in quantum physics e.g. [13]. It is currently accepted that the quantum vacuum spacetime is indeed characterized by fluctuations of electromagnetic fields with non-zero energy, called zero-point energy. It is the electromagnetic field that exists in the so-called vacuum that provides the perturbation energy necessary for these fluctuations and the interaction phenomena in quantum physics, and even potentially sources of matter in the matter-energy duality. In this sense, the electromagnetic field in vacuum is not only a static background of spacetime but a dynamic field that is an integral part of spacetime as is the gravitational field e.g. [72].

We aim to analyse the consequences of using Einstein-Cartan spacetime for investigating the coupling effects of relative gravitation with electromagnetism in the framework of classical physics. It is interesting to connect what Gonseth called microcosms [2] with the multiply connected Riemann manifold that was proposed by Rainich [8] and subsequently developed by Misner and Wheeler [7], in their quest to unify gravitation with electromagnetism and particularly to base their theory solely on vacuum spacetime geometry to model not only actions-at-a-distance, but also electric and magnetic charges as well as massive matter. We adopt Riemann-Cartan geometry for a synthetic approach e.g. [39], [250]. This chapter is devoted to the coupling of gravitation with electromagnetism for two different spacetimes: an Einstein spacetime with curvature but zero torsion and an Einstein-Cartan spacetime with curvature and non-zero torsion.

6.2 Einstein Gravitation and Electromagnetism

In this section we deal with the theory of electromagnetism and gravitation in the presence of electric charges and magnetic fields and explore two possibilities [251]:

(1) the case where the torsion vanishes everywhere (Einstein spacetime);
(2) the case with non-zero torsion (Einstein-Cartan spacetime).

We consider some special examples of spacetimes with spherical symmetry for illustration. The first example shows the influence of an electric charge on the spacetime geometry, without torsion. The second example illustrates the influence of torsion on the spacetime geometry. The two examples are merged into one.

The so-called Reissner-Nordström metric corresponds to the static solution of the Einstein-Maxwell equations allowing us to determine the gravitational field of a punctual spherical non-rotating charge Q, contained in a body of mass m. The torsion is assumed to be zero. The punctual electric charge Q is assumed to be at rest at the origin. The goal of this subsection is to calculate the gravitational effect of a body with electric charge. In astrophysics, most planets and stars are electrically neutral, so the Reissner-Nordström example may be considered as only an academic exercise, although interesting, rather than a realistic and relevant field of gravitation.

6.2.1 *Basic Electromagnetism: Maxwell's Equations*

Maxwell's equations of electromagnetism were discovered before Einstein's theory of special relativity. They hold in all inertial frames (group of Lorentz transformations) and are consistent with the laws of special relativity. In fact, the incompatibility of Newtonian mechanics, whose inertial frames are deduced from Galilean transformations, with the Maxwell equations inspired Einstein to develop special relativity theory with a finite speed of light. Historically, Minkowski formulated a four-dimensional version of special relativity (1908) soon after expressing the Maxwell equations in a four-dimensional form (1907). Then the initial coupling of electromagnetism and relativistic mechanics was advocated during this period. We recall hereafter the basic elements of electromagnetism. For the sake of simplicity, we limit to electromagnetism in a vacuum region, without electric charges and magnetic monopoles. In the framework of three-dimensional vectorial formulation, the electromagnetic variables are then the electric field **E**, the magnetic flux **B**, the electric displacement **B**, and magnetic induction **H** (the classical approach uses three-dimensional vectors for each of these electromagnetic variables). Physical interpretation of the Maxwell equations are respectively e.g. [252]:

(1) Gauss's law for static electric fields (conservation of electric charge),
(2) Gauss's law for static magnetic fields (classically the assumption that there are no magnetic fields),
(3) Faraday's law of induction,
(4) and Ampere's law.

Over the years, there have been several ways of formulating Maxwell's equations: the three-dimensional vector formulation (the best known for undergraduate level), the formulation with 1-forms and 2-forms in three or four dimensions, and finally the four-dimensional formulation in terms of potentials, which exploits the advantages of differential forms and appears to be the most appropriate for expressing a relationship with quantum physics. There are two equivalent formulations, namely the usual formulation and the dual formulation, corresponding respectively to the first and second pairs of Maxwell's equations, e.g. [252]. We focus on the four-dimensional formulation of the Maxwell equations in this book e.g. [253], with a short reminder of the axiomatic derivation with differential forms.

Remark 6.1. Physically, the role of light in cosmology is fundamental, since the emission of light from stars makes it possible to perform most observational measurements of galaxies and galaxy clusters. This immediately justifies the treatment of electromagnetic waves in the present book, as the study of light waves–particular cases of electromagnetic waves–is inseparable from the study of gravitation. Light can be described both as a wave and as a particle. When regarded as a collection of particles called photons, light consists of massless quanta, each carrying a discrete amount of energy. Conversely, light may be viewed as propagating electromagnetic waves, as exemplified by the Cosmic Microwave Background. Regardless of whether it behaves as a wave or as a particle, light always propagates in vacuum at a speed close to 300,000 km s^{-1}. The concept of a photon wave function plays an important role in bridging quantum electrodynamics and classical electromagnetism e.g. [73]. Photon behavior is usually modeled within the framework of quantum field theory, in which the electromagnetic field is quantized and interactions with charged particles are described in terms of creation and annihilation operators for particles and antiparticles. Maxwell's equations may therefore be interpreted either as governing the dynamics of macroscopic electromagnetic waves or as describing the dynamics of a quantum wave function for photons e.g. [75]. In this book, we retain the first interpretation, since our focus is on cosmology and thus on phenomena at large length scales.

6.2.1.1 *Axiomatic Formulations of Maxwell's Equations*

In parallel with the formulation of Newton's law in classical mechanics, let's recall how the axiomatic structure of the laws of classical electromagnetism is established using differential forms e.g. [252], [254]. The first fundamental notion of this theory is based on the existence of electric charges ρ modeled by a 3-form:

$$\rho_e = \rho_e \, dx^1 \wedge dx^2 \wedge dx^3, \quad Q := \int_{\mathscr{B}} \rho_e \tag{6.1}$$

where we use the canonical volume form $dx^1 \wedge dx^2 \wedge dx^3$ for the sake of simplicity. By abuse of language, we keep the same notation for the charge and its component on the volume form. Any electric charge is supposed to move in space and thus create an electric current which becomes a flow through the surface $\partial\mathscr{B}$:

$$\mathfrak{J} := \int_{\partial\mathscr{B}} \mathbf{j} \tag{6.2}$$

where the electric current density is a 2-form:

$$\mathbf{j} := \frac{1}{2!} \, \mathrm{j}_{ab} \, dx^a \wedge dx^b = \mathrm{j}_{12} \, dx^1 \wedge dx^2 + \mathrm{j}_{23} \, dx^3 \wedge dx^2 + \mathrm{j}_{31} \, dx^3 \wedge dx^1 \tag{6.3}$$

The use of a 2-form is not an artificial complication but due to the need of a variable which can be integrated over a surface, say $\partial\mathscr{B}$.

Axiom 1: Conservation of electric charge. The global electric charge within any domain of the continuum manifold $\mathscr{B}$ with a boundary $\partial\mathscr{B}$ is conserved:

$$\partial_0 \int_{\mathscr{B}} \rho_e + \int_{\partial\mathscr{B}} \mathbf{j} = 0, \quad \forall\mathscr{B} \tag{6.4}$$

By virtue of Stokes' theorem, changing the surface integral into a volume integral, we deduce:

$$\partial_0 \rho_e + d\mathbf{j} = 0 \tag{6.5}$$

where ∂_0 means a derivative with respect to $x^0 := ct$, c being the light speed in vacuum. Now, since the electric charge is a 3-form within a three-dimensional space, its exterior derivative vanishes, $d\rho_e = 0$. By a theorem of de Rham, it follows that there exists a 2-form, $\mathcal{D}$, such that the 3-form ρ_e is derived from this 2-form. The conservation of electric charge implies mathematically that:

$$\rho_e = d\mathcal{D} \tag{6.6}$$

where the 2-form $\mathcal{D}$ is called electric displacement, but probably more rigorously electric excitation. By introducing the expression (6.6) into (6.5), we can write $d(\mathbf{j} + \partial_0\mathcal{D}) = 0$, and once again by virtue of the de Rham theorem previously cited, there exists a 1-form $\mathcal{H}$ such that e.g. [254]:

$$\mathbf{j} + \partial_0\mathcal{D} = d\mathcal{H} \tag{6.7}$$

where the 1-form $\mathcal{H}$ is called magnetic excitation.

Gauss's law and the Oersted-Ampère law. The two equations (6.6) and (6.7) constitute Gauss's law and the Oersted-Ampère law respectively:

$$\begin{cases} d\mathcal{D} = \rho_e \\ d\mathcal{H} - \partial_0\mathcal{D} = \mathbf{j} \end{cases} \tag{6.8}$$

where ρ_e is the electric density 3-form, $\mathcal{D}$ the electric excitation 2-form, $\mathcal{H}$ the magnetic excitation 1-form and $\mathbf{j}$ the electric current 2-form.

Remark 6.2. Equations (6.8) were deduced from the conservation of electric charge and by using de Rham theorem on forms. They are in principle valid even in quantum physics since the elementary charge may be the electron charge e. Indeed, $\mathcal{D}$ is at the same length level, and so are $\mathcal{H}$ and $\mathbf{j}$. At this stage of development, there is no need for metric and connection tools on the manifold.

Axiom 2: Lorentz force of electromagnetic field. For the second axiom of electromagnetism, the existence of two fields is first assumed, namely the electric strength E and a magnetic strength B. The second axiom is then related to the motion of a test particle of electric charge e within a spacetime with an electromagnetic field $(\mathbf{E}, \mathbf{B})$ as background. With respect to an inertial frame, the test particle is moving with a vector velocity $\mathbf{v}$, and the Lorentz force expressed with a three-dimensional vector formulation is given by:

$$\mathbf{F} = e\left(\mathbf{E} + \mathbf{v} \times \mathbf{B}\right) \tag{6.9}$$

The total force, considered as a 1-form mathematically, has components $F := F_\mu \, dx^\mu$ whose line integral yields a scalar, the energy needed to transport the test particle along the line $\mathscr{L}$ (do not confuse with a Lagrangian). The definition of the interior product of a vector and a 2-form $i_{\mathbf{v}}\omega$ leads to an interesting interpretation of a vector product in a three-dimensional vector space. For a 2-form ω, it is shown that the interior product $-i_{\mathbf{v}}\omega$ is the covariant version of a vector product $\mathbf{v} \times \mathbf{w}$. In effect, we can use the vector product e.g. [120]:

$$\left(\mathbf{v} \times \mathbf{w}\right) \cdot \mathbf{u} \equiv \omega_n\left(\mathbf{v}, \mathbf{w}, \mathbf{u}\right), \quad n = 3, \quad \forall \mathbf{u} \in \mathcal{E}^3$$

To this end, we write:

$$\begin{aligned}
\left(\mathbf{v} \times \mathbf{w}\right) \cdot \mathbf{u} \equiv \omega_n\left(\mathbf{v}, \mathbf{w}, \mathbf{u}\right) &= -\omega_n\left(\mathbf{w}, \mathbf{v}, \mathbf{u}\right) \\
&= -\left[i_{\mathbf{w}}\omega_n\right]\left(\mathbf{v}, \mathbf{u}\right) = -\omega\left(\mathbf{v}, \mathbf{u}\right) \\
&= -\left[i_{\mathbf{v}}\omega\right](\mathbf{u}), \quad \forall \mathbf{u} \in \mathcal{E}^3
\end{aligned}$$

This shows that vector product $\mathbf{v} \times \mathbf{B}$, usually performed during the derivation of a Lorentz force in electromagnetism, is formally an interior product of a vector and a 2-form. The motion of a particle of electric charge e is governed by the Lorentz force, whose expression was determined from experimental measurements:

$$F = e\left(E - i_{\mathbf{v}}B\right) \tag{6.10}$$

where E is the electric strength 1-form, and B the magnetic strength 2-form. The term $i_{\mathbf{v}}B$ is the interior product of the velocity vector $\mathbf{v}$ with a 2-form B. Notice that the interior product decreases the rank of a form k to a rank $(k-1)$ e.g. [128].

Remark 6.3. The definition of Lorentz force (6.10) assumes that the electric charge is conserved. The two axioms do not stand alone—the second presupposes the first. The axiom of conservation of electric charge is very fundamental.

Let's consider a surface $\mathscr{S}$ and its boundary $\partial\mathscr{S}$. The second axiom is based on the magnetic flux conservation e.g. [254]:

$$\partial_0 \int_{\mathscr{S}} B + \int_{\partial\mathscr{S}} E = 0 \tag{6.11}$$

stating that the rate of the magnetic flux 2-form B across $\mathscr{S}$ plus the circulation of the electric 1-form field E along its closed boundary $\partial\mathscr{S}$ is equal to zero. By virtue of Stokes' theorem, we have $\int_{\partial\mathscr{S}} E = \int_{\mathscr{S}} dE$, giving us Faraday's law of induction, assuming a sufficient smoothness (differentiability) of fields:

$$\partial_0 B + dE = 0 \tag{6.12}$$

Historically, this law was inspired and verified by experimental measurement and was discovered by Faraday. However, its mathematical statement is due to Kelvin and Maxwell. The exterior derivative of (6.12) together with the closeness of the 1-form E allows us to write $\partial_0 dB = 0$ meaning that the exterior derivative dB is constant. Assuming that there is no magnetic monopole charge:

$$dB = 0 \tag{6.13}$$

Equation (6.13) is called Gauss's law of magnetic induction, and is based on the observation, at that time, of the absence of magnetic monopoles.

Remark 6.4. In the axiomatic derivation of Maxwell's equations, it is well stated that we have deliberately assumed that there is no magnetic monopole charge $\rho_m = 0$. More recent research studies e.g. [27], [255–258] have, however, shown the possibility of existence of these magnetic monopoles, the genesis of this trend beginning with Dirac's work [28].

Faraday's induction law and Gauss's induction law. The two equations (6.12) and (6.13) constitute the Faraday and Gauss induction laws respectively:

$$\begin{cases} dB = \rho_m \\ dE + \partial_0 B = 0 \end{cases} \tag{6.14}$$

where $\rho_m = 0$ is a hypothetical magnetic charge (monopole) assumed to not exist, B is the 2-form magnetic strength, and E is the 1-form electric strength.

Remark 6.5. The possibility of the case where $\rho_m \neq 0$ is nowadays more and more admitted e.g. [26] and interestingly, magnetic monopoles are also associated with spacetime geometrical defects e.g. [90].

6.2.1.2 *Yang-Mills Action in a Minkowski Spacetime*

The term *dual* may be a source of ambiguity and therefore requires clarification. It must be used with care. In electromagnetism, the dual formulation is not related to the constitution of matter, but rather to a formulation based on the Hodge operator. By contrast, in the theory of constitutive laws, the term *dual* refers to physical quantities that are associated with the causes of physical phenomena. For example, in thermomechanics, the primal variables are the metric (as well as its derivatives) and the temperature, while the dual variables are the mechanical stress and the entropy.

Let's move to a four-dimensional formulation of Maxwell's equations within a Minkowski spacetime endowed with a metric and connection which can be reduced, for the sake of simplicity, to:

$$\eta^{\mu\nu} = \begin{pmatrix} 1 & 0 & 0 & 0 \\ 0 & -1 & 0 & 0 \\ 0 & 0 & -1 & 0 \\ 0 & 0 & 0 & -1 \end{pmatrix} \tag{6.15}$$

with connection calculated as Christoffel's symbols (zero for the particular projection). The Faraday strength tensor, considered as a primal variable, is:

$$\mathcal{F}_{\mu\nu} = \begin{pmatrix} 0 & -E_1 & -E_2 & -E_3 \\ E_1 & 0 & B^3 & -B^2 \\ E_2 & -B^3 & 0 & B^1 \\ E_3 & B^2 & -B^1 & 0 \end{pmatrix} \tag{6.16}$$

The essence of constitutive law (it determines rather a model than a law) is that it defines a particular material. The vacuum spacetime constitutes an example. Assuming constitutive laws $\mathbf{D}\,[\mathbf{E}, \mathbf{B}]$ and $\mathbf{H}\,[\mathbf{E}, \mathbf{B}]$, the dual variables (in the sense of constitutive laws) are expressed as follows:

$$\mathcal{H}^{\mu\nu} = \begin{pmatrix} 0 & D^1 & D^2 & D^3 \\ -D^1 & 0 & H_3 & -H_2 \\ -D^2 & -H_3 & 0 & H_1 \\ -D^3 & H_2 & -H_1 & 0 \end{pmatrix} \tag{6.17}$$

Definition 6.1. By means of the Faraday tensor and its dual, the Yang-Mills action is defined with a Lagrangian of the electromagnetic field [259]:

$$\mathscr{S}_{YM} := \int_{\mathscr{B}} \mathscr{L}\ \omega_n, \quad \mathscr{L} := -\frac{1}{4}\ \mathcal{H}^{\mu\nu}\ \mathcal{F}_{\mu\nu} \tag{6.18}$$

which defines the material's constitutive laws, and particularly for vacuum spacetime. Constitutive laws may be formulated by tensorial relations $\mathcal{H}^{\mu\nu}\,[\mathcal{F}_{\alpha\beta}]$.

Remark 6.6. Here we use a classical (in the non-quantum sense) form of Yang-Mills action which was originally designed to include the quantum version on isotope spin conservation and isotopic gauge invariance in [259]. In fact, the classical version (6.18) allows us to derive with a variational principle the covariant Maxwell equations for macroscopic electromagnetism.

In a three-dimensional formulation, the Yang-Mills Lagrangian takes the usual form:

$$\mathscr{L} := -\frac{1}{4}\ \mathcal{H}^{\mu\nu}\ \mathcal{F}_{\mu\nu} = \frac{1}{2}\,(\mathbf{D}\cdot\mathbf{E} - \mathbf{B}\cdot\mathbf{H}) \tag{6.19}$$

Electromagnetic waves can travel through a vacuum space and transport energy (and, by extension, information). The simplest example of constitutive laws would be the linear relations of vacuum Minkowski spacetime, called Maxwell-Lorentz "axiom":

$$\mathbf{D} = \epsilon_0\ \mathbf{E}, \quad \mathbf{B} = \mu_0\ \mathbf{H}$$

where ϵ_0 and μ_0 are respectively the electric permittivity and the magnetic permeability of the vacuum. These two physical contants satisfy in our coordinate system convention the relation $\mu_0\epsilon_0 = c^2 = 1$. we can rewrite the particular Lagrangian of the vacuum spacetime electromagnetism by using the definition of the Faraday strength (6.16) within a flat spacetime such as that of the Minkowski metric. This Lagrangian is usually used to illustrate the electromagnetic action, but it should be more precisely defined if we would like to extend it to curved spacetime, as in relative gravitation and electromagnetism coupling. It is reminded that the tensor we are interested in is the constitutive dual variable $\mathcal{H}^{\mu\nu}$ which should be related to the primal variable $\mathcal{F}_{\lambda\sigma}$. Focusing again the interest on linear constitutive laws, we have e.g. [6], [260]:

$$\mathcal{H}^{\mu\nu} = \Xi^{\mu\nu\lambda\sigma}\,[\eta_{\alpha\beta}]\;\mathcal{F}_{\lambda\sigma}, \quad \Xi^{\mu\nu\lambda\sigma} = \epsilon_0\;\eta^{\mu\lambda}\eta^{\nu\sigma} \tag{6.20}$$

where the second relation of (6.20) defines a very special case of constitutive laws but represents a rich toy model of the electromagnetic field. Again, let us start with the Lagrangian which takes the general form of flat or hereafter curved continuum manifold. For the special constitutive laws (6.20) of Einstein vacuum spacetime, we merely have:

$$\mathscr{S}_{YM} := \int_{\mathscr{B}} \mathscr{L}\,\omega_n, \quad \mathscr{L} = -\frac{1}{4}\,\epsilon_0\;\eta^{\mu\lambda}\eta^{\nu\sigma}\;\mathcal{F}_{\lambda\sigma}\;\mathcal{F}_{\mu\nu} \tag{6.21}$$

from which we can derive the first pair of Maxwell's equations. The complete Maxwell's equations include two pairs of vectorial equations in three-dimensional formulation. The two pairs are not independent.

Remark 6.7. It should be noted that the geometrical structure of spacetime characterized by tensors such as the metric, curvature and possibly the torsion of the connection intervene only in the constitutive laws (for the example by means of the metric $\mathbf{g}$) and not in the definition of the electromagnetic excitation $\mathcal{H}$ and Faraday strength $\mathcal{F}$ tensors e.g. [248].

6.2.1.3 *Maxwell's Equations for Minkowski Vacuum Spacetime*

Before embarking on complex calculations based on quantum physics, or even quantum phenomena, it is more educational to link the equations in four dimensions to the usual Maxwell vector equations. It is known that the transition from macroscopic electromagnetism to quantum formulation is easier when we move on to the derivation of models via the four-dimensional electromagnetic potential $A^\mu, \mu = 0,1,2,3$ e.g. [252]. First, we introduce the primal variables as:

$$\mathcal{F}_{\mu\nu} = \nabla_\mu A_\nu - \nabla_\nu A_\mu$$

Starting again from the Yang-Mills action [259] previously introduced:

$$\mathscr{S}_{YM} := \int_{\mathscr{B}} \mathscr{L}\,\overline{\omega}_n, \quad \mathscr{L} := -\frac{1}{4}\,\mathcal{H}^{\mu\nu}\,\mathcal{F}_{\mu\nu} \tag{6.22}$$

where it is assumed that there is neither electric charge nor magnetic monopole distributions. The variational procedure induces:

$$\delta\mathscr{S} = -\frac{1}{4}\int_{\mathscr{B}} \delta\left(\mathcal{H}^{\mu\nu}\,\mathcal{F}_{\mu\nu}\right)\overline{\omega}_n + \mathcal{H}^{\mu\nu}\,\mathcal{F}_{\mu\nu}\,\delta\overline{\omega}_n$$

We have some obvious formulae:

$$\begin{aligned}\delta(\mathcal{H}^{\mu\nu}\,\mathcal{F}_{\mu\nu}) &= \delta(\mathcal{H}^{\mu\nu})\,\mathcal{F}_{\mu\nu} + \mathcal{H}^{\mu\nu}\,\delta\mathcal{F}_{\mu\nu}\\ &= \delta(\epsilon_0\eta^{\mu\lambda}\eta^{\nu\rho}\mathcal{F}_{\lambda\rho})\,\mathcal{F}_{\mu\nu} + \mathcal{H}^{\mu\nu}\,\delta\mathcal{F}_{\mu\nu}\\ &= \delta\mathcal{F}_{\lambda\rho}\,\mathcal{H}^{\lambda\rho} + \mathcal{H}^{\mu\nu}\,\delta\mathcal{F}_{\mu\nu}\end{aligned}$$

First, we consider that we are using a divergence-free potential (Lorentz gauge) together with the assumption that the volume-form is compatible with the connection, say $\delta\overline{\omega}_n[\delta\mathbf{A}] = (\nabla_\mu\delta A^\mu)\,\overline{\omega}_n = 0$. The second term vanishes accordingly. Second, the symmetry of the first term gives:

$$\delta\mathscr{S} = -\frac{1}{2}\int_{\mathscr{B}} \mathcal{H}^{\mu\nu}\,\delta\mathcal{F}_{\mu\nu}\,\omega_n = -\frac{1}{2}\int_{\mathscr{B}} \mathcal{H}^{\mu\nu}\,\left(\nabla_\mu\delta A_\nu - \nabla_\nu\delta A_\mu\right)\omega_n = 0 \tag{6.23}$$

Dropping the boundary term, we deduce the four-dimensional Maxwell equations of Minkowski vacuum spacetime (without sources):

$$\boxed{\nabla_\nu\mathcal{H}^{\mu\nu} = 0} \tag{6.24}$$

Introducing the tensor $\mathcal{H}^{\mu\nu}$ (6.17), we obtain the three-dimensional Maxwell equations (first pair):

$$\begin{cases}\nabla\cdot\mathbf{D} = 0\\ \partial_0\mathbf{D} - \nabla\times\mathbf{H} = 0\end{cases} \tag{6.25}$$

Remark 6.8. For free electromagnetism, it is typical to invoke the transformation $(\mathbf{E},\mathbf{B}) \to (\mathbf{B},-\mathbf{E})$ to derive the second pair of Maxwell's equations. Such a method is obvious when $\mathbf{D}$ is reduced to $\mathbf{E}$ and $\mathbf{H}$ to $\mathbf{B}$. The second pair of the three-dimensional Maxwell equations is:

$$\begin{cases} \nabla \cdot \mathbf{B} = 0 \\ \partial_0 \mathbf{B} + \nabla \times \mathbf{E} = 0 \end{cases} \tag{6.26}$$

which is valid for vacuum spacetime without source. The case of Einstein spacetime should be treated by accounting for metric dependent on the event-point e.g. [6], [261]. It is developed in the main part of this chapter. There is a need for more mathematical support to explain this transformation.

For completeness, when introducing the Faraday tensor in (6.24), we have the wave propagation equation for electromagnetism:

$$\begin{aligned} \nabla_\mu \mathcal{H}^{\mu\nu} &= \epsilon_0 \eta^{\mu\lambda} \eta^{\nu\rho} \nabla_\mu \left(\nabla_\lambda A_\rho - \nabla_\rho A_\mu \right) \\ &= \epsilon_0 \eta^{\mu\lambda} \; \nabla_\mu \nabla_\lambda A^\nu - \epsilon_0 \eta^{\nu\rho} \; \nabla_\mu \nabla_\rho A^\mu \end{aligned}$$

where we can use commutativity of the covariant derivative in the absence of curvature in a flat Minkowski spacetime, and where we can introduce the Lorenz jauge to show that the second term of the second row is equal to zero. We then deduce that:

$$\nabla_\mu \mathcal{H}^{\mu\nu} = 0 \Longrightarrow \Box \; A^\nu := \eta^{\mu\lambda} \; \nabla_\mu \nabla_\lambda A^\nu = 0 \tag{6.27}$$

This constitutes the electromagnetic wave propagation in flat vacuum spacetime.

6.2.1.4 *Classical Electromagnetism and Quantum Physics*

Maxwell's equations can be rearranged to represent the Schrödinger equations of a photon or a distribution of photons e.g. [75], [249]. For the sake of simplicity, let's consider a Minkowski spacetime with the time $x^0 := ct$ to allow us a comparison with the four-dimensional formulation[1]. Equations (6.25) and (6.26) are rewritten as for vacuum spacetime:

$$\begin{cases} \nabla \cdot \mathbf{D} = 0 \\ \mu_0 \; \partial_0 \mathbf{D} - \nabla \times \mathbf{B} = 0 \end{cases} \quad \text{and} \quad \begin{cases} \nabla \cdot \mathbf{B} = 0 \\ \epsilon_0 \; \partial_0 \mathbf{B} + \nabla \times \mathbf{D} = 0 \end{cases} \tag{6.28}$$

The introduction of the complex Weber vector:

$$\mathbf{W} := \frac{\mathbf{D}}{\sqrt{2\epsilon_0}} + i \; \frac{\mathbf{B}}{\sqrt{2\mu_0}} \tag{6.29}$$

[1]Quantum physics is described in a flat spacetime conversely to relative gravitation.

where constants ϵ_0 and μ_0 are introduced to match the dimensions of the real part and imaginary part of $\mathbf{W}$, permits us to formulate electromagnetism with complex variables [73]. First, by expressing the transversality condition of the Weber vector, we have:

$$\nabla \cdot \mathbf{W} = \frac{\nabla \cdot \mathbf{D}}{\sqrt{2\epsilon_0}} + i\, \frac{\nabla \cdot \mathbf{B}}{\sqrt{2\mu_0}} \tag{6.30}$$

Second, by writing the equation:

$$\frac{i}{c}\, \partial_0 \mathbf{W} = \nabla \times \mathbf{W} \tag{6.31}$$

we introduce the expression of the Weber vector:

$$\frac{i}{c}\, \partial_0 \left(\frac{\mathbf{D}}{\sqrt{2\epsilon_0}} + i\, \frac{\mathbf{B}}{\sqrt{2\mu_0}} \right) = \nabla \times \left(\frac{\mathbf{D}}{\sqrt{2\epsilon_0}} + i\, \frac{\mathbf{B}}{\sqrt{2\mu_0}} \right) \tag{6.32}$$

which gives after a straightforward distribution of derivative and curl:

$$\begin{cases} \mu_0\, \partial_0 \mathbf{D} - \nabla \times \mathbf{B} = 0 \\ \epsilon_0\, \partial_0 \mathbf{B} + \nabla \times \mathbf{D} = 0 \end{cases} \tag{6.33}$$

given that $c^2 \mu_0 \epsilon_0 = 1$. Therefore, the Maxwell equations in a Minkowski vacuum spacetime can be written in terms of the Weber vector:

$$\boxed{\begin{cases} i\, \partial_0 \mathbf{W} = c\, \nabla \times \mathbf{W} \\ \nabla \cdot \mathbf{W} = 0 \end{cases}} \tag{6.34}$$

which is similar to a Schrödinger equation with the general form e.g. [74]:

$$i\, \hbar\, \frac{\partial \psi}{\partial t} = \mathscr{H} \psi$$

where the vector $\mathbf{W}$ acts as a wave function whereas the Hamiltonian is related to the curl operator $\mathscr{H} \to c\, \nabla\, \times$ which is a 3×3 matrix. The term $\hbar$ is the Planck's constant which tends to zero for non-quantum physics. Gauss's law and the assumption of non-existence of magnetic monopoles represent the transversality conditions of the Weber vector. Classical electromagnetic equations describe a continuous distribution of large numbers of photons but the Dirac wave function ψ is rather limited to an electron originally. However, by defining first a density of photons using a Fourier transform in the space, and by defining accordingly a single photon wave function, it is possible to rigorously relate the Maxwell equations with quantum physics e.g. [75]. We do not go further since it is not the purpose of this book, which is focused on the phenomenological approach. However, despite the fact that Maxwell's equations were derived in the framework of classical physics, they bear a very tight resemblance to the quantum physics equation that governs the time evolution of wave functions in relativistic theory. Thus in some sense, Maxwell's equations in electromagnetism might be classified as a "slightly quantized" model associated with Schrödinger's equation for a photon [74].

Remark 6.9. Gravitation and electromagnetism are not at the same level in the hierarchy of physical theories. If gravitation is at the same level as geometric optics (similar to classical Newtonian mechanics), electromagnetism is one step above (in the sense of the quantization of physics). Without any pretense of wanting to propose any quantization of relative gravitation (we are very far from it), we must nevertheless be aware that the inclusion of electromagnetism constitutes *de facto* a step towards the direction of quantization in "continuous version" (first quantization).

6.2.2 *Covariant Maxwell and Einstein-Maxwell Equations*

Various phenomena may have effects on electromagnetic waves such as the presence of gravitational fields. The constitutive equations should be changed to account for the modification of spacetime in the presence of gravity for instance (curvature). Here, we thus consider the electromagnetic constitutive laws $\mathbf{D} = \epsilon\mathbf{E}$ and $\mathbf{B} = \mu\mathbf{H}$ where ϵ and μ are the electric and magnetic parameters of the spacetime replacing ϵ_0 and μ_0. The most comprehensive unification of Maxwell's electromagnetism with Einstein's gravitation goes back to Rainich's work in 1925 [8]. He essentially considered the metric $g_{\mu\nu}$ for gravitation and the anti-symmetric Faraday tensor $\mathcal{F}_{\mu\nu}$ for electromagnetism e.g. [8], [262]. It should be noted that these studies date from the same period as Friedmann's equations for cosmology (1922). In a more contemporary approach, the connection Rainich used was that of Levi-Civita. In the field equations he obtained, an electromagnetic stress-energy momentum was the only source of gravitation taken into account. The objective of this section is to revisit the variational formulation of equations developed in [8].

6.2.2.1 *Gravitation in an Einstein Spacetime*

To begin with, let us recall the basics of Einstein gravitation and the definition of an Einstein spacetime we adopt in the present book.

Definition 6.2. An Einstein spacetime is a four-dimensional manifold continuum (compacted and connected), endowed with a pseudo-Riemann hyperbolic metric $\mathbf{g}$ (verifying Hilbert causality) of signature $(1,3)$ and with a Levi-Civita connection $\overline{\nabla}$ without torsion field $\aleph \equiv 0$.

The covariance condition of the Lagrangian imposes that when the arguments of $\mathscr{L}$ are limited to the metric and its first and second derivatives, then it can only depend on the components of the metric and those of the curvature [110]. Remaining with the model of Einstein vacuum spacetime, let's consider the Hilbert-Einstein action.

Definition 6.3. The Hilbert-Einstein action of an Einstein vacuum spacetime is defined by:

$$\mathscr{S}_{HE} = \frac{1}{2\chi} \int_{\mathscr{B}} \overline{\mathcal{R}}\, \overline{\omega}_n, \quad \overline{\mathcal{R}} = \overline{\mathcal{R}}\left[g_{\alpha\beta}\right] \tag{6.35}$$

where curvature is calculated by means of the metric $g_{\mu\nu}$ and the connection $\overline{\Gamma}^{\gamma}_{\mu\nu}$, and importantly the Riemann volume-form $\overline{\omega}_n$.

To derive the field equations, the stationarity of the Hilbert-Einstein action (6.35) gives:

$$\begin{aligned}\delta\mathscr{S}_{HE} &= \frac{1}{2\chi} \int_{\mathscr{B}} \left(\overline{\mathcal{R}}_{\beta\lambda} - \frac{\overline{\mathcal{R}}}{2} g_{\beta\lambda}\right) \delta g^{\beta\lambda}\, \overline{\omega}_n \\ &+ \frac{1}{2\chi} \int_{\mathscr{B}} \left[\overline{\nabla}_{\gamma}\left(g^{\beta\lambda}\, \delta\overline{\Gamma}^{\gamma}_{\beta\lambda}\right) - \overline{\nabla}_{\beta}\left(g^{\beta\lambda}\, \delta\overline{\Gamma}^{\gamma}_{\gamma\lambda}\right)\right] \overline{\omega}_n\end{aligned}$$

given that the connection is assumed to be torsionless and metric compatible. The variation (2.104) was used. This gives us the classical Einstein field equation for vacuum spacetime, after dropping the boundary terms embodied by the second row where $\delta\overline{\Gamma}^{\gamma}_{\alpha\beta} \equiv 0$ on the boundary $\partial\mathscr{B}$:

$$\overline{\mathcal{R}}_{\beta\lambda} - \frac{\overline{\mathcal{R}}}{2} g_{\beta\lambda} = 0 \tag{6.36}$$

Starting with the Hilbert-Einstein action, the main conditions required to obtain the Einstein field equations (6.36) are the following:

(1) the connection $\overline{\nabla}$ is compatible with the metric $\mathbf{g}$;
(2) the torsion is equal to zero $\aleph = 0$;
(3) the connection $\overline{\nabla}$ is compatible with the volume-form $\overline{\omega}_n$.

All the three conditions are necessary (and sufficient). However, there is no reason, neither theoretical nor practical, which prevents *a priori* the generalization of spacetime from Riemann (Einstein spacetime) to that of Riemann-Cartan (Einstein-Cartan spacetime). For the sake of clarity, we sketch hereafter the divergence operator which was implicitly used for obtaining the field equations (6.36).

6.2.2.2 *Yang-Mills Action, Faraday Strength and Excitation Tensors*

Let's consider the example of action for a free electromagnetic field without sources and occurring within a Einstein curved spacetime.

Definition 6.4. The Lagrangian of an Einstein vacuum spacetime is defined by including both Einstein-Palatini (extension of Hilbert-Einstein Lagrangian) and Yang-Mills contributions e.g. [263]:

$$\mathscr{S} := \int_{\mathscr{M}} \mathscr{L}\, \varpi_n \quad \text{with} \quad \boxed{\mathscr{L} := \frac{1}{2\chi}\overline{\mathcal{R}} - \frac{1}{4}\mathcal{H}^{\mu\nu}\,\mathcal{F}_{\mu\nu}} \tag{6.37}$$

where no matter action is considered, and where the Faraday tensor and the scalar curvature are defined by the relationships:

$$\mathcal{F}_{\mu\nu} := \overline{\nabla}_\mu A_\nu - \overline{\nabla}_\nu A_\mu, \quad \overline{\mathcal{R}} := g^{\mu\nu}\overline{\mathfrak{R}}_{\mu\nu} \tag{6.38}$$

Contrary to Minkowski spacetime, Einstein spacetime has non-zero curvature. For more details, the Faraday strength tensor, which is the primal variable in the sense of constitutive laws of materials, is defined as:

$$\mathcal{F}_{\mu\nu} := \partial_\mu A_\nu - \partial_\nu A_\mu, \qquad \mathcal{F}_{\mu\nu} = \begin{bmatrix} 0 & -E_1 & -E_2 & -E_3 \\ E_1 & 0 & B^3 & -B^2 \\ E_2 & -B^3 & 0 & B^1 \\ E_3 & B^2 & -B^1 & 0 \end{bmatrix} \tag{6.39}$$

where the combined electromagnetic field $(\mathbf{E}, \mathbf{B})$ do not transform as 3-vectors but rather as the six components of the skew-symmetric tensor $\mathcal{F}_{\mu\nu}$. We conform here to the convention in e.g. [264]. The dual variables (here in the sense of constitutive laws) are merged into the tensor:

$$\mathcal{H}^{\mu\nu} = \begin{bmatrix} 0 & D^1 & D^2 & D^3 \\ -D^1 & 0 & H_3 & -H_2 \\ -D^2 & -H_3 & 0 & H_1 \\ -D^3 & H_2 & -H_1 & 0 \end{bmatrix} \tag{6.40}$$

The term (6.40) constitutes the general form of dual variables and the relation between (6.40) and (6.39) defines the *constitutive laws* of the continuum (here vacuum spacetime). *It should be emphasized that (6.40) and (6.39) do not depend on the geometric structure of the vacuum spacetime.*

Remark 6.10. In order to relate the expression of the Faraday strength tensor to the classical formulation in three dimensions, we can take term by term to obtain:

$$\begin{cases} \mathcal{F}_{01} = -E_1 = \overline{\nabla}_0 A_1 - \overline{\nabla}_1 A_0 \\ \mathcal{F}_{02} = -E_2 = \overline{\nabla}_0 A_2 - \overline{\nabla}_2 A_0 \\ \mathcal{F}_{03} = -E_3 = \overline{\nabla}_0 A_3 - \overline{\nabla}_3 A_0 \end{cases}, \quad \begin{cases} \mathcal{F}_{12} = B^3 = \overline{\nabla}_1 A_2 - \overline{\nabla}_2 A_1 \\ \mathcal{F}_{13} = -B^2 = \overline{\nabla}_1 A_3 - \overline{\nabla}_3 A_1 \\ \mathcal{F}_{23} = B^1 = \overline{\nabla}_2 A_3 - \overline{\nabla}_3 A_2 \end{cases}$$

which may be written as:

$$\mathbf{E} = -\partial_0 \mathbf{A} - \overline{\nabla} U, \quad \mathbf{B} = \overline{\nabla} \times \mathbf{A} \tag{6.41}$$

where $A_\mu := (-U, \mathbf{A})$ is the four-dimensional electromagnetic potential including the scalar and vector potential. The second equation may be interpreted as assuming the absence of magnetic monopoles since $\overline{\nabla} \cdot \mathbf{B} = \overline{\nabla} \cdot \overline{\nabla} \times \mathbf{A} \equiv 0$. In the absence of magnetic monopoles, Lorentz already noticed that the electromagnetic potential leading to a given field is not unique since the gauge transformations are:

$$\tilde{U} = U - \partial_0 g, \quad \tilde{\mathbf{A}} = \mathbf{A} + \overline{\nabla} g \tag{6.42}$$

where the arbitrary gauge function $g(x^0, x^1, x^2, x^3)$ assumed of class $\mathscr{C}^2$ leads to:

$$\tilde{\mathbf{E}} := -\partial_0 \tilde{\mathbf{A}} - \overline{\nabla} \tilde{U} = -\partial_0 \mathbf{A} - \overline{\nabla} U = \mathbf{E}, \quad \tilde{\mathbf{B}} := \overline{\nabla} \times \tilde{\mathbf{A}} = \overline{\nabla} \times \mathbf{A} = \mathbf{B} \tag{6.43}$$

in a simply connected (Riemannian) manifold. In an Einstein-Cartan spacetime, in the presence of magnetic monopoles, the gauge invariance is not so easy to interpret e.g. [27], [257].

6.2.2.3 *Maxwell and Einstein-Maxwell Equations*

In this subsection, we consider an Einstein spacetime for which the metric is point-event dependent. The linear constitutive laws take the special form:

$$\mathcal{H}^{\mu\nu} = \Xi^{\mu\nu\lambda\rho}\ \mathcal{F}_{\lambda\rho}, \quad \Xi^{\mu\nu\lambda\rho} = \varepsilon_0\ g^{\mu\lambda} g^{\nu\rho} \tag{6.44}$$

First, we have the variation of the gravitation contribution:

$$\delta\left(\mathscr{L}_G\ \overline{\omega}_n\right) = \frac{1}{2\chi}\left(\delta\overline{\Re}_{\mu\nu}\ g^{\mu\nu} + \overline{\Re}_{\mu\nu}\delta g^{\mu\nu}\right)\overline{\omega}_n + \frac{1}{2\chi}\ \overline{\Re}_{\mu\nu}\ g^{\mu\nu}\delta\overline{\omega}_n$$

The variation of the Ricci curvature may be dropped by using the divergence theorem when integrating and shifting the boundary terms. We deduce:

$$\delta\int_{\mathscr{B}}\mathscr{L}_G\ \overline{\omega}_n = \frac{1}{2\chi}\int_{\mathscr{B}}\left(\overline{\Re}_{\mu\nu} - \frac{\overline{\Re}}{2}\ g_{\mu\nu}\right)\delta g^{\mu\nu}\ \overline{\omega}_n \tag{6.45}$$

Remark 6.11. The variation of the Ricci curvature holds e.g. [49]:

$$\delta\int_{\mathscr{B}} g^{\mu\nu}(\delta\overline{\Re}_{\mu\nu})\ \overline{\omega}_n = \int_{\mathscr{B}} g^{\mu\nu}\left[\overline{\nabla}_\lambda\left(\delta\overline{\Gamma}^\lambda_{\mu\nu}\right) - \overline{\nabla}_\mu\left(\delta\overline{\Gamma}^\lambda_{\lambda\nu}\right)\right]\overline{\omega}_n = 0$$

by shifting the divergence terms at the boundary of the spacetime and assuming a zero divergence at this boundary.

Second, we consider the variation of the electromagnetic contribution:

$$\begin{aligned}\delta\int_{\mathscr{B}}\mathscr{L}_{EM}\ \overline{\omega}_n = &-\frac{1}{4}\int_{\mathscr{B}}\delta\mathcal{H}^{\mu\nu}\ \mathcal{F}_{\mu\nu}\ \overline{\omega}_n - \frac{1}{4}\int_{\mathscr{B}}\mathcal{H}^{\mu\nu}\delta\mathcal{F}_{\mu\nu}\ \overline{\omega}_n\\ &-\frac{1}{4}\int_{\mathscr{B}}\mathcal{H}^{\mu\nu}\mathcal{F}_{\mu\nu}\ \delta\overline{\omega}_n\end{aligned}$$

The last term of the integrand gives:

$$-\frac{1}{4}\int_{\mathscr{B}}\mathcal{H}^{\mu\nu}\mathcal{F}_{\mu\nu}\ \delta\overline{\omega}_n = \frac{1}{8}\int_{\mathscr{B}}\mathcal{H}^{\lambda\rho}\mathcal{F}_{\lambda\rho}\ g_{\mu\nu}\ \delta g^{\mu\nu}\ \overline{\omega}_n \tag{6.46}$$

The second term of the integrand becomes:

$$\begin{aligned}-\frac{1}{4}\int_{\mathscr{B}}\mathcal{H}^{\mu\nu}\delta\mathcal{F}_{\mu\nu}\ \overline{\omega}_n &= -\frac{1}{4}\int_{\mathscr{B}}\mathcal{H}^{\mu\nu}\left[\overline{\nabla}_\mu(\delta A_\nu) - \overline{\nabla}_\nu(\delta A_\mu)\right]\overline{\omega}_n\\ &= \frac{1}{2}\int_{\mathscr{B}}\mathcal{H}^{\mu\nu}\ \overline{\nabla}_\nu(\delta A_\mu)\ \overline{\omega}_n\end{aligned}$$

Integrating by parts and applying the divergence theorem give:

$$-\frac{1}{4}\int_{\mathscr{B}} \mathcal{H}^{\mu\nu}\delta\mathcal{F}_{\mu\nu}\,\varpi_n = \frac{1}{2}\int_{\mathscr{B}} \overline{\nabla}_\nu(\mathcal{H}^{\mu\nu}\delta A_\mu)\,\varpi_n - \frac{1}{2}\int_{\mathscr{B}} \overline{\nabla}_\nu(\mathcal{H}^{\mu\nu})\delta A_\mu\,\varpi_n \tag{6.47}$$

For variation of the Lagrangian (6.37), the independent variables are the metric and four-potential vector. Variation of the Faraday tensor takes the form of:

$$\delta\mathcal{F}_{\mu\nu} = \overline{\nabla}_\mu(\delta A_\nu) - \overline{\nabla}_\nu(\delta A_\mu) \tag{6.48}$$

This relation is directly obtained by writing:

$$\begin{aligned}\delta\mathcal{F}_{\mu\nu} &= \delta\left(\partial_\mu A_\nu - \overline{\Gamma}^\rho_{\mu\nu}A_\rho\right) - \delta\left(\partial_\nu A_\mu - \overline{\Gamma}^\rho_{\nu\mu}A_\rho\right)\\ &= \left(\partial_\mu\delta A_\nu - \overline{\Gamma}^\rho_{\mu\nu}\delta A_\rho - \delta\overline{\Gamma}^\rho_{\mu\nu}A_\rho\right) - \left(\partial_\nu\delta A_\mu - \overline{\Gamma}^\rho_{\nu\mu}\delta A_\rho - \delta\overline{\Gamma}^\rho_{\nu\mu}A_\rho\right)\end{aligned}$$

accounting for the fact that the connection variation $\delta\overline{\Gamma}^\rho_{\mu\nu}$ also contributes to variation of $\delta\mathcal{F}_{\mu\nu}$. The first term of the integrand may be obtained by introducing the constitutive laws:

$$\delta(\mathcal{H}^{\mu\nu})\mathcal{F}_{\mu\nu} = \varepsilon_0\left[\delta(g^{\mu\lambda})\;g^{\nu\rho}\;\mathcal{F}_{\lambda\rho} + g^{\mu\lambda}\;\delta(g^{\nu\rho})\;\mathcal{F}_{\lambda\rho} + g^{\mu\lambda}\;g^{\nu\rho}\;\delta\mathcal{F}_{\lambda\rho}\right]\mathcal{F}_{\mu\nu} \tag{6.49}$$

We observe that the last term gives the same quantity as previously. The first two terms are calculated as follows:

$$-\frac{\varepsilon_0}{4}\left[\delta(g^{\mu\lambda})\;g^{\nu\rho}\;\mathcal{F}_{\lambda\rho} + g^{\mu\lambda}\;\delta(g^{\nu\rho})\;\mathcal{F}_{\lambda\rho}\right]\mathcal{F}_{\mu\nu} = \frac{\varepsilon_0}{2}\left(\mathcal{F}_{\mu\lambda}\;g^{\lambda\rho}\;\mathcal{F}_{\rho\nu}\right)\delta g^{\mu\nu}$$

In sum, the two systems of conservation laws associated with the unknown primal variables (A_μ, and $g_{\mu\nu}$) are derived by varying the Lagrangian along the Lie-derivative variations $\mathcal{L}_\xi A_\mu$, and $\mathcal{L}_\xi g_{\mu\nu}$. We can rearrange the Lagrangian variation of the action to give:

$$\begin{aligned}\delta\mathscr{S} = \int_{\mathscr{M}} \left(\overline{\nabla}_\nu\mathcal{H}^{\mu\nu}\right)\delta A_\mu\,\varpi_n + \int_{\mathscr{M}} &\left[\frac{1}{2\chi}\left(\overline{\mathfrak{R}}_{\mu\nu} - \frac{\overline{\mathcal{R}}}{2}\,g_{\mu\nu}\right)\right.\\ &\left. + \frac{1}{8}\left(\mathcal{H}^{\lambda\rho}\mathcal{F}_{\lambda\rho}\right)g_{\mu\nu} + \frac{\varepsilon_0}{2}\left(\mathcal{F}_{\mu\lambda}\;g^{\lambda\rho}\;\mathcal{F}_{\rho\nu}\right)\right]\delta g^{\mu\nu}\,\varpi_n\end{aligned} \tag{6.50}$$

given that the Faraday tensor is *in fine* expressed in terms of the potential A_μ by means of (6.38).

Definition 6.5. We define the electromagnetic Maxwell stress-energy tensor:

$$\boxed{T^{\text{em}}_{\mu\nu} = \frac{1}{8}\left(\mathcal{H}^{\lambda\rho}\mathcal{F}_{\lambda\rho}\right) g_{\mu\nu} + \frac{1}{4}\mathcal{H}^{\lambda\rho}\left(g_{\nu\lambda}\mathcal{F}_{\rho\mu} + \mathcal{F}_{\mu\lambda}g_{\rho\nu}\right)} \tag{6.51}$$

Remark 6.12. It is easy to check that for the vacuum constitutive laws, the Maxwell stress-energy tensor is equal to:

$$T^{\text{em}}_{\mu\nu} = \frac{1}{8}\left(\mathcal{H}^{\lambda\rho}\mathcal{F}_{\lambda\rho}\right) g_{\mu\nu} + \frac{\varepsilon_0}{2}\,\mathcal{F}_{\mu\lambda}\,g^{\lambda\rho}\,\mathcal{F}_{\rho\nu} \tag{6.52}$$

but we prefer to use the expression (6.51) which is more convenient for later extension to more generalized spacetimes.

The stationarity $\delta\mathscr{S} \equiv 0$ for any compatible δA_ν and $\delta g^{\mu\nu}$ implies the Einstein-Maxwell equations:

$$\begin{cases} \overline{\nabla}_\nu \mathcal{H}^{\mu\nu} = 0 \\ \dfrac{1}{2\chi}\,\overline{G}_{\mu\nu} + T^{\text{em}}_{\mu\nu} = 0 \end{cases} \tag{6.53}$$

with the Einstein tensor $\overline{G}_{\mu\nu} := \overline{\mathfrak{R}}_{\mu\nu} - \left(\overline{\mathcal{R}}/2\right) g_{\mu\nu}$. The Mawxell stress-energy tensor (6.51) becomes a physical source for the gravitation. The unknowns are the metric and the electromagnetic potential.

The system of equations (6.53) is coupled:

(1) In a sense, the first equation shows that the curvature, via the coefficients of the connection and its derivatives, influences the propagation of electromagnetic waves (we will see in detail later);
(2) In the other direction, the second equation shows that the electromagnetic field via an isotropic component of the electromagnetic tensor but also by an anisotropic component modifies the geometric structure, in particular the Ricci curvature, of spacetime.
(3) Except in very special cases, the Schwarzschild metric is no longer automatically a solution for the metric due to this coupling.

Remark 6.13. The first equation is the covariant Maxwell equations in an Einstein spacetime. The second equation, which is the Einstein-Maxwell equation, governs the interaction of electromagnetism with gravitation. The unknowns are the spacetime metric and the electromagnetic potential.

Remark 6.14. The second equation can be rearranged as $\overline{G}_{\mu\nu} + 2\chi\, T^{\text{em}}_{\mu\nu} = 0$ where the first term represents the spacetime bending and generating gravitation whereas the energy-momentum second term represents the electromagnetic action which is the source of the spacetime bending.

6.2.2.4 *Maxwell Stress-Energy Tensor*

We report here below the components of the stress-energy tensor; some parts may also be found in e.g. [6]:

(1) for diagonal terms:

$$T^{\text{em}}_{00} = \frac{1}{4}\left(D^i E_i + B^i H_i\right) \tag{6.54}$$

$$T^{\text{em}}_{11} = \frac{1}{4}\left(-D^1 E_1 + D^2 E_2 + D^3 E_3\right) + \frac{1}{4}\left(-B^1 H_1 + B^2 H_2 + B^3 H_3\right) \tag{6.55}$$

$$T^{\text{em}}_{22} = \frac{1}{4}\left(+D^1 E_1 - D^2 E_2 + D^3 E_3\right) + \frac{1}{4}\left(+B^1 H_1 - B^2 H_2 + B^3 H_3\right) \tag{6.56}$$

$$T^{\text{em}}_{33} = \frac{1}{4}\left(+D^1 E_1 + D^2 E_2 - D^3 E_3\right) + \frac{1}{4}\left(+B^1 H_1 + B^2 H_2 - B^3 H_3\right) \tag{6.57}$$

The first component is related to the electromagnetic energy. The other components are classical e.g. [6]. Diagonal components constitute the backbone of the isotropic part of the electromagnetic field.

(2) for the first row (and, by extension, the first column):

$$T^{\text{em}}_{01} = T^{\text{em}}_{10} = \frac{1}{4}\left(D^3 B^2 - D^2 B^3\right) + \frac{1}{4}\left(E_3 H_2 - E_2 H_3\right) \tag{6.58}$$

$$T^{\text{em}}_{02} = T^{\text{em}}_{20} = \frac{1}{4}\left(D^1 B^3 - D^3 B^1\right) + \frac{1}{4}\left(E_1 H_3 - E_3 H_1\right) \tag{6.59}$$

$$T^{\text{em}}_{03} = T^{\text{em}}_{30} = \frac{1}{4}\left(D^2 B^1 - D^1 B^2\right) + \frac{1}{4}\left(E_2 H_1 - E_1 H_2\right) \tag{6.60}$$

Components of the Poynting vectors $\mathbf{D} \times \mathbf{B}$ and $\mathbf{E} \times \mathbf{H}$ are recognized in these equations. We further observe the mean of these two vectors resulting from an implicit symmetrization.

(3) for the remaining:

$$T^{\text{em}}_{12} = T^{\text{em}}_{21} = \frac{1}{4}\left[-\left(D^1E_2 + D^2E_1\right) + \left(B^1H_2 + B^2H_1\right)\right] \quad (6.61)$$

$$T^{\text{em}}_{23} = T^{\text{em}}_{32} = \frac{1}{4}\left[-\left(D^2E_3 + D^3E_2\right) + \left(B^2H_3 + B^3H_2\right)\right] \quad (6.62)$$

$$T^{\text{em}}_{31} = T^{\text{em}}_{31} = \frac{1}{4}\left[-\left(D^3E_1 + D^1E_3\right) + \left(B^3H_1 + B^1H_3\right)\right] \quad (6.63)$$

They are different from the non-symmetric stress-energy tensor of Minkowski. In fact components (6.61) are equal to the symmetrized Minkowski stress-energy tensor. In some sense, the stress-energy obtained in the present book is arising from a metrical variational principle and therefore is similar to that of Abraham; this was already remarked in e.g. [264]. The electromagnetic stress–energy tensor includes the flow of energy T^{em}_{00} and momentum of electromagnetic field in vacuum spacetime. For obtaining all these expressions, we have used the metric components in the tetrads $\vartheta^\alpha \otimes \vartheta^\beta$ which are $\eta_{\alpha\beta} = [+1, -1, -1, -1]$. Three-dimensional expressions of this stress-energy tensor may be also found in e.g. [265].

Remark 6.15. The electromagnetic stress-energy that we obtained is symmetric because it was obtained via factorization with the variation $\delta g^{\mu\nu}$. In this case, any anti-symmetric part is eliminated automatically. In fact, there are several methods to define this tensor, the best known of which are that of Minkowski (non-symmetric, 1908) and that of Abraham (symmetric, 1909). We do not go into the story regarding the Minkowski and Abraham controversy in conductive dielectric materials which lasted over a century e.g. [264]. Comparison of six theoretical models, including the two main originals of Minkowski and Abraham, have been the subject of more recent studies, namely their covariance properties and symmetry having been analyzed in e.g. [265]. In vacuum spacetime, the electromagnetic property that relates the electric permittivity ϵ_0 and the magnetic permeability μ_0 with the light speed avoids this controversial problem. Our stress-energy tensor is symmetric by construction.

6.2.2.5 *Gravito-Electromagnetism Equations and Rainich Conditions*

Let's consider an Einstein spacetime $\mathscr{B}$ endowed with a metric $g_{\mu\nu}$ satisfying Hilbert causality, and $\overline{\nabla}$ denotes the Levi-Civita connection.

A skew-symmetric tensor $\mathcal{F}_{\mu\nu}$ on $\mathscr{B}$ defines an electromagnetic field. First, we want to highlight in this paragraph the coupling of gravitation with electromagnetism without torsion. Predictably, given that the electromagnetic tensor field is skew-symmetric, it is unlikely to be easily related to the metric tensor and the Riemann curvature tensor. Nevertheless, it is necessary to mention approaches using curvature as mathematical variables of electromagnetism e.g. [266], while remaining within classical physics with a Riemann manifold [7], although the latter paper already advocated the use of multi-connected manifolds.

Gravitation and electromagnetism within Einstein spacetime. The gravitation and electromagnetism model of an Einstein vacuum spacetime is governed by the system of equations:

$$
\begin{aligned}
\overline{\nabla}_\nu \mathcal{H}^{\mu\nu} &= 0 \\
\overline{G}_{\mu\nu} &= -2\chi\, T^{\text{em}}_{\mu\nu} \\
T^{\text{em}}_{\mu\nu} &= \frac{1}{8}\left(\mathcal{H}^{\lambda\rho}\mathcal{F}_{\lambda\rho}\right) g_{\mu\nu} + \frac{1}{4}\mathcal{H}^{\lambda\rho}\left(g_{\nu\lambda}\mathcal{F}_{\rho\mu} + \mathcal{F}_{\mu\lambda} g_{\rho\nu}\right)
\end{aligned}
\tag{6.64}
$$

where $\overline{G}_{\mu\nu}$ is the Einstein tensor, $\mathcal{F}_{\mu\nu}$, and $\mathcal{H}^{\mu\nu}$ the excitation tensor.

Remark 6.16. For completeness, the Bianchi identity for the skew-symmetric tensor $\mathcal{F}_{\mu\nu}$ e.g. [262]:

$$
\overline{\nabla}_\lambda \mathcal{F}_{\mu\nu} + \overline{\nabla}_\mu \mathcal{F}_{\nu\lambda} + \overline{\nabla}_\nu \mathcal{F}_{\lambda\mu} = 0 \tag{6.65}
$$

is added to (6.64).

Given the assumptions for the derivation of (6.64) and (6.65), Rainich had asked the question about the necessary and sufficient conditions that allowed us to ensure the existence of a non-zero electromagnetic field for the metric tensor e.g. [8]. This is legitimate. Without going into the details that could be found for example in [267], one condition among four others is that the scalar curvature must be zero, $\mathcal{R} \equiv 0$. This condition is quite limitative since imposing a zero scalar curvature for the spacetime means that this theory should happen in a Minkowski spacetime. In the following, we attempt to extend the spacetime geometry to Riemann-Cartan manifolds rather than limiting the geometry to Einstein spacetime with zero curvature.

6.2.2.6 *Electromagnetic Waves and Gravitational Fields*

Most if not all known information about the Universe is more or less directly obtained by measuring and analyzing electromagnetic waves e.g. [64]. The modeling and analysis of electromagnetic waves is therefore a matter of concern. If the second line of (6.53) expresses the influence of the electromagnetic fields on gravitation (consider the right-hand side of this equation as the electromagnetic force field occurring in vacuum spacetime), the first line governs the influence of gravitation on electromagnetic wave propagation by means of metric $\mathbf{g}$ and connection $\overline{\nabla}$. Maxwell's equations (6.53) (first row) are used to analyze the electromagnetic wave within an Einstein spacetime. The first equation may be rewritten:

$$\begin{aligned}\overline{\nabla}_\nu \mathcal{H}^{\mu\nu} &= \varepsilon_0\, \overline{\nabla}_\nu \left(g^{\mu\alpha} g^{\alpha\nu} \mathcal{F}_{\alpha\beta}\right) = \varepsilon_0\, \overline{\nabla}_\nu \left[g^{\mu\alpha}\overline{\nabla}_\alpha A^\nu - g^{\nu\beta}\overline{\nabla}_\beta A^\mu\right] \\ &= \varepsilon_0\, g^{\mu\alpha}\left[\overline{\nabla}_\alpha \overline{\nabla}_\nu A^\nu + \Re^\nu_{\nu\alpha\gamma} A^\gamma\right] - \varepsilon_0\, g^{\nu\beta}\overline{\nabla}_\nu \overline{\nabla}_\beta A^\mu = 0\end{aligned}$$

where we have used the Schouten relations e.g. [49] with zero torsion. Maxwell's equations include a classical wave part, a divergence term, and the contribution of the Ricci curvature:

$$-\,g^{\nu\beta}\overline{\nabla}_\nu \overline{\nabla}_\beta A^\mu + g^{\mu\alpha}\overline{\nabla}_\alpha \overline{\nabla}_\nu A^\nu + g^{\mu\alpha}\overline{\Re}_{\alpha\gamma} A^\gamma = 0 \tag{6.66}$$

The first term expresses a D'Alembertian operator. The second term may be dropped if we assume a Lorenz gauge $\overline{\nabla}_\nu A^\nu = 0$. We then obtain in the absence of magnetic monopoles:

Electromagnetic wave within Einstein spacetime. The electromagnetic wave propagation equation within Einstein curved spacetime is given by:

$$g^{\nu\beta}\overline{\nabla}_\nu \overline{\nabla}_\beta A^\mu - g^{\mu\alpha}\overline{\Re}_{\alpha\gamma} A^\gamma = 0 \tag{6.67}$$

in which we notice the direct influence of the gravitation (represented by the Ricci curvature) on the electromagnetic wave propagation.

This is a fundamental equation for light wave propagation for observing gravitational lensing. This is also valid for e.g. optics [65], and electromagnetic fields in magnetohydrodynamics [64], [268]. (6.67) is not a wave-like equation and the method of variables separation does not hold in the general case. The form of the equation (6.67) resembles the wave equation in flat

Minkowski spacetime, except that the partial derivatives ∂_α are replaced by covariant derivatives $\overline{\nabla}_\alpha$ and there is also an additional term proportional to the Ricci curvature. Equation (6.67) is a classical model to study the electromagnetic wave propagation in different gravitation fields for instance in cosmology e.g. [64], or the vicinity of the Sun e.g. [94] although for the latter we rather use the three-dimensional formulation by introducing the metric in the Maxwell constitutive laws. In the following we will extend (6.67) to Einstein-Cartan spacetime.

Remark 6.17. For curved spacetime, the bending of light is modeled by these equations and the experimental confirmation was done by Eddington (1919) [79]. During a solar eclipse on 29 May 1919, Eddington and his co-workers measured the bending of light rays by seeing the outward shift of stars and found values near those predicted by Einstein's theory. This provided one among other evidences for Einstein gravitation. The Lorenz gauge condition can be incorporated in the Lagrangian to become:

$$\mathscr{L} := -\frac{1}{4}\mathcal{H}^{\mu\nu}\mathcal{F}_{\mu\nu} + \lambda\ \left(\overline{\nabla}_\nu A^\nu\right)^n$$

where λ is a Lagrange multiplier and $n = 1$ or $n = 2$ e.g. [64]. This method permits us to relax the solving of (6.67). Including a Lagrangian multiplier implicitly enforces the Lorenz gauge during resolution. The next step is to find the spacetime metric, which is necessary to define the Levi-Civita connection $\overline{\nabla}$.

6.2.3 *Reissner-Nordström Metric*

The simplest case of using the Einstein-Maxwell equation is that of spacetime due to an electrostatic charge with a spherical symmetry, which results in the Reissner-Nordström metric, originally discovered in 1918 by [269] and 1916 by [270], and studied more recently in the context of exact spherically symmetric solutions of the Einstein field equation e.g. [241], or in cosmology e.g. [271]. We suggest first the Lagrangian within vacuum spacetime where gravitation and electromagnetism occur. For the sake of simplicity, let's directly assume the existence of a radial electric field E, without magnetism and not depending on time, so that the Faraday electromagnetic field holds ("physical" values of the field are assumed) when projected onto an orthonormal spherical base:

$$\mathcal{F}_{(\mu\nu)} := E \begin{bmatrix} 0 & -1 & 0 & 0 \\ 1 & 0 & 0 & 0 \\ 0 & 0 & 0 & 0 \\ 0 & 0 & 0 & 0 \end{bmatrix} \Longrightarrow \mathcal{H}^{(\mu\nu)} := \epsilon_0 E \begin{bmatrix} 0 & 1 & 0 & 0 \\ -1 & 0 & 0 & 0 \\ 0 & 0 & 0 & 0 \\ 0 & 0 & 0 & 0 \end{bmatrix} \tag{6.68}$$

by using the Riemannian metric $g_{\mu\nu}$ to uprise the indices. In this example we consider a torsionless Riemann manifold $\mathscr{B}$. We can compute the electric contribution in the orthonormal coframe, expressed in terms of the Faraday strength:

$$\mathcal{F}_{(0r)} = -E = e^{-\nu} e^{-\mu}\, \mathcal{F}_{0r} \Longrightarrow \mathcal{H}^{0r} = \epsilon_0\, e^{-(\nu+\mu)}\, E,\ (\mathcal{F}_{0r})^2 = e^{2\nu+2\mu}\, (E)^2 \tag{6.69}$$

The electromagnetic energy-momentum tensor (6.51) gives us the components:

$$T_{00} = \frac{\epsilon_0}{4}\, g^{rr}\, (\mathcal{F}_{0r})^2 = -\frac{\epsilon_0}{4}\, e^{-2\mu}\, (\mathcal{F}_{0r})^2 \tag{6.70}$$

$$T_{rr} = \frac{\epsilon_0}{4}\, g^{00}\, (\mathcal{F}_{0r})^2 = \frac{\epsilon_0}{4}\, e^{-2\nu}\, (\mathcal{F}_{0r})^2 \tag{6.71}$$

$$T_{\theta\theta} = -\frac{\epsilon_0}{4}\, g^{00} g^{rr} g_{\theta\theta}\, (\mathcal{F}_{0r})^2 = -\frac{\epsilon_0}{4}\, e^{-2(\nu+\mu)}\, r^2\, (\mathcal{F}_{0r})^2 \tag{6.72}$$

$$T_{\varphi\varphi} = -\frac{\epsilon_0}{4}\, g^{00} g^{rr} g_{\varphi\varphi}\, (\mathcal{F}_{0r})^2 = -\frac{\epsilon_0}{4}\, e^{-2(\nu+\mu)}\, r^2 \sin^2\theta\, (\mathcal{F}_{0r})^2 \tag{6.73}$$

The second row of the system (6.53) which are the Einstein-Maxwell equations leads to the coupling of gravitation and electromagnetic fields:

$$\begin{cases} \left[e^{-2\mu}\left(\dfrac{2\mu'}{r} - \dfrac{1}{r^2}\right) + \dfrac{1}{r^2}\right] = +\dfrac{\epsilon_0 \chi}{2} E^2 \\ \left[e^{-2\mu}\left(\dfrac{2\nu'}{r} + \dfrac{1}{r^2}\right) - \dfrac{1}{r^2}\right] = -\dfrac{\epsilon_0 \chi}{2} E^2 \\ \left[\nu'' + (\nu')^2 - \nu'\mu' + \dfrac{\nu' - \mu'}{r}\right] = \dfrac{\epsilon_0 \chi}{2} E^2 \end{cases} \quad \text{and} \quad \begin{cases} \overline{\nabla}_r \mathcal{H}^{0r} = 0 \\ \overline{\nabla}_0 \mathcal{H}^{r0} = 0 \end{cases} \tag{6.74}$$

where unknowns are the three functions $\nu(r)$, $\mu(r)$ and $E(r)$. Summation of the first two equations (6.74) gives the necessary condition $\nu' + \mu' = 0$. Assuming a flat Minkowski spacetime at far distances we have, without losing the generality of our purpose, $\nu(r) + \mu(r) \equiv 0$.

6.2.3.1 *Alternative Expression of the Divergence*

For (pseudo)-Riemann manifolds, the divergence of tensor of order two $\mathcal{A}$ takes a particular form as follows:

$$\begin{aligned}\overline{\nabla}_\nu \mathcal{A}^{\mu\nu} &= \partial_\nu \mathcal{A}^{\mu\nu} + \overline{\Gamma}^{\mu}_{\nu\lambda}\mathcal{A}^{\lambda\nu} + \overline{\Gamma}^{\nu}_{\nu\lambda}\mathcal{A}^{\mu\lambda} \\ &= \overline{\Gamma}^{\mu}_{\nu\lambda}\mathcal{A}^{\lambda\nu} + \frac{1}{\sqrt{|\mathrm{Det}\mathbf{g}|}}\,\partial_\nu\left(\sqrt{|\mathrm{Det}\mathbf{g}|}\;\mathcal{A}^{\mu\nu}\right)\end{aligned}$$

We then deduce, for (pseudo)-Riemann manifold $\mathscr{B}$ endowed with a metric $g_{\alpha\beta}$, a volume form $\overline{\omega}_n$ and a Levi-Civita (symmetric) connection $\overline{\Gamma}^{\gamma}_{\alpha\beta}$:

(1) the divergence of skew-symmetric tensor of order 2

$$\overline{\nabla}_\nu \mathcal{A}^{\mu\nu} = \frac{1}{\sqrt{|\mathrm{Det}\mathbf{g}|}}\,\partial_\nu\left(\sqrt{|\mathrm{Det}\mathbf{g}|}\;\mathcal{A}^{\mu\nu}\right) \tag{6.75}$$

(2) the divergence of symmetric tensor of order 2

$$\overline{\nabla}_\nu \mathcal{A}^{\mu\nu} = \overline{\Gamma}^{\mu}_{\nu\lambda}\mathcal{A}^{\lambda\nu} + \frac{1}{\sqrt{|\mathrm{Det}\mathbf{g}|}}\,\partial_\nu\left(\sqrt{|\mathrm{Det}\mathbf{g}|}\;\mathcal{A}^{\mu\nu}\right) \tag{6.76}$$

For the example, we have to calculate the divergence of the skew-symmetric dual tensor $\mathcal{H}^{\mu\nu}$:

$$\overline{\nabla}_\nu \mathcal{H}^{\mu\nu} = \frac{1}{\sqrt{|\mathrm{Det}\mathbf{g}|}}\,\partial_\nu\left(\sqrt{|\mathrm{Det}\mathbf{g}|}\;\mathcal{H}^{\mu\nu}\right), \quad \text{with} \quad |\mathrm{Det}\mathbf{g}| = e^{2(\nu+\mu)} r^4 \sin^2\theta$$

Introduction of this particular electromagnetic field into the relations (6.53), and given that the Maxwell equations reduce to two components:

$$\begin{cases} \overline{\nabla}_0 \mathcal{H}^{r0} = \epsilon_0 \partial_0 E = 0 \\ \overline{\nabla}_r \mathcal{H}^{0r} = \dfrac{\epsilon_0}{r^2|\sin\theta|}\partial_r\left(r^2|\sin\theta| E\right) = 0 \end{cases}$$

The first row expresses that the field is stationary whereas the second allows us to explicitly obtain the electric field which is the case where the electric field surrounding a non-rotating charge Q (a constant of integration) depends on the distance r as: $E(r) := Q/r^2$.

6.2.3.2 *Reissner-Nordström Spacetime*

For the general solution of the first homogeneous equation of (6.74), as for the case where the electric field is set equal to zero, we rewrite the first row of Einstein-Maxwell equations by defining a variable change to have $e^{-2\mu} := 1 - V(r)$:

$$\left[e^{-2\mu}\left(\frac{2\mu'}{r} - \frac{1}{r^2}\right) + \frac{1}{r^2}\right] = \frac{V'(r)}{r} + \frac{V(r)}{r^2} = \frac{\epsilon_0 \chi}{2} \frac{Q}{r^4}$$

The homogeneous solution is obvious by defining a constant of integration A:

$$\frac{V'(r)}{r} + \frac{V(r)}{r^2} = 0 \quad \Longrightarrow \quad V(r) = \frac{A}{r}$$

Applying the Lagrange method of varying constant, the non-homogeneous equation holds:

$$\frac{V'(r)}{r} + \frac{V(r)}{r^2} = \frac{\epsilon_0 \chi}{2} \frac{Q}{r^4} \quad \Longrightarrow \quad \frac{A'}{r^2} - \frac{A}{r^3} + \frac{A}{r^3} = \frac{\epsilon_0 \chi}{2} \frac{Q}{r^4}$$

The solution for the "constant" A gives:

$$A(r) = -\frac{\epsilon_0 \chi}{2} \frac{Q}{r} + B \quad \Longrightarrow \quad V(r) = -\frac{\epsilon_0 \chi}{2} \frac{Q}{r^2} + \frac{B}{r}$$

We thus deduce:

$$e^{2\nu} = e^{-2\mu} = 1 - \frac{2m}{r} + \frac{q^2}{r^2}, \quad q^2 := \frac{\epsilon_0 \chi Q}{2} \tag{6.77}$$

where $2m$ is also a constant of integration ensuring that we obtain Minkowskian spacetime for long-distance point-event r far from the origin.

Remark 6.18. The physical dimension of m and q here is a length dimension.

From solution (6.77), we recognize components of the Reissner-Nordström metric. However, care should be taken in view of the entire system of equations (6.74) since we have four equations with three unknowns, thus the compatibility of all of them is mandatory. The resulting metric takes the form of:

$$\boxed{\begin{cases} g_{00} = \left(1 - \dfrac{2m}{r} + \dfrac{q^2}{r^2}\right) \\ g_{rr} = -\left(1 - \dfrac{2m}{r} + \dfrac{q^2}{r^2}\right)^{-1} \end{cases} \quad \begin{cases} g_{\theta\theta} = -r^2 \\ g_{\varphi\varphi} = -r^2 \sin^2\theta \end{cases}} \tag{6.78}$$

The line element is deduced accordingly:

$$ds^2 = \left(1 - \frac{2m}{r} + \frac{q^2}{r^2}\right)(dx^0)^2$$
$$- \left(1 - \frac{2m}{r} + \frac{q^2}{r^2}\right)^{-1} dr^2 - r^2 d\theta^2 - r^2 \sin^2\theta d\varphi^2 \qquad (6.79)$$

This gives the metric describing the spherically symmetric spacetime satisfying the Einstein field equation in a region with no matter but with a radial electric field generated by an equivalent charge q. Mathematically, it is shown that the only solution of the Einstein-Maxwell equation, in vacuum spacetime, reduces to the Reissner-Nordström metric under the conditions that we have previously given in this subsection e.g. [97]. an extension of the solution in the presence of the cosmological constant is not easy, and has been made in [97] where the solution uses a nonlinear Laplace equation.

Remark 6.19. Historically, the Reissner-Nordstrom metric was discovered independently by Reissner (1916) and Nordström (1918). Weyl also discovered this metric independently of Reissner and Nordström in 1917. This metric is a classical solution of the Einstein field equation for the case of an electrically q charged point of mass m without rotation in vacuum spacetime.

6.2.4 *Sketch for Resolution of Geodesics and Deviation*

The method used for establishing geodesic equations in Schwarzschild spacetime remains valid. We are not going into the details of the results that we can find in the references, but we just recall the main steps for this derivation. Without using the relations of (6.78) for calculating Christoffel's symbols, it is more direct to write directly the Levi-Civita connection:

$$\overline{\Gamma}^0_{0r} = \Gamma^0_{r0} = \nu' \qquad (6.80)$$

$$\overline{\Gamma}^r_{00} = \nu'\, e^{4\nu}, \quad \overline{\Gamma}^r_{rr} = \mu', \quad \overline{\Gamma}^r_{\theta\theta} = -r\, e^{2\nu}, \quad \overline{\Gamma}^r_{\varphi\varphi} = -r\, e^{2\nu}\sin^2\theta \qquad (6.81)$$

$$\overline{\Gamma}^\theta_{r\theta} = \Gamma^\theta_{\theta r} = \frac{1}{r}, \quad \overline{\Gamma}^\theta_{\varphi\varphi} = -\sin\theta\cos\theta \qquad (6.82)$$

$$\overline{\Gamma}^\varphi_{r\varphi} = \overline{\Gamma}^\varphi_{\varphi r} = \frac{1}{r}, \quad \overline{\Gamma}^\varphi_{\theta\varphi} = \overline{\Gamma}^\varphi_{\varphi\theta} = \cot\theta \qquad (6.83)$$

For the spherically symmetric spacetime of Schwarzschild, we can derive the geodesic equations with $e^{2\nu} = 1 - 2m/r + q^2/r^2$, which is similar to (5.138):

$$\boxed{\begin{cases} \dfrac{du^0}{d\tau} + 2\nu' \left(u^0 u^r\right) = 0 \\ \dfrac{du^r}{d\tau} + \nu' \, e^{4\nu}(u^0)^2 - \nu'(u^r)^2 - re^{2\nu}(u^\theta)^2 - r\sin^2\theta e^{2\nu}(u^\varphi)^2 = 0 \\ \dfrac{du^\theta}{d\tau} + \dfrac{2}{r}\left(u^r u^\theta\right) - \sin\theta\cos\theta \, (u^\varphi)^2 = 0 \\ \dfrac{du^\varphi}{d\tau} + \dfrac{2}{r}\left(u^r u^\varphi\right) + 2\cot\theta(u^\theta u^\varphi) = 0 \end{cases}} \tag{6.84}$$

From the connection coefficients, it is easy to obtain the curvature tensor $\overline{\mathfrak{R}}^{\gamma}_{\alpha\beta\lambda}$ and then derive the geodesic deviation equation. Results may be found in literature particularly for charged test particles along motion on circles e.g. [271], and for test particles in the presence of exotic matter e.g. [272]. The full set of analytical solutions for electrically and magnetically charged particles was derived in [42] in Reissner-Nordström spacetime.

6.2.4.1 *Cauchy and Event Horizons*

First of all, the causality principle, embodied by the relationships between events occurring in spacetime $\mathscr{B}$, constitutes the fundamental basis of relativity theories. The effective motion of particles can take place only within the spacetime light cone. It applies for both special and general relativity. For the former, global spacetime is considered whereas for relative gravitation, the motion of particles will always take place inside a local spacetime light cone. Hilbert causality is the key e.g. [6]. Analyses of spacetime structure is classically investigated by means of test particle motions and light wave propagation. For studying the geodesics of the Reissner-Nordström spacetime, let's consider the zeros of the function by calculating its discriminant Δ:

$$e^{2\nu} = 1 - \frac{2m}{r} + \frac{q^2}{r^2} = 0, \quad \Delta := 4\left(m^2 - q^2\right)$$

Solution for metric (6.78) is only valid for $\Delta \geq 0$, say $m^2 \geq q^2$. Such a case corresponds to a Reissner-Nordström black hole with one or two horizons:

$$r_C = m - \sqrt{m^2 - q^2}, \quad r_E = m + \sqrt{m^2 - q^2} \tag{6.85}$$

provided $m \geq q$, where r_C and r_E respectively are called the Cauchy (internal) horizon and event (or external) horizon. There are theoretically three possible configurations according to the relative values of the mass m and the charge q. For the limiting case when $m = q$, the two horizons merge and is referred to as an extremal case. For the case where $m < q$, we observe that the solution has a naked singularity. For that purpose, the line element of the Reissner-Nordström spacetime may then be rewritten as:

$$ds^2 = \left(1 - \frac{r_C}{r}\right)\left(1 - \frac{r_E}{r}\right)(dx^0)^2 - \frac{dr^2}{\left(1 - \frac{r_C}{r}\right)\left(1 - \frac{r_E}{r}\right)} - r^2\left(d\theta^2 + \sin^2\theta\, d\varphi^2\right) \tag{6.86}$$

Assuming that $m \geq q$ (and then $r_C < r_E$), we define three zones of the space according to the radius value r.

(1) For the two regions $r < r_C$ or $r > r_E$, the metric (6.86) holds, and the Hilbert causality condition is respected by the metric;
(2) For r between the horizons, say $r_C < r < r_E$, the radius coordinate becomes timelike with the coordinate x^0 spacelike;
(3) When approaching the horizons $r \to r_C$ or $r \to r_E$, the component $|g_{00}| \to 0$, and $|g_{rr}| \to \infty$.

Remark 6.20. Cauchy horizon r_C and external or event horizon r_E are important in relative gravitation theory, in the sense that the horizon is a borderline across which no information can pass [273]: (a) no light can cross the event horizon r_E, which indeed delimits a black hole; (b) Cauchy horizon r_C is a boundary of the future, and crossing this radius changes the type of equations of the spacetime field and the problem is no more of Cauchy type.

6.2.4.2 *Example of a Circular Motion in the Equatorial Plane*

For this case, we assume $\theta = \pi/2$ and $u^\theta := \dot{\theta} \equiv 0$ (derivative with respect to proper time τ). For defining the geodesic, the general method would be to consider the system of equations (6.84) deduced from the geodesic equation (5.76). Another method for describing the motion of relativistic test particle of mass m^* carrying an electric charge q^* uses the Lagrangian formalism. Here, we adopt this Lagrangian method. For geodesic determination, it is easier to consider the analogy of the Schwarzschild Lagrangian (5.147) and add the electromagnetic contribution:

$$\mathscr{L} = -m^* c \sqrt{g_{\mu\nu}\ u^\mu u^\nu} + q^*\ A_\nu u^\nu \tag{6.87}$$

where $A_\nu = (-\Phi, A_r, A_\theta, A_\varphi)$ is the vacuum electromagnetic potential. Assuming velocity to be much lower than light speed, the Lagrangian holds as:

$$\mathscr{L} \simeq -m^* c^2 + \frac{1}{2}\ g_{\mu\nu}\ u^\mu u^\nu + q^*\ A_\nu u^\nu \tag{6.88}$$

For the example, we then deduce, by dropping the constant term:

$$\mathscr{L} = \frac{m^*}{2}\left\{\left(1-\frac{r_C}{r}\right)\left(1-\frac{r_E}{r}\right)(\dot{x}^0)^2 - \left[\left(1-\frac{r_C}{r}\right)\left(1-\frac{r_E}{r}\right)\right]^{-1}\dot{r}^2 \right.$$
$$\left. - r^2\left(\dot{\theta}^2 + \sin^2\theta\ \dot{\varphi}^2\right)\right\} \tag{6.89}$$
$$-q^*\ \Phi(r)\ \dot{x}^0 \tag{6.90}$$

Similarly to Schwarzschild spacetime, no explicit dependence on x^0 and on φ occurs, then the derivative of the Lagrangian with respect to $\dot{x}^0$ and $\dot{\varphi}$ are motion constants:

$$\begin{cases} m^* \dfrac{dx^0}{d\tau}\left(1-\dfrac{r_C}{r}\right)\left(1-\dfrac{r_E}{r}\right) - q^*\Phi(r) = E_0 \\ m^* \dfrac{d\varphi}{d\tau}\ r^2 \sin^2\theta = L_0 \end{cases} \tag{6.91}$$

where the two conserved quantities are E_0 the energy and L_0 the angular momentum in relative gravitation. This allows us to obtain the velocity:

$$\begin{cases} u^0 = \dfrac{(E_0 + q^*\Phi(r))}{m^*}\left[\left(1-\dfrac{r_C}{r}\right)\left(1-\dfrac{r_E}{r}\right)\right]^{-1} \\ u^\varphi = \dfrac{L_0}{m^*}\dfrac{1}{r^2\sin^2\theta} \end{cases} \tag{6.92}$$

These equations conform to the previous Schwarzschild solutions e.g. [42], [271]. For calculating the radial component of velocity u^r, if any, we may consider the Euler-Lagrange equation for the radial coordinate:

$$\frac{\partial \mathscr{L}}{\partial r} - \frac{d}{d\tau}\left(\frac{\partial \mathscr{L}}{\partial \dot{r}}\right) = 0$$

However, for the particular case of a circular motion, we have $u^r = 0$. Therefore, the system of equations (6.84) can be used to give:

$$\frac{du^0}{d\tau} = 0, \quad \text{and} \quad \frac{du^\varphi}{d\tau} = 0 \tag{6.93}$$

which gives the solutions for a radius $r = a$:

$$\begin{cases} u_a^0 = \dfrac{(E_0 + q^*\Phi(a))}{m^*}\left[\left(1 - \frac{r_C}{a}\right)\left(1 - \frac{r_E}{a}\right)\right]^{-1} \\ u_a^\varphi = \dfrac{L_0}{m^*}\dfrac{1}{a^2} \end{cases} \tag{6.94}$$

where radius a is still unknown and to be calculated. For that purpose, we can again use the second equation of (6.94), which becomes necessary to write the equation giving the position of the test particle of charge q^* and mass m^*:

$$\nu'(a)\, e^{2\nu(a)} \left(u_a^0\right)^2 - a\left(u_a^\varphi\right)^2 = 0 \tag{6.95}$$

More explicitly, the equation to solve with unknown a takes the form of:

$$\frac{1}{2}\left[\left(1 - \frac{r_C}{a}\right)\left(1 - \frac{r_E}{a}\right)\right]' \left(\frac{E_0 + q^*\Phi(a)}{m^*}\right)^2 - \frac{L_0^2}{(m^*)^2\, a^3} = 0 \tag{6.96}$$

The prime means a derivative with respect to a as variable which is the unknown. To sum up, the Cauchy horizon r_C and exterior horizon r_E depend solely on the mass m and the electric charge q of the body at $r = 0$ which induces the gravitation and the electromagnetic field $\Phi(r)$. Energy E_0 and angular momentum L_0 are related to initial conditions of the test particle with respect to the body at the center. The test particle with mass m^* and electric charge q^* is assumed as usual to not influence the gravitation and electromagnetic field.

Remark 6.21. Considering Reissner-Nordström horizons (6.85), it is noticed that, given Hilbert causality, component $g_{00} > 0$. This is possible if and only if $r \geq r_e$.

Provided the function $e^{2\nu}$, we are able to calculate the curvature $\overline{\Re}$ and the associated tangent u^μ of geodesic curves; it is easy but cumbersome to introduce them in the deviation equation:

$$\frac{D^2\xi^\alpha}{D\tau^2} = \overline{\Re}^\alpha_{\delta\gamma\beta}\, u^\beta u^\delta\, \xi^\gamma$$

where $\mathbf{u} = (u^0, 0, 0, u^\varphi)$ from the particular case of circular orbit equation (6.94). In the general case of spherical symmetry, it is necessary to solve the system of differential equations (6.84).

Remark 6.22. Some other examples of electrovacuum spacetime may be found in the literature e.g. [241], and the interest in the role of magnetic fields in cosmology continues to motivate recent research e.g. [64], [94], [96].

6.3 Gravitation and Electromagnetism in Einstein-Cartan Spacetime

Analysis of electromagnetic fields in the presence of extremely massive gravitation remains a relevant topic in relativistic astrophysics. Propagation of electromagnetic waves governed by Maxwell's equations within a curved spacetime constitutes a fundamental basis for studying signals received from neutron stars and black holes to name but a few in astrophysics. Other methods consist in measuring signals due to gravitational waves.

6.3.1 *Context for Einstein-Cartan Spacetime*

In this section we consider the gravitation-electromagnetism interaction within a Riemann-Cartan spacetime endowed with metric $g_{\alpha\beta}(x^\mu)$ and connection $\Gamma^\gamma_{\alpha\beta}(x^\mu)$. *It should be noted that we first use in this section a Riemann volume-form $\overline{\omega}_n$ for integrating forms in the Einstein-Cartan spacetime.* We want to get back to the basics of Einstein-Cartan cosmology, knowing that torsion represents *in fine* in one way or another an electromagnetic radiation field (photon). Historically, introduction of torsion, in addition to curvature, was already made from the beginning in the development of relative gravitation theory. From this period, the consideration of a connection independent of the metric tensor was motivated to model the coupling of gravitation and electromagnetism; the Lagrangian proposed by Reichenbach, for example, included metric, connection and partial derivative of connection e.g. [58]. Since then, numerous models of the gravito-electromagnetic field of vacuum spacetime have been proposed, generalizing Einstein gravitation towards cases with spinning fluids e.g. [78]. Indeed, numerous investigations by including a non-zero torsion have led to the conclusion that this torsion field is necessarily zero in the absence of other physical sources, such as spin fluids or an ambient electromagnetic field e.g. [61], [137], [217]. The role of a magnetic monopole field is gaining increasing importance and the efforts to prove their existence remain relevant, for instance their creation might be associated to plasma e.g. [274]. In cosmology, what remains of prime importance is nevertheless research on dark matter and dark energy which play a primordial role in the acceleration of the Universe e.g. expansion [89], [54]. The link between torsion, itself related with spacetime defects, and magnetic monopoles is increasingly suggested in relative gravitation e.g. [90], [91] and also in cosmology to include in Friedman equations e.g. [89], [275]. The use of Einstein-Cartan spacetime therefore constitutes an approach which is gaining importance in cosmology e.g. [4], [51], [276]. In this work, we look at electromagnetic radiation, possibly from magnetic monopole fields, as external sources of the torsion field.

6.3.2 *Einstein-Cartan Spacetime and the Role of $\overline{\omega}_n$*

The spacetime that we are interested in is inspired from the work of Reichenbach [58] as early as 1929 which extended the Hilbert-Einstein action to define more general spacetimes including the connection coefficients $\Gamma^{\gamma}_{\alpha\beta}$, which are not necessarily Christoffel's symbols, and their partial derivatives $\partial_\lambda \Gamma^{\gamma}_{\alpha\beta}$ as arguments of the Lagrangian in addition to the metric.

6.3.2.1 *Einstein-Palatini Action for Gravitation*

As required by the presence of defect density within spacetime, used connection presents non-zero torsion and curvature that may depend on time $x^0 := ct$ in a four-dimensional formulation. Such a spacetime was originally proposed to unify gravitation and electromagnetism. There will be a need for external sources such as fluid with spin or electromagnetic field e.g. [78].

Definition 6.6. An Einstein-Cartan spacetime (generalized continuum) is a four-dimensional manifold (compacted and connected), endowed with a pseudo-Riemannian hyperbolic metric $\mathbf{g}$ (verifying Hilbert causality) of signature $(1, 3)$ and with a connection ∇ compatible with metric, say $\nabla \mathbf{g} \equiv 0$.

Torsion field characterizes density of defects (defined here by the dislocations between "microcosms" on the figure e.g. [49]). In brief, dislocation is a line defect, and the density is defined as the total length of dislocation per unit volume $[m/m^3] = [m^{-2}]$. It is the number of dislocation lines intersecting a unit area. For a loose comparison with metals, dislocation density is usually of order of $10^{10}[m^{-2}]$ in a metal, increasing up to $\simeq 10^{16}[m^{-2}]$ after work hardening and fatigue. In the present work, we expect a kind of smooth dust cloud (magnified in Figure 6.1), advocated and named "microcosms" in e.g. [2]. Einstein-Cartan spacetime may then be interpreted as a set of very small microcosms. Now, considering the same form of the action

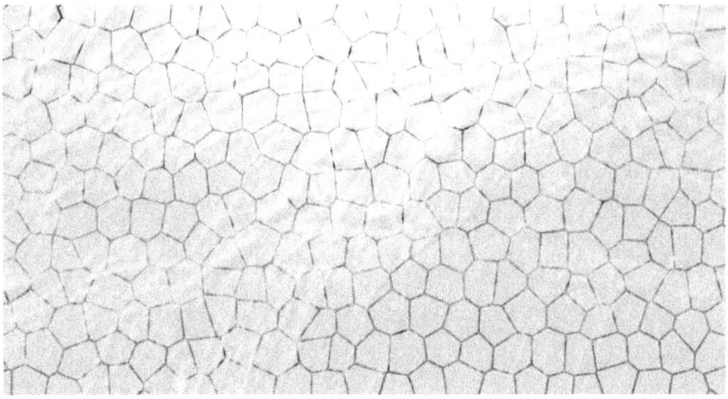

Fig. 6.1: Scheme of an affinely connected manifold with torsion. The Riemann-Cartan spacetime $\mathscr{RC}$ is assumed to be a set of infinitesimal "microcosms" linked to each other by metric-compatible connection $\Gamma^{\gamma}_{\ \alpha\beta} = D^{\gamma}_{\ \alpha\beta} + \mathfrak{T}^{\gamma}_{\ \alpha\beta}$. Metric compatibility means that $Q^{\gamma}_{\ \alpha\beta} := \nabla^{\gamma} g_{\alpha\beta} \equiv 0$ on $\mathscr{RC}$.

but using curvature in terms of metric and independent connection ∇ (that is called Einstein-Palatini action), we write:

$$\mathscr{S}_{EP} = \frac{1}{2\chi}\int_{\mathscr{B}} \mathcal{R}\,\overline{\omega}_n, \quad \mathcal{R} = \mathcal{R}[g_{\alpha\beta}, \Gamma^{\gamma}_{\ \alpha\beta}]$$

The stationarity of this action provides us the following variational equation:

$$\delta\mathscr{S}_{EP} = \frac{1}{2\chi}\int_{\mathscr{B}} \left(\mathcal{R}_{\beta\lambda} - \frac{\mathcal{R}}{2} g_{\beta\lambda}\right)\delta g^{\beta\lambda}\,\overline{\omega}_n \tag{6.97}$$

$$+\frac{1}{2\chi}\int_{\mathscr{B}} \left[\nabla_{\gamma}(g^{\beta\lambda}\delta\Gamma^{\gamma}_{\ \beta\lambda}) - \nabla_{\beta}(g^{\beta\lambda}\Gamma^{\gamma}_{\ \gamma\lambda}) + g^{\beta\lambda}\aleph^{\mu}_{\ \alpha\beta}\delta\Gamma^{\gamma}_{\ \mu\lambda}\right]\overline{\omega}_n \tag{6.98}$$

Riemann volume-form $\overline{\omega}_n$ compatible with $\overline{\nabla}$ is not compatible with ∇ as long as torsion is not zero. Let's assume that the variations of metric and connection are arbitrary, while remaining zero on the boundary, say $\delta g^{\beta\lambda} \equiv 0$ and $\delta\Gamma^{\gamma}_{\ \alpha\beta} \equiv 0$. We cannot deduce hastily the following two (false) equations in the vacuum:

$$\mathcal{R}_{\beta\lambda} - \frac{\mathcal{R}}{2} g_{\beta\lambda} = 0, \quad \aleph^{\gamma}_{\ \beta\lambda} = 0 \tag{6.99}$$

A direct use of the first equation of (6.99) and adding the energy-momentum of the external interactions is very tempting. But we will show that this method is not without ambiguity since ∇ is not compatible with $\overline{\omega}_n$ in (6.97).

Remark 6.23. The two equations (6.36) and (6.99) are similar although they can hide the subtleties and pitfalls in the derivation of field equations in relative gravitation. Indeed, if the first equation (6.36) is well established, the second, although exact, requires a finer interpretation. We will return to this later.

In accordance with Einstein-Cartan gravitation, it is necessary to consider the source terms: on the one hand for spacetime bending due to gravitation by the presence of matter (usually characterized by Einstein tensor $G_{\mu\nu}$); and on the other hand for spacetime distortion due to torsion field, usually attributed to the existence of matter spins e.g. [184]. Therefore, introduction of an independent connection requires the introduction of external physical interactions as sources with spin. Another question would be about the necessity of introducing a tensor other than only torsion to measure the spacetime distortion (the counterpart of Einstein tensor but for torsion).

Remark 6.24. As for the introduction of a perfect fluid matter to derive the Friedmann equations, within Einstein spacetime, cosmologists and astrophysicists have often considered fluids with spins as physical sources of torsion (otherwise the latter identically vanishes). In this book, we opt instead for electromagnetism, since it is the only other action-at-a-distance with an infinite radius of action as gravitation. This is inspired by the fact that the involved dimensions are textually astronomical.

6.3.2.2 *Divergence Operator and Field Equation*

During the variational calculus, the derivation of Einstein field equations uses the divergence operator as e.g. [128]:

$$\overline{\mathrm{Div}}\ \xi := \frac{1}{\sqrt{g}}\, \partial_\alpha \left(\sqrt{g}\ \xi^\alpha\right), \quad \alpha = 0, 1, 2, 3 \tag{6.100}$$

with the volume-form $\overline{\omega}_n := \sqrt{g}\ dx^0 \wedge \cdot \wedge dx^3$.[2] This is the original definition of the divergence operator in a general differentiable manifold $\mathscr{B}$; neither metric tensor $g_{\alpha\beta}$ nor affine connection $\Gamma^\gamma_{\alpha\beta}$ on the manifold are necessary. Provided a general volume-form ω_n on the manifold $\mathscr{B}$, it is first recalled

[2]Throughout this work, we set $g := |\mathrm{Det}\ \mathbf{g}|$ knowing that Det $\mathbf{g}$ can be positive or negative.

that the volume-form has only one component:

$$\omega_n := \Omega \; dx^0 \wedge \cdots \wedge dx^3 \tag{6.101}$$

Then, it is also worth recalling the vector $\xi = (\xi^\beta, \beta = 0, 1, 2, 3)$. The definition of a generalized divergence operator extending (6.100) is straightforward:

$$\text{Div } \xi := \frac{1}{\Omega} \, \partial_\alpha \left(\Omega \; \xi^\alpha\right), \quad \alpha = 0, 1, 2, 3 \tag{6.102}$$

At this first step, neither metric nor affine structure are necessary, only differentiability is required. Now, let's consider a Riemann-Cartan manifold and write the covariant derivative of an n-form (with only one component in a manifold of dimension n) and a contravariant vector:

$$\nabla_\alpha \Omega = \partial_\alpha \Omega - \Gamma^\beta_{\alpha\beta} \; \Omega, \quad \nabla_\alpha \xi^\gamma = \partial_\alpha \xi^\gamma + \Gamma^\gamma_{\alpha\beta} \; \xi^\beta$$

From the definition (6.102), the divergence of a vector field ξ on a Riemann-Cartan manifold holds:

$$\begin{aligned} \text{Div } \xi &= \frac{1}{\Omega} \left[\Omega \; \partial_\alpha \xi^\gamma + (\partial_\alpha \Omega) \; \xi^\gamma\right] \\ &= \frac{1}{\Omega} \left[\Omega \left(\nabla_\alpha \xi^\gamma - \Gamma^\gamma_{\alpha\beta} \; \xi^\beta\right) + \left(\nabla_\alpha \Omega + \Gamma^\beta_{\alpha\beta} \; \Omega\right) \xi^\gamma\right] \end{aligned} \tag{6.103}$$

We deduce the covariant expression of the divergence operator on a Riemann-Cartan manifold e.g. [131]:

$$\boxed{\text{Div } \xi = \frac{1}{\Omega} \; \nabla_\alpha \left(\Omega \xi^\alpha\right) + \aleph^\alpha_{\beta\alpha} \; \xi^\beta} \tag{6.104}$$

where we have the Riemann-Cartan volume-form $\omega_n = \Omega \; dx^0 \wedge \cdots \wedge dx^3$ on $\mathscr{B}$.

Remark 6.25. Riemann volume-form $\overline{\omega}_n$ is not covariantly constant with respect to connection ∇ with non-zero torsion. The action defined as integral using Riemann volume-form does not lead to the minimally coupled field equations in Riemann-Cartan manifolds. Practically, application of divergence theorem is not warranted. In the special case when the trace of the torsion tensor is derivable from a scalar potential, Mosna and Saa construct a covariantly constant volume element and use it to obtain minimally coupled field equations [134].

6.3.2.3 *Compatible Volume-Form* ω_n

Compatibility of ω_n with ∇ was studied for Einstein-Cartan spacetime e.g. [132] which made it possible to define an appropriate volume-form [134], and relate it to gauge invariance of electromagnetic fields in e.g. [132], [135] and also recently in e.g. [131]. Compatibility of ∇ with ω_n is expressed with a Lie derivative along an arbitrary vector field ξ:

$$\mathcal{L}_\xi \omega_n = (\nabla_\mu \xi^\nu)\, \omega_n$$

The expression of such a volume-form was already given [132], so we re-derive the proof with a more direct calculus. For this purpose, let's express the Lie derivative of the volume-form components:

$$\begin{aligned}\mathcal{L}_\xi \omega_{\alpha_1,\cdots\alpha_n} &= \xi^\mu \nabla_\mu \omega_{\alpha_1,\cdots\alpha_n} + (\nabla_{\alpha_1}\xi^\mu)\, \omega_{\mu,\alpha_2,\cdots\alpha_n} \\ &+ (\nabla_{\alpha_2}\xi^\mu)\, \omega_{\alpha_1,\mu,\alpha_3,\cdots\alpha_n} + \cdots + (\nabla_{\alpha_n}\xi^\mu)\, \omega_{\alpha_1,\cdots,\alpha_{n-1},\mu} \\ &+ \xi^\mu \aleph^\nu_{\mu\alpha_1}\ \omega_{\nu,\alpha_2,\cdots\alpha_n} + \cdots + \xi^\mu \aleph^\nu_{\mu\alpha_n}\ \omega_{\alpha_1,\cdots\alpha_{n-1},\nu}\end{aligned}$$

Since there is only one component $\Omega := \omega_{\alpha_1<\alpha_2\cdots<\alpha_n}$, the Lie derivative reduces to:

$$\mathcal{L}_\xi \Omega = \xi^\mu \nabla_\mu \Omega + (\nabla_\mu \xi^\mu)\, \Omega + \tau_\mu\ \xi^\mu\ \Omega, \quad \text{with} \quad \tau_\mu := \aleph^\nu_{\mu\nu} = \sum_{\nu=0}^{\nu=3} \aleph^\nu_{\mu\nu} \tag{6.105}$$

Imposing compatibility of ∇ with ω_n induces the relation:

$$\xi^\mu \nabla_\mu \Omega + \tau_\mu\ \xi^\mu = 0 \tag{6.106}$$

By definition Ω is always positive, so it is possible to make a change of variable $\Omega := e^\vartheta$ to obtain, ξ^μ being arbitrary:

$$\nabla_\mu \vartheta = -\tau_\mu \tag{6.107}$$

Coming back to the projection of the volume-form onto an arbitrary tensor base $dx^1 \wedge \cdots dx^n$, we may write:

$$\omega_n = e^\vartheta\ \omega_{12\cdots n}\ dx^1 \wedge \cdots \wedge dx^n = e^\vartheta\ \mathrm{Det}(\partial_1, \partial_2, \cdots, \partial_n)\ dx^1 \wedge \cdots \wedge dx^n \tag{6.108}$$

Particularly, we can suggest a compatible volume-form with the connection:

$$\omega_n = e^\vartheta\ \overline{\omega}_n, \quad \vartheta = \vartheta(\mathbf{g}, \aleph) \tag{6.109}$$

The introduction of the determinant is necessary to have an invariant component during a base transformation.

6.3.2.4 *Fundamental Forces*

Just as a brief reminder, there are four fundamental forces in the Universe, linked to the four fundamental interactions.

(1) *The gravitational force* is related to the gravitational interaction undergone by a mass m in a gravitational field involving a constant $G \simeq 6.674 \times 10^{-11}[m^3/(kg\, s^2)]$. Any material point A of mass m_A exerts on a point material B of mass m_B an attractive gravitational force or gravitation. The radius of action of the gravitational force is infinite, and it does not need material support to act—it can act in a vacuum.
(2) *The electromagnetic force* is the manifestation of the electromagnetic interaction which is exerted on an electric charge q in motion. The radius of action of the electromagnetic force is also infinite. Nor does it need material support to act—it can act in a vacuum.
(3) *The strong nuclear force* ensures the cohesion of the atomic nucleus (force between nuclei, i.e., between protons, neutrons and quarks). It keeps the protons together despite the electromagnetic repulsion. It is a short-range force because it does not act outside the core. It binds quarks to form neutrons and protons and binds protons between them. Simple considerations of orders of magnitude always impose neglecting the forces of gravitation and electromagnetism in the case of elementary particles. Its radius of action is of the order of $10^{-15}[m]$.
(4) *The weak nuclear force* is 10^6 (one million) times weaker than the strong nuclear force. Its range is the shortest since it acts at $10^{-18}[m]$, i.e., practically in contact with two particles. It is at the origin of thermonuclear reactions (in the stars) and radioactivity.

Gravitation and electromagnetic forces constitute the only *actions-at-a-distance* and unmediated forces among the fundamental ones (Fig. 6.2). For example, the gravitational force between the Earth and the Moon acts at a distance of nearly $3.84399 \times 10^8[m]$, and light, a special case of electromagnetic waves, travels nearly 150 million kilometers from the Sun to the Earth. Their range of action is infinite contrary to the nuclear forces which can be considered as "contact forces".

Remark 6.26. Among the two possible actions-at-a-distance, gravitation is weaker than electromagnetic interactions at the atomic scale. However, gravitation is thought to be the most important of the four fundamental forces for large-scale distances as is the case in cosmology. Indeed, even though electromagnetic forces are far stronger than gravitation for

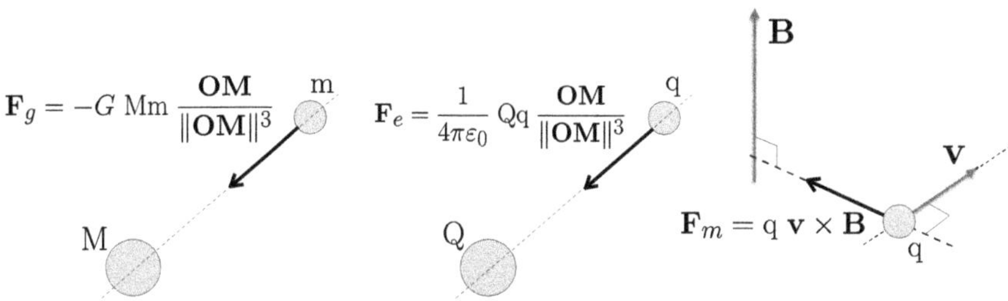

Fig. 6.2: The three forces acting at a distance: (a) Gravitational force $\mathbf{F}_g$ (the mass M is located at the origin O), (b) electrostatic force $\mathbf{F}_e$ (the charge Q is located at the origin O), and (c) magnetic force $\mathbf{F}_m$ ($\mathbf{B}$ is a prescribed magnetic induction and $\mathbf{v}$ the velocity of the charge q). The last two forces may be unified as electromagnetic force $\mathbf{F}_{em} = \mathbf{F}_e + \mathbf{F}_m$.

infinitesimally small distant objects, electrostatic attraction is not relevant for celestial bodies, such as planets, stars, and galaxies. In sum, only gravitation interaction is thought to matter on the large-scale structure of the Universe, in principle.

6.3.3 *Cosmological Constant* Λ *and* $\mathscr{A}d\mathscr{S}$ *Spacetime*

The so-called ΛCDM (Lambda-Cold Dark Matter) is nowadays more or less considered as the standard model for describing the evolution of the Universe. It includes three major ingredients: (1) an *ad hoc* cosmological constant Λ that describes a postulated dark energy responsible for the accelerated expansion of the Universe; (2) a postulated CDM which is thought to be necessary to prevent the matter in the Universe from expanding too rapidly; and finally (3) ordinary baryonic matter. Overall, the implicit hypothesis is nevertheless that the cosmological model assumes that relative gravitation is the correct theory of gravity at large galactic length scales.

One of the major problems in physics and cosmology theories concerns the presence and smallness of the cosmological constant Λ. Accounting for dark energy requires a cosmological constant $\Lambda \simeq\leq 10^{-52}[m^{-2}]$. It is proven that the static spherically symmetric solutions of Einstein's equation with cosmological constant $\Lambda < 0$ lead to anti-de Sitter spacetime in Einstein gravitation. To overcome this problem, we try a different path in this work by considering rather an Einstein-Cartan spacetime with uniform torsion (or not) to arrive at the $\mathscr{A}d\mathscr{S}$ spacetime.

Despite the increasingly pronounced importance of the quantum aspect in the analysis of the cosmological constant, it is not uninteresting to investigate the influence of torsion (distribution of spacetime defects), in the non-quantum sense, for determining this constant. This section is based on the Einstein-Palatini action:

$$\mathscr{S}_{EP} := \frac{1}{2\chi}\int_{\mathscr{B}} \mathcal{R}(\mathbf{g}, \nabla)\,\overline{\omega}_n \tag{6.110}$$

where the curvature is a function of the metric and independent connection. Since the discovery of the accelerated expansion of the Universe, several hypotheses have been proposed. We can cite at least four, namely: dark energy embodied by Λ; theory of quintessence; design and then use of a modified relative gravitation; and finally holographic theory. What follows comes from a method located between cosmological constant and modified relative gravitation, the simplest version of which consists of considering the extension of Einstein spacetime towards that of Einstein-Cartan based on action (6.110). This can be considered as a first step to a more complete theory in the next chapter.

6.3.3.1 *Torsion Field and Cosmological Constant*

The next particular example constitutes a concrete realization of a de Sitter spacetime and illustrates the close link between components of torsion $\aleph$ and cosmological constant Λ. First, the contortion tensor is defined as the difference:

$$\mathcal{T}^{\gamma}_{\alpha\beta} := \Gamma^{\gamma}_{\alpha\beta} - \overline{\Gamma}^{\gamma}_{\alpha\beta} \tag{6.111}$$

where $\overline{\Gamma}^{\gamma}_{\alpha\beta}$ are Christoffel's symbols associated with $g_{\alpha\beta}$. Contortion describes deviation from Riemann geometry. Let's calculate the curvature:

$$\begin{aligned}\mathfrak{R}^{\gamma}_{\alpha\beta\lambda} &= \partial_\alpha(\overline{\Gamma}^{\gamma}_{\beta\lambda} + \mathcal{T}^{\gamma}_{\beta\lambda}) - \partial_\beta(\overline{\Gamma}^{\gamma}_{\alpha\lambda} + \mathcal{T}^{\gamma}_{\alpha\lambda}) \\ &\quad - (\overline{\Gamma}^{\mu}_{\alpha\lambda} + \mathcal{T}^{\mu}_{\alpha\lambda})(\overline{\Gamma}^{\gamma}_{\beta\mu} + \mathcal{T}^{\gamma}_{\beta\mu}) + (\overline{\Gamma}^{\mu}_{\beta\lambda} + \mathcal{T}^{\mu}_{\beta\lambda})(\overline{\Gamma}^{\gamma}_{\alpha\mu} + \mathcal{T}^{\gamma}_{\alpha\mu})\end{aligned}$$

We arrive at the expression of curvature which includes the Riemann curvature and contortion contribution:

$$\mathfrak{R}^{\gamma}_{\alpha\beta\lambda} = \overline{\mathfrak{R}}^{\gamma}_{\alpha\beta\lambda} + \overline{\nabla}_\alpha \mathcal{T}^{\gamma}_{\beta\lambda} - \overline{\nabla}_\beta \mathcal{T}^{\gamma}_{\alpha\lambda} - (\mathcal{T}^{\gamma}_{\beta\mu}\mathcal{T}^{\mu}_{\alpha\lambda} - \mathcal{T}^{\gamma}_{\alpha\mu}\mathcal{T}^{\mu}_{\beta\lambda}) \tag{6.112}$$

For Ricci curvature we obtain:

$$\mathfrak{R}_{\beta\lambda} := \mathfrak{R}^{\alpha}_{\alpha\beta\lambda} = \overline{\mathfrak{R}}_{\beta\lambda} + \overline{\nabla}_\alpha \mathcal{T}^{\alpha}_{\beta\lambda} - \overline{\nabla}_\beta \mathcal{T}^{\alpha}_{\alpha\lambda} - (\mathcal{T}^{\alpha}_{\beta\mu}\mathcal{T}^{\mu}_{\alpha\lambda} - \mathcal{T}^{\alpha}_{\alpha\mu}\mathcal{T}^{\mu}_{\beta\lambda}) \tag{6.113}$$

We note the use of the Levi-Civita covariant derivative. In view of (6.113), we observe that the Ricci curvature may not be symmetric. The curvature (6.113) allows us to calculate the scalar curvature $\mathcal{R} := g^{\beta\lambda}\Re_{\beta\lambda}$. The symmetric $\Re^S_{\beta\lambda}$ and asymmetric $\Re^A_{\beta\lambda}$ parts of the Ricci curvature are obtained from (6.113):

$$\boxed{\begin{cases} \Re^S_{\beta\lambda} = \overline{\Re}_{\beta\lambda} \\ \Re^A_{\beta\lambda} = \overline{\nabla}_\alpha \mathfrak{T}^\alpha_{\beta\lambda} - \overline{\nabla}_\beta \mathfrak{T}^\alpha_{\alpha\lambda} - (\mathfrak{T}^\alpha_{\beta\mu}\mathfrak{T}^\mu_{\alpha\lambda} - \mathfrak{T}^\alpha_{\alpha\mu}\mathfrak{T}^\mu_{\beta\lambda}) \end{cases}} \tag{6.114}$$

The second line of (6.114) is not necessarily skew-symmetric. For Einstein-Cartan gravitation, the Einstein-Palatini action holds:

$$\begin{aligned} \mathscr{S}_{EP} &= \frac{1}{2\chi}\int_{\mathscr{B}} \Re\,\overline{\omega}_n = \frac{1}{2\chi}\int_{\mathscr{B}} g^{\beta\lambda}\left(\Re^S_{\beta\lambda} + \Re^A_{\beta\lambda}\right)\,\overline{\omega}_n \\ &= \frac{1}{2\chi}\int_{\mathscr{B}} g^{\beta\lambda}\,\overline{\Re}_{\beta\lambda}\,\overline{\omega}_n \\ &+ \frac{1}{2\chi}\int_{\mathscr{B}} g^{\beta\lambda}\left[\overline{\nabla}_\alpha \mathfrak{T}^\alpha_{\beta\lambda} - \overline{\nabla}_\beta \mathfrak{T}^\alpha_{\alpha\lambda} - (\mathfrak{T}^\alpha_{\beta\mu}\mathfrak{T}^\mu_{\alpha\lambda} - \mathfrak{T}^\alpha_{\alpha\mu}\mathfrak{T}^\mu_{\beta\lambda})\right]\overline{\omega}_n \end{aligned}$$

Compatibility of $\overline{\nabla}$ with $\mathbf{g}$ and shift of the divergence terms to boundary $\partial\mathscr{B}$ lead to a convenient version of the Einstein-Palatini action in vacuum spacetime:

$$\mathscr{S}_{EP} = \frac{1}{2\chi}\int_{\mathscr{B}} \overline{\Re}\,\overline{\omega}_n - \frac{1}{2\chi}\int_{\mathscr{B}} g^{\beta\lambda}(\mathfrak{T}^\alpha_{\beta\mu}\mathfrak{T}^\mu_{\alpha\lambda} - \mathfrak{T}^\alpha_{\alpha\mu}\mathfrak{T}^\mu_{\beta\lambda})\,\overline{\omega}_n \tag{6.115}$$

where the first integral denotes the Hilbert-Einstein action whereas the second represents the torsion contribution. Similar decomposition is found in [38].

Cosmological constant as averaged torsion field. Such a background might give new insights for explaining the smallness of the constant. From the decomposition (6.115), we can define an average value:

$$\mathscr{S}_{EP} = \frac{1}{2\chi}\int_{\mathscr{B}} \left(\overline{\Re} - 2\overline{\Lambda}\right)\,\overline{\omega}_n \tag{6.116}$$

where the averaged value of the cosmological constant is defined as:

$$\overline{\Lambda} := \frac{1}{2\int_{\mathscr{B}}\overline{\omega}_n}\int_{\mathscr{B}} g^{\beta\lambda}(\mathfrak{T}^\alpha_{\beta\mu}\mathfrak{T}^\mu_{\alpha\lambda} - \mathfrak{T}^\alpha_{\alpha\mu}\mathfrak{T}^\mu_{\beta\lambda})\,\overline{\omega}_n \tag{6.117}$$

Remark 6.27. Decomposition (6.115) shows that torsion may serve as an origin of cosmological constant Λ although the integrand in (6.115) is not a constant but may depend on x^μ. Using Friedman and Raychaudhuri equations, torsion was also shown to contribute to curvature and thus mimic at least in part the cosmological constant. e.g. [276].

The integrand is neither constant (time varying) nor uniform (space varying) in the whole Universe e.g. [38]. Sign of $\overline{\Lambda}$ is not *a priori* imposed since it depends on the torsion field, meaning that the Universe may include locally both de Sitter and anti-de Sitter bubbles generally in fluctuations e.g. [25].

Remark 6.28. In this work, we go a little further by linking torsion with electromagnetism, which allows us *in fine* to support the hypothesis that the spacetime background generating cosmological constant Λ would be also electromagnetic (at the length scale of relative gravitation). Definition (6.117) conforms to the idea that $\overline{\Lambda}$ is generated by spacetime defect distribution and that it is also closely related to magnetic monopoles e.g. [90], [91]. It may give new insights to explain geometrically the origin for dark matter and dark energy e.g. [54].

It is important to go back a little to the derivation of Einstein's field equations. Considering the Hilbert-Einstein action $\mathscr{S}_{HE}$, which we may extend to Einstein-Palatini action $\mathscr{S}_{EP}$ e.g. [277], we can write:

$$\mathscr{S}_{EP} := \frac{1}{2\chi} \int_{\mathscr{B}} \mathcal{R}\, \overline{\omega}_n, \quad \text{with} \quad \mathcal{R} := g^{\mu\nu}\, \mathfrak{R}_{\mu\nu} \tag{6.118}$$

for which the Ricci curvature is obtained by independent connection ∇. The idea is to start from an Einstein-Cartan spacetime whose geometry is defined by a metric $g_{\mu\nu}$, analogous to that of Schwarzschild, and an independent, but compatible, connection $\Gamma^\gamma_{\mu\nu}$ which can be determined by contortion $\mathcal{T}^\gamma_{\mu\nu}$:

$$\Gamma^\gamma_{\mu\nu} = \overline{\Gamma}^\gamma_{\mu\nu} + \mathcal{T}^\gamma_{\mu\nu} \tag{6.119}$$

Torsion may be considered as sourced by fermions (matter is made of fermions, and such is the case for fluids with spin), but can also be sourced by bosons and particularly photons for electromagnetic fields e.g. [88]. It is known that torsion may be relevant for the fermion-sourced case especially in cases of high-spin densities of fermionic matter. We limit to the macroscopic fields of gravitation and electromagnetism. Let's assume that contortion constitutes a particular case of uniform diffuse field acting all over the spacetime. Precisely we start from (6.116) and (6.117):

$$\mathscr{S}_{EP} = \frac{1}{2\chi} \int_{\mathscr{B}} \mathcal{R}\, \varpi_n = \frac{1}{2\chi} \int_{\mathscr{B}} \left(\overline{\Re} - 2\overline{\Lambda}\right) \varpi_n$$

where $\overline{\Lambda}$ becomes an *input* data of the problem. Assuming that torsion may be averaged, the field equations reduce to a well-known expression:

$$\overline{\Re}_{\alpha\beta} - \frac{\overline{\mathcal{R}}}{2}\, g_{\alpha\beta} + \overline{\Lambda}\, g_{\alpha\beta} = 0 \tag{6.120}$$

giving the same equation as for the de Sitter spacetime. The average value of the cosmological constant is defined by (6.117). We thus obtain the classical field equation of Schwarzschild-de Sitter gravitation with a cosmological constant $\overline{\Lambda}$.

In sum, for Einstein-Cartan spacetime, it is possible to define a cosmological constant by averaging the torsion. Equation (6.120) constitutes the basis of the ΛCDM model in which no mention of exotic matter with "new physics" is necessary. Returning to accelerated expansion of the Universe, we note that $\overline{\Lambda}$ can be associated with dark energy when it is negative, or with matter if positive. However, for the moment it is purely geometric but does not provide any physical explanation, even phenomenological. The next step would be the search for this physical explanation by considering the two possible actions-at-a-distance.

6.3.3.2 *Example of Spacetime with Spherical Symmetry*

Consider hereafter a Riemann-Cartan spacetime with a particular torsion $\aleph$ where the only non-vanishing components are $\aleph^0_{23} = 2\aleph_0(r) = -\aleph^0_{32}$ (other components are zero) as in e.g. [61], [92]. We do not consider the problem of stationary spinning fluid as in [61], but directly consider a vacuum and non-zero torsion within a Riemann-Cartan manifold.

Calculus of contortion tensor. We consider a Einstein-Cartan gravitation model with spherical symmetry. First, the metric is guessed to have the form of:

$$ds^2 := e^{2\nu}\, (dx^0)^2 - e^{2\mu}\, dr^2 - r^2\, d\theta^2 - r^2 \sin^2 \theta d\varphi^2 \tag{6.121}$$

Physically, the torsion field is interpreted as the consequence of spins of particle fluids (fermions) that are aligned in the radial direction and

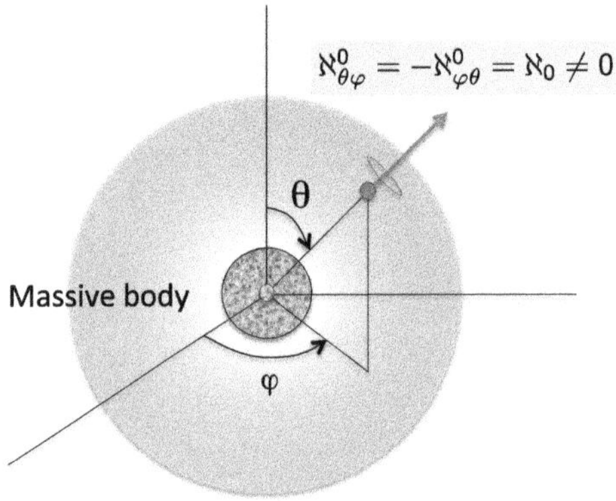

Fig. 6.3: A spherical massive body source of the curvature and torsion of the surrounding spacetime. Only $\aleph^0_{23} = \aleph_0(r) = -\aleph^0_{32}$ is non-zero whereas all other components vanish.

where particles are assumed to be static. The torsion decomposition holds (Fig. 6.3):

$$\aleph = \aleph^0_{23}\ \vartheta^2 \otimes \vartheta^3 \otimes \mathbf{e}_0 + \aleph^0_{32}\ \vartheta^3 \otimes \vartheta^2 \otimes \mathbf{e}_0 \tag{6.122}$$

Nevertheless, we have not related this torsion field with any fluid spin. We consider the torsion on its own being a geometric property of the spacetime, independent of the presence or not of any particles and matter (photons). The Einstein tensor $G_{\mu\nu}$ is calculated by means of contortion $\mathcal{T}^{\gamma}_{\alpha\beta}$ and curvature $\Re^{\gamma}_{\alpha\beta\lambda}$ calculated from (6.112). The general formula of contortion with symmetric and skew-symmetric parts holds e.g. [6]:

$$\mathcal{T}^{\gamma}_{\mu\nu} = \frac{1}{2}\aleph^{\gamma}_{\mu\nu} + \frac{g^{\gamma\sigma}}{2}\left(g_{\mu\epsilon}\aleph^{\epsilon}_{\sigma\nu} + g_{\nu\epsilon}\aleph^{\epsilon}_{\sigma\mu}\right) \tag{6.123}$$

where we distinguish the skew-symmetric contribution and the symmetric part. The non-zero components of the contortion are thus:

$$\mathcal{T}^0_{23} = \Omega^0_{23} = -\mathcal{T}^0_{23}, \quad \mathcal{T}^2_{03} = g^{22}g_{00}\ \Omega^0_{23} = \mathcal{T}^2_{30}, \quad \mathcal{T}^3_{02} = g^{33}g_{00}\ \Omega^0_{32} = \mathcal{T}^3_{20} \tag{6.124}$$

The first term is skew-symmetric whereas the last two are symmetric with respect to lower indices. Relations (6.124) remain valid for diagonal metrics.

In Prasanna's work, spin fluid is considered e.g. [61], which is not in our case where the component of torsion rather relates to magnetic monopoles, a part of the electromagnetic field. The goal in [61] was to understand all influences of Einstein-Cartan spacetime itself on the fluid spin surrounding a sphere. Spacetime is therefore not empty although it is suggested in [61] to assume the existence of radially outgoing magnetic flux generated by magnetic monopoles at the center of the sphere under consideration. Initially proposed by Dirac (1931), magnetic monopoles are now associated with the Grand Unification Theory of particle physics, which attempts to include fundamental forces at high energies into a single force e.g. [256]. According to this theory, monopoles are too massive to be detected in particle colliders. However, some extensions of the standard model postulate the existence of monopoles whose mass would be in a range accessible to colliders. To date, no evidence of magnetic monopoles has yet been found, despite advances in research in this area; see review in e.g. [26]. Recent studies tend towards the geometrization of magnetic monopoles which is represented by spacetime torsion, characterizing its defects e.g. [90] or an underlying electromagnetic field e.g. [96], [6], or as a mediator with cosmological constant Λ e.g. [38], and dark matter and energy e.g. [54], [89]. For his model of electric spin fluids, Prasanna already in 1975 concluded that there is a need to introduce a radial magnetic field, that is to say a magnetic monopole located at the sphere center [61]. With the most recent results, this hypothesis becomes increasingly plausible e.g. [26].

Different bases for projecting tensors and connection. Tensor components and connection coefficients depend on the base of projection. The coordinate base we usually use in the case of spherical symmetry is defined by:

$$(\partial_0, \partial_r, \partial_\theta, \partial_\varphi) \quad \text{and its dual } \left(dx^0, dr, d\theta, d\varphi\right) \tag{6.125}$$

From the line element, we define an orthonormal base $\left(\vartheta^0, \vartheta^1, \vartheta^2, \vartheta^3\right)$ as follows:

$$ds^2 = e^{2\nu}\ \left(dx_0\right)^2 - e^{-2\nu}\ dr^2 - r^2\ d\theta^2 - r^2 \sin^2\theta\ d\varphi^2$$

rearranging as:

$$ds^2 = \left(e^{\nu} dx_0\right)^2 - \left(e^{-\nu} dr\right)^2 - \left(r\ d\theta\right)^2 - \left(r \sin\theta\ d\varphi\right)^2$$

It shows that we can define the dual base as:

$$\vartheta^0 := e^{\nu} dx^0, \quad \vartheta^1 := e^{-\nu} dr, \quad , \vartheta^2 := r\ d\theta, \quad \vartheta^3 := r \sin\theta\ d\varphi \tag{6.126}$$

which is orthonormal but non-coordinate. Using duality $\vartheta^\alpha\left(\mathbf{e}_\beta\right) \equiv \delta^\alpha_\beta$, we obtain the orthonormal non-coordinate base:

$$\mathbf{e}_0 := e^{-\nu} \partial_0, \quad \mathbf{e}_1 := e^{\nu} \partial_r, \quad \mathbf{e}_2 := r^{-1}\ \partial_\theta, \quad \mathbf{e}_3 := \left(r \sin\theta\right)^{-1}\ \partial_\varphi \tag{6.127}$$

The assumed torsion is projected onto the coordinate base:

$$\aleph = \aleph_0 r^2 \, e^{-\nu} \, \sin\theta \, (d\theta \otimes d\varphi \otimes \partial_0 - d\varphi \otimes d\theta \otimes \partial_0) \tag{6.128}$$

From the Levi-Civita connection, we calculate the connection coefficients (without overline) since the corresponding contortion components are zero:

$$\Gamma^0_{01} = \Gamma^0_{10} = \nu'$$

$$\Gamma^1_{00} = \nu' \, e^{2\nu - 2\mu}, \quad \Gamma^1_{11} = \mu', \quad \Gamma^1_{22} = -r \, e^{-2\mu}, \quad \Gamma^1_{33} = -r \, e^{-2\mu} \sin^2\theta$$

$$\Gamma^2_{12} = \Gamma^2_{21} = \frac{1}{r}, \quad \overline{\Gamma}^2_{33} = -\sin\theta\cos\theta$$

$$\Gamma^3_{13} = \Gamma^3_{31} = \frac{1}{r}, \quad \Gamma^3_{23} = \Gamma^3_{32} = \cot\theta$$

projected onto the base $(\partial_0, \partial_r, \partial_\theta, \partial_\varphi)$ and its dual $(dx^0, dr, d\theta, d\varphi)$.

Remark 6.29. When the spin fluid is confined in a spherical ball, Einstein gravitation remains valid outside the sphere. Developments on static fluids have been made: by introducing torsion as an external environment to the gravitational field, torsion is considered as a source. For example, Prasanna assumed that spins of fluid particles are all aligned along the radial direction of the sphere. In our case, contortion $\mathfrak{T}^0_{32}$ is assumed to be an integral part of spacetime [61].

To Christoffel's symbols, we add the contortion $\mathfrak{T}^\gamma_{\alpha\beta}$ and obtain the following additional connection coefficients engendered by (6.124):

$$\Gamma^0_{23} = \frac{\aleph_0}{2} = \Gamma^0_{32}, \quad \Gamma^2_{03} = -\frac{\aleph_0}{2} \, e^\nu \sin\theta = \Gamma^2_{30}, \quad \Gamma^3_{02} = \frac{e^\nu}{\sin\theta} \frac{\aleph_0}{2} = \Gamma^3_{20} \tag{6.129}$$

projected on the base $(\partial_0, \partial_r, \partial_\theta, \partial_\varphi)$ and its dual $(dx^0, dr, d\theta, d\varphi)$. Fortunately, connection coefficients due to torsion are decoupled from those of the metric. This permits us to go further for this toy model.

Einstein field equation. Non-zero components of the Ricci curvature deduced from $\overline{\mathfrak{R}}_{\alpha\beta} := \overline{\mathfrak{R}}^\gamma_{\gamma\alpha\beta}$ hold:

$$\begin{cases} \overline{\mathfrak{R}}_{00} = e^{2\nu-2\mu} \left[\nu'' + (\nu')^2 - \nu'\mu' + (2\nu'/r)\right] \\ \overline{\mathfrak{R}}_{11} = -\left[\nu'' + (\nu')^2 - \nu'\mu' - (2\mu'/r)\right] \\ \overline{\mathfrak{R}}_{22} = e^{-2\mu} \left(-1 + r\,\mu' - r\,\nu'\right) + 1 \\ \overline{\mathfrak{R}}_{33} = e^{-2\mu} \left[(-1 + r\,\mu' - r\,\nu') + 1\right] \sin^2\theta \end{cases}$$

General expression of curvature is reminded:

$$\mathfrak{R}_{\beta\lambda} := \mathfrak{R}^{\alpha}_{\alpha\beta\lambda} = \overline{\mathfrak{R}}_{\beta\lambda} + \overline{\nabla}_{\alpha}\mathfrak{T}^{\alpha}_{\beta\lambda} - \overline{\nabla}_{\beta}\mathfrak{T}^{\alpha}_{\alpha\lambda} + \mathfrak{T}^{\alpha}_{\alpha\mu}\mathfrak{T}^{\mu}_{\beta\lambda} - \mathfrak{T}^{\alpha}_{\beta\mu}\mathfrak{T}^{\mu}_{\alpha\lambda}$$

A straightforward calculus gives the non-zero components:

$$\mathfrak{R}_{00} = \overline{\mathfrak{R}}_{00} - 2g^{22}g^{33}g^{2}_{00}\ \Omega^{0}_{23}\Omega^{0}_{32}, \tag{6.130}$$

$$\mathfrak{R}_{11} = \overline{\mathfrak{R}}_{11}, \quad \mathfrak{R}_{22} = \overline{\mathfrak{R}}_{22}, \quad \mathfrak{R}_{33} = \overline{\mathfrak{R}}_{33} \tag{6.131}$$

Projection onto basis $\{dx^0, dr, d\theta, d\varphi\}$ gives [61]: $\mathfrak{R}_{00} = \overline{\mathfrak{R}}_{00} + e^{2\nu}\ (\aleph_0^2/2)$ since $\vartheta^0 := e^{\nu}dx^0$, $\vartheta^2 := rd\theta$ and $\vartheta^3 := r\sin\theta d\varphi$. Only the time component is influenced by torsion. From definition $\mathcal{R} := g^{\alpha\beta}\mathfrak{R}_{\alpha\beta}$, we directly obtain:

$$\mathcal{R} = \overline{\mathcal{R}} + \frac{\aleph_0^2}{2} \tag{6.132}$$

This simple example highlights the role of torsion on spacetime bending.

We have seen previously for Schwarzschild spacetime that both the equation $\mathfrak{R}_{\alpha\beta} = 0$ and $G_{\alpha\beta} = 0$ lead exactly to the same result. To conform with models in literature we also write the field equation as $G_{\alpha\beta} := \mathfrak{R}_{\alpha\beta} - (\mathcal{R}/2)g_{\alpha\beta} = 0$ which is in fact the consequence of implicitly choosing the Riemann volume-form $\overline{\omega}_n$ in the Hilbert-Einstein action instead of ω_n. This may play an important role in the consistency of the model. We nevertheless conform to literature e.g. [4], [38], [278]. The Einstein tensor is obtained accordingly by considering this additional term in the field equations. Projection of the Einstein tensor $G_{\alpha\beta}$ onto co-basis gives from equation $G_{\alpha\beta} + \Lambda\ g_{\alpha\beta} = 0$:

$$\begin{cases} e^{2\nu}\left[e^{-2\mu}\left(\dfrac{2\mu'}{r} - \dfrac{1}{r^2}\right) + \dfrac{1}{r^2} + \dfrac{\aleph_0^2}{4}\right] = 0 \\ e^{2\mu}\left[e^{-2\mu}\left(\dfrac{2\nu'}{r} + \dfrac{1}{r^2}\right) - \dfrac{1}{r^2} - \dfrac{\aleph_0^2}{4}\right] = 0 \\ r^2\ e^{-2\mu}\left[\nu'' + (\nu')^2 - \nu'\mu' + \dfrac{\nu' - \mu'}{r} - \dfrac{\aleph_0^2}{4}\right] = 0 \\ r^2\ e^{-2\mu}\left[\nu'' + (\nu')^2 - \nu'\mu' + \dfrac{\nu' - \mu'}{r} - \dfrac{\aleph_0^2}{4}\right]\sin^2\theta = 0 \end{cases} \tag{6.133}$$

Adding the first two equations of (6.133) gives: $\nu' + \mu' = 0$ to give $e^{\mu} = e^{-\nu}$ by choosing again asymptotic behavior for $r \to \infty$. Taking the first equation and given that $e^{2\nu} \neq 0$, we have for a vacuum:

$$e^{-2\mu}\left(\frac{2\mu'}{r} - \frac{1}{r^2}\right) + \frac{1}{r^2} + \frac{\aleph_0^2}{4} = 0$$

by making a change of variable $U(r) := e^{-2\mu}$ the equation reduces to:

$$[rU(r)]' = 1 + \frac{\aleph_0^2}{4}(r)\, r^2$$

In a general case, the field $\aleph_0^2(r)$ is unknown and its resolution is coupled with other unknowns in the system (6.133). If we assume a uniform torsion field $\aleph_0(r) = \aleph_0$, for the sake of simplicity, we directly obtain the solution:

$$e^{2\nu} := U(r) = 1 - \frac{2m}{r} + \frac{1}{3}\left(\frac{\aleph_0^2}{4}\right) r^2 \tag{6.134}$$

by choosing the constant of integration $2m$. Constant $2m := 2GM/c^2$ reduces to the Schwarzschild radius of a massive body M. We recognize here a Schwarzschild-anti-de Sitter ($\mathscr{A}d\mathscr{S}$) spacetime metric where $\aleph_0^2(r)/4 := \Lambda > 0$ (see (5.154)) is a *cosmological constant*, which is a particular case of a Kottler spacetime metric with line element:

$$\boxed{\begin{cases} g_{00} = 1 - \dfrac{2m}{r} + \dfrac{\Lambda_\aleph}{3} r^2 \\ g_{rr} = -\left(1 - \dfrac{2m}{r} + \dfrac{\Lambda_\aleph}{3} r^2\right)^{-1} \end{cases} \quad \begin{cases} g_{\theta\theta} = -r^2 \\ g_{\varphi\varphi} = -r^2 \sin^2\theta \end{cases}} \tag{6.135}$$

where we have defined the cosmological constant:

$$\boxed{\Lambda_\aleph := \frac{\aleph_0^2}{4} > 0} \tag{6.136}$$

defining an anti-de Sitter spacetime. We have accordingly the line:

$$\begin{aligned} ds^2 = &\left(1 - \frac{2m}{r} + \frac{\Lambda_\aleph}{3} r^2\right)(dx^0)^2 - \left(1 - \frac{2m}{r} + \frac{\Lambda_\aleph}{3} r^2\right)^{-1} dr^2 \\ &- r^2 d\theta^2 - r^2 \sin^2\theta d\varphi^2 \end{aligned} \tag{6.137}$$

where care should be taken again about the sign convention.

Remark 6.30. Expression of $\Lambda_\aleph \geq 0$ confirms that torsion contributes to the cosmological constant. For this example, the positive sign illustrates that it induces a contraction of Einstein-de Sitter spacetime e.g. [275], [276]. In the general case, and this needs to be proven, nothing prevents torsion from being one of the variables which participate in dark matter or dark energy e.g. [25].

Remark 6.31. This example illustrates the role of torsion. Starting directly from an electromagnetic field for which only $\mathfrak{F}_{32} = -\mathfrak{F}_{23} \neq 0$ would have been equivalent to considering a magnetic monopole field e.g. [97]. Similarly, using a spin fluid present in vacuum spacetime causing a torsion field would also have been an option e.g. [61]. In the present study, we have shown putting the geometric structure of spacetime instead, so working within the Einstein-Cartan spacetime framework is a valuable path.

As a finding, the $\mathscr{A}\mathrm{d}\mathscr{S}$ spacetime constitutes a solution of the Einstein-Cartan field equations with an analogy of an attractive cosmological constant. This seems unphysical but could be considered as a regularization of the gravitational field at large distances which is not flat for $r \to \infty$. The horizon of the anti-de Sitter spacetime is merely related to torsion field $\ell^2 := 12/\aleph_0^2$, suggesting that the torsion amplitude is very large. By analogy, in continuum mechanics of defects, for physical interpretation of torsion as the density of defects e.g. [49] (as dislocations), it suggests that the spacetime has a very high defect density as it can be considered as a set of microcosms (see Figure 6.1 for a schematic picture of the spacetime).

Remark 6.32. We analyse Einstein-Cartan gravity in its standard form $\mathcal{R} = \overline{\mathcal{R}} + \mathcal{C}(\mathcal{T})$, where Einstein-Cartan curvature is equal to Einstein curvature plus a function of contortion. In such a way our study conforms to the findings of e.g. [38] which claims that torsion can serve as an origin for vacuum energy density, given by the cosmological constant or dark energy density in the Universe. Maybe such a finding supports the small value of the cosmological constant Λ, which remains a long-standing problem in cosmology e.g. [95].

6.3.3.3 *Cosmological Constant, Torsion, and $\mathscr{A}d\mathscr{S}$ Spacetime*

A quasi-exhaustive review is available in a recent paper [281] about the relevance of the non-zero cosmological constant, namely about the very unnatural small value of Λ and its introduction as a quasi-empirical constant to capture the dark energy in the framework of ΛCDM theory. The ΛCDM model relies upon the assumption of the existence of dark matter, dark energy and the inflation field, which induces some controversies.

On the one hand, this model turns out to be predictive and relatively robust with respect to old and recent cosmological experiments to a large extent (although very large length scales greater than the Solar System need further experimental observations, and very small length scales at the quantum level are out of reach). On the other hand, incorporation of an extremely low and unnatural value Λ, particularly in the Planck units, instead of a zero value, is not satisfactory from a theoretical point of view. Despite its success in explaining most cosmological observations, defining the cosmological constant as an *ad hoc* parameter without any clear connection to small-scale physical phenomena is also unsatisfactory from a quantum physicist's perspective. Estimates of Λ derived from quantum field theory are approximately 10^{120} times larger than those obtained from phenomenological models, which are themselves fitted to experimental measurements, e.g. [36].

Several approaches have been proposed to construct theories of gravitation that overcome the problem of Λ in the ΛCDM model. These include increasing the dimensionality of spacetime (beyond four dimensions), introducing non-local gravity, and developing modified theories of gravity. The accounting for local topology of spacetime defects, or more generally the accounting of the theory of microcosms e.g. [2], based on a non-vanishing torsion field $\aleph$ in the spacetime allows us to derive the Schwarzschild-anti-de Sitter spacetime by considering contortion as a primal variable in addition to the metric.

Some authors already suggested a relation between cosmological constant Λ and defects as torsion by considering Einstein-Cartan spacetime. They take into account a torsion field $\aleph \neq 0$ which leads to a negative pressure contribution and therefore to an accelerated expansion of the Universe. They stated that the presence of defects and torsion has the same effect as a cosmological constant e.g. [76], [279]. In some sense, a spin density acts as a source of torsion that can replace the dark matter concept. In Einstein-Cartan gravitation, a more recent work has nevertheless contradicted the previous results in [279], by confronting them with the supernovae data from Hubble [280].

Torsion and cosmological constant. For a non-zero torsion field, it is shown that the torsion field may be considered as a source and engenders a non-uniform cosmological constant $\Lambda(x^\mu)$.

$\Lambda > 0$ leads to a Universe re-collapsing rather than expansion, nevertheless the $\mathscr{A}d\mathscr{S}$/CFT conjecture is hopefully a guiding result to link gravitation with quantum physics [246]. The equality $\aleph_0^2/4 := \Lambda$ seems very interesting given that a torsion field $\aleph_{\alpha\beta}^{\gamma}$ on a manifold captures the local density of defects on a manifold e.g. [49]. The present work systematically attempts to show that torsion is related to the cosmological constant by generalizing the anti-de Sitter spacetime of (6.135). Extremely low values of Λ means a high density of defects.

Remark 6.33. Attractiveness of anti-de Sitter spacetime with metric (6.135) can also be interpreted as a vacuum spacetime with negative energy, which causes this spacetime to collapse in on itself. The interest of $\mathscr{A}d\mathscr{S}$ spacetime is mostly in the domain of quantum physics, thermodynamics of black holes, the correspondence between $\mathscr{A}d\mathscr{S}$ and Conformal Field Theory (CFT) , and high-energy processes. Nevertheless, we think that the use of Einstein-Cartan spacetime torsion may give new insights for future relation of $\mathscr{A}d\mathscr{S}$, the torsion field and quantum mechanics. Recently, starting from a compilation of experimental measurements in cosmology, Anjan and coworkers [95] compare numerous models and found that models with a negative Λ are either preferred over or perform equally as the ΛCDM model. This seems to support the presence of a negative Λ (anti-de Sitter spacetime regions) in our Universe. Such a fact can naturally arise in string theory, and is consistent with cosmological observations. This might be seen as a strong suggestive work for providing new insights in $\mathscr{A}d\mathscr{S}$/CFT conjectured by [246].

Remark 6.34. The recent trend to "geometrize" dark matter and dark energy by including torsion in relative gravitation models can be found in e.g. [90], [38], [54], [89], to name but a few. In particular, Milton proposed a theory which makes it possible to calculate the torsion field, using a hypothesis that he himself recognizes he has not justified, but makes it possible to close constitutive laws in the jargon of mechanics [54]. This hypothesis assumes that geodesics are equated with autoparallel curves. In view of equations for geodesic deviation (5.111) and autoparallel deviation (6.215), we believe that this interesting hypothesis can be avoided by considering other areas of physics such as electromagnetism.

Remark 6.35. Many studies have suggested the growing importance of torsion in cosmology, particularly through its combination with the FLRW framework. Its role is not limited to the resolution of singularity problems e.g. [4] but also extends to providing a more geometric interpretation of Λ and, above all, to the possibility of elucidating its contribution to the accelerated expansion of the Universe as part of dark matter and/or dark energy e.g. [54], [89], [276]. This is shown by relation (6.117). The question now is to know if, beyond the approach by the quantum theory of physical particles, it is possible to put a more classical physical meaning to this torsion field, linked in particular to one or more of the fundamental forces of the Universe linked to action-at-a-distance.

6.3.4 *Gravitation in an Einstein-Cartan Spacetime*

For this model, let us go back to the Hilbert-Einstein action by paying more attention to compatibility or not of ∇ and $\overline{\omega}_n$ that we have overlooked throughout this work.

Definition 6.7. The Palatini-Einstein action is the immediate extension of the Hilbert-Einstein action and writes:

$$\mathscr{S}_{PE} := \frac{1}{2\chi} \int_{\mathscr{B}} \mathcal{R}(\mathbf{g}, \nabla)\ \overline{\omega}_n \tag{6.138}$$

where Riemann volume-form $\overline{\omega}_n$ measures the spacetime volume element.

Einstein-Cartan cosmology has gained interest during the last decade, particularly to avoid an initial singularity e.g. [4]. But once again, we use Riemann volume-form for the moment. The reason is that most authors use this volume-form. The idea is to have the ability to compare various models. This will be a matter of concern later.

6.3.4.1 *Field Equations of Gravitation in Einstein-Cartan Spacetime*

The variation of Einstein-Palatini action concerns both metric and connection:

$$\delta\mathscr{S}_{PE} = \frac{1}{2\chi} \int_{\mathscr{B}} \left(\delta g^{\beta\lambda} \mathcal{R}_{\beta\lambda} + g^{\beta\lambda} \delta \mathcal{R}_{\beta\lambda} - \frac{\mathcal{R}}{2} g_{\beta\lambda} \delta g^{\beta\lambda} \right) \overline{\omega}_n$$

We will omit hereafter the factor $1/2\chi$ for convenience. The variation of curvature tensor has been developed in the chapter on geometry:

$$\delta \Re^{\gamma}_{\alpha\beta\lambda} = \nabla_\alpha \delta\Gamma^{\gamma}_{\beta\lambda} - \nabla_\beta \delta\Gamma^{\gamma}_{\alpha\lambda} + \aleph^{\mu}_{\alpha\beta} \delta\Gamma^{\gamma}_{\mu\lambda}$$

We have the term expressing the Ricci curvature variation:

$$g^{\beta\lambda} \delta \mathcal{R}_{\beta\lambda} = \delta^{\alpha}_{\gamma} g^{\beta\lambda} (\nabla_\alpha \delta\Gamma^{\gamma}_{\beta\lambda} - \nabla_\beta \delta\Gamma^{\gamma}_{\alpha\lambda} + \aleph^{\mu}_{\alpha\beta} \delta\Gamma^{\gamma}_{\mu\lambda})$$

After some tedious but straightforward calculus, we obtain the relation from factorizing the connection variation $\delta\Gamma^{\mu}_{\beta\lambda}$:

$$g^{\beta\lambda}\delta\mathcal{R}_{\beta\lambda} = \delta^{\alpha}_{\gamma}g^{\beta\lambda}\left(\overline{\nabla}_{\alpha}\delta\Gamma^{\gamma}_{\beta\lambda} - \overline{\nabla}_{\beta}\delta\Gamma^{\gamma}_{\alpha\lambda}\right) \tag{6.139}$$

$$+ \left(g^{\beta\lambda}\mathfrak{T}^{\gamma}_{\gamma\mu} - g^{\alpha\lambda}\mathfrak{T}^{\beta}_{\alpha\mu}\right)\delta\Gamma^{\mu}_{\beta\lambda} - \left(g^{\gamma\lambda}\mathfrak{T}^{\beta}_{\mu\gamma} - g^{\gamma\lambda}\mathfrak{T}^{\beta}_{\gamma\mu}\right)\delta\Gamma^{\mu}_{\beta\lambda} \tag{6.140}$$

$$- \left(g^{\beta\gamma}\mathfrak{T}^{\lambda}_{\mu\gamma} - \delta^{\beta}_{\mu}g^{\alpha\gamma}\mathfrak{T}^{\lambda}_{\alpha\gamma}\right)\delta\Gamma^{\mu}_{\beta\lambda} + g^{\gamma\lambda}\aleph^{\beta}_{\mu\gamma}\,\delta\Gamma^{\mu}_{\beta\lambda} \tag{6.141}$$

As in the case of Einstein cosmology (without torsion), the first row is a divergence thanks to compatibility between $\overline{\nabla}$ and $\overline{\omega}_n$. Now, $\overline{\nabla}$ is compatible with $\overline{\omega}_n$, so we can use the divergence theorem, which allows us to write:

$$\begin{aligned}\delta\mathscr{S}_{EP} = &\frac{1}{2\chi}\int_{\mathscr{M}}\left(\mathcal{R}_{\mu\nu} - \frac{\mathcal{R}}{2}g_{\mu\nu}\right)\delta g^{\mu\nu}\,\overline{\omega}_n\\ &+\frac{1}{2\chi}\int_{\mathscr{M}}\left[\left(g^{\beta\lambda}\mathfrak{T}^{\gamma}_{\gamma\mu} - g^{\alpha\lambda}\mathfrak{T}^{\beta}_{\alpha\mu}\right) - \left(g^{\gamma\lambda}\mathfrak{T}^{\beta}_{\mu\gamma} - g^{\gamma\lambda}\mathfrak{T}^{\beta}_{\gamma\mu}\right)\right.\\ &-\left.\left(g^{\beta\gamma}\mathfrak{T}^{\lambda}_{\mu\gamma} - \delta^{\beta}_{\mu}g^{\alpha\gamma}\mathfrak{T}^{\lambda}_{\alpha\gamma}\right) + g^{\gamma\lambda}\aleph^{\beta}_{\mu\gamma}\right]\delta\Gamma^{\mu}_{\beta\lambda}\,\overline{\omega}_n\end{aligned}$$

where we notice independent variations $\delta g^{\mu\nu}$ and $\delta\Gamma^{\mu}_{\beta\lambda}$. Divergence terms are dropped by assuming $\delta g^{\mu\nu} = 0$ and $\delta\Gamma^{\mu}_{\beta\lambda} = 0$ at $\partial\mathscr{M}$. The factor of connection variation $\delta\Gamma^{\mu}_{\beta\lambda}$ may be calculated and simplified as follows:

$$\boxed{\mathscr{D}^{\beta\lambda}_{\mu} = \left(g^{\beta\lambda}\,\mathfrak{T}^{\alpha}_{\alpha\mu} - g^{\alpha\lambda}\,\mathfrak{T}^{\beta}_{\alpha\mu}\right) - \left(g^{\beta\gamma}\,\mathfrak{T}^{\lambda}_{\mu\gamma} - \delta^{\beta}_{\mu}g^{\alpha\gamma}\,\mathfrak{T}^{\lambda}_{\alpha\gamma}\right)} \tag{6.142}$$

Tensor $\mathscr{D}^{\beta\lambda}_{\mu}$ is the counterpart of Einstein tensor $G_{\mu\nu}$ in the presence of torsion which geometrically measures the density of defects, or equivalently the subdivision of an otherwise continuous spacetime into a set of microcosms.

Field equations of Einstein-Cartan spacetime. In Einstein-Cartan vacuum spacetime, the stationarity of the Einstein-Palatini action makes it possible to obtain the system of two field equations:

$$\begin{cases} G_{\mu\nu} := \mathcal{R}_{\mu\nu} - \dfrac{\mathcal{R}}{2}\,g_{\mu\nu} = 0\\[2ex] \mathscr{D}^{\beta\lambda}_{\mu} := \left(g^{\beta\lambda}\,\mathfrak{T}^{\alpha}_{\alpha\mu} - g^{\alpha\lambda}\,\mathfrak{T}^{\beta}_{\alpha\mu}\right) - \left(g^{\beta\gamma}\,\mathfrak{T}^{\lambda}_{\mu\gamma} - \delta^{\beta}_{\mu}g^{\alpha\gamma}\,\mathfrak{T}^{\lambda}_{\alpha\gamma}\right) = 0 \end{cases} \tag{6.143}$$

where we recognize the classical Einstein field equation in the first row.

For the sake of clarity, we resume the variation of Einstein-Palatini action as:

$$\delta\mathscr{S}_{EP} = \frac{1}{2\chi}\int_{\mathscr{M}}\left(G_{\mu\nu}\,\delta g^{\mu\nu} + \mathscr{D}_{\mu}^{\beta\lambda}\,\delta\Gamma^{\mu}_{\beta\lambda}\right)\overline{\omega}_n = 0 \tag{6.144}$$

where the Einstein tensor and (Cartan) distortion tensor are defined by (6.143), which are not equal to zero in the presence of a physical source as we will develop hereafter. The two equations (6.143) give *a priori* vacuum field equations in the presence of non-zero curvature (cf. Einstein spacetime) and non-zero torsion (cf. Einstein-Cartan spacetime). Torsion and curvature measure the defect density in a continuum e.g. [6], [11]. This aspect would provide new insights into the understanding of spacetime as a set of grains called "microcosms" [2]. This opens the way to "discretized" spacetime at infinitely small dimensional scales, for instance at Planck length scale ℓ_P.

Remark 6.36. An elegant way to obtain (6.143) is to consider the expression (6.114) and introducing it into the Einstein-Palatini action:

$$\begin{aligned}\mathscr{S}_{EP} = &\frac{1}{2\chi}\int_{\mathscr{B}} g^{\beta\lambda}\left[\overline{\mathfrak{R}}_{\beta\lambda} + \left(\overline{\nabla}_\alpha \mathfrak{T}^{\alpha}_{\beta\lambda} - \overline{\nabla}_\beta \mathfrak{T}^{\alpha}_{\alpha\lambda}\right)\right]\overline{\omega}_n\\ &-\frac{1}{2\chi}\int_{\mathscr{B}} g^{\beta\lambda}\left(\mathfrak{T}^{\alpha}_{\beta\mu}\,\mathfrak{T}^{\mu}_{\alpha\lambda} - \mathfrak{T}^{\alpha}_{\alpha\mu}\,\mathfrak{T}^{\mu}_{\beta\lambda}\right)\overline{\omega}_n\end{aligned}$$

Variation of this action and separating terms in front of the two independent variations $\delta g^{\beta\lambda}$ and by $\delta\mathfrak{T}^{\mu}_{\beta\lambda}$ induces exactly (6.143). This may be considered as a checking of the calculus.

6.3.4.2 *Example 1: Distortion Example with Projective Transformation*

It is remarkable that when considering a contortion of the form $\mathfrak{T}^{\mu}_{\beta\lambda} := \delta^{\mu}_{\lambda}\,u_\beta$, for an arbitrary 4-vector u_β (zero or not), the second row identically vanishes, inducing $\mathscr{D}_{\mu}^{\beta\lambda} \equiv 0$. Such a contortion was already proposed in the literature e.g. [54], [276], [282] where the Einstein-Palatini action is compared to the Hilbert-Einstein action. In other words, the second equation of (6.143) has non-zero solutions in contortion. This means that introduction of an arbitrary non-zero vector $(u_\alpha, \alpha = 0, 1, 2, 3)$ makes it possible to construct a torsion field, and consequently, we would have the possibility of having an Einstein-Cartan vacuum spacetime with curvature and non-zero torsion $\aleph^{\mu}_{\beta\lambda} = \delta^{\mu}_{\lambda}u_\beta - \delta^{\mu}_{\beta}u_\lambda$. However, a straightforward calculus shows that the gauge transformation $\Gamma^{\gamma}_{\alpha\beta} \to \Gamma^{\gamma}_{\alpha\beta} + \delta^{\gamma}_{\alpha}\,u_\beta$ modifies neither the Ricci curvature (6.113) nor scalar curvature.

6.3.4.3 *Example 2: Distortion Example with Spherical Symmetry*

As an illustration, consider the example of a torsion field (6.122). It was inspired by a problem solved in [61] who considered the presence of a perfect spin fluid which is similar to the presence of a magnetic field, resulting from a magnetic monopole field at the center of a sphere. Spacetime is that of Einstein-Cartan. In this example, the torsion tensor is assumed to be:

$$\aleph = \aleph^0_{23}\, \vartheta^2 \otimes \vartheta^3 \otimes \mathbf{e}_0 + \aleph^0_{32}\, \vartheta^3 \otimes \vartheta^2 \otimes \mathbf{e}_0$$

which gave us the non-zero components of contortion (6.124):

$$\mathfrak{T}^0_{23} = \Omega^0_{23} = -\mathfrak{T}^0_{32}, \quad \mathfrak{T}^2_{03} = g^{22} g_{00}\, \Omega^0_{23} = \mathfrak{T}^2_{30}, \quad \mathfrak{T}^3_{02} = g^{33} g_{00}\, \Omega^0_{32} = \mathfrak{T}^3_{20}$$

All other components vanish. Calculus of $\mathscr{D}^{\beta\lambda}_{\mu}$ gives us the not-zero components:

$$\mathscr{D}^{23}_0 = -\left(g^{33}\mathfrak{T}^2_{30} + g^{22}\mathfrak{T}^3_{02}\right), \quad \mathscr{D}^{32}_0 = -\left(g^{22}\mathfrak{T}^3_{20} + g^{33}\mathfrak{T}^2_{03}\right) \tag{6.145}$$

$$\mathscr{D}^{30}_2 = -\left(g^{00}\mathfrak{T}^3_{02} + g^{33}\mathfrak{T}^0_{23}\right), \quad \mathscr{D}^{03}_2 = -\left(g^{33}\mathfrak{T}^0_{32} + g^{00}\mathfrak{T}^3_{20}\right) \tag{6.146}$$

$$\mathscr{D}^{20}_3 = -\left(g^{00}\mathfrak{T}^2_{03} + g^{22}\mathfrak{T}^0_{32}\right), \quad \mathscr{D}^{02}_3 = -\left(g^{22}\mathfrak{T}^0_{23} + g^{00}\mathfrak{T}^2_{30}\right) \tag{6.147}$$

We remark that these coefficients are equal to zero if and only if contortion is equal to zero, say $\aleph_0 = 0$. Once again, results confirm that without an external source such as spinning fluid matter or electromagnetic field, vacuum spacetime cannot produce torsion e.g. [217].

Remark 6.37. As for the variation of Einstein-Palatini action, we can introduce the contortion $\delta\aleph^\rho_{\mu\nu} = \delta\mathfrak{T}^\rho_{\mu\nu} - \delta\mathfrak{T}^\rho_{\nu\mu}$ and factorize by the variations $\delta g^{\mu\nu}$ and $\delta\mathfrak{T}^\gamma_{\mu\nu}$. Torsion is replaced by contortion.

6.3.5 *Gravito-Electromagnetism in an Einstein-Cartan Spacetime*

We have seen that the use of Einstein-Palatini action does not lead to a new vacuum spacetime model of relative gravitation when there is no external physical source (fluids with spin, electromagnetic fields or quantum physics). This leads us to propose the following example involving electromagnetism.

6.3.5.1 *Faraday Tensor in Einstein-Cartan Spacetime*

Formulation of Maxwell's equations with differential forms may not be equivalent to formulation by means of connection in Riemann-Cartan continuum e.g [283].

Definition 6.8. In an Einstein-Cartan spacetime, the Faraday tensor is defined as follows e.g. [284], [285]:

$$\mathfrak{F}_{\mu\nu} := \nabla_\mu A_\nu - \nabla_\nu A_\mu = \partial_\mu A_\nu - \partial_\nu A_\mu + \aleph^\rho_{\mu\nu} A_\rho \tag{6.148}$$

where A_ν is an electromagnetic four-vector potential.

The Faraday tensor is rather different if calculated by means of an exterior derivative of 1-form $\mathbf{A} = (A_\mu)$ e.g. [284]:

$$\mathfrak{F} := d\mathbf{A} \Longrightarrow \mathfrak{F}_{\mu\nu} = \partial_\mu A_\nu - \partial_\nu A_\mu \tag{6.149}$$

where, in such a case, we have exactly the same form of Faraday tensor in either Minkowski spacetime or Einstein spacetime. In this framework, two of the Maxwell equations $d\mathfrak{F} = 0$ would be expected since the Faraday tensor 2-form $\mathfrak{F}$ is exact, say $\mathfrak{F} := d\mathbf{A}$, and hence closed, $d\mathfrak{F} = d\,(d\mathbf{A}) = 0$.

To investigate electromagnetic waves within twisted and curved spacetime, it is assumed that the electromagnetic field is described by a Faraday electromagnetic 2-form $\mathfrak{F}_{\mu\nu}$. It constitutes an extended model of electromagnetism within curved continuum as in e.g. [260], and in the framework of differential forms e.g. [284]. [284] has derived Maxwell's equations in a Einstein-Cartan spacetime. It should be observed that the Faraday tensor (minimally coupled to gravitation in an Einstein-Cartan spacetime via torsion) and the scalar curvature are retained:

$$\mathfrak{F}_{\mu\nu} := \nabla_\mu A_\nu - \nabla_\nu A_\mu = \overline{\nabla}_\mu A_\nu - \overline{\nabla}_\nu A_\mu + \aleph^\rho_{\mu\nu} A_\rho, \quad \mathcal{R} := g^{\mu\nu} \mathfrak{R}_{\mu\nu} \tag{6.150}$$

where the contravariant components of $\mathfrak{F}^{\mu\nu}$ are calculated by means of connection with torsion e.g. [286]. Definition (6.149) in a Riemann continuum holds for both Euclidean and (pseudo)-Riemannian and also proposed in some post-Riemannian spacetimes e.g. [287]. As an extension the definition (6.150) is valid for both Euclidean, Riemannian and Riemann-Cartan continua. This again illustrates the fact that the extension of physical variables such as $\mathfrak{F}_{\mu\nu}$ can be done in many ways (as a 2-form in e.g. [287] or as a skew-symmetric part of the gradient in e.g. [286]). This shows that the

Lagrangian variation of this 2-form and curvature include both the variation of the potential A_μ, the variation of the Riemann metric $g_{\alpha\beta}$, and also the variation of the connection $\Gamma^\gamma_{\alpha\beta}$. In the following, we derive covariant Maxwell equations in an Einstein-Cartan spacetime $\mathscr{M}$.

6.3.5.2 *Yang-Mills Action and Its Variation*

Extension of the Einstein-Maxwell model for gravitation coupled with electromagnetism was proposed in e.g. [58] as early as 1929, and since then by numerous authors suggesting the introduction of connection other than the Levi-Civita connection where the Lagrangian method implicitly introduces constitutive laws of the medium e.g. [250]. Given that our interest is focused on long-distance phenomena within vacuum spacetime, we propose the working hypothesis which consists of considering only the Yang-Mills action as electromagnetic sources for gravitation. It is not necessary to re-develop everything in this chapter, we only have to consider again the Yang-Mills action. Consider the action for free electromagnetic field where, once more, the volume-form is assumed Riemannian as for most models in literature e.g. [4], [38], [278].

Definition 6.9. The action of the electromagnetic field within a Einstein-Cartan vacuum spacetime is defined by the Yang-Mills Lagrangian:

$$\mathscr{S}_{YM} := \int_{\mathscr{M}} \mathscr{L}\, \overline{\omega}_n \quad \text{with} \quad \boxed{\mathscr{L} := -\frac{1}{4}\mathcal{H}^{\mu\nu}\, \mathcal{F}_{\mu\nu}} \tag{6.151}$$

For the variation of (6.151), it is worth recalling that metric and torsion are independent primal variables in addition to the electromagnetic four-potential.

Remark 6.38. The Riemannian assumption may be questionable due to non-compatibility of $\overline{\omega}_n$ and ∇ with non-zero torsion. We will deepen this potential but subtle controversy in the next sections.

For the Lagrangian we consider the Yang-Mills electromagnetic part and the Einstein-Hilbert gravitational part e.g. [261].

Remark 6.39. Two principles are involved in the formulation of four-dimensional electromagnetism when defining the Faraday tensor $\mathcal{F}_{\mu\nu}$: the principle of gauge invariance, and the principle of minimal coupling. In Einstein spacetime, the use of Levi-Civita connection $\overline{\nabla}$ does not present any difficulties but the presence of a non-zero torsion $\aleph \neq 0$ in Einstein-Cartan spacetime raises an apparent contradiction between these two principles e.g. [349]. Defining the Faraday strength tensor as $\mathcal{F}_{\mu\nu} := \nabla_\mu A_\nu - \nabla_\nu A_\mu$ with non-zero torsion respects the principle of minimal coupling but seems to violate the principle of gauge invariance, at least in the absence of magnetic monopoles. Some authors e.g. [217] proposed to use the Levi-Civita connection to define the Faraday strength tensor even in Einstein-Cartan spacetime. Such a definition means that photons are decoupled from torsion, as we will see in the next chapter on cosmology. Derivation of gravitation theory coupled with electromagnetism which respects both the gauge invariance principle and the minimal coupling principle was suggested in e.g. [349] by proposing a restricted form of torsion generated by a scalar function.

For completeness, Maxwell's equations need the introduction of continuum constitutive laws. Again, we study the particular electromagnetic constitutive laws:

$$\mathcal{H}^{\mu\nu} = \Xi^{\mu\nu\lambda\rho}\, \mathcal{F}_{\lambda\rho}, \quad \Xi^{\mu\nu\lambda\rho} = \varepsilon_0\, g^{\mu\lambda} g^{\nu\rho} \tag{6.152}$$

Variation of the Faraday strength tensor is obtained by directly writing:

$$\begin{aligned}\delta\mathcal{F}_{\mu\nu} &= \delta\left(\partial_\mu A_\nu - \Gamma^\rho_{\mu\nu} A_\rho\right) - \delta\left(\partial_\nu A_\mu - \Gamma^\rho_{\nu\mu} A_\rho\right)\\ &= \left(\partial_\mu \delta A_\nu - \Gamma^\rho_{\mu\nu}\delta A_\rho - \delta\Gamma^\rho_{\mu\nu} A_\rho\right) - \left(\partial_\nu \delta A_\mu - \Gamma^\rho_{\nu\mu}\delta A_\rho - \delta\Gamma^\rho_{\nu\mu} A_\rho\right)\end{aligned}$$

This allows us to obtain:

$$\delta\mathcal{F}_{\mu\nu} = \nabla_\mu(\delta A_\nu) - \nabla_\nu(\delta A_\mu) - \delta\aleph^\rho_{\mu\nu} A_\rho \tag{6.153}$$

showing that variation of geometric structure, say $\delta\aleph^\rho_{\mu\nu}$, induces a variation $\delta\mathcal{F}_{\mu\nu}$. First, the variation of the Yang-Mills action takes the form of:

$$\delta\mathscr{S}_{YM} = -\frac{1}{4}\int_{\mathscr{M}} \delta\left(\mathcal{H}^{\mu\nu}\mathcal{F}_{\mu\nu}\right)\overline{\omega}_n + \left(\mathcal{H}^{\lambda\rho}\mathcal{F}_{\lambda\rho}\right)\delta\overline{\omega}_n \tag{6.154}$$

where we can easily calculate:

$$\delta\mathcal{H}^{\mu\nu} = \epsilon_0\left(\delta g^{\mu\alpha}\, g^{\nu\beta}\mathcal{F}_{\alpha\beta} + g^{\mu\alpha}\, \delta g^{\nu\beta}\mathcal{F}_{\alpha\beta} + g^{\mu\alpha}\, g^{\nu\beta}\delta\mathcal{F}_{\alpha\beta}\right) \tag{6.155}$$

accounting for spacetime constitutive laws $\mathcal{H}^{\mu\nu} = \epsilon_0 g^{\mu\lambda} g^{\nu\rho}\, \mathcal{F}_{\lambda\rho}$ and:

$$\delta\left(\mathcal{H}^{\mu\nu}\mathcal{F}_{\mu\nu}\right) = -2\epsilon_0\, \mathcal{F}_{\mu\lambda} g^{\lambda\rho}\mathcal{F}_{\rho\nu}\, \delta g^{\mu\nu} + 2\mathcal{H}^{\mu\nu}\delta\mathcal{F}_{\mu\nu} \tag{6.156}$$

By analogy to gravito-electromagnetism within an Einstein spacetime, we define the following Maxwell stress-energy momentum.

Definition 6.10. Maxwell electromagnetic energy-momentum is defined as:

$$T^{\text{em}}_{\mu\nu} := \frac{1}{8}\left(\mathcal{H}^{\beta\lambda}\mathcal{F}_{\beta\lambda}\right) g_{\mu\nu} + \frac{1}{4}\mathcal{H}^{\lambda\rho}\left(g_{\nu\lambda}\mathcal{F}_{\rho\mu} + \mathcal{F}_{\mu\lambda}g_{\rho\nu}\right) \tag{6.157}$$

where $\mathcal{F}_{\mu\nu} := \nabla_\mu A_\nu - \nabla_\nu A_\mu$ and the constitutive laws $\mathcal{H}^{\mu\nu} = \epsilon_0 g^{\mu\lambda} g^{\nu\rho}\, \mathcal{F}_{\lambda\rho}$.

From these relations, we may write the variation of Yang-Mills action:

$$\begin{aligned}\delta\mathscr{S}_{YM} = &\int_{\mathscr{M}} -T^{\text{em}}_{\mu\nu}\; \delta g^{\mu\nu}\; \overline{\omega}_n \\ &-\frac{1}{2}\int_{\mathscr{M}} \mathcal{H}^{\mu\nu}\left(\nabla_\mu(\delta A_\nu) - \nabla_\nu(\delta A_\mu) - \delta\aleph^{\rho}_{\mu\nu}A_\rho\right)\overline{\omega}_n\end{aligned}$$

To go further, let's express the variation of the Faraday tensor in terms of covariant derivatives using Levi-Civita connection:

$$\delta\mathcal{F}_{\mu\nu} = \overline{\nabla}_\mu\left(\delta A_\nu\right) - \overline{\nabla}_\nu\left(\delta A_\mu\right) - \aleph^{\rho}_{\mu\nu}\delta A_\rho - A_\rho\delta\aleph^{\rho}_{\mu\nu} \tag{6.158}$$

The variation of Yang-Mills action therefore takes the form of:

$$\delta\mathscr{S}_{YM} = \int_{\mathscr{M}} \left[T^{\text{em}}_{\mu\nu}\delta g^{\mu\nu} + \left(\overline{\nabla}_\mu\mathcal{H}^{\mu\nu} + \mathcal{H}^{\mu\nu}\mathcal{T}^{\rho}_{\mu\nu}\right)\delta A_\rho + \mathcal{H}^{\mu\nu}A_\rho\delta\mathcal{T}^{\rho}_{\mu\nu}\right]\overline{\omega}_n \tag{6.159}$$

where we have used the following relations:

$$\mathcal{T}^{\rho}_{\mu\nu} = \frac{1}{2}\aleph^{\rho}_{\mu\nu} + D^{\rho}_{\mu\nu}, \quad \mathcal{H}^{\mu\nu}\left(\frac{1}{2}\aleph^{\rho}_{\mu\nu} + D^{\rho}_{\mu\nu}\right) = \mathcal{H}^{\mu\nu}\mathcal{T}^{\rho}_{\mu\nu}$$

It is essential to note that the difference between consideration of an Einstein spacetime and that of Einstein-Cartan spacetime matters in the derivation of Maxwell's equations and those of Einstein-Maxwell.

6.3.5.3 *System of Field Equations*

Let's now consider the total action including gravitation and electromagnetism within an Einstein-Cartan spacetime:

$$\mathscr{S}_{GEM} := \int_{\mathscr{M}} \mathscr{L}\,\overline{\omega}_n, \quad \mathscr{L} := \mathscr{L}_{EP} + \mathscr{L}_{YM} \tag{6.160}$$

Remark 6.40. Again, we point out that the use of Riemann volume-form is quite typical in this domain. We will deviate from this rule later.

Thanks to the divergence theorem we have the following. The variation of total action is deduced from previous subsections:

$$\delta\mathscr{S} = \int_{\mathscr{M}} \left(\overline{\nabla}_\nu \mathcal{H}^{\mu\nu} + \mathcal{T}^{\nu}_{\mu\rho}\mathcal{H}^{\rho\mu}\right)\delta A_\nu\, \overline{\omega}_n \tag{6.161}$$

$$+ \int_{\mathscr{M}} \left[\frac{1}{2\chi}\left(\mathfrak{R}_{\mu\nu} - \frac{\mathcal{R}}{2} g_{\mu\nu}\right) + T^{\text{em}}_{\mu\nu}\right]\delta g^{\mu\nu}\, \overline{\omega}_n$$

$$+ \int_{\mathscr{M}} \left(\mathcal{H}^{\mu\nu}\, A_\rho + \frac{1}{2\chi}\, \mathscr{D}^{\mu\nu}_{\rho}\right)\delta\Gamma^{\rho}_{\mu\nu}\, \overline{\omega}_n \tag{6.162}$$

We should add effects of Ricci curvature variation including contortion (6.139).

Gravitation and electromagnetism in Einstein-Cartan spacetime. The system of field equations governing the Einstein-Cartan spacetime involving gravitation and electromagnetism, using Riemann volume-form $\overline{\omega}_n$, is:

$$\begin{cases} \overline{\nabla}_\mu \mathcal{H}^{\mu\nu} + \mathcal{T}^{\nu}_{\rho\mu}\, \mathcal{H}^{\mu\rho} = 0 \\ \dfrac{1}{2\chi}\, G_{\mu\nu} + T^{\text{em}}_{\mu\nu} = 0 \\ \dfrac{1}{2\chi}\, \mathscr{D}^{\beta\lambda}_{\mu} + \mathcal{H}^{\beta\lambda}\, A_\mu = 0 \end{cases} \tag{6.163}$$

with the Einstein tensor $G_{\mu\nu} := \mathfrak{R}_{\mu\nu} - (\mathcal{R}/2)\, g_{\mu\nu}$ and a spacetime distortion:

$$\mathscr{D}^{\beta\lambda}_{\mu} := \left[\left(g^{\beta\lambda}\mathcal{T}^{\alpha}_{\alpha\mu} - g^{\alpha\lambda}\mathcal{T}^{\beta}_{\alpha\mu}\right) - \left(g^{\beta\gamma}\mathcal{T}^{\lambda}_{\mu\gamma} - \delta^{\beta}_{\mu} g^{\alpha\gamma}\mathcal{T}^{\lambda}_{\alpha\gamma}\right)\right] \tag{6.164}$$

Unknowns are electromagnetic potential A_ν, metric $g_{\mu\nu}$ and contortion $\mathcal{T}^{\gamma}_{\mu\nu}$.

Remark 6.41. Faraday strength $\mathcal{F}_{\mu\nu}$, its dual $\mathcal{H}^{\mu\nu}$, and Ricci curvature $\mathfrak{R}_{\mu\nu}$ are defined by means of metric $\mathbf{g}$ and connection ∇.

Let's consider each row of the system of field equations.

(1) The first row of (6.163) expresses Maxwell's equations in Einstein-Cartan spacetime. The Lagrangian approach is equivalent to the differential form approach when the continuum is Riemannian without torsion ([283]). Maxwell's equations (6.163) show that the Lagrangian approach with torsion is assumed to be "naturally" deduced also from

a variation principle. However, the expression of Maxwell's equations does not fit the Minimal Coupling Principle.

(2) The second row of (6.163) gives the coupling equation of electromagnetism and gravitation. The electromagnetic terms act as a source-term for gravitation. Despite its apparent relative simplicity, the system (6.163) remains complex since the connection, and by extension the Ricci and total curvatures, includes both (gravitational) metric $g_{\mu\nu}$ and contortion $\mathcal{T}^{\gamma}_{\mu\nu}$.

(3) The third equation of system (6.163) may be rewritten in vectorial form after calculating the trace:

$$\mathcal{H}^{\mu\nu}\ A_{\nu} \rightarrow \begin{pmatrix} -\mathbf{D}\cdot\mathbf{A} \\ \mathbf{H}\times\mathbf{A} \end{pmatrix} \tag{6.165}$$

Once more, we notice that the choice of an Einstein-Cartan spacetime but using a Riemann volume-form $\overline{\omega}_n$ led us to a complicated (violating the Principle of Minimal Coupling) form between electromagnetism and torsion. Rigorously, we should not consider $\overline{\omega}_n$ but a volume-form ω_n compatible with the connection ∇ with non-zero torsion $\aleph \neq 0$. However, the interest of the present approach in this section is to be able to compare our results with those of authors who adopted the Riemann volume-form e.g. [52], [137].

6.3.6 *Dual Electromagnetism and Gravitation*

The third equation of (6.163) plays a special role when looking at its physical interpretation, especially in physical optics. However, for completeness, we must also consider the second pair of Maxwell's equations, which is more related to Gauss's law of magnetic induction **B**. This second pair cannot be deduced directly from the standard Yang-Mills Lagrangian.

6.3.6.1 *Hodge Transformation*

Three-dimensional vector equations might perturb uninformed readers who could confuse vectors with flux vectors normal to a surface. Moreover, covariant formulation is not easy to set up with the notions of "axial vectors" and "polar vectors" and so on. In this subsection we introduce the Hodge operator which allows us to better clarify these notions by working with 1-forms and 2-forms. In this way, the second pair is rigorously derived by means of dual formulation of electromagnetism needing the notion of the Hodge operator.

Vector product and Hodge operator. It finds its origin in the isomorphism between a 2-form and a 1-form in a three-dimensional space. To start with, consider two vectors $\mathbf{E}$ and $\mathbf{u}$ and determine their vector product:

$$\mathbf{E}\times\mathbf{u} = \begin{pmatrix} E^1 \\ E^2 \\ E^3 \end{pmatrix} \times \begin{pmatrix} u^1 \\ u^2 \\ u^3 \end{pmatrix} = \begin{pmatrix} E^2u^3 - E^3u^2 \\ E^3u^1 - E^1u^3 \\ E^1u^2 - E^2u^1 \end{pmatrix} \begin{pmatrix} 0 & -E^3 & E^2 \\ E^3 & 0 & -E^1 \\ -E^2 & E^1 & 0 \end{pmatrix} \begin{pmatrix} u^1 \\ u^2 \\ u^3 \end{pmatrix}$$

indicating that vector $\mathbf{E}$ is isomorph to a matrix ${}^*\mathbf{E}$ with the components:

$$^*\mathbf{E} = \begin{pmatrix} 0 & -E^3 & E^2 \\ E^3 & 0 & -E^1 \\ -E^2 & E^1 & 0 \end{pmatrix} \tag{6.166}$$

Let's now recall that the vector product may be defined by an coordinate-free relation involving the volume-form ω_n:

$$(\mathbf{E}\times\mathbf{u})\cdot\mathbf{w} := \omega_n\left(\mathbf{E},\mathbf{u},\mathbf{w}\right), \quad \text{for arbitrary } \mathbf{w}\in\mathscr{E}$$

The previous equation may be rewritten as follows:

$$*\mathbf{E}(\mathbf{u},\mathbf{w}) = \omega_n\left(\mathbf{E},\mathbf{u},\mathbf{w}\right), \quad \text{for arbitrary } \mathbf{u},\mathbf{w}\in\mathscr{E}$$

Projected onto any base $(\mathbf{e}_1,\mathbf{e}_2,\mathbf{e}_3)$ with its dual $(\mathbf{e}^1,\mathbf{e}^2,\mathbf{e}^3)$, we have:

$$\begin{aligned} *\mathbf{E}(\mathbf{u},\mathbf{w}) &= *E_{jk}\ \mathbf{e}^j\wedge\mathbf{e}^k(\mathbf{u},\mathbf{w}) = *E_{jk}\ u^j w^k = \omega_n(\mathbf{e}_i,\mathbf{e}_j,\mathbf{e}_k)\ E^i u^j w^k \\ &= \sqrt{|g|}\ \epsilon_{ijk}\ E^i u^j w^k \end{aligned}$$

We deduce the definition by means of components of the Hodge operator:

$$\boxed{*E_{jk} := \frac{\sqrt{|g|}}{1!}\ \epsilon_{ijk}\ E^i}\ , \quad \text{where 1 is the dimension of } \mathbf{E} \tag{6.167}$$

where ϵ_{ijk} is the Levi-Civita permutation symbol of rank 3. The Hodge operator is an application relating the 1-vector $\mathbf{E}$ to the 2-form ${}^*\mathbf{E}$. It depends on the volume-form and the dimension of the space.

Hodge operator of forms and applications. Let Λ^p be the set of p-forms on the Riemann manifold $\mathscr{B}$ of dimension n. Then Λ^p is isomorphic to Λ^{n-p}. Consider a Riemann manifold of dimension n and the p-forms

$\omega_{\mu_1\cdots\mu_p}$:

$$\omega = \omega_{\mu_1\cdots\mu_p}\, dx^{\mu_1} \wedge \cdots \wedge dx^{\mu_p} \tag{6.168}$$

where we assume a coordinate system $(x^1, \cdots, x^n)$ and with base of forms $(dx^1, \cdots, dx^n)$.

Definition 6.11. Consider a p-form on $\mathscr{B}$, endowed with a metric $\mathbf{g}$ and a volume-form ω_n. The Hodge operator $*$ is an isomorphism which assigns a p-form ω to a $n-k$-form denoted $*\omega$ and defined by e.g. [120]:

$$< \eta, \omega > \; \omega_n := (\eta \wedge *\omega)\,, \quad \forall \eta \in \Lambda^p \tag{6.169}$$

Projected onto a coordinate system, this definition takes the form of:

$$\eta_{\nu_1\cdots\nu_p}\, \omega_{\mu_1\cdots\mu_p}\, g^{\mu_1\nu_1} \cdots g^{\mu_p\nu_p}\, \omega_n \left(\partial_{\lambda_1}, \cdots \partial_{\lambda_p}, \partial_{\lambda_{p+1}}, \cdots \partial_{\lambda_n}\right)$$
$$=$$
$$\eta\left(\partial_{\lambda_1}, \cdots \partial_{\lambda_p}\right)(*\omega)\left(\partial_{\lambda_{p+1}}, \cdots \partial_{\lambda_n}\right) - (*\omega)\left(\partial_{\lambda_1}, \cdots \partial_{\lambda_p}\right)\eta\left(\partial_{\lambda_{p+1}}, \cdots \partial_{\lambda_n}\right)$$

The p-form η being arbitrary, we can choose $\eta = dx^{\lambda_1} \wedge \cdots \wedge dx^{\lambda_p}$. For a given and fixed set of indices $(\nu_1, \cdots \nu_p)$, we have:

$$\eta\left(\partial_{\nu_1}, \cdots \partial_{\nu_p}\right) = dx^{\lambda_1} \wedge \cdots \wedge dx^{\lambda_p}\left(\partial_{\nu_1}, \cdots \partial_{\nu_p}\right) = \delta^{\lambda_1}_{\nu_1} \cdots \delta^{\lambda_p}_{\nu_p} \tag{6.170}$$

We also have the two relations:

$$\eta\left(\partial_{\lambda_{p+1}}, \cdots \partial_{\lambda_n}\right) = 0 \tag{6.171}$$

and, remarking that we have a permutation of indices since we have a summation,

$$\eta\left(\partial_{\sigma(\lambda_1)}, \cdots \partial_{\sigma(\lambda_p)}\right) = \delta^{\lambda_1}_{\sigma(\lambda_1)} \cdots \delta^{\lambda_p}_{\sigma(\lambda_p)} = p! \tag{6.172}$$

This is checked by calculating for $p = 2$, $p = 3, \ldots$ and so on. We then deduce the Hodge operator which transforms p-form ω to $n-p$-form denoted $*\omega$:

$$*\omega = (*\omega)_{\mu_{p+1}\cdots\mu_n}\, dx^{\mu_{p+1}} \wedge \cdots \wedge dx^{\mu_n}$$

with the components of the $n-p$-form $*\omega$:

$$(*\omega)_{\mu_{p+1}\cdots\mu_n} := \frac{1}{p!}\omega_n\left(\partial_{\mu_1}, \cdots \partial_{\mu_p}, \partial_{\mu_{p+1}}, \cdots \partial_{\mu_n}\right)\, \omega_{\nu_1\cdots\nu_p}\; g^{\mu_1\nu_1} \cdots g^{\mu_p\nu_p} \tag{6.173}$$

For a Riemann manifold its components are particularized e.g. [5], [128]:

$$\boxed{(*\omega)_{\mu_{p+1}\cdots\mu_n} := \frac{\sqrt{|g|}}{p!}\varepsilon^{1\ 2\cdots n}_{\mu_1\ \mu_2\cdots\mu_n} \quad \omega_{\nu_1\cdots\nu_p} \ g^{\mu_1\nu_1} \ \cdots \ g^{\mu_p\nu_p}} \tag{6.174}$$

where $g := \det g_{\mu\nu}$, $\varepsilon^{1\ 2\cdots n}_{\mu_1\ \mu_2\cdots\mu_n}$ is the permutation symbol and $g^{\mu\nu}$ the inverse of the metric. As a reminder:

$$\varepsilon^{1\ 2\cdots n}_{\mu_1\ \mu_2\cdots\mu_n} : \begin{cases} +1 \text{ if } (\mu_1\ \mu_2\cdots\mu_n) & \text{cyclic permutation of} \quad (12\cdots n) \\ -1 \text{ if } (\mu_1\ \mu_2\cdots\mu_n) & \text{anticyclic permutation of} \quad (12\cdots n) \\ 0 \text{ if otherwise} & \end{cases}$$

We note the tensorial transformation:

$$\varepsilon^{\mu_1\ \mu_2\cdots\mu_n}_{1\ 2\cdots n} := g^{\mu_1\nu_1} \ \cdots \ g^{\mu_n\nu_n} \ \varepsilon^{1\ 2\cdots n}_{\nu_1\ \nu_2\cdots\nu_n}$$

which induces the relation:

$$\varepsilon^{\mu_1\ \mu_2\cdots\mu_n}_{1\ 2\cdots n} := g^{-1} \ \varepsilon^{1\ 2\cdots n}_{\nu_1\ \nu_2\cdots\nu_n}, \quad g := \det g_{\mu\nu} \tag{6.175}$$

Application of Hodge operator. Now we are interested in the dual formulation; such duality should not be confused with the duality of constitutive laws. Application of the Hodge operator is straightforward for Faraday strength tensor $\mathcal{H}^{\mu\nu}$ to become a dual variable $^*\mathcal{H}^{\mu\nu}$:

$$^*\mathcal{H}^{\mu_1\mu_2} = \frac{\sqrt{|g|}}{2!} \ \varepsilon^{\mu_1\mu_2\mu_3\mu_4}_{1234} \ g_{\mu_3\nu_3} g_{\mu_4\nu_4} \mathcal{H}^{\mu_3\mu_4} \tag{6.176}$$

where we have to recall the original concept of the Hodge operator to relate a vector field to a 1-form (vector product notion) and *vice versa*. Considering orthonormal base, we obtain the components of tensor $^*\mathcal{H}^{\mu_1\mu_2}$:

$$\mathcal{H}^{\mu\nu} = \begin{pmatrix} 0 & D^1 & D^2 & D^3 \\ -D^1 & 0 & H_3 & -H_2 \\ -D^2 & -H_3 & 0 & H_1 \\ -D^3 & H_2 & -H_1 & 0 \end{pmatrix} \qquad ^*\mathcal{H}^{\mu\nu} = \begin{pmatrix} 0 & H_1 & H_2 & H_3 \\ -H_1 & 0 & -D^3 & D^2 \\ -H_2 & D^3 & 0 & -D^1 \\ -H_3 & -D^2 & D^1 & 0 \end{pmatrix} \tag{6.177}$$

By applying again the Hodge operator, we can write:

$$\mathcal{F}_{\mu\nu} = \begin{pmatrix} 0 & -E_1 & -E_2 & -E_3 \\ E_1 & 0 & B^3 & -B^2 \\ E_2 & -B^3 & 0 & B^1 \\ E_3 & -B^2 & -B^1 & 0 \end{pmatrix} \quad {}^*\mathcal{F}_{\mu\nu} = \begin{pmatrix} 0 & B^1 & B^2 & B^3 \\ -B^1 & 0 & E_3 & -E_2 \\ -B^2 & -E_3 & 0 & E_1 \\ -B^3 & E_2 & -E_1 & 0 \end{pmatrix} \tag{6.178}$$

The following Lagrangian was investigated in the domain of light birefrigence in vacuum spacetime and was considered as an effective Lagrangian related to quantum electrodynamics, including the scalar invariant and pseudo-scalar invariant respectively:

$$\mathscr{L} := -\frac{1}{4}\mathcal{H}^{\mu\nu}\mathcal{F}_{\mu\nu}, \quad \mathscr{L}^* := -\frac{1}{4}\,{}^*\mathcal{H}^{\mu\nu}\mathcal{F}_{\mu\nu} \tag{6.179}$$

Given that $\mu_0\ \epsilon_0 = 1$ for vacuum, the pseudo-scalar invariant is equal to $\mathscr{L}^* = \mathbf{B}\cdot\mathbf{D}$. In the following, we focus on the Lagrangian $\mathscr{L}$.

6.3.6.2 *Dual Formulation of Electromagnetism*

To highlight the physical aspect, we use a classical dual formulation e.g. [288]. We first replace tensors $\mathcal{F}_{\mu\nu}$ and $\mathcal{H}^{\mu\nu}$ respectively by $*\mathcal{F}_{\mu\nu}$ and ${}^*\mathcal{H}^{\mu\nu}$ defined by the Hodge operator (6.176):

$$\mathcal{F}_{\mu\nu} = \begin{bmatrix} 0 & -E_1 & -E_2 & -E_3 \\ E_1 & 0 & B^3 & -B^2 \\ E_2 & -B^3 & 0 & B^1 \\ E_3 & B^2 & -B^1 & 0 \end{bmatrix}, \quad \mathcal{H}^{\mu\nu} = \begin{bmatrix} 0 & D^1 & D^2 & D^3 \\ -D^1 & 0 & H_3 & -H_2 \\ -D^2 & -H_3 & 0 & H_1 \\ -D^3 & H_2 & -H_1 & 0 \end{bmatrix}$$

Dual Faraday tensors in the sense that the Hodge operator is applied (it is by no means the duality in the sense of constitutive laws) are then:

$$*\mathcal{F}_{\mu\nu} = \begin{bmatrix} 0 & B^1 & B^2 & B^3 \\ -B^1 & 0 & E_3 & -E_2 \\ -B^2 & -E_3 & 0 & E_1 \\ -B^3 & E_2 & -E_1 & 0 \end{bmatrix}, \quad {}^*\mathcal{H}^{\mu\nu} = \begin{bmatrix} 0 & -H_1 & -H_2 & -H_3 \\ H_1 & 0 & D^3 & -D^2 \\ H_2 & -D^3 & 0 & D^1 \\ H_3 & D^2 & -D^1 & 0 \end{bmatrix}$$

Remark 6.42. This operation only reverses the position of the electric and magnetic variables in the Faraday tensor. To obtain the second pair of Maxwell's equations, it is classical to apply the transformation $(\mathbf{E}, \mathbf{B}) \rightarrow (\mathbf{B}, -\mathbf{E})$.

Let's recall the Maxwell equations in vacuum spacetime written in four dimensions are:

$$\nabla_\nu \mathcal{H}^{\mu\nu} = 0 \quad \Longrightarrow \quad \begin{cases} \nabla \cdot \mathbf{D} = 0 \\ \nabla \times \mathbf{H} - c\partial_0 \mathbf{D} = 0 \end{cases} \tag{6.180}$$

with its vectorial form e.g. [6] which represent Gauss's electrical law and Ampère's law respectively:

$$\begin{cases} \nabla \cdot \mathbf{D} = 0 \\ \nabla \times \mathbf{H} - c\partial_0 \mathbf{D} = 0 \end{cases}, \quad \begin{cases} \nabla \cdot \mathbf{B} = 0 \\ \nabla \times \mathbf{E} + c\partial_0 \mathbf{B} = 0 \end{cases}$$

The second group of Maxwell's equations is deduced from the dual theory of electromagnetism. It starts by defining a dual Yang-Mills Lagrangian. First it is necessary to define the following dual tensors from the previous ones obtained with the Hodge operator.

Definition 6.12. The dual Faraday strength tensors are given by:

$$\begin{cases} \tilde{\mathcal{H}}^{\mu\nu} := {}^*\mathcal{F}^{\mu\nu} = g^{\mu\alpha} \; * \mathcal{F}_{\alpha\beta} \; g^{\beta\nu}, \\ \tilde{\mathcal{F}}_{\mu\nu} := *\mathcal{H}_{\mu\nu} = g_{\mu\alpha} \; {}^*\mathcal{H}^{\alpha\beta} \; g_{\beta\nu} \end{cases} \tag{6.181}$$

Remark 6.43. Obtention of these dual variables is more than just intervening the electric and magnetic fields $(\mathbf{E}, \mathbf{B}) \to (\mathbf{B}, -\mathbf{E})$.

For the sake of clarity, the expression in a Minkowski spacetime of the two dual tensors are given by:

$$\tilde{\mathcal{H}}^{\mu\nu} = \begin{bmatrix} 0 & -B^1 & -B^2 & -B^3 \\ B^1 & 0 & E_3 & -E_2 \\ B^2 & -E_3 & 0 & E_1 \\ B^3 & E_2 & -E_1 & 0 \end{bmatrix}, \quad \tilde{\mathcal{F}}_{\mu\nu} = \begin{bmatrix} 0 & H_1 & H_2 & H_3 \\ -H_1 & 0 & D^3 & -D^2 \\ -H_2 & -D^3 & 0 & D^1 \\ -H_3 & D^2 & -D^1 & 0 \end{bmatrix}$$

Now the condition to have a physical interpretation of the Lagrangian governing the dual electromagnetism is to impose the same Lagrangian in three-dimensional formulation. Either form the covariant four-dimensional form or by remarking the expression in three-dimensional vectorial form, we have the following.

Definition 6.13. The Yang-Mills Lagrangian of a vacuum spacetime for the dual formulation of electromagnetism is defined by:

$$\boxed{\tilde{\mathscr{L}}_{YM} := \frac{1}{4}\, \tilde{\mathcal{H}}^{\mu\nu}\, \tilde{\mathfrak{F}}_{\mu\nu}}, \quad \tilde{\mathscr{S}}_{YM} := \int_{\mathscr{M}} \tilde{\mathscr{L}}_{YM}\, \overline{\omega}_n \tag{6.182}$$

where dual Faraday tensors are defined by (6.181).

Remark 6.44. It is easy to check the consistency of the definition, namely the sign was chosen to satisfy this condition:

$$\tilde{\mathscr{L}}_{YM} := \frac{1}{4}\, \tilde{\mathcal{H}}^{\mu\nu}\, \tilde{\mathfrak{F}}_{\mu\nu} = -\frac{1}{4}\, \mathcal{H}^{\mu\nu}\, \mathfrak{F}_{\mu\nu} = \mathscr{L} = \frac{1}{2}\,(\mathbf{D}\cdot\mathbf{E} - \mathbf{B}\cdot\mathbf{H})$$

meaning that the two Lagrangians (Yang-Mills and its dual) describe the same physical phenomenon.

Derivation of the second group of Maxwell's equations begins with the variation:

$$\delta\tilde{\mathscr{S}}_{YM} = \frac{1}{4}\int_{\mathscr{M}} \Big[(\delta\tilde{\mathcal{H}}^{\mu\nu})\, \tilde{\mathfrak{F}}_{\mu\nu} + \tilde{\mathcal{H}}^{\mu\nu}\, \delta\tilde{\mathfrak{F}}_{\mu\nu}\Big]\, \overline{\omega}_n + \tilde{\mathcal{H}}^{\lambda\rho}\tilde{\mathfrak{F}}_{\lambda\rho}\, \delta\overline{\omega}_n = 0 \tag{6.183}$$

where the primal variable is, by analogy, defined as:

$$\tilde{\mathfrak{F}}_{\mu\nu} := \nabla_\mu \tilde{A}_\nu - \nabla_\nu \tilde{A}_\mu \tag{6.184}$$

in which $\tilde{A}_\mu$ is a dual electromagnetic quadri-potential (we keep the same notation for avoiding a plethora of variables) e.g. [288], [289]. We consider the usual constitutive laws between $\mathcal{H}^{\mu\nu}$ and $\mathfrak{F}_{\mu\nu}$, and also the constitutive laws between the dual variables $\tilde{\mathcal{H}}^{\mu\nu}$ and $\tilde{\mathfrak{F}}_{\mu\nu}$.

Similar to standard electromagnetism, we have spacetime constitutive laws in dual formulation:

$$\tilde{\mathcal{H}}^{\mu\nu} = \frac{1}{\mu_0} g^{\mu\lambda}\, \tilde{\mathfrak{F}}_{\lambda\rho}\, g^{\rho\nu} \tag{6.185}$$

The notion of constitutive laws becomes essential in relative gravitation due to the spacetime curvature, even for vacuum without source distribution. It is essentially provoked by the non-homogeneous metric tensor. The variation of the integrand is given by:

$$\delta(\tilde{\mathcal{H}}^{\mu\nu}\tilde{\mathfrak{F}}_{\mu\nu}\, \overline{\omega}_n) = \Big[(\delta\tilde{\mathcal{H}}^{\mu\nu})\tilde{\mathfrak{F}}_{\mu\nu} + \tilde{\mathcal{H}}^{\mu\nu}\, \delta\tilde{\mathfrak{F}}_{\mu\nu}\Big]\, \overline{\omega}_n + \tilde{\mathcal{H}}^{\lambda\rho}\, \delta\tilde{\mathfrak{F}}_{\lambda\rho}\, \delta\overline{\omega}_n \tag{6.186}$$

with

$$\delta\tilde{\mathfrak{F}}_{\mu\nu} = \overline{\nabla}_\mu(\delta\tilde{A}_\nu) - \overline{\nabla}_\nu(\delta\tilde{A}_\mu) - \aleph^\rho_{\mu\nu}\, \delta\tilde{A}_\rho - \tilde{A}_\rho\, \delta\aleph^\rho_{\mu\nu} \tag{6.187}$$

and by analogy to Maxwell stress tensor $T^{\text{em}}_{\mu\nu}$, we remark its dual formulation.

Definition 6.14. The dual Maxwell stress energy tensor is defined as:

$$\tilde{T}^{\text{em}}_{\mu\nu} := \frac{1}{8}\tilde{\mathfrak{H}}^{\lambda\rho}\tilde{\mathfrak{F}}_{\lambda\rho}\, g_{\mu\nu} + \frac{1}{4}\tilde{\mathfrak{H}}_{\lambda\rho}\left(g_{\nu\lambda}\tilde{\mathfrak{F}}_{\rho\mu} + \tilde{\mathfrak{F}}_{\mu\lambda}g^{\rho\nu}\right) \tag{6.188}$$

where $\tilde{\mathfrak{F}}_{\mu\nu} = \nabla_\mu \tilde{A}_\nu - \nabla_\nu \tilde{A}_\mu$ with the constitutive laws $\tilde{\mathfrak{H}}^{\mu\nu} = (1/\mu_0)\, g^{\mu\lambda}g^{\nu\rho}\tilde{\mathfrak{F}}_{\lambda\rho}$.

Again, we check the equality of the two expressions in such a case:

$$\tilde{T}^{\text{em}}_{\mu\nu} := \frac{1}{2\mu_0}\tilde{\mathfrak{F}}_{\mu\lambda}\, g^{\lambda\rho}\, \tilde{\mathfrak{F}}_{\rho\mu} + \frac{1}{8}\tilde{\mathfrak{H}}^{\lambda\rho}\tilde{\mathfrak{F}}_{\lambda\rho}\, g_{\mu\nu} \tag{6.189}$$

Since $\tilde{\mathfrak{H}}^{\mu\nu}$ is skew-symmetric and $\mathfrak{T}^{\rho}_{\mu\nu}$ may be decomposed into skew-symmetric and symmetric parts in terms of the torsion $\aleph^{\rho}_{\mu\nu}$, we can write:

$$\tilde{\mathfrak{H}}^{\mu\nu}\, \delta\tilde{\mathfrak{F}}_{\mu\nu} = 2\tilde{\mathfrak{H}}^{\mu\nu}\, \overline{\nabla}_\mu(\delta\tilde{A}_\nu) - 2\tilde{\mathfrak{H}}^{\mu\nu}\, \mathfrak{T}^{\rho}_{\mu\nu}\, \delta\tilde{A}_\rho - 2\tilde{\mathfrak{H}}^{\mu\nu}\, \tilde{A}_\rho\delta\, \mathfrak{T}^{\rho}_{\mu\nu} \tag{6.190}$$

Again, by using the divergence theorem, we obtain the variation:

$$\delta\tilde{\mathscr{S}}_{YM} = \int_{\mathscr{B}} \left[-\left(\overline{\nabla}_\mu\tilde{\mathfrak{H}}^{\mu\nu} + \mathfrak{T}^{\nu}_{\mu\rho}\, \tilde{\mathfrak{H}}^{\mu\rho}\right)\delta\tilde{A}_\nu - \tilde{T}^{\text{em}}_{\mu\nu}\, \delta g^{\mu\nu} + \tilde{\mathfrak{H}}^{\mu\nu}\tilde{A}_\rho\, \delta\mathfrak{T}^{\rho}_{\mu\nu}\right]\overline{\omega}_n \tag{6.191}$$

that we will combine with the gravitational effect in the next subsection.

6.3.6.3 *Chern-Simons Current and Spin Angular Momentum*

Let's combine the two actions-at-a-distance of the Universe by assuming the Lagrangian $\mathscr{L}_G := \tilde{\mathscr{L}}_{YM} + \mathscr{L}_{EP}$.[3] The model of relative gravitation in the presence of electromagnetism is defined by the Lagrangian:

$$\boxed{\tilde{\mathscr{S}}_{GEM} := \int_{\mathscr{B}} \left(\frac{\mathcal{R}}{2\chi} + \frac{1}{4}\tilde{\mathfrak{H}}^{\mu\nu}\, \tilde{\mathfrak{F}}_{\mu\nu}\right)\overline{\omega}_n} \tag{6.192}$$

where we adopt a dual formulation of the Maxwell theory of vacuum space-time. The dual electromagnetic variation may be coupled with variation of

[3]The dual electromagnetic Lagrangian is equal to the classical one. It is obvious when it is formulated in three dimensions.

the Einstein-Palatini Lagrangian to give:

$$\delta\tilde{\mathscr{S}}_{GEM} = \int_{\mathscr{M}} -\left(\overline{\nabla}_\mu \tilde{\mathfrak{H}}^{\mu\nu} + \mathfrak{T}^{\nu}_{\mu\rho}\,\tilde{\mathfrak{H}}^{\mu\rho}\right)\delta\tilde{A}_\nu\,\overline{\omega}_n \tag{6.193}$$

$$+\int_{\mathscr{M}} \left[\frac{1}{2\chi}\left(\mathfrak{R}_{\mu\nu} - \frac{\mathcal{R}}{2}g_{\mu\nu}\right) - \tilde{T}^{\text{em}}_{\mu\nu}\right]\delta g^{\mu\nu}\,\overline{\omega}_n \tag{6.194}$$

$$+\int_{\mathscr{M}} \left(-\tilde{\mathfrak{H}}^{\mu\nu}\,\tilde{A}_\rho + \frac{1}{2\chi}\,\mathscr{D}^{\mu\nu}_{\rho}\right)\delta\Gamma^{\rho}_{\mu\nu}\,\overline{\omega}_n \tag{6.195}$$

Arbitrariness of the variations $\delta\tilde{A}_\nu$, $\delta g^{\mu\nu}$, and $\delta\Gamma^{\rho}_{\mu\nu}$ induces the field equations in dual formulation.

Gravitation and dual electromagnetism in Einstein-Cartan spacetime. The system of field equations governing the Einstein-Cartan spacetime involving gravitation and electromagnetism using the dual electromagnetism approach is:

$$\begin{cases} \overline{\nabla}_\mu \tilde{\mathfrak{H}}^{\mu\nu} + \mathfrak{T}^{\nu}_{\mu\rho}\,\tilde{\mathfrak{H}}^{\mu\rho} = 0 \\ \dfrac{1}{2\chi}\,G_{\mu\nu} - \tilde{T}^{\text{em}}_{\mu\nu} = 0 \\ \dfrac{1}{2\chi}\,\mathscr{D}^{\beta\lambda}_{\mu} - \tilde{\mathfrak{H}}^{\beta\lambda}\,\tilde{A}_\mu = 0 \end{cases} \tag{6.196}$$

with respectively the Einstein tensor and the distorsion tensor (6.164):

$$\begin{cases} G_{\mu\nu} := \mathfrak{R}_{\mu\nu} - \dfrac{\mathcal{R}}{2}g_{\mu\nu} \\ \mathscr{D}^{\beta\lambda}_{\mu} := : \left[\left(g^{\beta\lambda}\mathfrak{T}^{\alpha}_{\alpha\mu} - g^{\alpha\lambda}\mathfrak{T}^{\beta}_{\alpha\mu}\right) - \left(g^{\beta\gamma}\mathfrak{T}^{\lambda}_{\mu\gamma} - \delta^{\beta}_{\mu}g^{\alpha\gamma}\mathfrak{T}^{\lambda}_{\alpha\gamma}\right)\right] \end{cases}$$

These results suggest the contribution of the magnetic field (instead of electric) to the evolution of the Universe at large scale e.g. [290]. Equations (6.196) constitute a sound basis for this purpose. The trace with respect to indices ν and λ induces the equation:

$$\frac{1}{2\chi}\,\mathscr{D}^{(\text{m})\mu\nu}_{\nu}(\mathbf{g},\mathfrak{T}) - \tilde{\mathfrak{H}}^{\mu\nu}\,\tilde{A}_\nu = 0 \tag{6.197}$$

where we recognize the first term as the Chern-Simons current [319]. The dual formulation of electromagnetism is worth highlighting that the magnetic field of the Universe plays an fundamental role in the formation and evolution of large-scale structure such as galaxies, clusters of galaxies and importantly giant filaments and their spinning. Indeed, the vectorial form is given explicitly:

$$\tilde{\mathcal{H}}^{\mu\nu}\ \tilde{A}_\nu \to -\begin{pmatrix} \mathbf{B}\cdot\tilde{\mathbf{A}} \\ \mathbf{E}\times\tilde{\mathbf{A}} \end{pmatrix} \tag{6.198}$$

Component $\mathbf{B}\cdot\tilde{\mathbf{A}}$ is the magnetic helicity density and the three-dimensional vector $\mathbf{E}\times\tilde{\mathbf{A}}$ is the spin angular momentum e.g. [288], [291]. The index (m) is introduced because the induction magnetic vector is involved in the helicity.

Remark 6.45. First, experimental observations of galaxies, clusters of galaxies and giant filaments allow us to deduce that the Universe is magnetized and that magnetic fields, together with gravitational fields, are the macroscopic physical phenomena that shape the Universe's gigantic web structure e.g. [290]. This is a fact, although the primordial origin of magnetic fields remains an open field of research. Second, magnetic helicity, and also spin angular momentum, are important, if not essential, physical quantities in cosmology. They were related to problems of C and CP symmetry violations according to Peccei and Quinn e.g. [292] and with axion field evolution e.g. [293], who deduced the existence of chargeless pseudo-particles called axions. While this book is limited to phenomenological cosmological spacetime models (very large dimensions), it is nevertheless curious to note that taking into account local defects, quantified by torsion and curvature, leads us to highlight formal linking between electromagnetism, in this case magnetic helicity $\mathbf{B}\cdot\tilde{\mathbf{A}}$ and extended to Chern-Simons current $\tilde{\mathcal{H}}^{\mu\nu}\ \tilde{A}_\nu$, and the distortion $\mathscr{D}_\nu^{\mu\nu}(\mathbf{g},\aleph)$ of spacetime.

6.3.7 *Gravito-Electromagnetism and Birth of Distortion*

Electromagnetic helicity includes electrical and magnetic helicities. The use of the dual formulation of electromagnetism was necessary to have all the physical sources which influence the spacetime distortion.

6.3.7.1 *Electrical Helicity and Spacetime Distortion*

The third row of (6.163) gives an equation to calculate torsion. We observe an analogy with a suggestion of [99] by considering a particular Riemann-Cartan continuum and working with contortion. What is interesting is that the electromagnetic field allows us to calculate torsion. The third equation may be rearranged to express the Chern-Simmons current $\mathcal{H}^{\beta\lambda}A_\mu$ in terms of metric and torsion.

Theorem 6.1. *(Geometry and electromagnetism) Presence of an electromagnetic field induces a torsion field which changes an Einstein spacetime into an Einstein-Cartan spacetime [6]. The torsion field appears related to the geometric structure by:*

$$\boxed{\mathscr{D}_{\mu}^{\beta\lambda}(\mathbf{g},\mathfrak{T}) = -2\epsilon_0\,\chi\left(g^{\beta\rho}g^{\lambda\nu} - g^{\beta\nu}g^{\lambda\rho}\right)A_\mu\nabla_\rho A_\nu} \tag{6.199}$$

Proof. From the third equation of (6.163), we introduce the definition $\mathfrak{F}_{\mu\nu} := \nabla_\mu A_\nu - \nabla_\nu A_\mu$ where A_μ is the quadri-potential of the electromagnetic field. It may be noticed that the contribution of the electromagnetic potential which modifies torsion combined with metric is of second order "$A_\rho \nabla_\alpha A_\beta$". □

Remark 6.46. The idea that torsion is the source of magnetic monopoles was already proposed in e.g. [90], [222], knowing that torsion measures the non-closure of infinitely small loops in a spacetime manifold e.g. [6]. This theorem (6.199) then definitively confirms the link between electromagnetic and spacetime distortion e.g. [263]. More precisely, Chern-Simons current represents magnetic helicity and spin angular momentum of electromagnetic waves, which added to orbital angular momentum gives the total angular momentum e.g. [108].

Remark 6.47. Investigation of electromagnetism interacting with Einstein-Cartan spacetime was conducted by numerous authors for charged and spinning "static" dust (static means no displacement of mass center), and for perfect fluids with spin density e.g. [286]. By analyzing solutions of Maxwell's equations, torsion together with spin of Einstein-Cartan gravitation theory may be suggested as being produced by the electromagnetic field e.g. [295]. Paraphrasing these authors, it was concluded that in the absence of electromagnetism, a body has a vanishing spin density which itself is associated to torsion e.g. [37].

Remark 6.48. What could be relevant is that the torsion field of the Riemann-Cartan continuum can be in principle calculated. In other words and from a physical point of view, torsion is thought to be generated by the electromagnetic field demonstrated here by assuming a simple Lagrangian (6.151).

Remark 6.49. From (6.163), we notice that vacuum electromagnetic energy-momentum has also the same shape as energy-momentum within matter [296]:

$$- T_{\text{em}}^{\lambda\rho} = \frac{1}{4}\mathcal{H}^{\mu\nu}\mathcal{F}_{\mu\nu}\, g^{\lambda\rho} + \frac{\varepsilon_0}{2}\mathcal{F}_{\mu\nu}\left(g^{\mu\lambda}\mathcal{F}^{\rho\nu} + \mathcal{F}^{\mu\lambda}g^{\rho\nu}\right) \tag{6.200}$$

This is a Minkowski (canonical) energy-momentum tensor e.g. [264] for free electromagnetic field in spacetime. For electromagnetic field without sources within a Riemann-Cartan continuum, the Lagrangian (6.151) is slightly extended to:

$$\mathscr{S} := \int_{\mathscr{M}} \mathscr{L}\,\overline{\omega}_n \quad \text{with} \quad \mathscr{L} := -\frac{1}{4}\mathcal{H}^{\mu\nu}\,\mathcal{F}_{\mu\nu} + \frac{1}{2\chi}\mathcal{R} \tag{6.201}$$

with constitutive laws $\mathcal{H}^{\mu\nu}[\mathcal{F}_{\alpha\beta}]$ e.g. [6]. There is a controversy between the version of Minkowski and that of Abraham, not deduced from a Lagrangian. We do not enter into this long debate, which is in the past. We just have to remember that the Minkowski version is defined from Lagrange-Noether conforming to the invariance approach we adopt in this work. Obukhov & Hehl suggested the adoption of the Minkowski-like version (6.200) which is motivated by the Lagrangian axiomatic approach and by the experimental evidence conducted in the past by Walker & Walker (which is based on experimental measurements of a dielectric disk placed in a crossed oscillating radial electric and longitudinal magnetic fields) and James (which is based on a similar experimental jig but with radial electric field and azimuthal magnetic field) [296].

Remark 6.50. In previous equations, torsion seems not to propagate, but it is in fact a propagating variable. In order to account for this propagation, a well-known method would be to add a scalar bilinear term of the covariant derivatives of the torsion e.g. [263] where the trace of torsion $\aleph_\nu := \aleph^\mu_{\nu\mu}$ can be considered as an electromagnetic four-potential, and the skew-symmetric part of the Ricci curvature tensor as proportional to the Faraday strength tensor. For that purpose, the action: $\mathscr{S} := \int_{\mathscr{M}}(\frac{\mathcal{R}}{\chi} + a\,\mathscr{G}_{\mu\nu}\mathscr{G}^{\mu\nu})$ with $\mathscr{G}_{\mu\nu} := \partial_\mu\aleph_\nu - \partial_\nu\aleph_\mu$, where arbitrariness of metric and torsion variations hold, is considered. In this approach the electromagnetic variables are directly deduced from continuum geometry.

Remark 6.51. By proposing Einstein-Cartan spacetime, and working with contortion instead of torsion, a similar and interesting relation to the second row of system (6.163) is obtained in [99] without going through our deductive approach.

We define the Faraday tensor as $\mathscr{F}_{\alpha\beta} := \nabla_\alpha A_\beta - \nabla_\beta A_\alpha$ where the connection $\Gamma^\gamma_{\alpha\beta}$ has torsion. The $U(1)$ gauge invariance of Maxwell's equations may be violated without being cautious with this choice. Indeed by modifying the potential as $A_\beta \to A_\beta + g_{\beta\gamma}\nabla^\gamma\Lambda$ where $\Lambda(x^\mu)$ is an arbitrary function, we get: $\mathscr{F}_{\alpha\beta} = \nabla_\alpha A_\beta - \nabla_\beta A_\alpha - \aleph^\gamma_{\alpha\beta}\nabla_\gamma\Lambda$ where the last term vanishes if and only if the torsion is zero or the function Λ is covariantly uniform. Some previous authors proposed to define $\mathscr{F}_{\alpha\beta} := \overline{\nabla}_\alpha A_\beta - \overline{\nabla}_\beta A_\alpha$ as a Faraday tensor even in Riemann-Cartan continuum e.g. [50], [285]. Further analyses are required in the future. Results in the framework of Riemann-Cartan gravitation (e.g. [137]) may highlight some problems of gauge invariance since the electromagnetic tensor does not satisfy the Lorenz gauge invariance (say $U(1)$ gauge invariance) e.g. [287]. Choosing the Faraday tensor as $\mathscr{F} := d\mathbf{A}$ or equivalently $\mathscr{F}_{\mu\nu} := \overline{\nabla}_\mu A_\nu - \overline{\nabla}_\nu A_\mu = \nabla_\mu A_\nu - \nabla_\nu A_\mu - \aleph^\rho_{\mu\nu} A_\rho$ allows us to obtain a $U(1)$ invariant model but induces the following field equations: $\nabla_\nu \mathscr{F}^{\mu\nu} - 2\aleph^\mu_{\rho\nu}\mathscr{F}^{\rho\nu} = 0$ and $\aleph^\rho_{\mu\nu} = 0$, replacing the first Maxwell equation and the third equation coupling torsion and electromagnetic field, meaning that the torsion is identically zero within the continuum. A next question would be the eventual role of the electric charge on the creation of spacetime torsion that seems to be suggested by the third equation of the system (6.163). This is legitimate since we have assumed but not deduced that the torsion field is equal to zero for the classical Reissner-Nordström metric. Now we have to check if this is only the possible situation.

Remark 6.52. We observe that the system (6.163) governing both gravitation and electromagnetism is highly coupled in Riemann-Cartan physics. The third row of (6.163) is expected to allow us to determine the torsion field by means of the algebraic equation from electromagnetic variables.

6.3.7.2 *Magnetic Helicity and Spacetime Distortion*

By analogy to the electric development, the second pair of Maxwell's equations was obtained by means of the Lagrangian:

$$\mathscr{S} := \int_{\mathscr{M}} \mathscr{L}\,\overline{\omega}_n, \quad \mathscr{L} := \frac{1}{4}\tilde{\mathcal{H}}^{\mu\nu}\,\tilde{\mathfrak{F}}_{\mu\nu} + \frac{1}{2\chi}\mathcal{R} \tag{6.202}$$

where the dual Yang-Mills reduces in fact to the Yang-Mills one when formulated in the three dimensions. The difference is the choice of variable by successively using the Hodge operator and writing constitutive laws. Instead of using the transformation $(\mathbf{E} \to \mathbf{B}), (\mathbf{B} \to -\mathbf{E})$, we prefer to distinguish the two separate steps and therefore we do no have to simplify by $\mathbf{D} = \mathbf{E}$ and $\mathbf{B} = \mathbf{H}$. We think that extension to matter other than the vacuum might be more easily obtained with our method. In addition to previous electrical helicity, magnetic fields also induce a spacetime distortion when coupled with gravitation within an Einstein-Cartan spacetime (and even the magnetic fields are more involved in cosmology with very large structures).

Theorem 6.2. *(Geometry and dual electromagnetism) The presence of an electromagnetic field induces a torsion field and modifies an Einstein spacetime into an Einstein-Cartan spacetime. The torsion field appears related to the geometric structure by:*

$$\boxed{\mathscr{D}_{\mu}^{\beta\lambda}(\mathbf{g}, \mathfrak{T}) = 2\frac{\chi}{\mu_0}\left(g^{\beta\rho}g^{\lambda\nu} - g^{\beta\nu}g^{\lambda\rho}\right)\tilde{A}_{\mu}\nabla_{\rho}\tilde{A}_{\nu}} \tag{6.203}$$

The proof is straightforward as for electric helicity. Despite the fact that we have dealt with the dual formulation, it does not prevent us from noticing that the influence of magnetic fields is often at least as important, if not more so, in the field of cosmology, particularly in their influence on the formation and evolution of the large structures of the Universe e.g. [290], [297]. This will be developed in the next chapter.

Remark 6.53. Some remarks regarding the coupling between gravitation and electromagnetism should be made in the context of teleparallel gravity. A description based on teleparallel gravity in Weitzenböck spacetime, which has non-vanishing torsion but zero curvature, is not the only possible approach to coupling electromagnetism and gravitation as suggested in e.g. [50]. The presence of the torsion in the governing equations is essentially due to the Palatini relation (variation of the curvature), and its consequence is the presence of an additional skew-symmetric term in the variation of the Faraday tensor $\mathfrak{F}^{\nu\mu}$ e.g. [5]. This seems to conform to the finding of Andrade & Pereira [50] who have shown, by using the teleparallel description of gravitation, that an electromagnetic field is also able to produce torsion by reading (6.204) from right to left.

6.3.7.3 *Optical Helicity and Spacetime Distortion*

Electric helicity is related to the linking number of electric lines whereas magnetic helicity is related to the number of magnetic lines linked. What we have shown previously is that both of them influence the distortion of the Einstein-Cartan spacetime. We have seen not only the possibility of their coupling but also the principle of determination of torsion in terms of the electromagnetic strength of Faraday. The torsion field and by extension the spacetime distortions are respectively produced according to the relationship:

$$\mathscr{D}_{\mu}^{\beta\lambda(e)}(\mathbf{g},\mathcal{T}) = -2\chi\ \mathcal{H}^{\beta\lambda}\ A_{\mu}, \quad \mathscr{D}_{\mu}^{\beta\lambda(m)}(\mathbf{g},\mathcal{T}) = 2\chi\ \tilde{\mathcal{H}}^{\beta\lambda}\ \tilde{A}_{\mu} \tag{6.204}$$

within Einstein-Cartan vacuum spacetime. A systematic investigation of the 4-potential $A_{\mu} := (\phi, A_1, A_2, A_3)$ and the dual potential $\tilde{A}_{\mu}$ should be conducted in the future. Both of them are governed by the first and second pair of Maxwell's equations respectively.

To construct a Lagrangian formalism that includes the two pairs of Maxwell's equations with the two Lagrangians, that of Yang-Mills and its dual, some authors e.g. [288], [298] suggested a combination of the actions $\mathscr{S}_{YM}$ and $\tilde{\mathscr{S}}_{YM}$ for modeling the light propagation. It is interesting to use one Lagrangian to derive the two pairs of Maxwell's equations. We propose therefore the symmetrized Lagrangian combined with the Einstein-Palatini action. Symmetry between electric and magnetic fields is an intrinsic property of Maxwell's equations in vacuum spacetime free of sources. Taken separately, standard formulation leads to first pair of Maxwell's equations ("electric part"), whereas dual formulation arrives at the second pair ("magnetic part"). Considering the symmetrized Yang-Mills Lagrangian is not only cosmetically dictated by the inclusion of the two pairs of equations in one Lagrangian formulation. It is more motivated by the very basic meaning of Maxwell electromagnetism where the two fields are closely interweaved and form one electromagnetic field. Taken separately, the standard and dual Yang-Mills action present asymmetry. The seek of symmetry in the formulation is closely related to conservation of optical (electromagnetic) helicity and, as we will show here below, is closely related to the separation of spin and orbital degrees of freedom of the propagating light.

Definition 6.15. Within an Einstein-Cartan spacetime $\mathscr{M}$, we define the action $\mathscr{S}_{SYM}$ (symmetrized Yang-Mills action) for modeling the electromagnetic field in vacuum spacetime:

$$\mathscr{S}_{SYM} : -\int_{\mathscr{M}} \mathscr{L}\ \overline{\omega}_n, \quad \mathscr{L} = -\frac{1}{8}(\mathcal{H}^{\mu\nu}\ \mathfrak{F}_{\mu\nu} - \tilde{\mathcal{H}}^{\mu\nu}\ \tilde{\mathfrak{F}}_{\mu\nu}) \tag{6.205}$$

with the associated Faraday tensors and the constitutive laws:

$$\begin{cases} \mathcal{F}_{\mu\nu} = \nabla_\mu A_\nu - \nabla_\nu A_\mu \ , \ \mathcal{H}^{\mu\nu} = \epsilon_0 \ g^{\mu\lambda} g^{\nu\rho} \ \mathcal{F}_{\lambda\rho} \\ \tilde{\mathcal{F}}_{\mu\nu} = \nabla_\mu \tilde{A}_\nu - \nabla_\nu \tilde{A}_\mu \ , \ \tilde{\mathcal{H}}^{\mu\nu} = \dfrac{1}{\mu_0} \ g^{\mu\lambda} g^{\nu\rho} \ \tilde{\mathcal{F}}_{\lambda\rho} \end{cases} \tag{6.206}$$

Remark 6.54. We have two electromagnetic potentials A_μ and $\tilde{A}_\mu$. The Lagrangian reduces to classic Lagrangian in a three-dimensional formulation:

$$\mathscr{L} = \frac{1}{2} (\mathbf{D} \cdot \mathbf{E} - \mathbf{B} \cdot \mathbf{H}) \tag{6.207}$$

The variational method applied to symmetrized Yang-Mills action allows us to obtain the two pairs of Maxwell's equations within a Minkowski spacetime.

Definition 6.16. We define the action of symmetrized electromagnetic and gravitational fields of a vacuum Einstein-Cartan spacetime $\mathscr{M}$ which is given by:

$$\mathscr{S}_{GEMS} := \int_{\mathscr{M}} \mathscr{L} \ \overline{\omega}_n, \quad \mathscr{L} := \frac{\mathcal{R}}{2\chi} - \frac{1}{8} (\mathcal{H}^{\mu\nu} \ \mathcal{F}_{\mu\nu} - \tilde{\mathcal{H}}^{\mu\nu} \ \tilde{\mathcal{F}}_{\mu\nu}) \tag{6.208}$$

Since most terms were already calculated previously, it is straightforward to obtain, by means of the variation of total action, the field equations governing the gravito-electromagnetism of the vacuum spacetime.

$$\delta \mathscr{S}_{GEMS} = 0, \quad \forall \delta A_\mu, \delta \tilde{A}_\mu, \delta g^{\mu\nu}, \delta \mathcal{T}^{\lambda}_{\beta\lambda}$$

We deduce:

Gravitation and electromagnetism in Einstein-Cartan spacetime. The system of equations governing gravitation and electromagnetism within an Einstein-Cartan vacuum spacetime is given by:

$$\begin{cases} \overline{\nabla}_\mu \mathcal{H}^{\mu\nu} + \mathcal{T}^{\nu}_{\mu\rho} \ \mathcal{H}^{\mu\rho} = 0 \\ \overline{\nabla}_\mu \tilde{\mathcal{H}}^{\mu\nu} + \mathcal{T}^{\nu}_{\mu\rho} \ \tilde{\mathcal{H}}^{\mu\rho} = 0 \\ \dfrac{1}{2\chi} \ G_{\mu\nu} + T^{\text{em}}_{\mu\nu} = 0 \\ \dfrac{1}{2\chi} \mathscr{D}^{\beta\lambda}_{\mu} + \dfrac{1}{2} \left(\mathcal{H}^{\beta\lambda} \ A_\mu - \tilde{\mathcal{H}}^{\beta\lambda} \ \tilde{A}_\mu \right) = 0 \end{cases} \tag{6.209}$$

where unknowns are A_μ, $\tilde{A}_\mu$, $g^{\mu\nu}$, and $\mathcal{T}^{\lambda}_{\beta\lambda}$.

With this approach, we observe in the first and second rows that the two pairs of Maxwell's equations are deduced with one variation equation. Due to equality of the two Maxwell stress-energy tensors $T^{\text{em}}_{\mu\nu} = \tilde{T}^{\text{em}}_{\mu\nu}$, we obtain easily the second row expressing the Einstein-Cartan-Maxwell. The third row is obtained since the two Maxwell stress-energy tensors are equal and of opposite sign. The vacuum spacetime distortion includes the electric and magnetic helicity to give the optical helicity. The three-dimensional vector formulation is very illuminating for the physical interpretation (see e.g. [107]):

$$\frac{1}{2}\left(\mathcal{H}^{\beta\lambda}\ A_{\mu} - \tilde{\mathcal{H}}^{\beta\lambda}\ \tilde{A}_{\mu}\right) = \frac{1}{2}\begin{pmatrix} \mathbf{B}\cdot\tilde{\mathbf{A}} - \mathbf{D}\cdot\mathbf{A} \\ \mathbf{E}\times\tilde{\mathbf{A}} + \mathbf{H}\times\mathbf{A} \end{pmatrix} := \begin{pmatrix} \mathcal{H} \\ \mathcal{S} \end{pmatrix} \tag{6.210}$$

where $\mathcal{H}$ and $\mathcal{S}$ are optical helicity and the spin angular momentum respectively e.g. [288], [289].

Remark 6.55. For basic insights of electromagnetism, we have two families of physical lines e.g. [216], [252]. On the one hand, *electric lines* can be defined as spatial curves so that the tangent at any point to it gives the direction of the electric field intensity at that point. On the other hand, *magnetic lines* can be defined as spatial curves so that the tangent at any point to it gives the direction of the magnetic field intensity at that point. For a vacuum spacetime where the electric displacement and the magnetic induction are respectively proportional to the electric field and magnetic field, we can use these definitions to have a physical interpretation of the optical helicity $\mathcal{H}$ in (6.210): Magnetic helicity is related to the linkage of magnetic lines whereas electric helicity is related to the linking number of electric lines e.g. [291]. We remark that some authors suggest an opposite sign for the electric helicity but *in fine* merges to the same physical interpretation e.g. [299].

Both magnetic and electric helicities are classically conserved quantities of electromagnetic fields where the domains of interest (portions of space) are governed by vector fields whose evolutions are described by volume-preserving diffeomorphims, and more precisely by topological transformations under typical boundary conditions. This aspect was extensively studied and fundamental results have been obtained since the original findings of Moreau [300] and independently earlier by [351] about the helicity invariance of any vector field under some regular transformations. For our case, the question is whether helicity invariance can be extended to topological volume-preserving homeomorphisms. Some authors recently answered such a question affirmatively for flows without rest points e.g. [301]. In the present work, we are working within a Riemann-Cartan manifold and these conditions may change results. Indeed, this is a key point since the evolution of the Universe may not be governed solely by diffeomorphisms due to the nucleation and development of spacetime defects during cosmic evolution. This aspect might be approached by the third linking equation relating distortion $\mathscr{D}_{\mu}^{\beta\lambda}$ with helicities $\mathcal{H}$ and spin angular momentum $\mathcal{S}$.

Spacetime distortion, optical helicity, and spin angular momentum. With this new equation (6.209) we have a new fact: optical helicity (including magnetic and electric helicity) and the spin angular momentum of electromagnetic radiation are directly linked to the distortion (torsion) of vacuum spacetime.

6.3.7.4 *Electromagnetic Fields Bend and Twist the Spacetime*

The third row of system (6.53) is the field equation which extends Einstein's equation to Einstein-Cartan spacetime, where we recognize the Maxwell energy-momentum (6.51):

$$T_{\mu\nu}^{\text{em}} = \frac{1}{8}\left(\mathcal{H}^{\lambda\rho}\mathcal{F}_{\lambda\rho}\right) g_{\mu\nu} + \frac{1}{4}\mathcal{H}^{\lambda\rho}\left(g_{\mu\lambda}\mathcal{F}_{\rho\mu} + \mathcal{F}_{\mu\lambda}g_{\rho\nu}\right) = \tilde{T}_{\mu\nu}^{\text{em}} \tag{6.211}$$

It is the Minkowski energy-momentum tensor due to the electromagnetic field. It modifies the gravitational field as a source whereas the spacetime modifies the electromagnetic field according to (6.67). The temporal component of the energy-momentum (6.211) holds: $T^{00} = \frac{1}{2}\left(D^i E_i + B^i H_i\right)$, which is exactly the electromagnetic energy density

$$T^{00} = (1/2)\left(\mathbf{D}\cdot\mathbf{E} + \mathbf{B}\cdot\mathbf{H}\right) := \mathscr{E}$$

in a three-dimensional formulation. By introducing the electromagnetic tensors into the expression of the energy-momentum tensor, we have the Minkowski tensor $T^{\mu\nu}$ which is not symmetric when considering the time index 0. It is worth expressing the Minkowski energy momentum as:

$$T_M^{\mu\nu} = \begin{bmatrix} \mathscr{E} & \mathbf{E}\times\mathbf{H} \\ \mathbf{D}\times\mathbf{B} & \mathbf{T}^M \end{bmatrix} \tag{6.212}$$

where $\mathscr{E}$ is the energy and $\mathbf{T}^M$ is the three-dimensional Maxwell tensor.

Remark 6.56. The Poynting vector $\mathbf{S} := \mathbf{E} \times \mathbf{H}$ (originally discovered by JH Poynting in 1884) represents the vector rate of energy. However, it is known that Maxwell's equations admit only the fundamental fields $\mathbf{E}$ and $\mathbf{B}$, without considering any model of medium where electromagnetic fields occur. In such a case, the Poynting vector is rigorously defined as $\mathbf{S} = \frac{1}{\mu_0}\mathbf{E}\times\mathbf{B}$, where μ_0 is the vacuum permeability (Minkowski spacetime). This is considered as its general expression. The moment of the Poynting vector $\mathbf{J} := \int \mathbf{x}\times\mathbf{S}dv$ is defined as the angular momentum including orbital and spin angular momentum e.g. [109].

We relate electromagnetic fields $T^{\lambda\rho}$ as sources of spacetime bending, so we can multiply by the covariant components of the metric $g_{\lambda\rho}$ to obtain without difficulty the Ricci curvature and then spacetime curvature:

$$-\overline{\mathcal{R}} = \mathscr{T} := g_{\lambda\rho}T_M^{\lambda\rho} = -2\chi\mathcal{F}_{\lambda\rho}\mathcal{H}^{\lambda\rho} = \chi\left(\mathbf{D}\cdot\mathbf{E} - \mathbf{B}\cdot\mathbf{H}\right) \qquad (6.213)$$

which is exactly χ times twice of Yang-Mills Lagrangian.

Remark 6.57. However, the goal is not reached because electromagnetic variables should be introduced as additional variables in the action. They are not directly described by spacetime geometry. We should add the electromagnetic potential $\mathbf{A}$ (or the electromagnetic Faraday strength $\mathscr{F}$) as arguments of the Yang-Mills action.

6.4 Geodesics and Deviations in Einstein-Cartan Spacetime

In relative gravitation, the Jacobi equation makes it possible to calculate the geodesic deviation, for example the evolution of the separation of two test particles moving in a gravitational field (see (5.111)). It is implicitly accepted that each of these particles with their respective masses does not have other structures such as spin or even electric charges. The ambient spacetime therefore remains that of Einstein. The next step is the inclusion of electromagnetism in relative gravitation. Generalization of geodesic deviation from Riemann to Riemann-Cartan manifold was already done in the past e.g. [302], without reference to any electromagnetic fields, but in a post-Riemannian geometric structure of spacetime. The aim of this section is to derive equations of geodesics deviation within Einstein-Cartan spacetime.

6.4.1 *Gravitation and Electromagnetism at Human Length Scale*

Similarly to mass which creates gravitation interaction, electric charge and its motion create an electromagnetic field and then generates forces.[4] Attempts to model gravito-electromagnetism suggested the use of curved spacetime with non-zero torsion e.g. [6], [99], [100]. Einstein-Cartan spacetime was already used to model this coupling e.g. [50], [100], [284], [286], to name but a few. Introduction of torsion as an additional variable extends the field of applications to complex spacetime in cosmology e.g. [4]. Motivated by recent discoveries on acceleration of the Universe expansion, where dark matter and dark energy are suggested to be connected with torsion e.g. [38], [89], some authors lean upon the concept of geodesics and autoparallel curves to establish their theory e.g. [54]. As reminded in e.g. [248], interrelation between gravitation and electromagnetism by means of their unified background geometry as metric and connection is an age-old problem, where Reichenbach (1929) already introduced connection coefficients and their partial derivatives as arguments of the Lagrangian. In the present work, we focus again on the use of Riemann-Cartan manifolds to model the coupling of gravitation and electromagnetism e.g. [6], [39].

6.4.2 *Autoparallel Deviation Equation*

Research of particle trajectory within gravitational and electromagnetic fields has been longstanding since the first developments of relativity. Accordingly, geodesic deviation including the influence of electric charge on test particle motions was studied e.g. [105]. Physical interest may thus be also found in studying the separation of two small electrically charged particles subject to gravitation and electromagnetism e.g. [6], [99]; the idea is now to detect the relative acceleration of two nearby particles when the spacetime is curved with torsion.

First of all, it is worth recalling that in any spacetime the deviation from an autoparallel curve is obtained from the definition of the deviation equation of two autoparallel curves (5.70):

$$\frac{D^2\xi^\alpha}{D\tau^2} = u^\gamma \nabla_\gamma \left(u^\beta \nabla_\beta \xi^\alpha\right) \tag{6.214}$$

[4]The term autoparallels may be replaced by geodesics but with caution in this work. Sometimes they are also called worldlines.

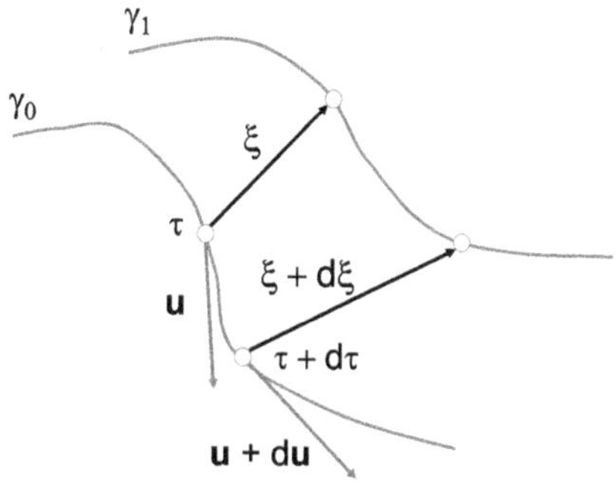

Fig. 6.4: Autoparallel curves. The gap vector ξ separating autoparallel timelike curves respectively γ_0 and γ_1, defined by $\nabla_{\mathbf{u}}\mathbf{u} = 0$.

In a previous work, [104] proposed a deviation equation by defining *a priori* a deviation operator of the connection in the presence of torsion and curvature $\mathcal{L}\Gamma(\xi, \mathbf{u}) := [\mathcal{L}_x i, \nabla_{\mathbf{u}}] - \nabla_{[\xi,\mathbf{u}]}$ to extend the deviation equation to Riemann-Cartan spacetime. Intrinsic definition (6.220) can be used to calculate the deviation within either Riemann (5.111) or Riemann-Cartan manifold.

Theorem 6.3. *Let $(\mathscr{M}, \mathbf{g}, \nabla)$ be a Einstein-Cartan spacetime with $u^\beta := dx^\beta/d\tau$, four-vector velocity (timelike vector), and ξ^α the separation between two autoparallel curves γ_0, and γ_1 (Fig. 6.4). We assume the Lie derivative of u^α along ξ vanishes (as for classical assumption in relative gravitation) $\mathcal{L}_\xi u^\alpha \equiv 0$. Then the acceleration of the separation between γ_0 and γ_1 takes the form of [6]:*

$$\boxed{\frac{D^2\xi^\alpha}{D\tau^2} = \tilde{\Re}^\alpha_{\gamma\beta\rho} u^\gamma \xi^\beta u^\rho + \left(\tilde{\aleph}^\rho_{\beta\gamma} \tilde{\nabla}_\rho u^\alpha - \tilde{\aleph}^\alpha_{\gamma\rho} \tilde{\nabla}_\beta u^\rho\right) u^\gamma \xi^\beta} \tag{6.215}$$

where the connection $\tilde{\Gamma}^\gamma_{\alpha\beta} := \Gamma^\gamma_{\alpha\beta} - \aleph^\gamma_{\alpha\beta}$ is defined from the connection $\Gamma^\gamma_{\alpha\beta}$ by substracting its proper torsion.

Proof. First, we express the two assumptions of the theorem. The vanishing Lie derivative in a Riemann-Cartan manifold writes:

$$\mathcal{L}_\xi u^\alpha \equiv 0 \quad \Longrightarrow \quad \xi^\beta \nabla_\beta = \xi^\beta \nabla_\beta u^\alpha - \xi^\beta \aleph^\alpha_{\beta\mu} u^\mu \tag{6.216}$$

Second, the definition of autoparallel curves holds (for any index α):

$$u^\gamma \nabla_\gamma u^\alpha = 0 \tag{6.217}$$

Introducing these two equations into the definition (6.220) gives:

$$\frac{D^2\xi^\alpha}{D\tau^2} = \xi^\gamma \nabla_\gamma u^\beta \left(\nabla_\beta u^\alpha - \aleph^\alpha_{\beta\mu} u^\mu\right) + \xi^\gamma \aleph^\beta_{\gamma\rho} u^\rho \left(\aleph^\alpha_{\beta\mu} u^\mu - \nabla_\beta u^\alpha\right) + u^\gamma \xi^\beta \nabla_\gamma \nabla_\beta u^\alpha - u^\gamma \xi^\beta \nabla_\gamma \left(\aleph^\alpha_{\beta\mu} u^\mu\right)$$

Factorizing the covariant derivative allows us to rewrite the equation as:

$$\frac{D^2\xi^\alpha}{D\tau^2} = \xi^\gamma \nabla_\gamma u^\beta \left(\nabla_\beta u^\alpha - \aleph^\alpha_{\beta\mu} u^\mu\right) + \xi^\gamma \aleph^\beta_{\gamma\rho} u^\rho \left(\aleph^\alpha_{\beta\mu} u^\mu - \nabla_\beta u^\alpha\right) + u^\gamma \xi^\beta \nabla_\gamma \left(\nabla_\beta u^\alpha - \aleph^\alpha_{\beta\mu} u^\mu\right)$$

Let's now define a new connection $\tilde{\nabla} := \nabla - \aleph$, which remains a connection on $\mathscr{M}$. Assuming metric compatibility of the connection, we can write:

$$\tilde{\Gamma}^\gamma_{\alpha\beta} := \overline{\Gamma}^\gamma_{\alpha\beta} + \mathfrak{T}^\gamma_{\alpha\beta} - \aleph^\gamma_{\alpha\beta} = \overline{\Gamma}^\gamma_{\alpha\beta} + D^\gamma_{\alpha\beta} - \Omega^\gamma_{\alpha\beta}$$

The skew-symmetric part of the new connection is merely $\tilde{\aleph}^\gamma_{\alpha\beta} = -\aleph^\gamma_{\alpha\beta}$. We rewrite the separation acceleration as:

$$\frac{D^2\xi^\alpha}{D\tau^2} = \xi^\gamma \tilde{\nabla}_\gamma u^\beta \tilde{\nabla}_\beta u^\alpha + u^\gamma \xi^\beta \tilde{\nabla}_\gamma \tilde{\nabla}_\beta u^\alpha - \tilde{\aleph}^\alpha_{\gamma\rho} (\tilde{\nabla}_\beta u^\rho) u^\gamma \xi^\beta \tag{6.218}$$

Let's recall the relation about second-order derivatives for either ∇ or $\tilde{\nabla}$:

$$\nabla_\gamma \nabla_\beta u^\alpha - \nabla_\beta \nabla_\gamma u^\alpha = \mathfrak{R}^\alpha_{\gamma\beta\rho} u^\rho + \aleph^\mu_{\beta\mu} \nabla_\mu u^\alpha \tag{6.219}$$

It is straightforward to deduce the relation, by exploiting the Leibniz relation for derivative of products and by rearranging the autoparallel equation:

$$\frac{D^2\xi^\alpha}{D\tau^2} = \tilde{\mathfrak{R}}^\alpha_{\gamma\beta\rho} u^\gamma \xi^\beta u^\rho + (\tilde{\aleph}^\rho_{\beta\gamma} \tilde{\nabla}_\rho u^\alpha - \tilde{\aleph}^\alpha_{\gamma\rho} \tilde{\nabla}_\beta u^\rho) u^\gamma \xi^\beta \qquad \square$$

Remark 6.58. Curvature and torsion occur in a linear fashion in (6.215). The first term is proportional to curvature while the second term is proportional to torsion. However, in this form, the correlation with gravitation and electromagnetism still remains unclear.

6.4.3 *Geodesics in the Presence of Gravitation and Electromagnetism*

The component form of the geodesic deviation equation does not allow us to point out the coupled effects of gravitation and electromagnetism. It remains difficult to give an easy physical interpretation to (6.215).

6.4.3.1 *Coordinate-Free Expression of the Jacobi Equation*

The idea is now to derive an intrinsic formulation. First of all, the deviation from an autoparallel curve is obtained from the intrinsic definition (5.70):

$$\frac{D^2\xi}{D\tau^2} := \nabla_{\mathbf{u}}\left(\nabla_{\mathbf{u}}\xi\right) \tag{6.220}$$

Let's now particularize to Riemann-Cartan spacetime.

Theorem 6.4. *Let* $(\mathscr{M}, \mathbf{g}, \nabla)$ *be a Riemann-Cartan spacetime with* $\mathbf{u} := d\mathbf{x}/d\tau$*, four-vector velocity (timelike vector), and* ξ *the separation between two autoparallel curves* γ_0*, and* γ_1*. We assume the Lie derivative of* ξ *along* $\mathbf{u}$ *vanishes (as for classical assumption in relative gravitation)* $\mathcal{L}_{\mathbf{u}}\xi \equiv 0$*. Then the acceleration of the separation between two autoparallel curves* γ_0 *and* γ_1 *holds:*

$$\frac{D^2\xi}{D\tau^2} = \Re\left(\mathbf{u}, \xi, \mathbf{u}\right) + \aleph\left(\frac{D\xi}{D\tau}, \mathbf{u}\right) + \nabla_{\mathbf{u}}\aleph\left(\xi, \mathbf{u}\right) \tag{6.221}$$

extending (5.111) from Riemann to Riemann-Cartan spacetime.

Proof. Relation (5.109) induces $\nabla_{\mathbf{u}}\nabla_{\xi}\mathbf{u} = \Re\left(\mathbf{u}, \xi, \mathbf{u}\right)$. Vanishing of the Lie derivative also holds $\nabla_{\mathbf{u}}\xi = \nabla_{\xi}\mathbf{u} + \aleph\left(\xi, \mathbf{u}\right)$. We deduce accordingly:

$$\begin{aligned}\frac{D^2\xi}{D\tau^2} &= \nabla_{\mathbf{u}}\left(\nabla_{\mathbf{u}}\xi\right) = \nabla_{\mathbf{u}}\left(\nabla_{\xi}\mathbf{u} + \aleph\left(\xi, \mathbf{u}\right)\right) \\ &= \Re\left(\mathbf{u}, \xi, \mathbf{u}\right) + \nabla_{\mathbf{u}}\left[\aleph\left(\xi, \mathbf{u}\right)\right]\end{aligned}$$

For the sake of clarity, we calculate separately the last term:

$$\nabla_{\mathbf{u}}\left[\aleph\left(\xi, \mathbf{u}\right)\right] = \left(\nabla_{\mathbf{u}}\aleph\right)\left(\xi, \mathbf{u}\right) + \aleph\left(\nabla_{\mathbf{u}}\xi, \mathbf{u}\right) + \aleph\left(\xi, \nabla_{\mathbf{u}}\mathbf{u}\right)$$

where the third last term vanishes since $\mathbf{u}$ is tangent to the geodesic. Given that $\nabla_{\mathbf{u}}\xi := D\xi/D\tau$, we get the result. □

Remark 6.59. Coordinate-free equation constitutes a convenient alternative expression for autoparallel deviation curves. Conversely to (6.215), the covariant derivative of torsion occurs in a coordinate-free equation (6.221). We will see hereafter that this alternative way may be very helpful to relate results to gravitational and electromagnetic fields.

Remark 6.60. In the 1980s, another coordinate-free form of the Jacobi equation for autoparallel curves (including curvature and torsion) was proposed [103]. Their results can be applied to various gravitation theories such as Einstein-Cartan, but (6.221) seems more tractable for practical applications, and uses exclusively curvature and torsion as operators.

6.4.3.2 *Geodesic Deviation and Electromagnetism*

Each term in (6.221) has its physical interpretation when identified with results of [34], [105], where torsion $\aleph$ which is proportional to Faraday tensor $\mathscr{F}$ appears as well as its covariant derivative $\nabla\aleph$ related to $\nabla\mathscr{F}$. Assuming a Riemann-Cartan manifold allows us to entirely include in the geometric spacetime gravitation and electromagnetism. [99] suggested that the spacetime torsion is due to the spin of angular momentum. Indeed, by considering the relation between torsion and electromagnetism as deduced in the book [6], it is hopefully guessed that the last two terms of (6.221) represent forces due to electromagnetism [105]. As an illustration, the geodesic deviation equation in the presence of gravitation and electrically charged particles is given by [105]:

$$\frac{D^2\xi^\alpha}{D\tau^2} = \overline{\Re}^\alpha_{\gamma\beta\rho}\ u^\gamma\xi^\beta u^\rho + \frac{q}{m}\ \mathfrak{F}^\alpha_\nu \frac{D\xi^\nu}{D\tau} + \frac{q}{m}\ \overline{\nabla}_\rho\mathfrak{F}^\alpha_\nu\ u^\nu\xi^\rho \tag{6.222}$$

where m, q, and $\mathfrak{F}$ are the mass and electric charge of particles and $\mathfrak{F}$ is the Faraday strength field. The similarity of (6.222) with (6.221) is striking. It obviously provides a strong support to utilize the Einstein-Cartan spacetime in the presence of electromagnetism, instead of Einstein spacetime.

Remark 6.61. Despite the encouraging similarity of (6.222) and (6.221), it is however troubling that (6.221) does not include the value of mass m nor the value of charge q of the test particle. It confirms that this relation is of a purely geometric character of Einstein-Cartan spacetime. The duality of light-matter probably has its words to say.

6.4.4 *Geodesics in an Anti-de Sitter Spacetime*

A uniform distribution of torsion $\aleph_0$ leads to an anti-de Sitter spacetime with metric (6.135) with nevertheless a difference that the obtained spacetime has non-vanishing torsion, acting as a vacuum energy (cosmological constant).

6.4.4.1 *Geodesic in Einstein-Cartan Spacetime*

For calculating the four-vector u^γ, we recall the definition of autoparallel curves which is also valid for Einstein-Cartan spacetime. Autoparallel curves are the integral curves of the differential equations e.g. [216]:

$$\frac{Du^\gamma}{D\tau} := \frac{du^\gamma}{d\tau} + \Gamma^\gamma_{\mu\nu} u^\mu u^\nu = 0 \tag{6.223}$$

The skew-symmetric part of the connection is not involved in (6.223). The system, after decomposing the torsion into Levi-Civita connection and contortion tensor e.g. [5], becomes:

$$\frac{Du^\gamma}{D\tau} := \frac{du^\gamma}{d\tau} + \overline{\Gamma}^\gamma_{\mu\nu} u^\mu u^\nu + D^\gamma_{\mu\nu} u^\mu u^\nu = 0 \tag{6.224}$$

where $D^\gamma_{\mu\nu}$ is the symmetric part of the contortion $\mathcal{T}^\gamma_{\mu\nu}$ with respect to the two lower indices. This relation involves 40 coefficients $\overline{\Gamma}^\gamma_{\mu\nu}$ and another 40 components for $D^\gamma_{\mu\nu}$. Symmetries may reduce the number of components.

Remark 6.62. Extension of the deviation equation to Riemann-Cartan spacetime was already done in the past e.g. [104]. In the present work, we slightly modify the formulation by highlighting a second term which is expressed explicitly in terms of torsion. Basically (6.221) includes two terms: the first corresponds to bending of spinless particles moving under gravitation, whereas the second term acts on particles as a consequence of spacetime torsion. It is observed that for a non-curved spacetime, but with torsion, there is a separation of autoparallel curves. At least from a theoretical point of view, the possibility to detect torsion appears in (6.221). This possibility was sketched as a perspective in [101] where the covariant form of the relativistic top deviation was stated, in the presence of spin. Extension of the deviation equation explicitly demonstrates that even in a non-curved spacetime, torsion influences the path of spinning particles. This gives an answer to the third argument advocated by Manoff's paper [104].

The Lagrangian method we previously used is not applicable for Einstein-Cartan spacetime since the symmetric part of the contortion does not vanish. Integral curves are defined by connection coefficients from both the metric and the symmetric part of contortion (6.124):

$$\begin{cases} g_{00} = \left(1 - \dfrac{2m}{r} + \dfrac{\aleph_0^2}{12} r^2\right) \\ g_{11} = -\left(1 - \dfrac{2m}{r} + \dfrac{\aleph_0^2}{12} r^2\right)^{-1} \end{cases}, \quad \begin{cases} g_{22} = -r^2 \\ g_{33} = -r^2 \sin^2\theta \end{cases} \tag{6.225}$$

and all the connection coefficients calculated as follows.

(1) From the Levi-Civita connection analogous to previous developments, we use again the coefficients:

$$\Gamma^0_{01} = \Gamma^0_{10} = \nu'$$

$$\Gamma^1_{00} = \nu' \, e^{2\nu - 2\mu}, \quad \Gamma^1_{11} = \mu', \quad \Gamma^1_{22} = -r \, e^{-2\mu}, \quad \Gamma^1_{33} = -r \, e^{-2\mu} \sin^2\theta$$

$$\Gamma^2_{12} = \Gamma^2_{21} = \frac{1}{r}, \quad \overline{\Gamma}^2_{33} = -\sin\theta\cos\theta$$

$$\Gamma^3_{13} = \Gamma^3_{31} = \frac{1}{r}, \quad \Gamma^3_{23} = \Gamma^3_{32} = \cot\theta$$

in which we introduce the function:

$$e^{2\nu} = \left(1 - \frac{2m}{r} + \frac{\Lambda}{3} r^2\right) = e^{-2\mu}$$

where we have exactly the same shape of function than for an anti-de Sitter metric if we replace Λ by $\aleph_0^2/4$;

(2) From the ansatz torsion, we calculate the remaining coefficients:

$$\Gamma^0_{23} = \frac{\aleph_0}{2} - \Gamma^0_{32}, \quad \Gamma^2_{03} = -\frac{e^{2\nu}}{r^2}\frac{\aleph_0}{2} = \Gamma^2_{30}, \quad \Gamma^3_{02} = \frac{e^{2\nu}}{r^2 \sin^2\theta}\frac{\aleph_0}{2} = \Gamma^3_{20} \tag{6.226}$$

Conversely to anti-de Sitter spacetime, connection is non-symmetric in the present spacetime which couples gravitation with electromagnetism.

Non-symmetric coefficients of the connection are introduced into the relation (6.224) to derive the geodesic equation:

$$\frac{Du^\gamma}{D\tau} := \frac{du^\gamma}{d\tau} + \overline{\Gamma}^\gamma_{\mu\nu} u^\mu u^\nu + D^\gamma_{\mu\nu} u^\mu u^\nu = 0 \tag{6.227}$$

Identification of skew-symmetric part $\Omega^\gamma_{\mu\nu}$ and symmetric part $D^\gamma_{\mu\nu}$ is obvious from relations (6.226). Only the skew-symmetric contribution of the contortion was dropped since we remark that the symmetric part plays a role in addition to the Levi-Civita connection. In general, relation (6.227) shows that geodesics should not be confused with autoparallel curves due to the presence of the symmetric part of the contortion $D^\gamma_{\mu\nu} u^\mu u^\nu$. This seems to contradict the assumption in e.g. [54] for the explanation of dark energy and dark matter by torsion fields. Illustration with an example will be given hereafter.

6.4.4.2 *Geodesic Equations*

Given the complete set of connection coefficients, we are now able to derive the system of equations governing the geodesic in this example with the function previously obtained $e^{2\nu} = 1 - (2m/r) + (\aleph_0^2/3)\, r^2$. Expression (6.227) may be nevertheless incomplete and may induce errors, and should be replaced by a first-order differential system (directly deduced from the original definition of geodesics $\nabla_{\mathbf{u}}\mathbf{u} = 0$):

$$\boxed{\left\{\begin{aligned} \frac{du^0}{d\tau} + 2\nu'\,(u^0u^1) &= 0 \\ \frac{du^1}{d\tau} + \nu'\, e^{4\nu}(u^0)^2 - \nu'(u^1)^2 - re^{2\nu}(u^2)^2 - r\sin^2\theta e^{2\nu}(u^3)^2 &= 0 \\ \frac{du^2}{d\tau} + \frac{2}{r}\,(u^1u^2) - \sin\theta\cos\theta\,(u^3)^2 - \aleph_0 e^{\nu}\sin\theta\,(u^0u^3) &= 0 \\ \frac{du^3}{d\tau} + \frac{2}{r}\,(u^1u^3) + 2\cot\theta\,(u^2u^3) + \frac{\aleph_0 e^{\nu}}{\sin\theta}(u^0u^2) &= 0 \end{aligned}\right.} \tag{6.228}$$

Autoparallel curves are solutions of system (6.228), which are an extension of systems (5.138), (6.84) and (5.166). For the last three equations, only function $e^{2\nu}$ differs. Moreover, it should be pointed out that geodesic equation (5.166) is not the same as (6.228) despite the fact that both of them are associated with $\mathscr{A}d\mathscr{S}$ spacetime. Contribution of the electromagnetic field is highlighted in the last system. The difference between geodesics and autoparallel curves comes from $\aleph_0$.

6.4.4.3 *Geodesic Deviation Equations*

Considering an Einstein-Cartan spacetime, the tidal forces include two major contributions. Since all the (asymmetric) connection coefficients are available, we can recall the autoparallel deviation equation (6.221) as:

$$\frac{D^2\xi}{D\tau^2} = \underbrace{\Re\,(\mathbf{u},\xi,\mathbf{u})}_{\text{Dominant RG}} + \underbrace{\aleph\left(\frac{D\xi}{D\tau},\mathbf{u}\right) + \nabla_{\mathbf{u}}\aleph\,(\xi,\mathbf{u})}_{\text{Dominant EM}} \tag{6.229}$$

Two groups of physical phenomena govern the deviation autoparallel curves in (6.229). The first term is gravitationally dominated, while the second and third terms are mainly due to electromagnetism. It is clear that in the first term if we isolate the Riemann curvature, the contribution of the contortion

$\mathfrak{T}^{\gamma}_{\alpha\beta}$ also includes electromagnetic phenomena implicitly (see (6.114)), and in the last two terms, the implicit presence of the metric in the connection shows the influence of gravitation. That is the reason the two groups are called "dominant relative gravitation" and "dominant electromagnetism". The last two terms vanish in an Einstein spacetime, and only appear when we extend the spacetime by including the torsion. This enlargement has also many other implications in cosmology e.g. [4]. Equation (6.229) states that geodesics are different from autoparallel curves, and an attempt to compare the two notions, for instance in e.g. [54], must involve more precision in their respective definitions.

Remark 6.63. A complete resolution of geodesic structures in anti-de Sitter spacetime can be found in e.g. [44]. In particular, they study a black hole where radial and non-radial trajectories were obtained. Namely, it has also been shown that geodesics of anti-de Sitter spacetimes present differences compared to Schwarzschild geodesics e.g. [41].

Considering the influence of torsion in spacetime models remains an open question even in classical physics despite a large number of published works e.g. [4], [39], [53], [156], [217], [218], [303], [304], and particularly by suggesting a relation between torsion and the cosmological constant e.g. [38], [280]. Focusing on gravitation and electromagnetic interaction, it was already observed by e.g. [216] that the third equation of the system (6.163) would imply that background electromagnetic radiation (microwave) of the Universe would create a non-propagating torsion field. This suggests that spacetime might be better modeled by Riemann-Cartan rather than Riemann manifolds.

Geometric interpretation of torsion and curvature, via the Schouten equations e.g. [11], is essentially based on Cartan parallelogram disclosure which is a version of what we can define as a loop in the theory of loop quantum gravitation. Of course, the continuous version of spacetime can be used when dimensions of considered regions are large enough without going in the domain of physical particles, given that the Planck length is about $1.616255 \times 10^{-35}[m]$ which is the length scale at which quantum gravitation becomes relevant. Roughly speaking, the radius at which the gravitation of the interaction energy is as important as the interaction energy itself is the Planck length (divided by 11.7, the square root of 137). It is sometimes conjectured that vacuum spacetime becomes a quantum foam at Planck length scale. Nevertheless, the loop diagram via Cartan parallelogram is interesting for generalizing relative gravitation to the conceptual set of microcosms constituting spacetime e.g. [2] (Figure 6.1).

On the other hand, torsion amplitude (density of spacetime defects) would be extremely "unnaturally" small since it is equal to the cosmological constant, and even be out of reach of experimental measurements as reminded by e.g. [216], [217], [281]. However, the finding (6.134) that torsion $\aleph_0$ may be directly related to cosmological constant Λ may give interesting insights for further indirect measurement of background torsion fields, becoming a link between local spacetime properties and the large-scale Universe. Dark energy can be described in terms of an unknown quantity such as the cosmological constant Λ, and it has been suggested in the past that geometric arguments might be used to address the dark energy problem e.g. [305]. In this last reference, the existence of gravitational waves other than those based on Einstein spacetime was investigated by means of teleparallel gravitation theory. Authors have shown that gravitational waves in teleparallel gravity were the same as those in Einstein gravitation. In the present book, we have a slightly different approach since we have shown that cosmological constant Λ may be approached in an Einstein-Cartan spacetime. *In fine*, cosmological dark energy may arise from both unknown physical contributions and geometric effects.

Key results. Combining the results of the present work, we may suggest that the electromagnetic field is potentially a source of torsion, and that torsion is itself potentially a source of the cosmological constant, dark matter and dark energy. It is necessary to develop Maxwell electromagnetism with its dual approach to show that both electric helicity $\mathcal{H}^{\mu\nu} A_\nu$ and magnetic helicity $\tilde{\mathcal{H}}^{\mu\nu} A_\nu$ are sources of vacuum spacetime distortion. These two suggestions are supported by simple and explicit examples.

Remark 6.64. In view of (6.229), it is obvious that defining an inertial frame of reference remains an asymptotic point of view since we can never completely eliminate curvature and torsion, and make spacetime locally free of physical interactions—gravitation and/or electromagnetism or others. Of course, when an observer's measurements are sufficiently sensitive enough, deviations from Galileo's law of inertia—that means straight trajectories of all test particles (by assumption, they cannot induce interaction fields)—can in general be detected. Relation (6.229) merely shows that the concept of inertial frames of reference is not exclusively related to gravitational fields but also to electromagnetic fields. The approach with Einstein-Cartan spacetime allows us to highlight this property with a pure geometric model.

6.4.4.4 *Some Comments on Anti-De Sitter Spacetime*

De Sitter spacetime is an exact solution of Einstein's field equations, void of matter but which includes a repulsive force embodied by cosmological constant $\Lambda > 0$. If the sign of cosmological constant Λ is changed, the repulsion force becomes attractive and the model is transformed into an anti-de Sitter spacetime. In this work, we have clearly found that repulsive force may be deduced from torsion of the Einstein-Cartan spacetime. The anti-de Sitter spacetime acquires a hyperbolic spatial geometry (i.e., of negative curvature) and, even if it is infinite, it has a well-defined boundary. To represent this boundary, we use the Poincaré representation of the hyperbolic disk which, using a conformal transformation preserving the angles but not the distances, brings infinity to a finite distance. Combining these results suggests a link between general relativity, electromagnetism, and anti–de Sitter spacetime with negative curvature. The role of anti–de Sitter spacetime in theories of cosmological bubble expansion and collapse has recently attracted renewed interest in the physics of large-scale structure and high-energy phenomena. The correspondence between higher-dimensional $\mathscr{A}d\mathscr{S}$ spacetime and CFT—proposed as a pathway toward a quantum theory of gravity—was conjectured by [246]. This is beyond the scope of this book.

The putative existence of $\mathscr{A}d\mathscr{S}$ spacetime with a negative value of cosmological constant Λ may seem to contradict the commonly accepted theory that the Universe is constantly expanding. Nevertheless, the study of such spacetime models remains of great interest, not only because of their fundamental role in the Maldacena conjecture, which opens avenues toward unifying gravitation and quantum mechanics, but also because anti–de Sitter spacetime plays a crucial role in recent theories of bubble fluctuations (regions of spacetime that alternately expand and then contract) e.g. [25].

Chapter 7

Continuum Spacetime Models for Cosmology

This chapter presents the Friedman equations and their extension under different conditions in cosmology. For that purpose, we re-derive the field equations showing the relation between the gravito-electromagnetic field and the spacetime geometric structure. Clearly, the presence of gravitation and electromagnetism interweave with bending and twisting of the spacetime. Different expressions of spacetime curvature and torsion tensor are obtained, and it was shown that the compatibility of the volume-form plays an important role in the derivation of the field equations. The common thread that links all the models developed in this book is the search for a formal expression of the cosmological constant Λ. The presence of torsion results in a non-homogeneous and non-isotropic "cosmological field", meaning a violation of the cosmological principle. However, the presence of a diffuse cosmic electromagnetic field associated with a non-zero torsion field is shown to be a serious phenomenological candidate for modeling dark matter and dark energy.

7.1 Introduction

Any reflection on the notion of spacetime in cosmology, as suggested by the title of this book, cannot ignore scenarios of Universe formation in terms of its constituents and their organization. The Universe is structured as a hierarchical assembly of (baryonic) matter, forming an intricate network of galaxies, galaxy clusters, and giant filaments—collectively known as the cosmic web.

The term cosmological model in its modern sense dates back to the physicist Alexander Friedmann, who was the first to formulate it in this form in 1922 [306]. During the development of the theory of general relativity, Einstein proposed in 1917 the first known cosmological solution. Although highly idealized, this solution was obtained from Einstein's field equations and can be regarded as the first truly scientific basis for a cosmological model. On the one hand, general relativity is nowadays considered a well-established and plausible theory, at least on large length scales. Its experimental tests are highly accurate, including, among others, the precession of Mercury's perihelion, the bending of light in gravitational fields, and the gravitational redshift. On the other hand, cosmological theories still rely mainly on general relativity, together with quantum physics at small scales, and most of the time remain extrapolations beyond experimentally tested regimes. The characteristic length scale of experimental tests of general relativity is below 10^{11} $[m]$, whereas for cosmological theories it is of the order of 10^{26} $[m]$. This disparity motivates the search for appropriate spacetime models. Furthermore, quantum effects become relevant at length scales smaller than the Planck length $\ell_{\mathrm{P}} = 10^{-35}$ $[m]$.

While gravitation plays an essential role in the formulation of cosmological equations, the existence of magnetic fields has also proven to be significant. In particular, the evolution of the large-scale structure of the cosmic web owes much to electromagnetic fields. A review of the role and possible influence of magnetic fields in the early Universe can be found in e.g. [56]. Recent observational results showing the alignment of giant filaments and the concentration of diffuse matter around them with magnetic fields constitute a major advance in understanding the role of electromagnetism in the evolution and structuring of the cosmic web e.g. [290]. The growing recognition of a "magnetic Universe" leaves no doubt as to the role of electromagnetism in the development of vacuum spacetime models. Moreover, this seems logical since the only two universal interactions with an infinite radius are gravitational and electromagnetic forces.

Currently, one widely accepted assumption in cosmological theories is that the Universe is composed predominantly of vacuum, but is permeated by matter–referred to as baryonic matter—whose mean density is very low $\rho \simeq 10^{-28}[kg/m^3]$. Baryonic matter is made up of protons and neutrons. However, most of the matter content of the Universe, called dark matter, remains unknown to this day. The existence of dark matter, as well as dark energy, remains a central concept of modern cosmology. These hypotheses are essential for explaining the major observational evidence of cosmic evolution, in particular the accelerated expansion of the Universe [18, 19].

In a modest way, this chapter is devoted to proposing some basic features of the vacuum spacetime by including classical interactions such as gravitation and electromagnetism within the model. The chapter will attempt to provide some elements to our understanding of cosmic web formation and the accelerated expansion of spacetime. In addition to the "visible" large-scale structure, discovery of dark matter and dark energy to explain the accelerated expansion implies new unknowns and challenges. To our knowledge, the addition of magnetic fields therefore seems mandatory to derive reliable spacetime models in cosmology.

7.2 Spacetime Models Based on Einstein Gravitation

It is worth recalling some basic principles of cosmology. Historically, the Copernican principle assumes that any observer studying the evolution of the Universe may be located anywhere within it, and the direction of observation would not influence the results. In a slightly extended form, this principle states that humans—or any experimental devices on Earth, in the Solar System, or on any other planets or satellites—are not privileged observers of the evolution of the Universe. When properly calibrated, observations made from any location may be regarded as representative of those made from an average position in the Universe. From a pragmatic point of view, cosmological theory is therefore based on the assumption that the cosmological principle is approximately, though not exactly, valid on the largest scales. In other words, under the assumption of the Copernican principle, the Universe appears homogeneous (having the same properties everywhere) and isotropic (having the same properties in all directions). In reality, astronomical observations show that the Universe exhibits non-uniform structures on scales up to galaxies, galaxy clusters, giant filaments, and possibly even interstellar or intergalactic regions. Nevertheless, the underlying spacetime of the Universe is assumed to become increasingly homogeneous and isotropic on progressively larger scales. This suggests that homogeneity and isotropy may be locally violated, but are still assumed to hold statistically at large scales. Like other principles in physics, the principles of cosmology have never been formally proven—and, in the most general sense, cannot be proven—but they remain implicit assumptions in many modern physical theories.

Principles of Cosmology. Derivation of cosmology theory is essentially based on the properties of homogeneity (in the sense of constant curvature) and isotropy of the spacetime at very large scale.

There are numerous publications which report on the different theories underlying the various major international projects whose main objective is, for example, to observe and thus clarify the notion of dark energy e.g. [20]. This chapter makes no claim to summarize or discuss the state-of-the-art of the various vacuum spacetime models in cosmology. Herein, we only propose a simple comparative analysis of various spacetime models which take up Friedmann's equations by considering relative gravitation and electromagnetism.

Remark 7.1. The first quantitative cosmological models based on general relativity date back to the work of Einstein and de Sitter (1917). These models describe the Universe as a whole, on very large scales. In contrast, the fundamental physical laws—particularly those governing gravitation—are formulated locally. It is also curious that the consistent classical counterpart, based on Newtonian mechanics, of these cosmological models had emerged around 1934 (McCrea and Milne). The difficulty with classical mechanics essentially stemmed from the problems of divergent boundary conditions e.g. [307].

7.2.1 *FLRW Metric in a Flat Spacetime*

From the outset, the construction of FLRW spacetime was based on homogeneity and isotropy. We know the construction of Schwarzschild spacetime through the Hilbert-Einstein action by variational procedure. In this section we limit ourselves to a direct construction of the Roberston-Walker metric.

This specific spacetime results from the independent works of Friedmann [306], Lemaître [308], Robertson [309], and Walker [310]. FLRW constitutes a class of exact solutions where the search of metric form may be done using translation (homogeneity) and rotation (isotropy) symmetry. With some calculus details, the expression of FLRW metric is obtained as follows. We start by a classical line element:

$$ds^2 = e^{2\nu}\,(dx^0)^2 - e^{2\mu}\,dr^2 - r^2\left(d\theta^2 + \sin^2\theta\,d\varphi^2\right) \tag{7.1}$$

where (ν, μ) are functions of the radius r. It is the same as for all spherically symmetric spacetimes we have studied previously. The complete derivation of this line element is given in the Appendix.

Remark 7.2. In relative gravitation, the Birkhoff theorem states that any spherically symmetric solution of the Einstein vacuum field equations must be static and asymptotically flat e.g. [125]. The theorem then ensures that the metric, which is a solution of the vacuum spacetime outside of a spherical, non-rotating, massive body, is reduced to the Schwarzschild metric. The converse of the Birhkoff theorem is true and is called Israel's theorem [311].

7.2.1.1 *Method of Killing Vectors*

To derive a spacetime model, it is generally assumed that the evolution of the Universe proceeded from a nearly uniform initial state—both homogeneous and isotropic—to a globally irregular cosmic web. For that purpose, we need to consider six tangent directions ξ^λ. The Killing vector method allows us to write the Lie derivative of the metric leading to the equations governing the invariance group:

$$\mathcal{L}_\xi g_{\mu\nu} = \xi^\lambda \partial_\lambda g_{\mu\nu} + g_{\mu\lambda}\partial_\nu \xi^\lambda + g_{\lambda\nu}\partial_\mu \xi^\lambda \equiv 0 \tag{7.2}$$

which is valid for the Riemann manifold. The idea is to apply this invariance equation on the metric components to account for translations and rotations. To do this, we will use a spherical coordinate system for space which does not mean that there is a center of the Universe. It's simply by convenience. To begin with, let us write the relations between Cartesian and spherical coordinates:

$$x^1 = r\sin\theta\,\cos\varphi, \quad x^2 = r\sin\theta\,\sin\varphi, \quad x^3 = r\cos\theta \tag{7.3}$$

The tangent vectors of the local spherical base are directly expressed with the tangent base associated with the Cartesian coordinates $(\partial_1, \partial_2, \partial_3)$:

$$\partial_r = \begin{pmatrix} \sin\theta\,\cos\varphi \\ \sin\theta\,\sin\varphi \\ \cos\theta \end{pmatrix}, \quad \frac{\partial_\theta}{r} = \begin{pmatrix} \cos\theta\,\cos\varphi \\ \cos\theta\,\sin\varphi \\ -\sin\theta \end{pmatrix}, \quad \frac{\partial_\varphi}{r\sin\theta} = \begin{pmatrix} -\sin\varphi \\ \cos\varphi \\ 0 \end{pmatrix} \tag{7.4}$$

This gives a system of three independent equations as follows:

$$\begin{cases} \partial_r = \sin\theta\cos\varphi\ \partial_1 + \sin\theta\sin\varphi\ \partial_2 + \cos\theta\ \partial_3 \\ \dfrac{1}{r}\partial_\theta = \cos\theta\cos\varphi\ \partial_1 + \cos\theta\sin\varphi\ \partial_2 - \sin\theta\ \partial_3 \\ \dfrac{1}{r\sin\theta}\partial_\varphi = -\sin\varphi\ \partial_1 + \cos\varphi\ \partial_2 \end{cases} \tag{7.5}$$

The Cartesian local tangent base is the unknown; expressing the Cartesian local base in terms of the spherical local base allows us to have e.g. [125]:

$$\begin{cases} \partial_1 = \sin\theta\cos\varphi\ \partial_r + \dfrac{\cos\theta}{r}\cos\varphi\ \partial_\theta - \dfrac{\sin\varphi}{r\sin\theta}\ \partial_\varphi \\ \partial_2 = \sin\theta\sin\varphi\ \partial_r + \dfrac{\cos\theta}{r}\sin\varphi\ \partial_\theta + \dfrac{\cos\varphi}{r\sin\theta}\ \partial_\varphi \\ \partial_3 = \cos\theta\ \partial_r - \dfrac{\sin\theta}{r}\ \partial_\theta \end{cases} \tag{7.6}$$

7.2.1.2 *Isotropy and Homogeneity*

It is worth noticing that the principles of cosmology are in fact a statement about space only. It means that the space-like slices of a four-dimensional spacetime are homogeneous and isotropic. Consider vectors ξ associated with rotation and translation symmetry. For the three translations we have the Killing vectors:

$$\xi_4 = \partial_1, \quad \xi_5 = \partial_2, \quad \xi_6 = \partial_3 \tag{7.7}$$

where (7.7) are projected onto a Cartesian frame (x^1, x^2, x^3). For translation, projected onto $(\partial_0, \partial_r, \partial_\theta, \partial_\varphi)$ in four dimensions from (7.6), we write:

$$\begin{cases} \xi_1 = \left(0, \sin\theta\cos\varphi, \dfrac{1}{r}\cos\theta\cos\varphi, \dfrac{1}{r}\dfrac{\sin\varphi}{\sin\theta}\right) \\ \xi_2 = \left(0, \sin\theta\sin\varphi, \dfrac{1}{r}\cos\theta\sin\varphi, \dfrac{1}{r}\dfrac{\cos\varphi}{\sin\theta}\right) \\ \xi_3 = \left(0, \cos\theta, -\dfrac{1}{r}\sin\theta, 0\right) \end{cases} \tag{7.8}$$

For rotation, the three infinitesimal Killing vectors of the spacetime are:

$$\xi_4 = x^2\partial_3 - x^3\partial_2, \quad \xi_5 = x^3\partial_1 - x^1\partial_3, \quad \xi_6 = x^1\partial_2 - x^2\partial_1 \tag{7.9}$$

Projecting vectors onto spherical coordinate base $(\partial_0, \partial_r, \partial_\theta, \partial_\varphi)$, we obtain:

$$\begin{cases} \xi_4 = (0, 0, -\sin\varphi, -\cot\theta\cos\varphi) \\ \xi_5 = (0, 0, \cos\varphi, -\cot\theta\sin\varphi) \\ \xi_6 = (0, 0, 0, 1) \end{cases} \tag{7.10}$$

Introducing successively the six symmetry vectors $\xi_i, i = 1, 6$ in the Lie derivative (7.2) allows us to obtain six partial differential equations which can be solved. For the Killing vectors (7.10) associated with rotation, we have from (7.2):

(1) for ξ_6

$$\partial_\varphi g_{\mu\nu} = 0, \quad \forall \mu, \nu \tag{7.11}$$

(2) for ξ_4 with $(\mu\nu) = (00)$

$$\begin{cases} \dfrac{\partial g_{00}}{\partial \theta}(-\sin\varphi) + \dfrac{\partial g_{00}}{\partial \varphi}\left(-\dfrac{\cos\theta}{\sin\theta}\cos\varphi\right) = 0 \\ \dfrac{\partial g_{00}}{\partial \theta}\ \sin^2\varphi + \dfrac{\partial g_{00}}{\partial \varphi}\ \dfrac{\cos\theta}{\sin\theta}\cos\varphi\sin\varphi = 0 \end{cases} \tag{7.12}$$

(3) for $(\mu\nu) = (01)$, for $(\mu\nu) = (11)$, since ξ_3 and ξ_4 are independent of x^0 and r:

$$\partial_\theta g_{r0} = 0, \quad \partial_\theta g_{rr} = 0 \tag{7.13}$$

(4) for $(\mu\nu) = (22)$ and $(\mu\nu) = (23)$ and $(\mu\nu) = (33)$

$$\begin{cases} \dfrac{\partial g_{\theta\theta}}{\partial \theta}\ \sin\varphi + 2g_{\theta\varphi}\ \dfrac{\cos\varphi}{\sin^2\theta} = 0 \\ \left(\dfrac{\partial g_{\theta\varphi}}{\partial \theta} - g_{\theta\varphi}\dfrac{\cos\theta}{\sin\theta}\right)\sin\varphi + \left(g_{\theta\theta} - \dfrac{g_{\varphi\varphi}}{\sin^2\theta}\right)\cos\varphi = 0 \\ \left(\dfrac{\partial g_{\varphi\varphi}}{\partial \theta} - g_{\varphi\varphi}\ \dfrac{\cos\theta}{\sin\theta}\right)\sin\varphi + 2g_{\theta\varphi}\cos\varphi = 0 \end{cases} \tag{7.14}$$

(5) for $(\mu\nu) = (12)$ and $(\mu\nu) = (13)$

$$\begin{cases} \dfrac{\partial g_{r\theta}}{\partial \theta}\ \sin\varphi + 2g_{r\varphi}\ \dfrac{\cos\varphi}{\sin^2\theta} = 0 \\ \left(-\dfrac{\partial g_{r\varphi}}{\partial \theta} + g_{r\varphi}\ \dfrac{\cos\theta}{\sin\theta}\right)\sin\varphi - g_{r\theta}\ \cos\varphi = 0 \end{cases} \tag{7.15}$$

(6) for $(\mu\nu) = (02)$ and $(\mu\nu) = (03)$

$$\begin{cases} \dfrac{\partial g_{0\theta}}{\partial \theta}\ \sin\varphi - g_{0\varphi}\ \dfrac{\cos\varphi}{\sin^2\theta} = 0 \\ \left(\dfrac{\partial g_{r\varphi}}{\partial \theta} - g_{0\varphi}\ \dfrac{\cos\theta}{\sin\theta}\right)\sin\varphi - g_{0\theta}\ \cos\varphi = 0 \end{cases} \tag{7.16}$$

For the rotation associated with vector ξ_5 we have to replace $\sin\varphi$ by $-\cos\varphi$ and $\cos\varphi$ by $\sin\varphi$ to have the following:

(1) for $(\mu\nu) = (22)$ and $(\mu\nu) = (23)$ and $(\mu\nu) = (33)$

$$\begin{cases} -\dfrac{\partial g_{\theta\theta}}{\partial\theta}\cos\varphi + 2g_{\theta\varphi}\dfrac{\sin\varphi}{\sin^2\theta} = 0 \\ -\left(\dfrac{\partial g_{\theta\varphi}}{\partial\theta} - g_{\theta\varphi}\dfrac{\cos\theta}{\sin\theta}\right)\cos\varphi + \left(g_{\theta\theta} - \dfrac{g_{\varphi\varphi}}{\sin^2\theta}\right)\sin\varphi = 0 \\ \left(\dfrac{\partial g_{\varphi\varphi}}{\partial\theta} - g_{\varphi\varphi}\dfrac{\cos\theta}{\sin\theta}\right)\cos\varphi - 2g_{\theta\varphi}\sin\varphi = 0 \end{cases} \tag{7.17}$$

(2) for $(\mu\nu) = (12)$ and $(\mu\nu) = (13)$

$$\begin{cases} -\dfrac{\partial g_{r\theta}}{\partial\theta}\cos\varphi + 2g_{r\varphi}\dfrac{\sin\varphi}{\sin^2\theta} = 0 \\ \left(\dfrac{\partial g_{r\varphi}}{\partial\theta} - g_{r\varphi}\dfrac{\cos\theta}{\sin\theta}\right)\cos\varphi - g_{r\theta}\sin\varphi = 0 \end{cases} \tag{7.18}$$

(3) for $(\mu\nu) = (02)$ and $(\mu\nu) = (03)$

$$\begin{cases} \dfrac{\partial g_{0\theta}}{\partial\theta}\cos\varphi + g_{0\varphi}\dfrac{\sin\varphi}{\sin^2\theta} = 0 \\ \left(\dfrac{\partial g_{0\varphi}}{\partial\theta} + g_{0\varphi}\dfrac{\cos\theta}{\sin\theta}\right)\cos\varphi - g_{0\theta}\sin\varphi = 0 \end{cases} \tag{7.19}$$

These systems of equations (which mean space isotropy) put together allow us to calculate the general expression of the metric:

$$g_{\mu\nu} = \begin{pmatrix} g_{00}(t,r) & g_{0r}(t,r) & 0 & 0 \\ g_{r0}(t,r) & g_{rr}(t,r) & 0 & 0 \\ 0 & 0 & g_{\theta\theta}(t,r) & 0 \\ 0 & 0 & 0 & g_{\varphi\varphi}(t,r) \end{pmatrix} \tag{7.20}$$

with the line element:

$$\begin{aligned} ds^2 &= g_{00}(t,r)\,dt^2 - 2g_{0r}(t,r)\,dtdr - g_{rr}(t,r)\,dr^2 \\ &\quad - g_{\theta\theta}(t,r)\,d\theta^2 - g_{\theta\theta}\,\sin^2\,d\varphi^2 \end{aligned} \tag{7.21}$$

which is the most general isotropic metric with spherical symmetry. Now we have to impose the space homogeneity by using translational vectors (ξ_1, ξ_2, ξ_3):

(1) for $(\mu\nu) = (00)$ and $(\mu\nu) = (22)$ (with ξ_2)

$$\partial_r g_{00} = 0, \quad \text{and} \quad g_{\theta\theta} = A(t)\,r^2 \tag{7.22}$$

(2) for $(\mu\nu) = (12)$ and $(\mu\nu) = (02)$

$$g_{rr} = \frac{1}{r^2}\, g_{\theta\theta} = A(t), \quad \text{and} \quad g_{0r} = 0 \tag{7.23}$$

The resulting line element is then given by:

$$ds^2 = g_{00}(t)\, dx_0^2 - A^2(t)\left[dr^2 + r^2 d\theta^2 + r^2 \sin^2\theta d\varphi^2\right]$$

where t is merely a universal parameter defined over the whole spacetime. Hilbert causality requires that $g_{00}(t) > 0$ which permits us to make a change of variable $dx_0'^2 := g_{00}(t)\, dx_0^2$ and then write, by dropping the prime:

$$\boxed{ds^2 = dx_0^2 - S^2(t)\left[dr^2 + r^2 d\theta^2 + r^2 \sin^2\theta d\varphi^2\right]} \tag{7.24}$$

where the positive function $S^2(t)$ is an expansion factor of the Universe. This is a form of the so-called Friedmann-Lemaître-Robertson-Walker metric in a flat spacetime. We have to recall the notion of space with constant curvature.

7.2.2 *Space with Constant Curvature*

The idea is now to introduce the Friedmann-Lemaître-Robertson-Walker metric for curved spacetime. To begin with, consider a three-dimensional Riemann manifold endowed with a metric g_{ij} with the curvature $\Re^k_{ijl}$, modeling the space contribution.

Definition 7.1. The Riemann curvature $\mathcal{R}$ on the manifold $\mathscr{B}$ is defined by:

$$\mathcal{R}(\mathbf{v}_1, \mathbf{v}_2, \mathbf{v}_3, \mathbf{v}_4) := \mathbf{g}\left(\mathbf{v}_1, \nabla_{\mathbf{v}_3}\nabla_{\mathbf{v}_4}\mathbf{v}_2 - \nabla_{\mathbf{v}_4}\nabla_{\mathbf{v}_3}\mathbf{v}_2 - \nabla_{[\mathbf{v}_3,\mathbf{v}_4]}\mathbf{v}_2\right) \tag{7.25}$$

for arbitrary vector fields $\mathbf{v}_i \in \mathscr{T}_{\mathbf{x}}\mathscr{B}$ on the manifold.

Theorem 7.1. *The Riemann curvature tensor satisfies the following relations*:

$$\begin{cases} \mathcal{R}(\mathbf{v}_1, \mathbf{v}_2, \mathbf{v}_3, \mathbf{v}_4) + \mathcal{R}(\mathbf{v}_2, \mathbf{v}_1, \mathbf{v}_3, \mathbf{v}_4) = 0 \\ \mathcal{R}(\mathbf{v}_1, \mathbf{v}_2, \mathbf{v}_3, \mathbf{v}_4) + \mathcal{R}(\mathbf{v}_1, \mathbf{v}_2, \mathbf{v}_4, \mathbf{v}_3) = 0 \\ \mathcal{R}(\mathbf{v}_1, \mathbf{v}_2, \mathbf{v}_3, \mathbf{v}_4) - \mathcal{R}(\mathbf{v}_3, \mathbf{v}_4, \mathbf{v}_1, \mathbf{v}_2) = 0 \\ \mathcal{R}(\mathbf{v}_1, \mathbf{v}_2, \mathbf{v}_3, \mathbf{v}_4) + \mathcal{R}(\mathbf{v}_1, \mathbf{v}_3, \mathbf{v}_4, \mathbf{v}_2) + \mathcal{R}(\mathbf{v}_1, \mathbf{v}_4, \mathbf{v}_2, \mathbf{v}_3) = 0 \end{cases} \tag{7.26}$$

Proof. It may be found in textbooks on geometry e.g. [128] and cosmology e.g. [241]. □

Definition 7.2. Let $\mathbb{R}^n$ be a pseudo-Riemann manifold endowed with the pseudo-metric (defined with its natural coordinates):

$$\mathbf{g}_0(\mathbf{u}, \mathbf{v}) = u^0 v^0 - u^1 v^1 - u^2 v^2 - u^3 v^3 \tag{7.27}$$

(it is easily transposed to manifolds of higher dimensions with different signatures). The manifold $(\mathscr{B}, \mathbf{g})$ is locally isometric to $(\mathbb{R}^n, \mathbf{g}_0)$ if any point $\mathbf{x} \in \mathscr{B}$ has an open neighbour which is isometric to a sub-manifold of $(\mathbb{R}^n, \mathbf{g}_0)$.

We thus have the keypoint proposition on the (pseudo)-Riemann manifold.

Theorem 7.2. *Let* $(\mathscr{B}, \mathbf{g})$ *be a pseudo-Riemann manifold. Then,* $(\mathscr{B}, \mathbf{g})$ *is locally isometric to* $(\mathbb{R}^n, \mathbf{g}_0)$ *if and only if* $\mathcal{R} = 0$ *where* $\mathcal{R}$ *is the scalar curvature (not the curvature tensor).*

Proof. The quite lengthy proof may be found in classical advanced textbooks on Riemann geometry e.g. [128]. □

Lemma 7.1. *Let* $\mathscr{E}$ *be a vector space of dimension* n *and* $\mathcal{R}$ *a tensor satisfying the following relations:*

$$\begin{cases} \mathcal{R}(\mathbf{v}_1, \mathbf{v}_2, \mathbf{v}_3, \mathbf{v}_4) + \mathcal{R}(\mathbf{v}_2, \mathbf{v}_1, \mathbf{v}_3, \mathbf{v}_4) = 0 \\ \mathcal{R}(\mathbf{v}_1, \mathbf{v}_2, \mathbf{v}_3, \mathbf{v}_4) + \mathcal{R}(\mathbf{v}_1, \mathbf{v}_2, \mathbf{v}_4, \mathbf{v}_3) = 0 \\ \mathcal{R}(\mathbf{v}_1, \mathbf{v}_2, \mathbf{v}_3, \mathbf{v}_4) + \mathcal{R}(\mathbf{v}_1, \mathbf{v}_3, \mathbf{v}_4, \mathbf{v}_2) + \mathcal{R}(\mathbf{v}_1, \mathbf{v}_4, \mathbf{v}_2, \mathbf{v}_3) = 0 \end{cases} \tag{7.28}$$

Then, the tensor $\mathcal{R}$ *is entirely defined by the knowledge of* $\mathcal{R}(\mathbf{v}_1, \mathbf{v}_2, \mathbf{v}_1, \mathbf{v}_2)$ *when* $(\mathbf{v}_1, \mathbf{v}_2)$ *span the entire vector space* $\mathscr{E}$.

Lemma 7.2. *Let* $\mathcal{R}$ *be the same tensor as in Lemma* 7.1. *For any two-dimensional subspace* $\mathscr{F}$ *of* $\mathscr{E}$*, we consider an orthonormal base* $(\mathbf{e}_1, \mathbf{e}_2)$*. Let's define the scalar:*

$$K(\mathscr{F}) := \mathcal{R}(\mathbf{e}_1, \mathbf{e}_2, \mathbf{e}_1, \mathbf{e}_2) \tag{7.29}$$

Then the scalar $K(\mathscr{F})$ *does not depend on the chosen base* $(\mathbf{e}_1, \mathbf{e}_2)$ *but only on the subspace* $\mathscr{F}$.

For any other base $(\mathbf{v}_1, \mathbf{v}_2)$ of the subspace $\mathscr{F}$, not necessarily orthonormal, we have:

$$K(\mathscr{F}) = \frac{\mathcal{R}(\mathbf{v}_1, \mathbf{v}_2, \mathbf{v}_1, \mathbf{v}_2)}{\mathbf{g}(\mathbf{v}_1, \mathbf{v}_1)\ \mathbf{g}(\mathbf{v}_2, \mathbf{v}_2) - [\mathbf{g}(\mathbf{v}_1, \mathbf{v}_2)]^2} \tag{7.30}$$

Definition 7.3. (Manifold with constant curvature) The scalar $K_{\mathbf{x}}(\mathscr{F})$ at each point $\mathbf{x}$ is called sectional curvature along the subspace (direction) $\mathscr{F}$

at the point $\mathbf{x}$ of $\mathscr{E}$. Whenever $K_{\mathbf{x}}(\mathscr{F})$ does not depend on the subspace (direction) $\mathscr{F}$ but depends only on the point $\mathbf{x}$, the manifold $\mathscr{B}$ is said to be a manifold with constant curvature.

We deduce:

Theorem 7.3. *Let $\mathscr{B}$ be a Riemann manifold (dimension n) endowed with the metric $\mathbf{g}$ and say $\mathbf{x} \in \mathscr{B}$. Then projected onto local coordinates, a necessary and sufficient condition of $\mathscr{B}$ to be a manifold with constant curvature at $\mathbf{x}$ is given by:*

$$\mathcal{R}_{ijkl}(\mathbf{x}) = \frac{\mathcal{R}(\mathbf{x})}{n(n-1)} \left[g_{ik}(\mathbf{x})\ g_{jl}(\mathbf{x}) - g_{il}(\mathbf{x})\ g_{jk}(\mathbf{x})\right] \tag{7.31}$$

In such a case, we also note $K_{\mathbf{x}} = \frac{\mathcal{R}(\mathbf{x})}{n(n-1)}$.

Proof. It may be found in textbooks on differential geometry e.g. [128] or reserach monographs e.g. [241]. □

This proposition gives a necessary and sufficient condition for a Riemann manifold. It is still valid for pseudo-Riemann manifolds. From the result (7.31), we can deduce easily the Ricci curvature:

$$\boxed{\mathcal{R}_{ij}(\mathbf{x}) := g^{hk}(\mathbf{x})\ \mathcal{R}_{hikj}(\mathbf{x}) = \frac{\mathcal{R}(\mathbf{x})}{n}\ g_{ij}(\mathbf{x})} \tag{7.32}$$

By using the identity conservation (on a four-dimensional Einstein spacetime for instance), we have:

$$\nabla_j \left(\mathcal{R}_{ij}(\mathbf{x}) - \frac{\mathcal{R}(\mathbf{x})}{2} g_{ij}(\mathbf{x}) \right) = 0 \tag{7.33}$$

given that we utilize here a metric-compatible connection. It is therefore deduced that the curvature is constant $\mathcal{R}(\mathbf{x}) = \mathcal{R}$. Finally, for a manifold with constant curvature we can write:

$$\begin{aligned} \mathcal{R}_{ijkl}(\mathbf{x}) &= \frac{\mathcal{R}}{n(n-1)} \left[g_{ik}(\mathbf{x})\ g_{jl}(\mathbf{x}) - g_{il}(\mathbf{x})\ g_{jk}(\mathbf{x})\right] \\ &= K \left[g_{ik}(\mathbf{x})\ g_{jl}(\mathbf{x}) - g_{il}(\mathbf{x})\ g_{jk}(\mathbf{x})\right] \end{aligned} \tag{7.34}$$

In summary, a Riemann manifold $\mathscr{B}$ endowed with metric $\mathbf{g} = (g_{ij})$ is said to be of constant curvature if all of its curvature components at any point of $\mathscr{B}$ are determined by the same single quantity K which is point-independent:

$$\mathcal{R}_{ijkl}(\mathbf{x}) = K \left[g_{ik}(\mathbf{x})\ g_{jl}(\mathbf{x}) - g_{il}(\mathbf{x})\ g_{jk}(\mathbf{x})\right] \tag{7.35}$$

Such a manifold is isotropic and homogeneous since the same relation is valid for all points and all directions. Because we have the Ricci curvature $\mathcal{R}_{ij} = K/(n-1)\, g_{ij}(\mathbf{x})$, the manifold is an Einstein-like manifold with the scalar curvature $\mathcal{R} = n(n-1)\, K$.

7.2.2.1 *Three-Dimensional Manifold of Constant Curvature*

To begin with, let's consider a three-dimensional manifold with coordinate system (x^1, x^2, x^3). In such a case, Riemann geometry results assess that there exists at least a suitable coordinate system allowing us to express the line element as:

$$ds^2 = \frac{(dx^1)^2 + (dx^2)^2 + (dx^3)^2}{\left[1 + \frac{1}{4}K\left((x^1)^2 + (x^2)^2 + (x^3)^2\right)\right]^2}$$

Depending on the sign of the curvature we can define the quantity K as:

$$K := \frac{k}{R^2}, \quad \text{with} \quad k = -1, 0, +1 \tag{7.36}$$

stating the three cases of negative, zero (flat) and positive curvature, and where R is a radius of curvature. Rescaling the metric line element by $1/R^2$, it may be rewritten as follows:

$$ds^2 = R^2 \frac{\left(dx^1/R\right)^2 + \left(dx^2/R\right)^2 + \left(dx^3/R\right)^2}{[1 + (k/4)((x^1/R)^2 + (x^2/R)^2 + (x^3/R)^2)]^2}$$

which reduces to $(x^i/R \to x^i)$:

$$ds^2 = \frac{(dx^1)^2 + (dx^2)^2 + (dx^3)^2}{\left[1 + \frac{1}{4}k\left((x^1)^2 + (x^2)^2 + (x^3)^2\right)\right]^2}$$

where coordinates are dimensionless since they are divided by R. Back to the case of spherical symmetry, we use a change of variables:

$$x^1 = \tilde{r}\sin\theta\cos\varphi, \quad x^2 = \tilde{r}\sin\theta\sin\varphi, \quad x^3 = \tilde{r}\cos\theta, \quad r := \frac{\tilde{r}}{(1 + k\tilde{r}^2/4)}$$

We recover a generic form of the standard metric (space contribution) of the Friedmann-Lemaître-Roberston-Walker metric in cosmology:

$$ds^2 = R^2\left(\frac{dr^2}{1 - kr^2} + r^2\, d\theta^2 + r^2\sin^2\theta\, d\varphi^2\right), \quad k = -1, 0, +1 \tag{7.37}$$

According to the sign of the curvature we recover the three cases e.g. [241]:

(1) $k = -1$. The manifold is a Lobatchevski space which is infinite, with a constant curvature (hyperbolic 3-space $\mathscr{H}^3$ with a change of variable $r = \sinh\chi$)

$$ds^2 = R^2 \left(d\chi^2 + \sinh^2\chi \; (d\theta^2 + \sin^2\theta \; d\varphi^2)\right) \tag{7.38}$$

(2) $k = 0$. The manifold is flat which is infinite and can be covered by a Cartesian coordinate system or by non-spherical coordinates with a renaming of variable $r = \chi$

$$ds^2 = R^2 \left(d\chi^2 + \chi^2 \; (d\theta^2 + \sin^2\theta \; d\varphi^2)\right) \tag{7.39}$$

(3) $k = +1$. The manifold is a 3-sphere which is finite with a constant positive curvature with the change of variable $r = \sin\chi$

$$ds^2 = R^2 \left(d\chi^2 + \sin^2\chi \; (d\theta^2 + \sin^2\theta \; d\varphi^2)\right) \tag{7.40}$$

They have nearly the same form although they represent different manifolds. The coordinates θ and φ have the same domains, but χ varies from zero to infinity for $k = 0$ and $k = -1$. For $k = +1$ (3-sphere), we have $\chi \in [0, \pi]$.

7.2.2.2 *FLRW Spacetimes in Cosmology*

For four-dimensional spacetime, we merely add the time $x^0 := ct$ to give the three situations and then obtain the general form of standard model spacetime in cosmology from (7.37):

$$\boxed{ds^2 = (dx^0)^2 - R^2(x^0) \left(\frac{dr^2}{1 - kr^2} + r^2 \; d\theta^2 + r^2 \sin^2\theta \; d\varphi^2\right), \quad k = -1, 0, +1} \tag{7.41}$$

The coordinates (r, θ, φ) are co-moving coordinates, meaning that galaxies keep these coordinates fixed, whereas the expansion or contraction of the Universe is defined by the function $R(x^0)$ e.g. [30]. Therefore, we obtain the following line elements e.g. [241]:

(1) $k = -1$. The manifold is a Lobatchevski space, with a constant curvature (hyperbolic 3-space $\mathscr{H}^3$) for space

$$ds^2 = (dx^0)^2 - R^2(x^0) \left(d\chi^2 + \sinh^2\chi \; (d\theta^2 + \sin^2\theta \; d\varphi^2)\right) \tag{7.42}$$

(2) $k = 0$. The manifold is flat and can be covered by a Cartesian coordinate system or by non-spherical coordinates

$$ds^2 = (dx^0)^2 - R^2(x^0) \left(d\chi^2 + \chi^2 \; (d\theta^2 + \sin^2\theta \; d\varphi^2)\right) \tag{7.43}$$

(3) $k = +1$. The manifold is a 3-sphere with a constant positive curvature for space

$$ds^2 = (dx^0)^2 - R^2(x^0)\left(d\chi^2 + \sin^2\chi\,(d\theta^2 + \sin^2\theta\;d\varphi^2)\right) \tag{7.44}$$

The metrics (7.42)–(7.44) define homogeneous and isotropic spacetimes and constitute the basics of cosmological models introduced by Friedmann (1922, 1924), Lemaître (1927) (cases of sphere and hyperbolic spaces), Robertson (1929), and Walker (1935). All the three cases can be summarized in one formulation as:

$$ds^2 = (dx^0)^2 - R^2(x^0)\left[d\chi^2 + S^2_{(k)}(\chi)\;(d\theta^2 + \sin^2\theta\;d\varphi^2)\right] \tag{7.45}$$

where:

$$S^2_{(k)}(\chi) := \begin{cases} \sinh\chi & \text{for } k = -1 \\ \chi & \text{for } k = 0 \\ \sin\chi & \text{for } k = +1 \end{cases} \tag{7.46}$$

corresponding to hyperbolic, flat and spherical spacetimes respectively.

7.2.2.3 *Metric and Redshift*

Most measurements in cosmology are based on electromagnetic waves, such as light emitted by planets and galaxies. Some elementary physical insights should be briefly recalled for understanding the Hubble parameter and redshift phenomenon. Given an electromagnetic radiation (such as light) emitted by a source, a redshift is an increase in the wavelength (or equivalently decrease of the frequency) due mainly to the motion of the source. Blueshift is the opposite phenomenon. Redshift is denoted by the letter z, corresponding to the fractional change in wavelength (positive for redshifts, negative for blueshifts), and by the wavelength ratio $1 + z$ (which is greater than 1 for redshifts and less than 1 for blueshifts).

$$1 + z(x^0) = \frac{\lambda_r}{\lambda_e} = \frac{f_e}{f_r} \tag{7.47}$$

where λ_e and λ_r are the emitted and received wavelength of the signals, and f_e and f_r the frequencies, respectively. Physically, absolute motions of planets and galaxies cannot be directly measured. Measurement results are based on the analysis of electromagnetic waves received from the moving sources. The motions of radiation sources give rise to the Doppler effect. Roughly speaking, Doppler blueshifts ($z < 0$) are associated with sources approaching the receiver whereas Doppler redshifts ($z > 0$) are associated with sources moving away from the receiver. It is observed that all sufficiently distant light sources show redshift that corresponds to recession speeds proportional to their distances from Earth. This is Hubble's law and shows that the Universe is expanding. It is then convenient to make a change of variable:

$$\frac{1}{1+z(x^0)} := \frac{R(x^0)}{R_0} \quad \Longleftrightarrow \quad R(x^0) := \frac{R_0}{1+z(x^0)} \tag{7.48}$$

by introducing the function $z(x^0)$ which is called redshift. First, this allows us to define the Hubble parameter as $H := \dot{R}/R$ which measures the rate of Universe expansion (dot means derivative with respect to proper spacetime). From this Hubble parameter, we thus easily express the proper spacetime as:

$$dx^0 = H^{-1}(R)\left(\frac{dR}{R}\right)$$

The line element takes the form of:

$$\begin{aligned} ds^2 &= H^{-2}(R)\left(\frac{dR}{R}\right)^2 - R^2\left[d\chi^2 + S^2_{(k)}(\chi)\ (d\theta^2 + \sin^2\theta\ d\varphi^2)\right], \\ R &= R(x^0) \end{aligned}$$

We deduce a formulation of the line element showing that in a Friedmann Universe model, the Hubble parameter is the only non-trivial function defining the spacetime, in addition to the redshift z, e.g. [16]:

$$\boxed{ds^2 = \frac{1}{(1+z)^2}\left\{\frac{dz^2}{H^2(z)} - R_0^2\left[d\chi^2 + S^2_{(k)}(\chi)\ (d\theta^2 + \sin^2\theta\ d\varphi^2)\right]\right\}} \tag{7.49}$$

The determination of the Hubble parameter $H[z(x^0)]$ thus constitutes an essential key step for cosmological theories and experimental measurements.

Remark 7.3. The discovery of the Doppler effect represents a cornerstone in the development of cosmological theory. Measurements of spectral shifts are essential in observational cosmology, as they allow the determination of the relative motions of celestial sources, such as stars and galaxies, via the Doppler effect. Indeed, the redshifts and blueshifts observed by Slipher, Wirtz, and others, later analyzed by Lemaître (1927) and Hubble (1929), made it possible to deduce that the Universe is expanding—leading to the formulation of the famous Hubble's law e.g. [12].

7.2.2.4 *(Anti)-de Sitter Metric vs. FLRW Metric*

Let's now use the hologram concept to relate (anti)-de Sitter and FLRW spacetimes. For that purpose, consider again a five-dimensional space with coordinates $(y^0, y^1, y^2, y^3, y^4)$ and let us apply particular transformations to embed the manifold, for de Sitter spacetime:

$$\begin{cases} y^0 = a \sinh\left(x^0/a\right) \\ y^1 = a \cosh\left(x^0/a\right) \cos\chi \end{cases}, \quad \begin{cases} y^2 = a \cosh\left(x^0/a\right) \sin\chi \cos\theta \\ y^3 = a \cosh\left(x^0/a\right) \sin\chi \sin\theta \cos\varphi \\ y^3 = a \cosh\left(x^0/a\right) \sin\chi \sin\theta \sin\varphi \end{cases} \tag{7.50}$$

given that the line element of de Sitter spacetime is:

$$ds^2 := (dy^0)^2 - (dy^1)^2 - (dy^2)^2 - (dy^3)^2 - (dy^4)^2$$

Introducing the transformations and variables $(x^0, \chi, \theta, \varphi)$ into the de Sitter metric gives the following after some lengthy calculus:

$$ds^2 = (dx^0)^2 - a^2 \cosh^2\left(x^0/a\right) \left[d\chi^2 + \sin^2\chi \left(d\theta^2 + \sin^2\theta \; d\varphi^2\right)\right] \tag{7.51}$$

It is remarkable that the expression of the de Sitter line element (7.51) belongs to the class of spherical FLRW spacetimes (7.44). For anti-de Sitter spacetime, we consider the special transformations:

$$\begin{cases} y^0 = a \sin\left(x^0/a\right) \\ y^1 = a \cos\left(x^0/a\right) \cosh\chi \end{cases}, \quad \begin{cases} y^2 = a \cos\left(x^0/a\right) \sinh\chi \cos\theta \\ y^3 = a \cos\left(x^0/a\right) \sinh\chi \sin\theta \cos\varphi \\ y^3 = a \cos\left(x^0/a\right) \sinh\chi \sin\theta \sin\varphi \end{cases} \tag{7.52}$$

given that the line element of anti-de Sitter spacetime is (be cautious with the sign of the second term):

$$ds^2 := (dy^0)^2 + (dy^1)^2 - (dy^2)^2 - (dy^3)^2 - (dy^4)^2$$

Introducing the transformations and variables $(x^0, \chi, \theta, \varphi)$ into anti-de Sitter metric gives the following after again some lengthy calculus:

$$ds^2 = (dx^0)^2 - a^2 \cos^2\left(x^0/a\right) \left[d\chi^2 + \sinh^2\chi \left(d\theta^2 + \sin^2\theta \; d\varphi^2\right)\right] \tag{7.53}$$

The de Sitter line element (7.53) belongs to the class of spherical FLRW spacetimes (7.42). At this step, the FLRW can be considered as holograms. Both de Sitter and anti-de Sitter spacetimes in five dimensions are transformed into four-dimensional FLRW spacetimes by special mappings. This is the Holography Principle proposed 25 years ago by [246] to relate relative gravitation and quantum field theory. In the present work, we recall that the expanding four-dimensional spacetime FLRW models, mainly utilized for cosmology, are holograms of de Sitter and anti-de Sitter five-dimensional spaces.

Remark 7.4. The search for extended Friedmann equations with a non-zero torsion would be deduced from two invariance equations: the zero Lie derivative of metric, the one that allowed us to obtain the FLRW metric, and additionally the zero Lie derivative of torsion. We will not, however, take this path; we return directly to field equations to study Einstein-Cartan spacetime.

7.2.3 *Derivation of Friedmann's Equations*

Briefly, Friedmann developed cosmology equations from solutions of Einstein's field equations as early as 1922, with a time-dependent scale factor $R(x^0)$, assuming perfect isotropic fluid as matter, with a energy density ρ and pressure p with these equations, Friedmann discovered that Einstein's relative gravitation predicts a non-stationary expansion of the Universe.[1] The aim of this subsection is to recall the derivation of Friedmann's equations.

7.2.3.1 *Einstein's Field Equations*

The action of vacuum spacetime $\mathscr{B}$ is given by $\mathscr{L}_{HE} := (1/2\chi)\int_{\mathscr{B}} \overline{\mathfrak{R}}\, \overline{\omega}_n$ where the curvature is now defined with the FLRW metric. A perfect fluid model is assumed to be the source of gravitation. To account for the perfect fluid acting force, authors introduced the energy-momentum:

$$T^{\text{fl}}_{\mu\nu} = (\rho + p)\, u_\mu u_\nu - p\, \eta_{\mu\nu} \tag{7.54}$$

within a flat Minkowski spacetime with metric $\eta_{\mu\nu} := \text{diag}\,[+1, -1, -1, -1]$ in special relativity. ρ and p are respectively the energy density and the pressure of the fluid. Since the pioneering works [312] and many others e.g. [313–315], several variational principles have been proposed, based on a Lagrangian for a perfect fluid $\mathscr{L}^{\text{fl}}$. In this discussion, we do not address issues related to the derivation of a fluid Lagrangian, including its limitations at the quantum level e.g. [316]; instead, we simply assume the

[1]At the time of the discovery of relative gravitation, the acceptance that the Universe is dynamic was not considered. Einstein himself had refuted this idea when he read Friedmann's mathematical results of May 1922 which showed that Einstein's field equations necessarily implied the non-stationarity of spacetime. This initially hindered the acceptance of Friedmann's discovery. However, Einstein ultimately acknowledged in May 1923 that Friedmann's solutions were indeed correct.

energy-momentum tensor as given. To begin with, we recall the related Einstein field equation:

$$\overline{\mathcal{R}}_{\mu\nu} - \frac{\overline{\mathcal{R}}}{2}\, g_{\mu\nu} + \Lambda\, g_{\mu\nu} = 8\pi G\, T^{\text{fl}}_{\mu\nu}$$

G is the universal gravity constant, introduced to fit with Newtonian gravitation in a flat spacetime and low velocity. For relative gravitation, the energy-momentum tensor is given by:

$$T^{\text{fl}}_{\mu\nu} := (\rho + p)\, u_\mu u_\nu - p\, g_{\mu\nu} \tag{7.55}$$

ρ and p are functions of only time. Conforming to the definition of a perfect isotropic fluid, the macroscopic velocity has only a time component $u_\mu = (1, 0, 0, 0)$. Accordingly, the energy-momentum tensor reduces to:

$$T^{\text{fl}}_{00} = \rho\, g_{00}, \quad T^{\text{fl}}_{rr} = -p\, g_{rr}, \quad T^{\text{fl}}_{\theta\theta} = -p\, g_{\theta\theta}, \quad T^{\text{fl}}_{\varphi\varphi} = -p\, g_{\varphi\varphi} \tag{7.56}$$

using a spherical coordinate system for space.

7.2.3.2 *Levi-Civita Connection, Ricci Curvature, and Einstein Tensor*

The first step for deriving the Friedmann equations is to express the Einstein tensor. Assuming spacetime homogeneity and isotropy, we use the FLRW metric. For completeness, we recall the line element formulation (7.41):

$$ds^2 = \left(dx^0\right)^2 - R^2(x^0)\left(\frac{dr^2}{1-kr^2} + r^2\, d\theta^2 + r^2 \sin^2\theta\, d\varphi^2\right), \quad k = -1, 0, +1$$

The time component of the metric is given by:

$$g_{00} = 1, \quad g^{00} = 1 \tag{7.57}$$

For the three spatial components, we have:

$$g_{rr} = -\frac{R^2}{1-kr^2}, \quad g^{rr} = -\frac{1-kr^2}{R^2} \tag{7.58}$$

$$g^{\theta\theta} = -r^2\, R^2, \quad g^{\theta\theta} = -\frac{1}{r^2\, R^2} \tag{7.59}$$

$$g_{\varphi\varphi} = -r^2 \sin^2\theta\, R^2, \quad g^{\varphi\varphi} = -\frac{1}{r^2 \sin^2\theta\, R^2} \tag{7.60}$$

Christoffel's symbols are calculated from the FLRW metric (7.37):

$$\overline{\Gamma}^0_{rr} = \frac{R\dot{R}}{1-kr^2}, \quad \overline{\Gamma}^0_{\theta\theta} = r^2\, R\dot{R}, \quad \overline{\Gamma}^0_{\varphi\varphi} = r^2\, R\dot{R}\, \sin^2\theta \tag{7.61}$$

$$\overline{\Gamma}^r_{rr} = \frac{kr}{1-kr^2}, \quad \overline{\Gamma}^r_{\theta\theta} = -r\left(1-kr^2\right), \tag{7.62}$$

$$\overline{\Gamma}^r_{\varphi\varphi} = -r\left(1-kr^2\right)\sin^2\theta$$

$$\overline{\Gamma}^r_{0r} = \overline{\Gamma}^r_{r0} = \overline{\Gamma}^\theta_{0\theta} = \overline{\Gamma}^\theta_{\theta 0} = \overline{\Gamma}^\varphi_{0\varphi} = \overline{\Gamma}^\varphi_{0\varphi} = \frac{\dot{R}}{R} \tag{7.63}$$

$$\overline{\Gamma}^\theta_{r\theta} = \overline{\Gamma}^\theta_{\theta r} = \overline{\Gamma}^\varphi_{r\varphi} = \overline{\Gamma}^\varphi_{\varphi r} = \frac{1}{r} \tag{7.64}$$

$$\overline{\Gamma}^\theta_{\varphi\varphi} = -\sin\theta\, \cos\theta, \quad \overline{\Gamma}^\varphi_{\theta\varphi} = \overline{\Gamma}^\varphi_{\varphi\theta} = \frac{\cos\theta}{\sin\theta} \tag{7.65}$$

Ricci curvature $\overline{\mathfrak{R}}_{\beta\lambda} := \overline{\mathfrak{R}}^\gamma_{\gamma\beta\lambda}$ whose components are:

$$\begin{cases} \overline{\mathcal{R}}_{00} = -3\, \dfrac{\ddot{R}}{R} \\ \overline{\mathcal{R}}_{rr} = \dfrac{R\ddot{R}}{1-kr^2} + \dfrac{2\dot{R}^2}{1-kr^2} + \dfrac{2k}{1-kr^2} \\ \overline{\mathcal{R}}_{\theta\theta} = r^2 R\ddot{R} + 2r^2\dot{R}^2 + 2kr^2, \quad \text{and} \quad \overline{\mathcal{R}}_{\varphi\varphi} = \overline{\mathcal{R}}_{\theta\theta}\, \sin^2\theta \end{cases} \tag{7.66}$$

It is worth observing that the spatial components write:

$$\overline{\mathfrak{R}}_{ii} = -g_{ii}\left(\frac{\ddot{R}}{R} + 2\,\frac{\dot{R}^2}{R^2} + 2\,\frac{k}{R^2}\right), \quad (ii) = (rr), (\theta\theta), (\varphi\varphi) \tag{7.67}$$

showing that the Ricci curvature is proportional to the metric tensor, for the space part. We deduce accordingly the scalar curvature:

$$\overline{\mathcal{R}} := g^{\alpha\beta}\, \overline{\mathfrak{R}}_{\alpha\beta} = -6\left(\frac{\ddot{R}}{R} + \frac{\dot{R}^2}{R^2} + \frac{k}{R^2}\right) \tag{7.68}$$

This expression conforms to the initial assumption that the Universe has a uniform curvature at each instant x^0, but this curvature evolves with time. From the Ricci curvature we deduce the Einstein tensor $\overline{G}_{\mu\nu} := \overline{R}_{\mu\nu} - (\overline{\mathcal{R}}/2)\, g_{\mu\nu}$, then:

$$\overline{G}_{00} = 3\left(\frac{\dot{R}^2}{R^2} + \frac{k}{R^2}\right), \quad \overline{G}_{ii} = \left(2\frac{\ddot{R}}{R} + \frac{\dot{R}^2}{R^2} + \frac{r}{R^2}\right) g_{ii} \tag{7.69}$$

where spatial components have the same shape. Equations (7.69) are consistent with (7.32) which mentions that the Ricci curvature is proportional to the metric. In the absence of matter and physical sources, field equations write $\overline{G}_{\beta\lambda} = 0$ with $\beta, \lambda = 0, 3$.

7.2.3.3 *Friedmann's Equations*

Derivation of Friedmann's equations starts with Einstein's field equations including a source term representing to a perfect fluid. The Hilbert-Einstein action with an *ad hoc* cosmological constant Λ is reminded:

$$\mathscr{S}_{HE} := \frac{1}{2\chi} \int_{\mathscr{B}} \left(\overline{\mathcal{R}} - 2\Lambda\right) \overline{\omega}_n$$

with volume-form $\overline{\omega}_n := \sqrt{|\mathrm{Det}\mathbf{g}|}\, dx^0 \wedge dx^1 \wedge dx^2 \wedge dx^3$. Curvature is calculated with the FLRW metric. Compatibility of connection with volume-form allows us to satisfy the divergence theorem:

$$\mathcal{L}_\xi \overline{\omega}_n = \left(\overline{\nabla}_\alpha \xi^\alpha\right) \overline{\omega}_n$$

Compatibilities of metric, connection and volume-form allow us to drop the Ricci curvature variation:

$$\begin{aligned}\int_{\mathscr{B}} g^{\beta\lambda} \left(\delta^\alpha_\gamma \delta \Re_{\beta\lambda}\right) \overline{\omega}_n &= \int_{\mathscr{B}} g^{\beta\lambda}\, \delta^\alpha_\gamma \left(\overline{\nabla}_\alpha \delta \overline{\Gamma}^\gamma_{\beta\lambda} - \overline{\nabla}_\beta \delta \overline{\Gamma}^\gamma_{\alpha\lambda}\right) \overline{\omega}_n \\ &= \int_{\mathscr{B}} \left[\overline{\nabla}_\alpha \left(g^{\beta\lambda}\, \delta^\alpha_\gamma \overline{\Gamma}^\gamma_{\beta\lambda}\right) - \overline{\nabla}_\beta \left(g^{\beta\lambda}\, \delta^\alpha_\gamma \delta \overline{\Gamma}^\gamma_{\alpha\lambda}\right)\right] \overline{\omega}_n\end{aligned}$$

Divergence theorem implies that the integral vanishes when assuming $\delta\Gamma^\gamma_{\alpha\beta} = 0$ at the boundary $\partial\mathscr{B}$. The classical Einstein field equations then hold. Perfect fluid energy-momentum is given by $T^{\mathrm{fl}}_{\mu\nu} = (\rho + p) u_\mu u_\nu - p g_{\mu\nu}$. First, starting from the temporal component of the Einstein field equations, we get:

$$\overline{G}_{00} + \Lambda\, g_{00} = 8\pi G\, T^{\mathrm{fl}}_{00} \implies 3\left(\frac{\dot{R}^2}{R^2} + \frac{k}{R^2}\right) = 8\pi G\, \rho - \Lambda \tag{7.70}$$

Considering a perfect isotropic fluid implies that the field equation reduces to its diagonal part $(\beta\lambda) = (ii)$, which is an important consequence of isotropy. Second, for the spatial part, we have:

$$\overline{G}_{ii} + \Lambda\, g_{ii} = 8\pi G\, T^{\text{fl}}_{ii}, \quad \{ii\} = (\{rr\}, \{\theta\theta\}, \{\varphi\varphi\})$$

We can factorize and then drop the metric component g_{ii} to obtain[2]:

$$2\,\frac{\ddot{R}}{R} + \frac{\dot{R}^2}{R^2} + \frac{k}{R^2} + \Lambda = -8\pi G\, p \tag{7.71}$$

Eliminating $\dot{R}^2/R^2$ by a linear combination of (7.70) and (7.71) gives the Friedmann equation for relative gravitation.

Friedmann's equations within Einstein spacetime (1929). Friedmann's equations constitute the basic model of an expanding Universe [306]:

$$\begin{cases} \left(\dfrac{\dot{R}}{R}\right)^2 = \dfrac{8\pi G}{3}\rho - \dfrac{\Lambda}{3} - \dfrac{k}{R^2} \\ \dfrac{\ddot{R}}{R} = -\dfrac{4\pi G}{3}(\rho + 3p) - \dfrac{\Lambda}{3} \end{cases} \tag{7.72}$$

where $k = -1, 0, +1$ and where ρ and p are respectively the energy density and the pressure of the fluid filling the vacuum spacetime.

(7.72) highlight that the Universe is not static in general. There are only two independent equations. Most solutions for various situations can be found in e.g. [241] when $k = -1, 0, +1$ or $\Lambda = 0, \Lambda \neq 0$. Solutions of $R(t)$ are also provided accordingly.

Remark 7.5. Equations (7.72) cover numerous contributions such as the fluid matter, the background radiation and "the energy of the vacuum". It is worth noting that these equations are independent of the equation of state of the fluid $p(\rho)$. The term k characterizing the Universe curvature (spatially open, flat, or spatially closed) is constant throughout a particular solution, but may vary from one solution to another. Following the discovery of the accelerated expansion of the Universe [18, 19], and remaining within the framework of Einstein's theory of gravitation, one proposed explanation is the existence of exotic—or at least hypothetical—forms of matter for which

[2]This is a four-dimensional generalization of three-dimensional Pascal fluid, whose internal stress reduces to a equal scalar pressure p on all cut-off surfaces in the fluid. The Cauchy stress tensor reduces to a hydrostatic pressure having the same value $\sigma := -p\,\mathbf{i}$ in all directions e.g. [133].

$\rho + 3p < 0$, giving rise to the so-called dark energy. The sign of the guessed cosmological constant Λ plays also a key role in the process. Since then, the determination of dark energy remains a crucial challenge for physics and mathematics of cosmology.

Remark 7.6. This subsection aims to recall these classic equations in order to highlight the key role of cosmological constant Λ in cosmology theory:

(1) $\Lambda < 0$ contributes to accelerated expansion;
(2) $\Lambda = 0$ plays no role;
(3) $\Lambda > 0$ contributes to contraction.

Remark 7.7. Despite the two Friedmann equations, published in 1922 and 1924, almost all researchers believed that the Universe was static until 1929, when Hubble discovered that it is expanding [12]. Even Einstein accepted the idea of expansion only much later. It was Lemaître, with a paper published in 1927, who shifted the opinion of the majority of researchers toward accepting cosmic expansion [308].

Remark 7.8. This book focuses on the macroscopic scale and is limited to relative gravitation based on continuum physics e.g. [6], leaving aside the important aspects of quantum physics of elementary particles e.g. [15] and historical perspectives e.g. [32]. More specifically, the effort is dedicated to the cosmological constant Λ and its possible origins from electromagnetism, particularly the magnetic component e.g. [27, 64, 90, 94, 258]. It should also be noted that the discussion is limited to Einstein gravitation without torsion. The introduction of a weak torsion field into Friedmann's equations has already been addressed e.g. [275, 276], where, in the first reference, the two equations are treated as a dynamical system with a focus on stability analysis. This approach allowed the authors to study the influence of the torsion vector on the system's behavior.

Remark 7.9. Phenomenologically, the use of a perfect fluid model filling vacuum spacetime is common and classic. At the quantum level and steady-state regime, and taking into account its particle constituents which have no interaction with each other but only with the gravitational field, the addition of the behaviors of these particles is incoherent in the sense that an equal *a priori* probability is assumed for individual quantum states. No dynamical justification for the appropriateness of such an assumption seems to exist, but this procedure is often taken as self-evident. Such incoherence of averaging in a curved spacetime was discussed in e.g. [316].

7.3 Cosmology with Torsion as Background Field

In deriving the Friedmann equations, only gravitation is considered. However, given that cosmology studies the Universe at very large scales, it makes sense to focus on both long-range forces—gravitation and electromagnetism—which act over infinite distances. We have seen that measurements of the Universe's expansion are based on the propagation of electromagnetic waves emitted by stars and galaxies moving away from each other (redshift). However, there is another fundamental reason to consider the electromagnetic field in cosmology: the Cosmic Microwave Background (CMB). The CMB is the residual electromagnetic radiation from the Big Bang which Dicke, Peebles, and their collaborators theoretically predicted in 1948 and which Penzias and Wilson accidentally discovered in 1965. Since then, several major international projects such as the Cosmic Background Explorer, Wilkinson Microwave Anisotropy Probe, Euclid, Planck and others have confirmed and refined theoretical models. The CMB exhibits great uniformity on large scales and near-isotropy, although local inhomogeneities and anisotropies can be observed (e.g., baryonic acoustic oscillations and diffusion damping). These phenomena are multi-parametric and multi-physical e.g. [20], which naturally motivates the consideration of electromagnetic waves in a book devoted to modeling vacuum spacetime.

Initially, spacetime models with electromagnetic fields based solely on a metric have been developed without introducing torsion. Such is the case for Reissner-Nordström which is a non-homogeneous spacetime due to a static electric charge. Another example of spacetime was proposed independently by Bertotti (1959) and Robinson (1959) by expressing a diagonal metric tensor e.g. [241]:

$$ds^2 = \frac{e^2}{r^2}\left(d\tau^2 - dr^2 - r^2 d\theta^2 - r^2 \sin^2\theta \; d\varphi^2\right) \tag{7.73}$$

using coordinate system $(\tau, r, \theta, \varphi)$. e is an electric charge with adapted dimensions. It is obvious that this spacetime is conformally flat; the terms in parentheses present the flat Minkowski metric. Taking advantage of the change of variables, without going into details:

$$\tau = \frac{e\sqrt{e^2 + R^2}\; \sin(t/e)}{R + \sqrt{e^2 + R^2}\; \cos(t/e)}, \quad r = \frac{e^2}{R + \sqrt{e^2 + R^2}\; \cos(t/e)}$$

The line element of Bertotti-Robinson takes the form of:

$$ds^2 = \left(1 + \frac{R^2}{e^2}\right) dt^2 - \left(1 + \frac{R^2}{e^2}\right)^{-1} dR^2 - e^2\left(d\theta^2 + \sin^2\theta \; d\varphi^2\right)$$

within the coordinate system (T, R, θ, φ) showing that the first two terms represent the metric of a two-dimensional $\mathscr{A}$de$\mathscr{S}$ spacetime of radius e. Despite the interest of such a spacetime, we will focus on spacetime with torsion.

7.3.1 *Method of Averaged Torsion*

Let's consider the Einstein-Palatini action modeling the vacuum Einstein-Cartan spacetime. In such a case, curvature depends on both metric via the Levi-Civita connection and on contortion tensor. *For the sake of clarity, this first step proposes a method based on the averaged value of torsion to define a cosmological constant* $\Lambda_{\aleph}$. In this subsection, the torsion field is assumed as a background field of the spacetime and does not constitute a dynamical geometric variable.

7.3.1.1 *Friedmann's Equations with Averaged Torsion Background*

We deal with a torsion field which remains uniform in space. It is a model of a diffuse magnetic monopole field to be more precise. We then find ourselves in the framework of the Einstein-Palatini action which reduces for a uniform torsion field to (see (6.116)):

$$\mathscr{S}_{EP} := \frac{1}{2\chi}\int_{\mathscr{B}} \mathcal{R}\,\overline{\omega}_n = \frac{1}{2\chi}\int_{\mathscr{B}} \left(\overline{\mathcal{R}} - 2\overline{\Lambda}_{\aleph}\right)\overline{\omega}_n$$

in which the averaged cosmological constant is derived from particular properties of the torsion field. In the general case, the following development should be improved and may become very complex. The Ricci curvature may be decomposed by calculating with the Levi-Civita connection and the part due to the contortion tensor, namely with a quadratic term (6.114):

$$\Re_{\beta\lambda} = \overline{\Re}_{\beta\lambda} + \overline{\nabla}_{\alpha}\mathcal{T}^{\alpha}_{\beta\lambda} - \overline{\nabla}_{\beta}\mathcal{T}^{\alpha}_{\alpha\lambda} - \left(\mathcal{T}^{\alpha}_{\beta\mu}\mathcal{T}^{\mu}_{\alpha\lambda} - \mathcal{T}^{\alpha}_{\alpha\mu}\mathcal{T}^{\mu}_{\beta\lambda}\right) \tag{7.74}$$

Compatibility of $\overline{\omega}_n$ and $\overline{\nabla}$ implies that the terms under covariant derivative tend to zero at the boundaries provided kinematical compatibility of the connection at $\partial\mathscr{B}$. The simplified hypothesis is then to consider that the averaged cosmological constant $\overline{\Lambda}_{\aleph}$ is none other than the contribution of the contortion tensor (or equivalently of the torsion tensor, (6.117)), given formally with the electromagnetic field (6.199):

$$\overline{\Lambda}_{\aleph} := \frac{1}{2\int_{\mathscr{B}}\overline{\omega}_n}\int_{\mathscr{B}} g^{\beta\lambda}\left(\mathcal{T}^{\alpha}_{\beta\mu}\mathcal{T}^{\mu}_{\alpha\lambda} - \mathcal{T}^{\alpha}_{\alpha\mu}\mathcal{T}^{\mu}_{\beta\lambda}\right)\overline{\omega}_n \tag{7.75}$$

The averaging method might be drastic but it allows us, via a phenomenological method, to more or less take into account the influence of the diffuse radiation of the electromagnetic field in the Universe on its evolution. This hypothesis also has the advantage of being very clear and can be justified if the torsion field is uniform on very large scales. Under these assumptions, we deduce Einstein's field equations (with the classical "overlines") with $\overline{\Lambda}_{\aleph}$. This does not allow us to completely appreciate the role of the electromagnetic field in cosmology but the classical shape of Einstein's field equation holds:

$$\overline{G}_{\mu\nu} + \overline{\Lambda}_{\aleph}\ g_{\mu\nu} = 0 \tag{7.76}$$

By analogy, adding the energy-momentum of a perfect fluid $T^{\text{fl}}_{\mu\nu}$, we deduce Friedmann's equations (7.72), where the cosmological constant is defined by torsion:

$$\begin{cases} \left(\dfrac{\dot{R}}{R}\right)^2 = \dfrac{8\pi G}{3}\rho - \dfrac{\overline{\Lambda}_{\aleph}}{3} - \dfrac{k}{R^2} \\ \dfrac{\ddot{R}}{R} = -\dfrac{4\pi G}{3}(\rho + 3p) - \dfrac{\overline{\Lambda}_{\aleph}}{3} \end{cases} \tag{7.77}$$

We see that a negative cosmological constant $\overline{\Lambda}_{\aleph} < 0$ contributes to the accelerated expansion of the Universe. It supports the idea that averaging the torsion can be considered as a source of the cosmological constant, and by extension the dark energy.

7.3.1.2 *Example of Torsion Vector Field*

The torsion tensor can be decomposed into two components: (a) the first is defined by its trace allowing us to calculate the torsion vector, (2) the second is the component defined by the torsion from which the trace has been removed. The present case will be introduced with more details later; at this step, we just take it as an example of application. We consider a particular case of a torsion field e.g. [282]:

$$\aleph^{\lambda}_{\mu\nu} = \aleph_{\mu}\ \delta^{\lambda}_{\nu} - \aleph_{\nu}\ \delta^{\lambda}_{\mu} \tag{7.78}$$

in which $\aleph_{\mu}$ is a 1-form we call a torsion vector or torsion trace vector. By calculating the trace, we obtain the expression of the vector:

$$\aleph^{\nu}_{\mu\nu} = \aleph_{\mu}\ \delta^{\nu}_{\nu} - \aleph_{\nu}\ \delta^{\nu}_{\mu} = 4\ \aleph_{\mu} - \aleph_{\mu} \quad \Longrightarrow \quad \boxed{\aleph_{\mu} = \frac{1}{3}\ \aleph^{\nu}_{\mu\nu}} \tag{7.79}$$

To express the contortion tensor in terms of the torsion vector, we use the formula:

$$\mathcal{T}^{\lambda}_{\mu\nu} := \frac{1}{2}\aleph^{\lambda}_{\mu\nu} + \frac{1}{2}g^{\lambda\sigma}\left(g_{\mu\epsilon}\aleph^{\epsilon}_{\nu\sigma} + g_{\nu\epsilon}\aleph^{\epsilon}_{\mu\sigma}\right) \tag{7.80}$$

By introducing the relation (7.79) we deduce:

$$\mathcal{T}^{\lambda}_{\mu\nu} = -\delta^{\lambda}_{\mu}\,\aleph_{\nu} + g^{\lambda\sigma}g_{\mu\nu}\aleph_{\sigma} = \left(g^{\lambda\sigma}g_{\mu\nu} - \delta^{\lambda}_{\mu}\delta^{\sigma}_{\nu}\right)\aleph_{\sigma} \tag{7.81}$$

For expressing the curvature in terms of the Riemann curvature and the torsion vector, intermediate calculus gives:

$$\mathcal{T}^{\alpha}_{\beta\mu}\mathcal{T}^{\mu}_{\alpha\lambda} = \left(\delta^{\sigma}_{\lambda}\delta^{\epsilon}_{\beta} - g^{\sigma\epsilon}g_{\beta\lambda}\right)\aleph_{\sigma}\aleph_{\epsilon}$$

and:

$$\mathcal{T}^{\alpha}_{\alpha\mu}\mathcal{T}^{\mu}_{\beta\lambda} = -3\left(g^{\sigma\epsilon}g_{\beta\lambda} - \delta^{\sigma}_{\beta}\delta^{\epsilon}_{\lambda}\right)\aleph_{\sigma}\aleph_{\epsilon}$$

As an intermediate step, we may write:

$$\mathcal{T}^{\alpha}_{\beta\mu}\mathcal{T}^{\mu}_{\alpha\lambda} - \mathcal{T}^{\alpha}_{\alpha\mu}\mathcal{T}^{\mu}_{\beta\lambda} = -2\left(\delta^{\sigma}_{\beta}\delta^{\epsilon}_{\lambda} - g^{\sigma\epsilon}g_{\beta\lambda}\right)\aleph_{\sigma}\aleph_{\epsilon} \tag{7.82}$$

$$g^{\beta\lambda}\left(\mathcal{T}^{\alpha}_{\beta\mu}\mathcal{T}^{\mu}_{\alpha\lambda} - \mathcal{T}^{\alpha}_{\alpha\mu}\mathcal{T}^{\mu}_{\beta\lambda}\right) = 6g^{\sigma\epsilon}\aleph_{\sigma}\aleph_{\epsilon} \tag{7.83}$$

After calculations, the cosmological constant is given by torsion contribution:

$$\boxed{\overline{\Lambda}_{\aleph} = g^{\beta\lambda}\left[-\aleph_{\beta}\,\aleph_{\lambda} + \left(g^{\mu\nu}\aleph_{\mu}\,\aleph_{\nu}\right)g_{\beta\lambda}\right] = 3\ g^{\beta\lambda}\aleph_{\beta}\aleph_{\lambda}} \tag{7.84}$$

We can distinguish numerous situations, decomposed onto the tetrad (ϑ^0, $\vartheta^1, \vartheta^2, \vartheta^3$), where the metric components reduce to diag$[+1, -1, -1, -1]$:

(1) For a timelike torsion vector $\aleph_{\mu}(\aleph_0, 0, 0, 0)$, we have:

$$\overline{\Lambda}_{\aleph} = 3\ \aleph_0^2 > 0$$

According to Friedmann's equations (7.77), torsion contributes to cosmic contraction (it acts as dark matter).

(2) For a spacelike torsion vector, $\aleph_{\lambda}(0, \aleph_{\text{sp}}, \aleph_{\text{sp}}, \aleph_{\text{sp}})$, the constant becomes:

$$\overline{\Lambda}_{\aleph} = -9\ \aleph_{\text{sp}}^2 < 0$$

Again from (7.77), the torsion vector contributes to the accelerating expansion (it acts as dark energy).

Remark 7.10. The two types of torsion vector (temporal and spatial) keep the isotropy of the Universe. Such is not the case in the general form of the torsion vector.

7.3.1.3 *Summary on the Averaged Torsion Method*

In these subsections, we considered an Einstein-Palatini Lagrangian and the integration over spacetime using a Riemann volume-form $\overline{\omega}_n$. Then, we decomposed curvature into a part depending only on Levi-Civita connection $\overline{\nabla}$ and a quadratic scalar function obtained using contortion $\mathcal{T}^{\gamma}_{\alpha\beta}$. Compatibility of Riemanian volume-form and Levi-Civita connection allowed us first to drop covariant gradient terms, and then to define by the average method a cosmological constant $\overline{\Lambda}_{\aleph}$ and thus obtain the associated Friedmann equations.

> **Averaged torsion method and Universe expansion.** The method of averaged torsion is certainly an approximate way to still be able to work with the FLRW metric and respect the two principles of cosmology. On the one hand, it is an approximate method for proposing geometrically $\Lambda_{\aleph}$ as the source of the acceleration of the Universe expansion, but on the other hand, this approach does not indicate the physical nature of torsion.

7.3.2 *FLRW Spacetime with a Particular Torsion Field*

For Einstein-Cartan spacetime, the influence of a uniform torsion has already been studied to generalize Friedmann's equations with torsion. Authors mostly assumed that the non-zero component of contortion $\mathcal{T}$ depending only on time x^0 is proportional to the radius of the Universe $\mathcal{T} = \alpha\ R(x^0)$ with $\alpha \in \mathbb{R}^+$ e.g. [89, 317]. In the same way, numerous works investigated the relevance of using characteristics of Einstein-Cartan spacetime, particularly the influence of torsion on the behavior of singularities e.g. [4], the role played by torsion to introduce the cosmological constant e.g. [38], the influence of torsion on the dynamic behavior of Friedmann spaces e.g. [275, 276], and the acceleration of these Friedmann spaces in the presence of torsion e.g. [318]. Conforming to these approaches, we give in this subsection some introductory elements to Einstein-Cartan cosmology where the common thread is to consider a FLRW metric to which we add a particular torsion tensor $\aleph$ that we have already seen before. *Again, the Riemann volume-form $\overline{\omega}_n$ is used in this section.* By the way, we already have shown that considering a particular example of a homogeneous torsion within a Schwarzschild spacetime leads to an anti-de Sitter spacetime, and permits us to relate the cosmological constant and the square

$\Lambda \propto \aleph_0^2$. Here, the difference is that we consider a FLRW spacetime from the outset.

Remark 7.11. Knowing that there is no encroachment on the part of affine connection deduced from metric, and that deduced directly by torsion, we can consider without any problem the components of the resulting connection.

A swift review of literature shows that generalizations of Friedmann's equations have been carried out by considering either modified gravitation, in which Einstein-Palatini is used instead of Hilbert-Einstein action, by using a function of scalar curvature $f(\mathcal{R})$, or by using FLRW metric but adding a torsion $\aleph^{\gamma}_{\alpha\beta}(x^\lambda)$ additionally to metric $g_{\alpha\beta}(x^\lambda)$ e.g. [51] (for a homogeneous torsion field) [52, 62, 137, 282]. Salient from these different studies is the reduction of torsion to torsion vector. Another finding is the need for matter or physical interactions to create torsion, thus the presence of torsion in vacuum spacetime is a matter of concern.

7.3.2.1 *Levi-Civita Connection*

Derivation of Friedmann's equations for Einstein gravitation requires the presence of a perfect fluid with a density ρ and a pressure p. Here, we also consider a fluid and additionally a particular torsion field $\aleph$ as previously (6.122):

$$\aleph = \aleph^0_{\theta\varphi}\, \vartheta^\theta \otimes \vartheta^\varphi \otimes \mathbf{e}_0 + \aleph^0_{\varphi\theta}\, \vartheta^\varphi \otimes \vartheta^\theta \otimes \mathbf{e}_0$$

assumed to be covariantly constant. It is worth not forgetting that the torsion field is mathematically linked to the electromagnetic field as shown in (6.199). It was deduced from an action combining the Hilbert-Einstein and the Yang-Mills Lagrangians. Before embarking on the calculus of Ricci curvature, it is worth recalling some tensor projection. Given that the metric is decomposed as follows onto the tetrads $(\vartheta^0, \vartheta^r, \vartheta^\theta, \vartheta^\varphi)$:

$$\mathbf{g} = \vartheta^0 \times \vartheta^0 - \vartheta^r \times \vartheta^r - \vartheta^\theta \times \vartheta^\theta - \vartheta^\varphi \times \vartheta^\varphi \tag{7.85}$$

we deduce their projection onto the standard base of 1-form $(dx^0, dr, d\theta, d\varphi)$:

$$\vartheta^0 = dx^0, \quad \vartheta^r = \frac{R}{\sqrt{1 - k\, r^2}}\, dr, \quad \vartheta^\theta = rR\, d\theta, \quad \vartheta^\varphi = rR \sin\theta\, d\varphi \tag{7.86}$$

We recall that the curvature parameter $k = -1$ stands for Lobatchevski hyperbolic manifold, $k = 0$ for a flat space, and $k = +1$ for spherical manifold. Therefore, the imposed torsion field has the components:

$$\aleph = \aleph^0_{\theta\varphi}\ r^2R^2\sin\theta\ d\theta\otimes d\varphi\otimes\partial_0 - \aleph^0_{\varphi\theta}\ r^2R^2\sin\theta\ d\varphi\otimes d\theta\otimes\partial_0 \tag{7.87}$$

showing that, practically, the component of the torsion $\aleph_0$ on the tetrad should be replaced by $\aleph_0\ r^2R^2\sin\theta$ when we are working on the standard base. The contortion is reminded for Riemann-Cartan manifolds:

$$\mathfrak{T}^{\gamma}_{\alpha\beta} = \Omega^{\gamma}_{\alpha\beta} + g^{\gamma\rho}\left(g_{\alpha\epsilon}\ \Omega^{\epsilon}_{\rho\beta} + g_{\beta\epsilon}\ \Omega^{\epsilon}_{\rho\alpha}\right), \quad \Omega := \frac{1}{2}\ \aleph$$

Non-zero components of $\mathfrak{T}$ associated with the prescribed torsion (6.124) are (they were established for Schwarzschild metric but they remain also valid for FLRW metric):

$$\begin{cases} \mathfrak{T}^0_{\theta\varphi} = \Omega^0_{\theta\varphi}, & \mathfrak{T}^0_{\varphi\theta} = -\mathfrak{T}^0_{\theta\varphi} \\ \mathfrak{T}^{\theta}_{0\varphi} = g^{\theta\theta}g_{00}\ \Omega^0_{\theta\varphi}, & \mathfrak{T}^{\theta}_{\varphi 0} = \mathfrak{T}^{\theta}_{0\varphi} \\ \mathfrak{T}^{\varphi}_{0\theta} = g^{\varphi\varphi}g_{00}\ \Omega^0_{\varphi\theta}, & \mathfrak{T}^{\varphi}_{\theta 0} = \mathfrak{T}^{\varphi}_{0\theta} \end{cases} \tag{7.88}$$

where the metric components $g_{\mu\nu}$ and their inverse $g^{\mu\nu}$ are those of the FLRW spacetime of classical cosmology (components (7.57)–(7.60)). For the time component we have respectively:

$$g_{00} = 1, \quad g^{00} = 1 \tag{7.89}$$

where we use the time coordinate $x^0 := ct$. For the spatial components, we obtain:

$$g_{rr} = -\frac{R^2}{1-kr^2}, \quad g^{rr} = -\frac{1-kr^2}{R^2} \tag{7.90}$$

$$g_{\theta\theta} = -r^2\ R^2, \quad g^{\theta\theta} = -\frac{1}{r^2\ R^2} \tag{7.91}$$

$$g_{\varphi\varphi} = -r^2\sin^2\theta\ R^2, \quad g^{\varphi\varphi} = -\frac{1}{r^2\sin^2\theta\ R^2} \tag{7.92}$$

Let's calculate the connection coefficients by the formula:

$$\Gamma^{\gamma}_{\alpha\beta} = \frac{1}{2}g^{\gamma\lambda}\left(\frac{\partial g_{\lambda\beta}}{\partial x^{\alpha}} + \frac{\partial g_{\alpha\lambda}}{\partial x^{\beta}} + \frac{\partial g_{\alpha\beta}}{\partial x^{\lambda}}\right) + \mathfrak{T}^{\gamma}_{\alpha\beta}$$

For the following components, Christoffel's symbols are the same as for the FLRW metric (7.37) without torsion; the "overline bar" is merely dropped:

$$\Gamma^0_{rr} = \frac{R\dot{R}}{1-kr^2}, \quad \Gamma^0_{\theta\theta} = r^2\, R\dot{R}, \quad \Gamma^0_{\varphi\varphi} = r^2\, R\dot{R}\, \sin^2\theta \tag{7.93}$$

$$\begin{aligned} \Gamma^r_{rr} &= \frac{kr}{1-kr^2}, \quad \Gamma^r_{\theta\theta} = -r\left(1-kr^2\right), \\ \Gamma^r_{\varphi\varphi} &= -r\left(1-kr^2\right)\sin^2\theta \end{aligned} \tag{7.94}$$

$$\Gamma^r_{0r} = \Gamma^r_{r0} = \Gamma^\theta_{0\theta} = \Gamma^\theta_{\theta 0} = \Gamma^\varphi_{0\varphi} = \overline{\Gamma}^\varphi_{0\varphi} = \frac{\dot{R}}{R} \tag{7.95}$$

$$\Gamma^\theta_{r\theta} = \Gamma^\theta_{\theta r} = \Gamma^\varphi_{r\varphi} = \Gamma^\varphi_{\varphi r} = \frac{1}{r} \tag{7.96}$$

$$\Gamma^\theta_{\varphi\varphi} = -\sin\theta\, \cos\theta, \quad \Gamma^\varphi_{\theta\varphi} = \Gamma^\varphi_{\varphi\theta} = \frac{\cos\theta}{\sin\theta} \tag{7.97}$$

There is no overlapping between the Christoffel symbols and the connection components engendered by the torsion. Calculus of Levi-Civita connection coefficients does not need the use of torsion tensor. This is why we can drop the "overline bar" on these coefficients without ambiguity.

7.3.2.2 *Connection Generated by the ad hoc Torsion*

Now let's introduce a uniform *ad hoc* tensor field characterized by a scalar $\aleph_0$. For the remaining connection coefficients due to torsion, we can use the relation $\Gamma^\gamma_{\alpha\beta} = \overline{\Gamma}^\gamma_{\alpha\beta} + \mathcal{T}^\gamma_{\alpha\beta}$ to give, considering the components of the contortion tensor on the base $(dx^0, dr, d\theta, d\varphi)$:

$$\begin{cases} \Gamma^0_{\theta\varphi} = \dfrac{\aleph_0}{2}\, r^2 R^2 \sin\theta, & \Gamma^0_{\varphi\theta} = -\dfrac{\aleph_0}{2}\, r^2 R^2 \sin\theta \\ \Gamma^\theta_{0\varphi} = -\dfrac{\aleph_0}{2}\, \sin\theta, & \Gamma^\theta_{\varphi 0} = -\dfrac{\aleph_0}{2}\, \sin\theta \\ \Gamma^\varphi_{0\theta} = -\dfrac{1}{\sin\theta}\dfrac{\aleph_0}{2}, & \Gamma^\varphi_{\theta 0} = -\dfrac{1}{\sin\theta}\dfrac{\aleph_0}{2} \end{cases} \tag{7.98}$$

All other coefficients are equal to zero. Knowledge of these coefficients allows us to derive geodesic equations as a first-order differential system.

Remark 7.12. In principle, derivation of Friedmann's equations for Einstein-Cartan spacetime will have to pass through the system of Einstein-Cartan-Maxwell equations, given that an electromagnetic field causes a torsion of spacetime (from (6.199)).[3] The first step is to introduce into this system the particular form of the FLRW metric, the second step is to also introduce the torsion (7.87) (compatibility should be checked carefully) that we wish to solve, and finally solution of the system must be carried out. However, resolution in metric $g_{\alpha\beta}$, torsion $\aleph^{\gamma}_{\alpha\beta}$ and electromagnetic potential A_α remains extremely difficult or even impossible in the current state, in its generality, namely because of the coupling of equations together with their nonlinearities. Some simplifying but nevertheless relevant assumptions are necessary to go further.

7.3.2.3 *Ricci Curvature Tensor After ad hoc Torsion Field*

We choose this example taken from Prasanna's paper [61] to illustrate the method which starts from a FLRW metric and then additionally introduces a torsion field supposed to be generated either by a spin fluid or a magnetic monopole. The Ricci curvature is accordingly obtained (6.114):

$$\mathfrak{R}_{\beta\lambda} = \overline{\mathfrak{R}}_{\beta\lambda} + \overline{\nabla}_\alpha \mathfrak{T}^{\alpha}_{\beta\lambda} - \overline{\nabla}_\beta \mathfrak{T}^{\alpha}_{\alpha\lambda} - (\mathfrak{T}^{\alpha}_{\beta\mu}\mathfrak{T}^{\mu}_{\alpha\lambda} - \mathfrak{T}^{\alpha}_{\alpha\mu}\mathfrak{T}^{\mu}_{\beta\lambda}) \tag{7.99}$$

We consider an uniform torsion field $\aleph = \aleph_0$ whose non-zero components are decomposed onto the tetrad basis $(\vartheta^0, \vartheta^1, \vartheta^2, \vartheta^3)$, with $\aleph_0 \in \mathbb{R}$ are $\aleph^0_{\theta\varphi} = \aleph_0$, and $\aleph^0_{\varphi\theta} = -\aleph_0$. We already calculated Ricci curvature $\overline{\mathfrak{R}}_{\alpha\beta}$ associated with the Levi-Civita connection by using the FLRW metric to which we add the part due to torsion $\aleph$:

$$\mathfrak{R}_{00} = \overline{\mathfrak{R}}_{00} - 2g^{\theta\theta}g^{\varphi\varphi}g^2_{00}\,\Omega^0_{\theta\varphi}\,\Omega^0_{\varphi\theta},$$

$$\mathfrak{R}_{rr} = \overline{\mathfrak{R}}_{rr}, \quad \mathfrak{R}_{\theta\theta} = \overline{\mathfrak{R}}_{\theta\theta}, \quad \mathfrak{R}_{\varphi\varphi} = \overline{\mathfrak{R}}_{\varphi\varphi}$$

showing that for this case, only the time-like component of the Ricci curvature is influenced by torsion. We need hereafter only the contortion components:

$$\Omega^0_{\theta\varphi} = \frac{\aleph_0}{2}\, r^2 R^2 \sin\theta, \quad \Omega^0_{\varphi\theta} = -\frac{\aleph_0}{2}\, r^2 R^2 \sin\theta \tag{7.100}$$

At this stage, we make a restrictive hypothesis, probably to be justified in more depth in the future, that despite the *ad hoc* introduction of an additional torsion, the use of the FLRW metric remains valid. In such a

[3]The angular momentum density of the electromagnetic field can be covariantly decomposed into orbital and spin parts. The torsion is exactly equal to the spin angular momentum e.g. [222], which is in fact a Chern-Simons current associated with the electromagnetic potential e.g. [99, 319]

case, we directly deduce the timelike Ricci curvature in the presence of this particular torsion:

$$\mathcal{R}_{00} = \overline{\mathcal{R}}_{00} + \frac{\aleph_0^2}{2} = -3\,\frac{\ddot{R}}{R} + \frac{\aleph_0^2}{2} \tag{7.101}$$

From relation (7.67), we recall that the spatial components may be written as:

$$\Re_{ii} = \overline{\Re}_{ii} = -g_{ii}\left(\frac{\ddot{R}}{R} + 2\,\frac{\dot{R}^2}{R^2} + 2\,\frac{k}{R^2}\right), \qquad (ii) = (rr), (\theta\theta), (\varphi\varphi)$$

and the scalar curvature is modified accordingly:

$$\mathcal{R} := g^{\alpha\beta}\,\Re_{\alpha\beta} = \overline{\mathcal{R}} + \frac{\aleph_0^2}{2} = -6\left(\frac{\ddot{R}}{R} + \frac{\dot{R}^2}{R^2} + \frac{k}{R^2}\right) + \frac{\aleph_0^2}{2}$$

This example highlights the role of torsion on the spacetime behavior. We note for this example that connection coefficients deduced from the FLRW metric and those deduced from torsion *ad hoc* do not overlap. This situation is fortuitous and does not represent the generality of cases. A potential problem with an *ad hoc* torsion is the loss of spacetime isotropy.

7.3.2.4 *Field Equations in the Presence of Perfect Fluids*

As a slightly extended model, we recall the field equations (6.163) and (7.141) with homogeneous torsion field:

$$\frac{1}{2\chi}\left(\Re_{\mu\nu} - \frac{\mathcal{R}}{2} g_{\mu\nu}\right) = T^{\mathrm{fl}}_{\mu\nu} \tag{7.102}$$

The perfect fluid filling the Universe, with a given energy density ρ and pressure p, admits a momentum-energy tensor:

$$T^{\mathrm{fl}}_{\mu\nu} := (\rho + p)\,u_\mu u_\nu - p\;g_{\mu\nu}, \quad u_\mu = (1,0,0,0)$$

which is isotropic. Again, the usual method is to introduce the Ricci curvature calculated from the FLRW metric into the Einstein field equations. The Einstein tensor of the Riemann contribution is given by (7.69).

(1) First, the temporal part of the Einstein field equations holds:

$$\Re_{00} - \frac{\mathcal{R}}{2}\,g_{00} = 8\pi G\,T_{00} = 8\pi G \rho g_{00}$$

Introducing the temporal Ricci curvature and scalar curvature writes:

$$\overline{G}_{00} + \frac{\aleph_0^2}{4}\, g_{00} = 8\pi G\rho\, g_{00}$$

giving us after simplification (details may be found in the previous section):

$$3\left(\frac{\dot{R}^2}{R^2} + \frac{k}{R^2}\right) = 8\pi G\rho - \frac{\aleph_0^2}{4} \tag{7.103}$$

where the added torsion appears in a very simple form. Torsion is not involved in other components of the Ricci curvature.

(2) Second, for the spatial part, we have the analogous equation to the torsionless spacetime:

$$\overline{\mathfrak{R}}_{ii} - \frac{\mathcal{R}}{2}\, g_{ii} = 8\pi G\, T^{\text{fl}}_{ii}, \quad \{ii\} = (\{rr\}, \{\theta\theta\}, \{\varphi\varphi\})$$

where it was shown that the Ricci curvature is proportional to the metric component. We have the intermediate equation after introduction of scalar curvature:

$$\overline{G}_{ii} + \frac{\aleph_0^2}{4}\, g_{ii} = 8\pi G\, T^{\text{fl}}_{ii}, \quad \{ii\} = (\{rr\}, \{\theta\theta\}, \{\varphi\varphi\}) \tag{7.104}$$

The same equation holds for the three spatial coordinates. It is then obvious that we can drop the metric component which allows us to write:

$$2\,\frac{\ddot{R}}{R} + \frac{\dot{R}^2}{R^2} + \frac{k}{R^2} = -8\pi G\, p + \frac{\aleph_0^2}{4} \tag{7.105}$$

Eliminating the square $\dot{R}^2/R^2$ by a linear combination of (7.103) and (7.105) gives the Friedmann equation in the presence of torsion.

They constitute the basic model of the expanding Universe accounting for the torsion background field and within an Einstein-Cartan spacetime [306]:

$$\begin{cases} \left(\dfrac{\dot{R}}{R}\right)^2 = \dfrac{8\pi G}{3}\,\rho - \dfrac{\aleph_0^2}{12} - \dfrac{k}{R^2} \\[2ex] \dfrac{\ddot{R}}{R} = -\dfrac{4\pi G}{3}\,(\rho + 3p) + \dfrac{\aleph_0^2}{6} \end{cases} \tag{7.106}$$

with constant $k = -1, 0, +1$. The main conclusion that we can draw is that torsion contribution $\aleph_0^2/6$ acts as dark energy that favors the acceleration of cosmic expansion, conforming to major findings of [19] and [18]. This also conforms to results on the role of torsion in e.g. [54, 275, 276, 318] and electromagnetic fields e.g. [62, 96] together with the cosmological constant e.g. [23, 30]. Some questions might nevertheless remain on the use of torsion tensor as an alternative to the cosmological constant e.g. [280]. The system of two equations covers numerous contributions such as the fluid matter and the background torsion representing "the energy of the vacuum". The curvature k of the Universe (spatially open, flat, or spatially closed) is constant throughout a particular solution, but may vary from one solution to another.

Remark 7.13. Despite the fact that the torsion might play the role of dark energy, these equations seem not to have firm background not only because it is a very particular torsion field, but also because the isotropy hypothesis was accepted without any real development. The latter merits further study subsequently.

7.3.2.5 *Necessity of Physical Sources in Einstein-Cartan Spacetime*

First, let's consider an Einstein-Cartan vacuum spacetime. However, let's use a Riemann volume-form as a guess. The Einstein-Palatini action holds:

$$\mathscr{S}_{EP} = \frac{1}{2\chi} \int_{\mathscr{B}} \mathcal{R}\, \overline{\omega}_n = \frac{1}{2\chi} \int_{\mathscr{B}} \left(\overline{\mathcal{R}} - 6\, g^{\beta\lambda} \aleph_\beta \aleph_\lambda \right) \overline{\omega}_n \tag{7.107}$$

where torsion is reduced to a torsion vector. It is worth mentioning that the Levi-Civita derivatives of contortion do not appear since $\overline{\nabla}$ is compatible with $\overline{\omega}_n$. To derive the field equation, the variation is calculated by using Euler variation:

$$\delta \mathscr{S}_{EP} = \frac{1}{2\chi} \int_{\mathscr{B}} \left(\overline{\mathfrak{R}}_{\beta\lambda}\, \delta g^{\beta\lambda} + g^{\beta\lambda} \delta \overline{\mathfrak{R}}_{\beta\lambda} - 6 \aleph_\beta \aleph_\lambda\, \delta g^{\beta\lambda} \right) \overline{\omega}_n \tag{7.108}$$

$$+ \frac{1}{2\chi} \int_{\mathscr{B}} \left(-12 g^{\beta\lambda} \aleph_\beta\, \delta \aleph_\lambda \right) \overline{\omega}_n \tag{7.109}$$

$$+ \frac{1}{2\chi} \int_{\mathscr{B}} \left(\overline{\mathcal{R}} - 6\, g^{\beta\lambda} \aleph_\beta \aleph_\lambda \right) \left(-\frac{1}{2} g_{\beta\lambda}\, \delta g^{\beta\lambda} \right) \overline{\omega}_n \tag{7.110}$$

Due to the arbitrariness of metric and torsion vector variations, we deduce the two field equations:

$$\boxed{\begin{cases} \overline{G}_{\beta\lambda} + \lambda_{\aleph}\ g_{\beta\lambda} - 6\ \aleph_\beta \aleph_\lambda = 0 \\ -12\ g^{\beta\lambda}\aleph_\beta = 0 \end{cases} , \quad \Lambda_{\aleph} := 3\ g^{\sigma\epsilon}\aleph_\sigma \aleph_\epsilon} \tag{7.111}$$

where the variation of the Ricci curvature can be dropped after application of the divergence theorem. We conclude that:

(1) From (7.111), without physical sources as gravitating fluid with spin or electromagnetic fields, the second row vanishes and torsion is identically equal to zero e.g. [278]. The system reduces to the classical field equation of Einstein.
(2) Isotropy is lost due to $6\ \aleph_\beta \aleph_\lambda \neq 0$ in the general case. There is not a single cosmological constant but a "cosmological field". The existence of zones with different accelerations of cosmic expansion was observed very recently with baryon acoustic oscillations e.g. [320].

In the case where there exists physical sources, as we will develop in the following, we remark that we recover a local cosmological "constant" $\Lambda_{\aleph} := 3\ g^{\mu\nu}\aleph_\mu \aleph_\nu$ (the same as previously) which tends to contract the Universe when negative (spacelike torsion vector) or to accelerate the expansion if positive (timelike torsion vector). In addition it is observed that the tensor product of the torsion vector breaks the spacetime isotropy. *As a key point, the Einstein-Cartan spacetime model needs to add physical sources.* The Friedmann equations (7.77) present a model as simple as the original ones. As a reminder, they were obtained by adding averaged torsion as an external source. They were thus deduced from Einstein's field equations:

$$\overline{G}_{\mu\nu} + \Lambda_{\aleph}\ g_{\mu\nu} - 6\aleph_\mu \aleph_\nu = 8\pi G\ T^{\text{fl}}_{\mu\nu} \tag{7.112}$$

in which a fairly strong hypothesis was to assume that spacetime remains isotropic even in the presence of torsion. Despite the interesting role of torsion in accelerating the expansion, via a "cosmological field", the model using $\Lambda_{\aleph}$ remains very incomplete since the torsion itself is intimately linked to the presence of an electromagnetic field as an external source as we have shown (6.199). The gravito-electromagnetic model is then appropriate, particularly by accepting that the electromagnetic field is not a background field but constitutes also a dynamical field.

Remark 7.14. Beyond the founding role of perfect fluids in cosmology, the existence and importance of magnetic fields, particularly in interstellar regions and around gigantic filaments in the neighborhood of galaxies, and clusters of galaxies, are currently considered major sources in the formation of the cosmic web and structure of the Universe. Taking into account the electromagnetic field, both primordial and amplified by magneto-hydrodynamics of plasmas, is therefore essential in cosmology and astrophysics e.g. [66].

For the classical approach, the Einstein-Maxwell field equations are written:

$$\overline{G}_{\mu\nu} + \Lambda_{\aleph}\, g_{\mu\nu} - 6\aleph_\mu \aleph_\nu = 8\pi G \left(T^{\mathrm{fl}}_{\mu\nu} - T^{\mathrm{em}}_{\mu\nu}\right) \tag{7.113}$$

From the outset, consideration of a perfect fluid obviously reduces the problem to an isotropic symmetry, which is not the case when we consider an the electromagnetic field, concretized by tensor $T^{\mathrm{em}}_{\mu\nu}$ which is rarely isotropic. Then, the fact that the electromagnetic field is generally temporally variable strongly suggests the use of Maxwell's equations. Equation (7.113) contains Einstein tensor $G_{\mu\nu}$ described with a connection ∇ (with non-zero torsion) and a Riemann volume-form $\overline{\omega}_n$ or a compatible ω_n. It also includes an additional cosmological constant Λ_{nem}, arising from non-electromagnetic interactions and assumed to be independent of any other physical sources, in particular elementary particles. For these reasons, we also include an energy-momentum contribution from a perfect cosmological fluid and an electromagnetic source.

Einstein-Cartan vacuum spacetime and physical interaction sources. (7.113) is not complete if we want to be rigorous because we would then have to return to the full Einstein-Cartan-Maxwell equations. It should be accompanied by electromagnetic and torsion field equations.

7.3.2.6 *Friedmann's Equations and Einstein-Cartan Spacetime*

The present book is based on continuum physics, leaving aside the important aspect of quantum physics of elementary particles e.g. [15, 36]. Effort is dedicated to the cosmological constant Λ and its possible origins in the field from electromagnetism and especially the magnetic component e.g. [27, 64, 90, 94, 258]. This chapter is limited to gravitation with torsion which may be compared to results with weak torsion fields in Friedmann's equations e.g. [275, 276]. Without being exhaustive, models of

Einstein-Cartan cosmology are reviewed by Medina *et al.* [4] where the interesting properties of the absence of initial singularity of cosmology and the possibility of expansion at the very beginning just after the Big Bang are pointed out.

Key results. With this slightly extended version of Friedmann's equations, we propose that:

(1) First of all, in the second equation of (7.106), the torsion term is always positive $\aleph_0^2 > 0$, which suggests that it contributes to the acceleration of cosmic expansion, conforming to the discoveries of [18] and [19].
(2) Second, we have rigorously shown that torsion is linked mathematically to the electromagnetic field. This implies that the diffuse radiation of the magnetic field is a source of the acceleration of cosmic expansion.

Remark 7.15. Several studies have shown that torsion can be considered as a source of the cosmological constant and, by extension, of dark energy e.g. [6, 38] and source of dark matter e.g. [38, 89]. Different authors have shown the essential role of torsion in accelerating the cosmic expansion e.g. [318, 321]. Still considering the existence of cosmic fluids, some works have suggested phenomena which are generated by torsion such as the helicity of galaxies, which are generally in the shape of spirals, and the Faraday effect, that is to say the rotation of the polarization of light relative to the axis of the emitting source in space e.g. [317].

In (7.106), we can distinguish several contributions:

(1) the contribution of perfect fluids and the introduction *ad hoc* of a cosmological constant Λ which can have several physical sources including quantum particles.
(2) the contribution of a homogenous torsion field. Extension to the non-homogeneous case is expected although this introduces some other difficulties.

The example we give here is a simple approach considering Einstein-Cartan spacetime. These results should be considered as complementary to the role of quantum physics on the expansion of the Universe, namely by considering the cosmological constant Λ_{nem} as a good representation of the quantum energy of the vacuum spacetime.

7.4 Spacetime with Torsion and Riemann Volume-Form $\overline{\omega}_n$

Numerous studies agree with the use of FLRW spacetime for phenomenological modeling of cosmology, the ΛCDM being accepted as the standard model. The problem of introducing torsion is the loss of isotropy. *In what follows, we would like to respond to two questions: (a) Why should we use an Einstein-Cartan spacetime, and (b) why introduce electromagnetism as an external source coupled with gravity? And incidentally, what would be the consequences of this choice to the derivation of cosmological equations?*

7.4.1 *Choice of Einstein-Cartan Spacetime*

Several reasons have been already given to justify the use of Einstein-Cartan spacetime to build phenomenological models of cosmology e.g. [4]. One of these reasons is linked to the evolution of the Universe itself, which is supposed to move from an extremely hot state to lower temperature situations, passing through several phases of expansion and cooling.

7.4.1.1 *Defects and Evolution of the Universe*

By analogy with the behavior of condensed matter processes e.g. [322], it is highly probable that during certain phases of the Universe's cooling during inflation, mismatches occurred that led to the formation of defects at higher densities e.g. [190, 215, 220]. The thermomechanical cooling of the Universe, which accompanied its expansion from nearly 100–200 [MeV] at the origin of the Big Bang ($10^{-5}[s]$) to temperatures around 10^{-4}[eV], could have been a source of such defects. Assuming that vacuum spacetime behaves as a continuum, the main types of defects observed in condensed matter–point, line, and surface defects—correspond geometrically to non-metricity, torsion, and curvature fields, respectively e.g. [11, 229] or may be linked to magnetic monopoles e.g. [91]. The earliest moments of the Universe were essentially governed by particle physics at extremely small length scales, i.e., quantum physics. Conversely, cosmology deals with very large scales, from astronomical distances to galactic dimensions. What is therefore interesting is the introduction of continuum variables to model defect fields in cosmological models, even though the defects themselves may exist at particle physics scales. As in condensed matter, the hypothetical proliferation of defects—such as translational and rotational dislocations (line and surface mismatches)—is accompanied by a redistribution in space and time, forming microstructures or cell-like patterns with filamentary networks.

Past and recent studies have highlighted the interplay between defects and electromagnetism e.g. [39]. On the one hand, recent works have shown the role of the electromagnetic field in the acceleration of cosmic expansion e.g. [21, 62]. On the other hand, numerous studies have focused on the influence of spacetime defects, via torsion, on this acceleration e.g. [38, 54, 276]. Given the gravito-electromagnetic field equations that we developed in the previous chapter, clearly and mathematically showing the coupling between torsion and electromagnetism, we consider again the Einstein-Cartan spacetime for this section.

7.4.1.2 *Gravito-Electromagnetic Model with Riemann Volume-Form*

Currently, there is a consensus supported by numerous experimental observations that magnetic fields are pervading in the Universe, and assumption of a primordial magnetic field remains one hypothesis suggested by several authors e.g. [56]. Regardless of the origin of magnetic fields, it is clear that their role should not be neglected in the formation and evolution of the Universe e.g. [290]. As a reminder the considered action includes the Einstein-Palatini Lagrangian for gravitation and the symmetrized Yang-Mills Lagrangian for electromagnetism:

$$\mathscr{S}_{GSEM} := \int_{\mathscr{B}} \left[\frac{1}{2\chi} \Re_{\mu\nu}(\nabla)\ g^{\mu\nu} - \frac{1}{8} \mathcal{H}^{\mu\nu} \mathfrak{F}_{\mu\nu} + \frac{1}{8} \tilde{\mathfrak{H}}^{\mu\nu} \tilde{\mathfrak{F}}_{\mu\nu} \right] \overline{\omega}_n$$

with $\mathcal{R} := \Re_{\mu\nu}\ g^{\mu\nu}$. The Ricci curvature is formulated explicitly only in terms of the connection coefficients. *As a reminder, throughout this section, we use the Riemann volume form* $\overline{\omega}_n$. This subsection is an extended model of the gravito-electromagnetism. The electromagnetic field is not reduced to a “decorative” background but may evolve and constitutes a dynamical element of the spacetime.

7.4.2 *Isotropy Condition and Torsion Vector*

A general study of the action governing gravitation with an integrand $f(\mathcal{R})$ was made in which the curvature is calculated with a metric and an independent connection with torsion e.g. [52, 137]. However, few studies have addressed the compatibility of the connection with the volume-form. This problem was noted indirectly and to compensate, some authors introduced an integral term with a Lagrange multiplier e.g. [137]. In this section we wish to propose the decomposition of the Ricci curvature $\mathcal{R}_{\beta\lambda}$ into a purely Riemannian Ricci curvature $\overline{R}_{\beta\lambda}$ to which is added the contribution of torsion (6.113). We consider in this section a particular case of a torsion field as follows e.g. [52, 137, 282]:

$$\aleph^{\lambda}_{\mu\nu} = \tau_\mu\ \delta^{\lambda}_{\nu} - \tau_\nu\ \delta^{\lambda}_{\mu} \tag{7.114}$$

in which τ_μ is what we called a torsion vector. This vector is given by means of the trace of the torsion as follows:

$$\aleph_\mu := \tau_\mu = \frac{1}{3}\ \aleph^{\nu}_{\mu\nu} = \frac{1}{3}\sum_{\nu=0}^{\nu=3} \aleph^{\nu}_{\mu\nu} \tag{7.115}$$

Non-zero components of the torsion vector $\aleph_\mu(x^0)$ depend only on the time x^0 e.g. [276]. Previous equations are deduced from contortion decomposition into skew-symmetric and symmetric parts:

$$\mathcal{T}^{\lambda}_{\mu\nu} = \frac{1}{2}\aleph^{\lambda}_{\mu\nu} + \frac{1}{2}g^{\lambda\sigma}\left(g_{\mu\epsilon}\ \aleph^{\epsilon}_{\sigma\nu} + g_{\nu\epsilon}\ \aleph^{\epsilon}_{\sigma\mu}\right)$$

A previous calculus gave us the expression of the contortion in such a case:

$$\mathcal{T}^{\lambda}_{\mu\nu} = \left(g^{\lambda\sigma} g_{\mu\nu} - \delta^{\lambda}_{\mu}\delta^{\sigma}_{\nu}\right) \aleph_\sigma \tag{7.116}$$

We have already calculated the term:

$$\mathcal{T}^{\alpha}_{\beta\mu}\ \mathcal{T}^{\mu}_{\alpha\lambda} - \mathcal{T}^{\alpha}_{\alpha\mu}\ \mathcal{T}^{\mu}_{\beta\lambda} = -2\ \aleph_\beta\ \aleph_\lambda + 2\left(g^{\mu\nu}\aleph_\mu\aleph_\nu\right) g_{\beta\lambda}$$

For completeness, the following equations can be be easily deduced:

$$\begin{aligned} \overline{\nabla}_\alpha \mathcal{T}^{\alpha}_{\beta\lambda} - \overline{\nabla}_\beta \mathcal{T}^{\alpha}_{\alpha\lambda} &= \left(g^{\alpha\sigma}\overline{\nabla}_\alpha \aleph_\sigma\right) g_{\beta\lambda} \\ &\quad + \left(\overline{\nabla}_\beta \aleph_\lambda + \overline{\nabla}_\lambda \aleph_\beta\right) + \left(\overline{\nabla}_\beta \aleph_\lambda - \overline{\nabla}_\lambda \aleph_\beta\right) \end{aligned} \tag{7.117}$$

where the first term is analogous to an isotropic dilation, the second a deformation and the third a rotation generated by the torsion vector. The Ricci curvature in an Einstein-Cartan spacetime then takes the form of:

$$\boxed{\begin{aligned} \mathcal{R}_{\beta\lambda} = \overline{\mathcal{R}}_{\beta\lambda} &+ \left(g^{\alpha\sigma}\overline{\nabla}_\alpha \aleph_\sigma\right) g_{\beta\lambda} \\ &+ \left(\overline{\nabla}_\beta \aleph_\lambda + \overline{\nabla}_\lambda \aleph_\beta\right) + \left(\overline{\nabla}_\beta \aleph_\lambda - \overline{\nabla}_\lambda \aleph_\beta\right) \\ &+ 2\ \aleph_\beta \aleph_\lambda - 2\left(g^{\mu\nu}\aleph_\mu\aleph_\nu\right) g_{\beta\lambda} \end{aligned}} \tag{7.118}$$

where the second term of the first row as well as the entire second and third rows are obtained from (7.117). The scalar curvature is deduced:

$$\boxed{\mathcal{R} := g^{\beta\lambda}\,\mathcal{R}_{\beta\lambda} = \overline{\mathcal{R}} + 6g^{\beta\lambda}\overline{\nabla}_\beta \aleph_\lambda - 6g^{\beta\lambda}\aleph_\beta\aleph_\lambda} \tag{7.119}$$

Coming back to Einstein-Palatini action we thus write:

$$\mathscr{S}_{EP} = \frac{1}{2\chi}\int_{\mathscr{B}} \left(\overline{\mathcal{R}} + 6g^{\mu\nu}\overline{\nabla}_\mu \aleph_\nu - 6g^{\mu\nu}\aleph_\mu\aleph_\nu\right)\overline{\omega}_n$$

which shows that the second term of the integrand is merely the divergence of torsion-vector $\overline{\nabla}_\mu \aleph^\mu$. We may define an *effective* Einstein-Palatini action with non-zero torsion vector:

$$\mathscr{S}_{EP} := \frac{1}{2\chi}\int_{\mathscr{B}} \left(\overline{\mathcal{R}} - 6g^{\mu\nu}\aleph_\mu\aleph_\nu\right)\overline{\omega_n} \tag{7.120}$$

where we have dropped the covariant derivatives of contortion ($\overline{\nabla}$ is compatible with $\overline{\omega}_n$) after application of the divergence theorem *under assumption that the flux of the torsion vector vanishes at the boundary $\partial\mathscr{M}$*. We have the same action which induced the field equation (7.111). The field equations stand, by putting apart the physical sources:

$$\begin{cases} \overline{G}_{\mu\nu} + \Lambda_\aleph\, g_{\mu\nu} - 6\aleph_\mu\aleph_\nu = \text{sources along } \mathbf{g}, \quad \Lambda_\aleph := 3g^{\sigma\epsilon}\aleph_\sigma\aleph_\epsilon \\ -12g^{\beta\lambda}\aleph_\beta = \text{sources along } \aleph \end{cases} \tag{7.121}$$

The presence of the tensor product $6\aleph_\mu\aleph_\nu$ makes the spacetime anisotropic. Let's now consider the particular torsion deduced from this vector:

$$\aleph^\lambda_{\mu\nu} := \aleph_\mu\;\delta^\lambda_\nu - \aleph_\nu\;\delta^\lambda_\mu, \quad \aleph_\nu = \aleph_\nu\left(x^0\right) \tag{7.122}$$

where the symbol $\aleph$ is used without confusion with the tensor itself. This special class of torsion field may satisfy the isotropy of the Universe. This should be checked nevertheless all along the study.

7.4.3 *Gravito-Electromagnetism in Einstein-Cartan Spacetime*

Now, we would like to find torsion contributions on $\mathscr{D}^{\beta\lambda}_\mu$ which represent the crossing point of external physical phenomena such as matter or electromagnetic fields on distortion of vacuum spacetime.

7.4.3.1 *Equation Governing Spacetime Distortion*

Let us go back to field equations with physical sources such as the electromagnetic field. From the system (6.209), let's consider the third row as:

$$\frac{1}{2\chi}\,\mathscr{D}_{\mu}^{\beta\lambda}+\frac{1}{2}\left(\mathcal{H}^{\beta\lambda}A_{\mu}-\tilde{\mathcal{H}}^{\beta\lambda}\tilde{A}_{\mu}\right)=0$$

where "defects" engendered (or the converse) by electromagnetism is given by (6.142) we previously established:

$$\mathscr{D}_{\mu}^{\beta\lambda}=\left[\left(g^{\beta\lambda}\mathcal{T}_{\alpha\mu}^{\alpha}-g^{\alpha\lambda}\mathcal{T}_{\alpha\mu}^{\beta}\right)-\left(g^{\beta\gamma}\mathcal{T}_{\mu\gamma}^{\lambda}-\delta_{\mu}^{\beta}g^{\alpha\gamma}\mathcal{T}_{\alpha\gamma}^{\lambda}\right)\right]$$

By introducing the particular torsion (7.122) within the general expression of $\mathscr{D}_{\mu}^{\beta\lambda}$, we obtain after a tedious but straightforward calculus:

$$\frac{1}{2\chi}\left(g^{\beta\lambda}\,\aleph_{\mu}+2\,\delta_{\mu}^{\beta}g^{\lambda\epsilon}\aleph_{\epsilon}\right)+\frac{1}{2}\left(\mathcal{H}^{\beta\lambda}\,A_{\mu}-\tilde{\mathcal{H}}^{\beta\lambda}\,\tilde{A}_{\mu}\right)=0$$

where it is assessed that the torsion vector $\aleph_{\mu}$ is directly related to the Chern-Simons current. For the sake of simplicity, we slightly change the expression:

$$\frac{1}{2\chi}\left(g^{\beta\lambda}\delta_{\mu}^{\sigma}+2\,\delta_{\mu}^{\beta}g^{\lambda\sigma}\right)\aleph_{\sigma}+\frac{1}{2}\left(\mathcal{H}^{\beta\lambda}\,A_{\mu}-\tilde{\mathcal{H}}^{\beta\lambda}\,\tilde{A}_{\mu}\right)=0 \qquad (7.123)$$

7.4.3.2 *System of Field Equations*

For completeness, the set of field equations as a reminder takes the form of:

$$\begin{cases}\overline{\nabla}_{\mu}\mathcal{H}^{\mu\nu}+\mathcal{H}^{\mu\nu}\,\aleph_{\mu}=0\\[4pt] \overline{\nabla}_{\mu}\tilde{\mathcal{H}}^{\mu\nu}+\tilde{\mathcal{H}}^{\mu\nu}\,\aleph_{\mu}=0\\[4pt] \dfrac{1}{2\chi}\,G_{\mu\nu}+T_{\mu\nu}^{\text{em}}=0\\[4pt] \dfrac{1}{2\chi}\left(g^{\beta\lambda}\delta_{\mu}^{\sigma}+2\,\delta_{\mu}^{\beta}g^{\lambda\sigma}\right)\aleph_{\sigma}+\dfrac{1}{2}\left(\mathcal{H}^{\beta\lambda}\,A_{\mu}-\tilde{\mathcal{H}}^{\beta\lambda}\,\tilde{A}_{\mu}\right)=0\end{cases} \qquad (7.124)$$

where unknowns are A_{μ}, $\tilde{A}_{\mu}$, $g_{\mu\nu}$, and $\aleph_{\nu}$. Resolution of (7.124) remains a great challenge although the three equations are highly coupled. By adding torsion, the Ricci curvature calculated with Riemann curvature becomes (6.114).

$$\begin{cases}\mathfrak{R}_{\beta\lambda}^{S}=\overline{\mathfrak{R}}_{\beta\lambda}\\ \mathfrak{R}_{\beta\lambda}^{A}=\overline{\nabla}_{\alpha}\mathcal{T}_{\beta\lambda}^{\alpha}-\overline{\nabla}_{\beta}\mathcal{T}_{\alpha\lambda}^{\alpha}-\left(\mathcal{T}_{\beta\mu}^{\alpha}\,\mathcal{T}_{\alpha\lambda}^{\mu}-\mathcal{T}_{\alpha\mu}^{\alpha}\,\mathcal{T}_{\beta\lambda}^{\mu}\right)\end{cases}$$

For the second row, we can omit terms with (Levi-Civita)-covariant derivative by assuming a compatibility at the boundary, say $\delta \mathcal{T}^{\gamma}_{\mu\nu} = 0, \forall x^{\alpha} \in \partial\mathscr{B}$. Assuming that flux at the boundary vanishes, a simpler expression of the Einstein tensor holds:

$$G_{\beta\lambda} = \overline{G}_{\beta\lambda} + 2\, \aleph_{\beta}\aleph_{\lambda} + (g^{\mu\nu}\, \aleph_{\mu}\aleph_{\nu})\, g_{\beta\lambda}$$

Remark 7.16. Scalar curvature has two other contributions: Riemann scalar curvature plus a tensor product of the torsion vector. Depending on the sign of the norm of the tensor product, its influence may be repulsive, neutral or attractive.

Remark 7.17. For Einstein-Cartan spacetime, a linking equation, similar to this third equation, was assumed in [99]. In our approach, the linking equation is logically deduced from the variational procedure.

Gravito-electromagnetism with torsion vector. The system of field equations take this particular form when adding a vector torsion:

$$\begin{cases} \overline{\nabla}_{\mu}\mathcal{H}^{\mu\nu} + \mathcal{H}^{\mu\nu}\, \aleph_{\mu} = 0 \\ \overline{\nabla}_{\mu}\tilde{\mathcal{H}}^{\mu\nu} + \tilde{\mathcal{H}}^{\mu\nu}\, \aleph_{\mu} = 0 \\ \dfrac{1}{2\chi}\, G_{\mu\nu} + T^{\text{em}}_{\mu\nu} = 0 \\ \dfrac{1}{2\chi}\left(g^{\beta\lambda}\delta^{\sigma}_{\mu} + 2\, \delta^{\beta}_{\mu} g^{\lambda\sigma}\right)\aleph_{\sigma} + \dfrac{1}{2}\left(\mathcal{H}^{\beta\lambda}\, A_{\mu} - \tilde{\mathcal{H}}^{\beta\lambda}\, \tilde{A}_{\mu}\right) = 0 \end{cases} \tag{7.125}$$

with the Einstein tensor and the distortion tensor being:

$$\begin{cases} G_{\beta\lambda} := \overline{G}_{\beta\lambda} + \Lambda_{\aleph}\, g_{\beta\lambda} + 2\, \aleph_{\beta}\aleph_{\lambda} \\ \mathscr{D}^{\beta\lambda}_{\mu} := \left(g^{\beta\lambda}\, \aleph_{\mu} + 2\, \delta^{\beta}_{\mu} g^{\lambda\epsilon}\aleph_{\epsilon}\right) \end{cases} \tag{7.126}$$

and $\Lambda_{\aleph} := g^{\mu\nu}\, \aleph_{\mu}\aleph_{\nu}$. The unknowns are A_{μ}, $\tilde{A}_{\mu}$, $g_{\mu\nu}$ and $\aleph_{\mu}$.

Remark 7.18. A brief analysis of this system leads us to make a few remarks:

(1) Equations (7.125) resume the coupling of relative gravitation and electromagnetism when torsion is added as a geometric variable.
(2) In Einstein-Cartan-Maxwell equations (second line), the torsion vector occurs quadratically. When $\aleph_{\mu} \equiv 0$, we find a classic case. Contribution of torsion is analogous to that of cosmological constant $\Lambda_{\aleph} := g^{\lambda\rho}\aleph_{\lambda}\aleph_{\rho}$,

plus a part intervenes as tensor product $\aleph \otimes \aleph$. Despite the apparent harmonicity of (7.125), tensor product $\aleph \otimes \aleph$ generates spacetime anisotropy. *This precludes the possibility of using FLRW metric.*

(3) Introduction of generalized action with $f(\mathcal{R})$ could be considered, but we did not go in this way. $f(\mathcal{R}) = \mathcal{R}^2$ was proposed e.g. [52, 137]. This seems illusory in view of all three equations because the third equation clearly indicates that when $\mathcal{H}^{\mu\nu}, \tilde{\mathcal{H}}^{\mu\nu} \equiv 0$, torsion is necessarily zero unless other external sources are introduced. We fall back into the case without torsion.

(4) The third equation (7.125) deserves to be underlined once again, because it shows that optical helicity and spin angular momentum of light are sources of torsion, and conversely. In addition to fermions which are known to be sources of torsion e.g. [282], these equations confirm that photons also induce torsion e.g. [90, 99, 222].

Theorem 7.4. *Let $\mathscr{B}$ be an Einstein-Cartan spacetime where gravitation is defined by Einstein-Palatini Lagrangian $\mathscr{L}_{EP}$ and electromagnetic field defined by symmetrized Yang-Mills Lagrangian $\mathscr{L}_{SYM}$. On this spacetime is defined a metric field $g_{\alpha\beta}$, an independent particular connection whose torsion is reduced to a torsion vector $\aleph_\mu$, and a Riemann volume-form field $\overline{\omega}_n$. Then the torsion is determined by the Chern-Simons current:*

$$\aleph_\mu = -\frac{1}{3}\chi \, g_{\mu\alpha} \left(\mathcal{H}^{\alpha\beta} A_\beta - \tilde{\mathcal{H}}^{\alpha\beta} \tilde{A}_\beta \right) \tag{7.127}$$

where χ is the gravitation constant, $\mathcal{H}^{\alpha\beta}$ the electromagnetic excitation (resp. dual $\tilde{\mathcal{H}}^{\alpha\beta}$), and A_β the four-vector electromagnetic potential (resp. dual $\tilde{A}_\beta$).

Proof. From the third equation (7.125), we may write:

$$\mathscr{D}_\lambda^{\beta\lambda} = g^{\beta\lambda}\aleph_\lambda + 2\delta_\mu^\beta g^{\lambda\epsilon}\aleph_\epsilon = 3g^{\beta\lambda}\aleph_\lambda$$

We deduce:

$$\frac{3}{2\chi} g^{\beta\lambda}\aleph_\lambda + \frac{1}{2}(\mathcal{H}^{\beta\lambda} A_\lambda - \tilde{\mathcal{H}}^{\beta\lambda}\tilde{A}_\lambda) = 0$$

□

Remark 7.19. As advocated for a long time by the pioneering works of Einstein and Weyl to geometrize electromagnetism [323], (7.125) can be considered as linking geometry with electromagnetism, and by extension with dark energy e.g. [324].

Given that the use of $\overline{\omega}_n$ could be subject to inconsistencies if not used judiciously for Einstein-Cartan spacetime, the result is correct mathematically (by deduction) and the torsion is explicitly determined by this theorem. Again, it allows us to deduce that torsion is generated by spin angular momentum of electromagnetic waves, that is to say photons e.g. [108, 288, 325]. This result makes more precise the suggestion of e.g. [222], completing results of the previous chapter on Einstein-Cartan spacetime. Nevertheless, the formulation of Maxwell's equations remains satisfactory.

Remark 7.20. The problem comes from the use of volume-forms not compatible with connection, which poses some inconsistency with respect to the use of the divergence theorem. To avoid too much redundancy, we will not add at this stage details for influence of magnetic helicity, which can be developed without any problem. However, in the following section, we will give more detailed results concerning both electric and magnetic helicity.

7.5 Spacetime with Torsion and Compatible Volume-Form ω_n

In the previous section, we used $\overline{\omega}_n$ to define action $\mathscr{S}$ that encompasses relative gravitation and electromagnetism. Again, torsion as well as the electromagnetic field are not static but constitute dynamical components of spacetime, and physical sources in spacetime.

7.5.1 *Action for Gravitation and Electromagnetism*

7.5.1.1 *Motivation for Using Compatible Volume-Form ω_n*

There remains a concern regarding Maxwell's equations, which do not respect the Minimal Coupling Principles[4] since the generic form of

[4]The choice of a volume-form (measure on the Riemann-Cartan manifold) might seem arbitrary because there are apparently no scientific criteria to decide on it. Moreover, authors relatively often use the Riemann volume-form $\overline{\omega}_n$ to define the action of their model. Even if it does not serve as a scientific principle, the *Minimal Coupling Principle* still has the advantage of satisfying the covariance of the conservation equations. We therefore stick to the principle as a consistency criterion rather than a physical principle. This is why we present the two possibilities in the present book: using the Riemann volume-form (compatible in the case of Einstein gravitation), and the Riemann-Cartan

Maxwell's equations of special relativity do not match with that of general relativity. In relativity and electromagnetism, the rule of thumb to render field equations covariant is to change partial derivatives ∂_α to covariant derivatives ∇_α, and to change Minkowski metric $\eta_{\alpha\beta}$ into Riemann metric $g_{\alpha\beta}$. While not a major problem (at first sight), it could be nevertheless annoying. In brief, incompatibility of ∇ with $\overline{\omega}_n$ introduces artifices in Maxwell's equations. To obviate such an inconsistency, compatibility of ω_n with ∇ was enforced for Einstein-Cartan spacetime e.g. [132] by defining an appropriate volume-form for any manifold with torsion and curvature. This requirement results in a particular n-form of the sort: $\omega_n := e^{\vartheta(\mathbf{g},\aleph)}\,\overline{\omega}_n$. Details are available in articles written mainly by Saa and Mosna between 1995 and 2005 e.g. [134], related to the gauge invariance of electromagnetic fields in e.g. [132, 135] and also recently in e.g. [131]. These results present the motivation for using a volume-form ω_n compatible with ∇.

7.5.1.2 *Action for Gravito-Electromagnetism in Vacuum*

The following action is an extension of the Einstein-Palatini action with contribution from the extended version of the Yang-Mills action. The total action is then given by:

$$\mathscr{S}_{GEM} := \frac{1}{2\chi}\int_{\mathscr{M}} \mathcal{R}(\mathbf{g},\nabla)\,\omega_n - \frac{1}{4}\int_{\mathscr{M}} \mathcal{H}^{\mu\nu}\mathcal{F}_{\mu\nu}\,\omega_n, \quad \omega_n := e^{\vartheta}\,\overline{\omega}_n \quad (7.128)$$

where $\mathscr{M}$ is a Riemann-Cartan continuum manifold endowed with a Lorentzian metric $\mathbf{g}$ (Hilbert causality). For a Riemann-Cartan manifold, a modified volume-form compatible with connection was already proposed, which is formulated as $\omega_n := e^{\vartheta}\,\overline{\omega}_n$ such that the trace of the torsion tensor is a gradient $\aleph^{\mu}_{\mu\nu} = \nabla_\nu \vartheta$ with $\vartheta(\mathbf{g},\aleph)$ e.g. [132]. We have developed the equations that govern the coupling of Einstein gravitation and Maxwell electromagnetism by using ∇ compatible with $\mathbf{g}$ but not with $\overline{\omega}_n$. The equations are explicitly completed but there remains one point which is not judicious concerning the Principle of Minimal Coupling because the electromagnetic wave equation (resulting from Maxwell's equation) does not have the same form of equation as for the case without torsion. *There is a lack of covariance in some sense.* We must develop a model where connection is compatible with both metric and volume form.

volume-form $\omega_n := e^{\vartheta(\mathbf{g},\aleph)}$ for the Einstein-Cartan spacetime. Determination of the function ϑ is another problem that we will sketch later.

7.5.2 Field Equations within Einstein-Cartan Spacetime

Extending Friedmann's equations to Einstein-Cartan spacetime is not simple. Most authors have paid attention to metric compatibility of connection, but not that of connection with volume-form *a priori*. Let's start from action (7.128).

7.5.2.1 Variation Equation

We focus on vacuum spacetime in the presence of gravitational and electromagnetic fields. Most of the details of the calculus have already been performed in the previous subsections; we just have to put them together. The variation of the action holds:

$$\delta\mathscr{S}_{GEM} = \frac{1}{2\chi}\int_{\mathscr{M}} (\delta g^{\mu\nu}\mathcal{R}_{\mu\nu} + g^{\mu\nu}\delta\mathcal{R}_{\mu\nu})\,\omega_n + g^{\mu\nu}\mathcal{R}_{\mu\nu}\,\delta\omega_n \quad (7.129)$$

$$-\frac{1}{4}\int_{\mathscr{M}} \delta\left(\mathcal{H}^{\mu\nu}\mathcal{F}_{\mu\nu}\right)\omega_n + \mathcal{H}^{\mu\nu}\mathcal{F}_{\mu\nu}\,\delta\omega_n \quad (7.130)$$

in which primal variables are respectively metric, connection (or torsion), and electromagnetic potential.

The volume-form is expressed as $\omega_n := e^{\vartheta(\mathbf{g},\aleph)}\overline{\omega}_n$. Its variation is explicitly derived, since $e^{\vartheta(\mathbf{g},\aleph)} > 0$:

$$\delta\omega_n = \left(\frac{\partial\vartheta\,(\aleph)}{\partial\aleph^{\gamma}_{\mu\nu}}\,\delta\aleph^{\gamma}_{\mu\nu} - \frac{1}{2}g_{\mu\nu}\,\delta g^{\mu\nu} + \frac{\partial\vartheta}{\partial g^{\mu\nu}}\,\delta g^{\mu\nu}\right)\omega_n \quad (7.131)$$

For convenience, we express this variation in terms of $\delta g^{\mu\nu}$ and $\delta\Gamma^{\gamma}_{\alpha\beta}$:

$$\boxed{\delta\omega_n = \left[\left(-\frac{1}{2}g_{\mu\nu} + \frac{\partial\vartheta}{\partial g^{\mu\nu}}\right)\delta g^{\mu\nu} + \left(\frac{\partial\vartheta\,(\aleph)}{\partial\aleph^{\gamma}_{\mu\nu}} - \frac{\partial\vartheta\,(\aleph)}{\partial\aleph^{\gamma}_{\nu\mu}}\right)\delta\Gamma^{\gamma}_{\mu\nu}\right]\omega_n} \quad (7.132)$$

The choice of the form of ω_n is dictated by the concern to be able to connect the equations of the model with those of Einstein gravitation. For the curvature, we recall the formula e.g. [158]:

$$\delta\mathfrak{R}^{\gamma}_{\alpha\beta\lambda} = \nabla_{\alpha}(\delta\Gamma^{\gamma}_{\beta\lambda}) - \nabla_{\beta}(\delta\Gamma^{\gamma}_{\alpha\lambda}) + \aleph^{\mu}_{\alpha\beta}\,\delta\Gamma^{\gamma}_{\mu\lambda} \quad (7.133)$$

As per previously, the variation of the curvature may be written as follows:

$$\delta\mathcal{R} = \delta g^{\mu\nu}\,\mathcal{R}_{\mu\nu} + \delta^{\alpha}_{\gamma}\,g^{\beta\lambda}[\nabla_{\alpha}(\delta\Gamma^{\gamma}_{\beta\lambda}) - \nabla_{\beta}(\delta\Gamma^{\gamma}_{\alpha\lambda}) + \aleph^{\mu}_{\alpha\beta}\delta\Gamma^{\gamma}_{\mu\lambda}]$$

Now for the electromagnetic part, i.e., the Faraday strength tensor e.g. [6]:

$$\delta \mathcal{F}_{\mu\nu} = \nabla_\mu (\delta A_\nu) - \nabla_\nu (\delta A_\mu) - A_\rho \, \delta \aleph^\rho_{\mu\nu} \tag{7.134}$$

where we retain the affine connection ∇ which is assumed to be compatible with the volume-form ω_n. In addition, we have calculated the variation:

$$\delta (\mathcal{H}^{\mu\nu} \mathcal{F}_{\mu\nu}) = -2\epsilon_0 \mathcal{F}_{\mu\alpha} g^{\alpha\beta} \mathcal{F}_{\beta\nu} \, \delta g^{\mu\nu} + 2\mathcal{H}^{\mu\nu} \delta \mathcal{F}_{\mu\nu} \tag{7.135}$$

It has the same shape as before. By extension, we recall the expression of the Maxwell energy-momentum (6.157):

$$T^{\text{em}}_{\mu\nu} := \frac{1}{8} (\mathcal{H}^{\beta\lambda} \mathcal{F}_{\beta\lambda}) g_{\mu\nu} + \frac{1}{4} \mathcal{H}^{\lambda\rho} (g_{\nu\lambda} \mathcal{F}_{\rho\mu} + \mathcal{F}_{\mu\lambda} g_{\rho\nu})$$

The main advantage of using a connection compatible with the volume-form ω_n is to introduce directly the ∇ operator. As we will see hereafter, the divergence theorem is therefore applicable without difficulty. In summary, the variation of the Einstein-Palatini action is given by:

$$\delta \mathscr{S}_{EP} = \frac{1}{2\chi} \int_{\mathscr{M}} \left[\left(\mathcal{R}_{\mu\nu} - \frac{\mathcal{R}}{2} g_{\mu\nu} \right) + \mathcal{R} \frac{\partial \vartheta}{\partial g^{\mu\nu}} \right] \delta g^{\mu\nu} \, \omega_n \tag{7.136}$$

$$+ \frac{1}{2\chi} \int_{\mathscr{M}} \left[\aleph^\mu_{\gamma\lambda} g^{\lambda\nu} + \mathcal{R} \left(\frac{\partial \vartheta}{\partial \aleph^\gamma_{\mu\nu}} - \frac{\partial \vartheta}{\partial \aleph^\gamma_{\nu\mu}} \right) \right] \delta \Gamma^\gamma_{\mu\nu} \, \omega_n \tag{7.137}$$

The variation of the Yang-Mills action holds:

$$\delta \mathscr{S}_{YM} = \int_{\mathscr{M}} \nabla_\mu \mathcal{H}^{\mu\nu} \, \delta A_\nu \, \omega_n \tag{7.138}$$

$$+ \int_{\mathscr{M}} \left[T^{\text{em}}_{\mu\nu} - \frac{1}{4} \mathcal{H}^{\alpha\beta} \mathcal{F}_{\alpha\beta} \frac{\partial \vartheta}{\partial g^{\mu\nu}} \right] \delta g^{\mu\nu} \, \omega_n \tag{7.139}$$

$$+ \int_{\mathscr{M}} \left[\mathcal{H}^{\mu\nu} A_\gamma - \frac{1}{4} \mathcal{H}^{\alpha\beta} \mathcal{F}_{\alpha\beta} \left(\frac{\partial \vartheta}{\partial \aleph^\gamma_{\mu\nu}} - \frac{\partial \vartheta}{\partial \aleph^\gamma_{\nu\mu}} \right) \right] \delta \Gamma^\gamma_{\mu\nu} \, \omega_n \tag{7.140}$$

During this calculation, two points deserve to be highlighted:

(1) the influence of the function $\vartheta (\mathbf{g}, \aleph)$ characterizing ω_n, which tends to complicate the equations since it adds another term;

(2) the fact that we no longer need to introduce the Levi-Civita connection $\overline{\nabla}$, which fortunately removes a lot of terms regarding the variation of curvature.

7.5.2.2 *Field Equations of Vacuum Spacetime*

Due to arbitrariness of δA_ν, $\delta g^{\mu\nu}$ and $\delta\Gamma^\gamma_{\mu\nu}$, we have the following equation which can be considered as an important step of this book. The system of field equations for Einstein-Cartan spacetime holds:

$$\boxed{\begin{cases} \nabla_\mu \mathcal{H}^{\mu\nu} = 0 \\ \dfrac{1}{2\chi}\, G_{\mu\nu} + T^{\text{em}}_{\mu\nu} + \mathscr{L}_{GEM} \dfrac{\partial \vartheta}{\partial g^{\mu\nu}} = 0 \\ \dfrac{1}{2\chi}\, \mathscr{D}^{\mu\nu}_{\gamma} + \mathcal{H}^{\mu\nu}\, A_\gamma + \mathscr{L}_{GEM} \left(\dfrac{\partial \vartheta}{\partial \aleph^\gamma_{\mu\nu}} - \dfrac{\partial \vartheta}{\partial \aleph^\gamma_{\nu\mu}} \right) = 0 \end{cases}} \tag{7.141}$$

with the Einstein tensor and the spacetime distortion tensor:

$$\boxed{G_{\mu\nu} := \Re_{\mu\nu} - \frac{\mathcal{R}}{2}\, g_{\mu\nu}, \quad \mathscr{D}^{\mu\nu}_{\gamma} := g^{\beta\nu}\, \aleph^\mu_{\gamma\beta}} \tag{7.142}$$

where $\aleph^\nu_{\nu\mu} \equiv \nabla_\mu \vartheta$. The system with three equations are highly coupled.

Remark 7.21. Terms in divergence can be set to zero by imposing the boundary conditions $\delta\Gamma^\gamma_{\mu\nu} \equiv 0$ and $\delta A_\nu \equiv 0$. The integral term:

$$\int_{\mathscr{M}} \delta^\alpha_\gamma\, g^{\beta\lambda} \left[\nabla_\alpha (\delta\Gamma^\gamma_{\beta\lambda}) - \nabla_\beta\, (\delta\Gamma^\gamma_{\alpha\lambda}) \right] \omega_n \to \int_{\mathscr{M}} \nabla_\alpha \xi^\alpha\, \omega_n$$

becomes a 3-surface integral, and so is the case for the two integrals, after applying integration by parts to the two integrands of the fourth row:

$$\int_{\mathscr{M}} \nabla_\mu\, (\mathcal{H}^{\mu\nu} \delta A_\nu)\, \omega_n \quad \text{and} \quad \int_{\mathscr{M}} \nabla_\nu\, (\mathcal{H}^{\mu\nu} \delta A_\mu)\, \omega_n \to \int_{\mathscr{M}} \nabla_\alpha \xi^\alpha\, \omega_n$$

Indeed, the use of compatibility (2.71) allows us to change these quantities into a 3-surface integral.

We have three equations whose unknowns are electromagnetic potential A_μ, metric $g_{\mu\nu}$ and torsion $\aleph^\gamma_{\mu\nu}$. Starting from actions including Einstein-Palatini and Yang-Mills Lagrangians, this model does not introduce new physical phenomena other than those contained in these actions. But from their tight coupling we aim to show that one cannot use parts of these equations carelessly without impunity. Additional coupling terms appear in the Einstein-Cartan-Maxwell equation, which makes the application in cosmology even more difficult if we still want to respect the principles of isotropy and homogeneity.

Remark 7.22. Equations seem "smooth" but remember that ∇ includes metric and torsion (2.69): $\Gamma^\gamma_{\alpha\beta} = (1/2)\, \aleph^\gamma_{\alpha\beta} + (1/2) g^{\sigma\gamma} (g_{\alpha\mu}\, \aleph^\mu_{\gamma\beta} + \aleph^\mu_{\gamma\alpha}\, g_{\mu\beta})$.

Practical and namely numerical implementation remains therefore difficult. However, one advantage of this model is ease of interpretation. The real conceptual difficulty is choosing (finding) the function $\vartheta(\mathbf{g}, \aleph)$, even if the calculus method allows us to concentrate the use of this function in the third row, where the Chern-Simons current appears explicitly.

Considering this model (7.141), we see that the electromagnetic field acts not only on curvature which bends spacetime but also on the distortion which "tears" spacetime to obtain an infinite number of small grains of spacetime (named microcosms by [2]) of very small length scales (7.142). As in condensed materials, the thermal evolution of the Universe would not have no consequences on its structure and its contents. By analogy, huge temperature changes starting from the estimated temperature at $T_0 \simeq 10^{31}\,[K]$ at the initial time $t \simeq 10^{-43}\,[s]$ and which drops to the currently measured temperature of $T_{CMB} \simeq 2.72528 \pm 0.000057\,[K]$ e.g. [326] can be sources of nucleation and proliferation of defects in the vacuum spacetime of the Universe. These defects include dislocation density (torsion) and disclination density, or even production of spacetime grains (curvature), which (7.142) would suggest.

Remark 7.23. Without torsion, field equations of an Einstein-Cartan vacuum reduce to $G_{\mu\nu} = 0$ and $\mathscr{D}_{\gamma}^{\mu\nu} = 0$ where unknowns are metric and connection (in the absence of the electromagnetic field). Tensor $\mathscr{D}_{\gamma}^{\mu\nu}$ is the counterpart of the Einstein tensor for torsion. In the case of non-dependence of function ϑ on torsion it implies that torsion is necessarily equal to zero, and by extension, the model reduces to Einstein gravitation.

Remark 7.24. In addition to the Chern-Simons current in (6.163) we note an additional contribution in (7.142). We also recognize the Lagrangian $\mathscr{L}_{GEM}$ as particularly interesting. In some sense, we derive here by a deductive manner a relationship between electromagnetism and spacetime distortion and bending. Change of spacetime is directly related to the Chern-Simons operator $A_\gamma \to \mathcal{H}^{\mu\nu} A_\gamma \to \mathcal{H}^{\mu\nu} A_\nu$ (after application of trace). By extension, we notice that keeping an isotropic universe would be possible for only very particular cases.

7.5.3 *Gravitation with Symmetrized Electromagnetism*

This time we consider the empty Einstein-Cartan spacetime with a symmetrized Yang-Mills Lagrangian coupled with the Einstein-Palatini Lagrangian.

7.5.3.1 *Action for Gravitation and Symmetrized Electromagnetism*

Action for gravitation and symmetrized electromagnetism. The action takes the form of:

$$\mathscr{S}_{GSEM} := \int_{\mathscr{M}} \frac{\mathcal{R}}{2\chi}\,\omega_n + \frac{1}{2}\left[\int_{\mathscr{M}} -\frac{1}{4}\mathfrak{H}^{\mu\nu}\mathfrak{F}_{\mu\nu}\,\omega_n + \int_{\mathscr{M}} \frac{1}{4}\tilde{\mathfrak{H}}^{\mu\nu}\tilde{\mathfrak{F}}_{\mu\nu}\,\omega_n\right] \tag{7.143}$$

where the definitions were given in the previous chapter.

We recall that after translation into a three-dimensional formulation the two electromagnetic integrals of the electromagnetic Lagrangian is exactly equal to:

$$\mathscr{L}_{YM} = \frac{1}{2}\left(\mathbf{D}\cdot\mathbf{E} - \mathbf{B}\cdot\mathbf{H}\right)$$

where $\mathbf{E}$, $\mathbf{H}$ are electric and magnetic fields respectively, whereas $\mathbf{D}$ and $\mathbf{B}$ are electric and magnetic inductions respectively.

7.5.3.2 *Field Equations of Einstein-Cartan Spacetime*

Since most of the development details were already elaborated on in the previous chapter, we have just to recall the results. The variation of the action includes three parts:

- for gravitation

$$\begin{aligned}\delta\mathscr{S}_{EP} &= \frac{1}{2\chi}\int_{\mathscr{M}} \left(\mathcal{R}_{\mu\nu}\,\delta g^{\mu\nu}\right)\omega_n \\ &+ \frac{1}{2\chi}\int_{\mathscr{M}} \left(g^{\lambda\nu}\aleph^{\mu}_{\gamma\lambda}\delta\Gamma^{\gamma}_{\mu\nu}\right)\omega_n + \text{flux on } \partial\mathscr{M} \\ &+ \frac{1}{2\chi}\int_{\mathscr{M}} \mathcal{R}\left[\left(-\frac{1}{2}g_{\mu\nu} + \frac{\partial\vartheta}{\partial g^{\mu\nu}}\right)\delta g^{\mu\nu} + \frac{\partial\vartheta}{\partial\aleph^{\gamma}_{\mu\nu}}\,\delta\aleph^{\gamma}_{\mu\nu}\right]\omega_n\end{aligned} \tag{7.144}$$

- for the standard electromagnetic action

$$\begin{aligned}\delta\mathscr{S}_{YM} &= -\frac{1}{4}\int_{\mathscr{M}} -\mathfrak{H}^{\lambda\rho}\left(g_{\nu\lambda}\mathfrak{F}_{\rho\mu} + \mathfrak{F}_{\mu\lambda}g_{\rho\nu}\right)\delta g^{\mu\nu}\,\omega_n \\ &- \frac{1}{4}\int_{\mathscr{M}} 2\mathfrak{H}^{\mu\nu}\left[\nabla_{\mu}(\delta A_{\nu}) - \nabla_{\nu}(\delta A_{\mu}) - A_{\gamma}\,\delta\aleph^{\gamma}_{\mu\nu}\right]\omega_n \\ &- \frac{1}{4}\int_{\mathscr{M}} \mathfrak{H}^{\lambda\rho}\mathfrak{F}_{\lambda\rho}\left[\left(-\frac{1}{2}g_{\mu\nu} + \frac{\partial\vartheta}{\partial g^{\mu\nu}}\right)\delta g^{\mu\nu} + \frac{\partial\vartheta}{\partial\aleph^{\gamma}_{\mu\nu}}\,\delta\aleph^{\gamma}_{\mu\nu}\right]\omega_n\end{aligned} \tag{7.145}$$

- for dual electromagnetic action

$$\delta\tilde{\mathscr{S}}_{YM} = \frac{1}{4}\int_{\mathscr{M}} -\tilde{\mathcal{H}}^{\lambda\rho}\left(g_{\nu\lambda}\tilde{\mathfrak{F}}_{\rho\mu} + \tilde{\mathfrak{F}}_{\mu\lambda}g_{\rho\nu}\right)\delta g^{\mu\nu}\ \omega_n$$
$$+\frac{1}{4}\int_{\mathscr{M}} 2\tilde{\mathcal{H}}^{\mu\nu}\left[\nabla_\mu(\delta\tilde{A}_\nu) - \nabla_\nu(\delta\tilde{A}_\mu) - \tilde{A}_\gamma\ \delta\aleph^\gamma_{\mu\nu}\right]\omega_n$$
$$+\frac{1}{4}\int_{\mathscr{M}} \tilde{\mathcal{H}}^{\lambda\rho}\tilde{\mathfrak{F}}_{\lambda\rho}\left[\left(-\frac{1}{2}g_{\mu\nu} + \frac{\partial\vartheta}{\partial g^{\mu\nu}}\right)\delta g^{\mu\nu} + \frac{\partial\vartheta}{\partial\aleph^\gamma_{\mu\nu}}\ \delta\aleph^\gamma_{\mu\nu}\right]\omega_n \tag{7.146}$$

In these relations, we have to observe that:

$$\mathcal{H}^{\mu\nu}\mathfrak{F}_{\mu\nu} = -\tilde{\mathcal{H}}^{\mu\nu}\tilde{\mathfrak{F}}_{\mu\nu}$$

Now, we define some variables, extended from classical ones. The Einstein tensor is extended as follows:

$$G_{\mu\nu} := \mathcal{R}_{\mu\nu} - \frac{\mathcal{R}}{2}\ g_{\mu\nu} + \mathcal{R}\ \frac{\partial\vartheta}{\partial g^{\mu\nu}} \tag{7.147}$$

where the last term represents the influence of the compatible volume-form ω_n. The distortion tensor of the vacuum spacetime is defined as:

$$\mathcal{D}^{\mu\nu}_\gamma := \frac{1}{2}g^{\lambda\nu}\ \aleph^\mu_{\gamma\lambda} + \frac{\mathcal{R}}{2\chi}\left(\frac{\partial\vartheta}{\partial\aleph^\gamma_{\mu\nu}} - \frac{\partial\vartheta}{\partial\aleph^\gamma_{\nu\mu}}\right) \tag{7.148}$$

where the last term is obtained by factorizing with $\delta\Gamma^\gamma_{\mu\nu}$ instead of $\delta\aleph^\gamma_{\mu\nu}$. We also extend the Maxwell energy-momentum to include torsion:

$$T^{\rm em}_{\mu\nu} := \frac{1}{8}\mathcal{H}^{\beta\lambda}\mathfrak{F}_{\beta\lambda}g_{\mu\nu} + \frac{1}{4}\mathcal{H}^{\lambda\rho}\left(g_{\nu\lambda}\mathfrak{F}_{\rho\mu} + \mathfrak{F}_{\mu\lambda}g_{\rho\nu}\right) - \frac{1}{4}\mathcal{H}^{\beta\lambda}\mathfrak{F}_{\beta\lambda}\frac{\partial\vartheta}{\partial g^{\mu\nu}} \tag{7.149}$$

where again the last term gives the influence of the torsion of the vacuum spacetime. For completeness, the following dual Maxwell energy-momentum is also defined:

$$\tilde{T}^{\rm em}_{\mu\nu} := \frac{1}{8}\tilde{\mathcal{H}}^{\beta\lambda}\tilde{\mathfrak{F}}_{\beta\lambda}g_{\mu\nu} + \frac{1}{4}\tilde{\mathcal{H}}^{\lambda\rho}\left(g_{\nu\lambda}\tilde{\mathfrak{F}}_{\rho\mu} + \tilde{\mathfrak{F}}_{\mu\lambda}g_{\rho\nu}\right) - \frac{1}{4}\tilde{\mathcal{H}}^{\beta\lambda}\tilde{\mathfrak{F}}_{\beta\lambda}\frac{\partial\vartheta}{\partial g^{\mu\nu}} \tag{7.150}$$

The two Maxwell energy momenta are fortunately opposite. Arbitrariness of the variations δA, $\delta\tilde{A}$, $\delta g^{\mu\nu}$ and $\delta\Gamma^{\gamma}_{\mu\nu}$ allows us to obtain the system of field equations governing the Einstein-Cartan vacuum spacetime.

Gravitation and symmetrized electromagnetism equations. For an Einstein-Cartan vacuum spacetime, the coupled gravitation and electromagnetic fields are governed by the system of partial differential equations:

$$\begin{cases} \nabla_\mu \mathcal{H}^{\mu\nu} = 0 \\ -\nabla_\mu \tilde{\mathcal{H}}^{\mu\nu} = 0 \\ \dfrac{1}{2\chi}\, G_{\mu\nu} + T^{\text{em}}_{\mu\nu} = 0 \\ \dfrac{1}{2\chi}\, \mathscr{D}^{\mu\nu}_{\gamma} + \dfrac{1}{2}\left(\mathcal{H}^{\mu\nu}\, A_\gamma - \tilde{\mathcal{H}}^{\mu\nu}\, \tilde{A}_\gamma\right) - \mathscr{L}_{GSEM}\left(\dfrac{\vartheta}{\partial \aleph^{\gamma}_{\mu\nu}} - \dfrac{\vartheta}{\partial \aleph^{\gamma}_{\nu\mu}}\right) = 0 \end{cases} \tag{7.151}$$

where variables are defined by (7.147), (7.149), (7.150), and (7.143).

Remark 7.25. It is remarkable that the Yang-Mills Lagrangian allows us to obtain both the Gauss laws for electric induction, the Ampère-Oersted law for current, the Gauss law for magnetic induction, and the Faraday law.

7.5.4 *Particular Case of Torsion Vector and Field Equations*

Development of relative gravitation models including torsion vector has already been the subject of advanced studies [52]. We take up this aspect with the model derived in this study. One main advantage to use torsion vector instead of torsion is the reduction of the number of unknowns.

7.5.4.1 *Variation of Compatible ω_n*

The torsion tensor may be represented as, where we have introduced the 1-form $\aleph_\mu$ as the trace of the torsion tensor: $\aleph^{\gamma}_{\mu\nu} = \aleph_\mu \delta^{\gamma}_{\nu} - \aleph_\nu \delta^{\gamma}_{\mu}$, with $\aleph_\mu = \frac{1}{3}\ \aleph^{\nu}_{\mu\nu}$. Let's consider the particular case where we have the dependence:

$$\omega_n := e^{\vartheta(\tau)}\,\overline{\omega}_n, \quad \tau := g^{\mu\nu}\aleph_\mu\aleph_\nu \in \mathbb{R} \tag{7.152}$$

This particular expression τ is dictated by the seeking of formulation invariance of the scalar function $\vartheta(\aleph)$, by virtue of the Cauchy theorem on scalar-valued function e.g. [49]. The Euler variation of the volume-form takes the form of:

$$\delta\omega_n = -\frac{1}{2}g_{\mu\nu}\,\delta g^{\mu\nu}\,\omega_n + \vartheta'(\tau)\delta\tau\,\omega_n \tag{7.153}$$

where $\delta\tau = \delta\left(g^{\mu\nu}\aleph_\mu\aleph_\nu\right) = \aleph_\mu\aleph_\nu\,\delta g^{\mu\nu} + 2\,g^{\mu\nu}\aleph_\nu\,\delta\aleph_\mu$. Variation is taken as the directional derivative along the geometric variables $\delta g^{\mu\nu}$ and $\delta\aleph_\mu$. The variation may be expressed in terms of 1-form $\aleph_\mu$, $\mu = 0, 1, 2, 3$:

$$\boxed{\delta\omega_n = \left(-\frac{1}{2}g_{\mu\nu} + \vartheta'(\tau)\,\aleph_\mu\aleph_\nu\right)\delta g^{\mu\nu}\,\omega_n + 2\,\vartheta'(\tau)g^{\mu\nu}\aleph_\nu\delta\aleph_\mu\,\omega_n} \tag{7.154}$$

This expression is used when we directly utilize the function $\vartheta(\tau)$ and the variations are formulated taking $\delta g^{\mu\nu}$ and $\delta\aleph_\mu$ as independent variables.

7.5.4.2 *Variation of Einstein-Palatini Action*

Take a vacuum spacetime with gravitation and electromagnetic fields, and let's consider the particular case of torsion vector. Let's remind the gravitational action of vacuum spacetime:

$$\mathscr{S}_{EP} := \frac{1}{2\chi}\int_{\mathscr{M}} \mathcal{R}\left(\mathbf{g}, \nabla\right)\omega_n \tag{7.155}$$

where the curvature is a function of the metric and the torsion, as independent variables. Since $\mathcal{R} := g^{\beta\lambda}\mathcal{R}_{\beta\lambda}$, the variation of the action is given by:

$$\delta\mathscr{S}_{EP} = \frac{1}{2\chi}\left(\int_{\mathscr{M}} \delta g^{\mu\nu}\mathcal{R}_{\mu\nu}\,\omega_n + \int_{\mathscr{M}} g^{\beta\lambda}\delta\mathcal{R}_{\beta\lambda}\,\omega_n + \int_{\mathscr{M}} \mathcal{R}\,\delta\omega_n\right)$$

where we have:

$$g^{\beta\lambda}\delta\mathcal{R}_{\beta\lambda} = \delta^\alpha_\gamma g^{\beta\lambda}\left[\nabla_\alpha(\delta\Gamma^\gamma_{\beta\lambda}) - \nabla_\beta(\delta\Gamma^\gamma_{\alpha\lambda}) + \aleph^\mu_{\alpha\beta}\,\delta\aleph^\gamma_{\mu\lambda}\right] \tag{7.156}$$

in which the term $\delta\Gamma^\gamma_{\mu\lambda}$ may be replaced by $\delta\aleph^\gamma_{\mu\lambda}$. The only term that could be non-zero after integration is:

$$\delta^\alpha_\gamma g^{\beta\lambda}\,\aleph^\mu_{\alpha\beta}\,\delta\aleph^\gamma_{\mu\lambda} = g^{\beta\lambda}\aleph^\mu_{\gamma\beta}\,\delta\aleph^\gamma_{\mu\lambda} = 3\,g^{\beta\lambda}\aleph_\beta\delta\aleph_\lambda$$

Application of the metric compatibility and the divergence theorem implies:

$$\delta\mathscr{S}_{EP} = \frac{1}{2\chi}\int_{\mathscr{M}}\left(\mathcal{R}_{\mu\nu} - \frac{\mathcal{R}}{2}g_{\mu\nu}\right)\delta g^{\mu\nu}\ \omega_n + \int_{\mathscr{M}}\frac{\mathcal{R}}{2\chi}\ \vartheta'(\tau)\ \aleph_\mu\aleph_\nu\delta g^{\mu\nu}\ \omega_n$$
$$+\frac{1}{2\chi}\int_{\mathscr{M}} 3\ g^{\mu\nu}\aleph_\nu\ \delta\aleph_\mu\ \omega_n + \int_{\mathscr{M}}\frac{\mathcal{R}}{2\chi}\ 2\ \vartheta'(\tau)\ g^{\mu\nu}\aleph_\nu\delta\aleph_\mu\ \omega_n$$

We have to go back to the rigorous expression of the extended Einstein tensor by writing first the Ricci curvature tensor (by introducing the Levi-Civita connection):

$$\Re_{\mu\nu} = \overline{\Re}_{\mu\nu} + \left(\overline{\nabla}_\alpha\mathfrak{T}^\alpha_{\mu\nu} - \overline{\nabla}_\mu\mathfrak{T}^\alpha_{\alpha\nu}\right) - \left(\mathfrak{T}^\alpha_{\mu\rho}\mathfrak{T}^\rho_{\alpha\nu} - \mathfrak{T}^\alpha_{\alpha\rho}\mathfrak{T}^\rho_{\mu\nu}\right)$$

and the scalar curvature:

$$\mathcal{R} = \overline{\mathcal{R}} + g^{\beta\lambda}\left(\overline{\nabla}_\alpha\mathfrak{T}^\alpha_{\beta\lambda} - \overline{\nabla}_\beta\mathfrak{T}^\alpha_{\alpha\lambda}\right) - g^{\beta\lambda}\left(\mathfrak{T}^\alpha_{\beta\rho}\mathfrak{T}^\rho_{\alpha\lambda} - \mathfrak{T}^\alpha_{\alpha\rho}\mathfrak{T}^\rho_{\beta\lambda}\right)$$

For completeness, we recall the details of the involved terms, for covariant derivative of contortion terms:

$$\overline{\nabla}_\alpha\mathfrak{T}^\alpha_{\mu\nu} = \left(g^{\alpha\sigma}g_{\mu\nu} - \delta^\alpha_\mu\delta^\sigma_\nu\right)\overline{\nabla}_\alpha\aleph_\sigma \tag{7.157}$$
$$\delta^\beta_\alpha\overline{\nabla}_\mu\mathfrak{T}^\alpha_{\beta\nu} = \delta^\beta_\alpha\left(g^{\alpha\sigma}g_{\beta\nu} - \delta^\alpha_\beta\delta^\sigma_\nu\right)\overline{\nabla}_\mu\aleph_\sigma = -3\ \overline{\nabla}_\mu\aleph_\nu \tag{7.158}$$

and for the contortion terms:

$$-\left(\mathfrak{T}^\alpha_{\mu\rho}\mathfrak{T}^\rho_{\alpha\nu} - \mathfrak{T}^\alpha_{\alpha\rho}\mathfrak{T}^\rho_{\mu\nu}\right) = -2\left(g^{\sigma\epsilon}g_{\mu\nu} - \delta^\sigma_\mu\delta^\epsilon_\nu\right)\aleph_\sigma\aleph_\epsilon \tag{7.159}$$
$$-g^{\beta\lambda}\left(\mathfrak{T}^\alpha_{\mu\rho}\mathfrak{T}^\rho_{\alpha\nu} - \mathfrak{T}^\alpha_{\alpha\rho}\mathfrak{T}^\rho_{\mu\nu}\right) = -6\ g^{\sigma\epsilon}\ \aleph_\sigma\aleph_\epsilon \tag{7.160}$$

It is worth recalling that only the symmetric part of the Ricci curvature should be accounted for, due to the scalar product with $\delta g^{\mu\nu}$, which induces the expressions:

$$\mathcal{R}_{\mu\nu} = \overline{\mathcal{R}}_{\mu\nu} + g^{\sigma\epsilon}\left[\frac{1}{2}\left(\overline{\nabla}_\sigma\aleph_\epsilon + \overline{\nabla}_\epsilon\aleph_\sigma\right) - 2\aleph_\sigma\aleph_\epsilon\right]g_{\mu\nu}$$
$$+\left(\overline{\nabla}_\mu\aleph_\nu + \overline{\nabla}_\nu\aleph_\mu\right) + 2\ \aleph_\mu\aleph_\nu \tag{7.161}$$
$$\mathcal{R} = \overline{\mathcal{R}} + 3g^{\sigma\epsilon}\left(\overline{\nabla}_\sigma\aleph_\epsilon + \overline{\nabla}_\epsilon\aleph_\sigma\right) - 6g^{\sigma\epsilon}\ \aleph_\sigma\aleph_\epsilon \tag{7.162}$$

By introducing the expression of the Einstein-like tensor, we obtain:

$$G_{\mu\nu} = \overline{G}_{\mu\nu} + g^{\sigma\epsilon}\left[\aleph_\sigma\aleph\epsilon - \left(\overline{\nabla}_\sigma\aleph_\epsilon + \overline{\nabla}_\epsilon\aleph_\sigma\right)\right]g_{\mu\nu}$$
$$+\left(\overline{\nabla}_\mu\aleph_\nu + \overline{\nabla}_\nu\aleph_\mu\right) + 2\ \aleph_\mu\aleph_\nu \tag{7.163}$$

where we recognize the Einstein tensor $\overline{G}_{\mu\nu}$ of classical relative gravitation. Finally, we have the variation:

$$\delta \mathscr{S}_{EP} = \frac{1}{2\chi} \int_{\mathcal{M}} [G_{\mu\nu} + \mathcal{R}\, \vartheta'(\tau)\, \aleph_\mu \aleph_\nu]\, \delta g^{\mu\nu} \omega_n + \frac{1}{2\chi} \int_{\mathcal{M}} [3 + 2\, \mathcal{R}\, \vartheta'(\tau)]\, g^{\mu\nu} \aleph_\nu\, \delta\aleph_\mu\, \omega_n$$

where, for convenience, we define the quantity $\Lambda_\aleph$ similar to the cosmological constant, for homogeneous field $\overline{\nabla}_\sigma \aleph_\epsilon \simeq 0$:

$$\Lambda_\aleph := g^{\alpha\beta} \aleph_\alpha \aleph_\beta = \tau \tag{7.164}$$

Remark 7.26. We notice that without physical sources the second integrand vanishes $\aleph_\beta = 0$, therefore the second line implies a zero torsion vector. In such a case, the variation reduces to the classical Einstein field equation $\overline{G}_{\mu\nu} = 0$.

7.5.4.3 *Variation of Yang-Mills Action*

Let's again consider the electromagnetic action of the vacuum spacetime action where the torsion is reduced to the torsion vector:

$$\mathscr{S}_{YM} := -\frac{1}{4} \int_{\mathcal{M}} \mathcal{H}^{\mu\nu} \mathcal{F}_{\mu\nu}\, \omega_n \tag{7.165}$$

with $\mathcal{F}_{\mu\nu}$ the Faraday strength tensor and $\mathcal{H}^{\mu\nu}$ the electromagnetic excitation tensor. The variation of this action writes:

$$\delta \mathscr{S}_{YM} := -\frac{1}{4} \int_{\mathcal{M}} (\delta \mathcal{H}^{\mu\nu} \mathcal{F}_{\mu\nu} + \mathcal{H}^{\mu\nu} \delta \mathcal{F}_{\mu\nu})\, \omega_n - \frac{1}{4} \int_{\mathcal{M}} \mathcal{H}^{\mu\nu} \mathcal{F}_{\mu\nu}\, \delta\omega_n$$

Each term of this variation can be formulated as:

$$\begin{aligned}\delta \mathscr{S}_{YM} = &-\frac{1}{4} \int_{\mathcal{M}} \left(-2\epsilon_0 \mathcal{F}_{\mu\alpha} g^{\alpha\beta} \mathcal{F}_{\beta\nu}\, \delta g^{\mu\nu} + \mathcal{H}^{\mu\nu} \delta \mathcal{F}_{\mu\nu}\right) \omega_n \\ &-\frac{1}{4} \int_{\mathcal{M}} \mathcal{H}^{\mu\nu} \delta \mathcal{F}_{\mu\nu} \omega_n \\ &-\frac{1}{4} \int_{\mathcal{M}} \mathcal{H}^{\sigma\epsilon} \mathcal{F}_{\sigma\epsilon} \Big[\Big(-\frac{1}{2} g_{\mu\nu} + \vartheta'(\tau)\, \aleph_\mu \aleph_\nu\Big)\, \delta g^{\mu\nu} \\ &+ 2\, \vartheta'(\tau)\, g^{\mu\nu} \aleph_\nu \delta\aleph_\mu\Big]\, \omega_n\end{aligned}$$

with $\delta \mathcal{F}_{\mu\nu} = \nabla_\mu(\delta A_\nu) - \nabla_\nu(\delta A_\mu) - A_\gamma \delta\aleph^\gamma_{\mu\nu}$, and for the case of torsion vector:

$$\delta \mathcal{F}_{\mu\nu} = \nabla_\mu(\delta A_\nu) - \nabla_\nu(\delta A_\mu) - (\delta\aleph_\mu A_\nu - \delta\aleph_\nu A_\mu) \tag{7.166}$$

After some rearrangements, we obtain the following expression:

$$\begin{aligned}\delta\mathscr{S}_{YM} &= \int_{\mathscr{M}} \left[\frac{1}{8}\mathcal{H}^{\sigma\epsilon}\mathcal{F}_{\sigma\epsilon}\, g_{\mu\nu} + \frac{1}{4}\mathcal{H}^{\lambda\rho}\left(g_{\mu\lambda}\mathcal{F}_{\rho\nu} + \mathcal{F}_{\mu\lambda}g_{\rho\nu}\right)\right]\delta g^{\mu\nu}\ \omega_n \\ &+ \int_{\mathscr{M}} \left(-\frac{1}{4}\mathcal{H}^{\sigma\epsilon}\mathcal{F}_{\sigma\epsilon}\right)\vartheta'(\tau)\ \aleph_\mu\aleph_\nu\ \delta g^{\mu\nu}\ \omega_n - \int_{\mathscr{M}} \mathcal{H}^{\mu\nu}\nabla_\mu(\delta A_\nu) \\ &+ \int_{\mathscr{M}} \mathcal{H}^{\mu\nu}A_\nu\delta\aleph_\mu\ \omega_n + \int_{\mathscr{M}} \left(-\frac{1}{4}\mathcal{H}^{\sigma\epsilon}\mathcal{F}_{\sigma\epsilon}\right) 2\ \vartheta'(\tau)\ g^{\mu\nu}\aleph_\nu\delta\aleph_\mu\ \omega_n\end{aligned}$$

where we arrive at the classical case of electromagnetic energy-momentum:

$$T^{\text{em}}_{\mu\nu} := \frac{1}{8}\mathcal{H}^{\sigma\epsilon}\mathcal{F}_{\sigma\epsilon}\, g_{\mu\nu} + \frac{1}{4}\mathcal{H}^{\lambda\rho}\left(g_{\mu\lambda}\mathcal{F}_{\rho\nu} + \mathcal{F}_{\mu\lambda}g_{\rho\nu}\right) \tag{7.167}$$

Remark 7.27. We recover the electromagnetic energy-momentum e.g. [296]. The function $\vartheta(\tau)$ is related to the so-called dilation for Brans-Dicke-like theory of gravitation e.g. [327].

7.5.4.4 *Variation of Dual Yang-Mills Action*

Without going into details which are not necessary at this stage, we obviously obtain the following relationships for the dual electromagnetism:

$$\begin{aligned}-\delta\tilde{\mathscr{S}}_{YM} &= \int_{\mathscr{M}} \left[\frac{1}{8}\tilde{\mathcal{H}}^{\sigma\epsilon}\tilde{\mathcal{F}}_{\sigma\epsilon}\, g_{\mu\nu} + \frac{1}{4}\tilde{\mathcal{H}}^{\lambda\rho}\left(g_{\mu\lambda}\tilde{\mathcal{F}}_{\rho\nu} + \tilde{\mathcal{F}}_{\mu\lambda}g_{\rho\nu}\right)\right]\delta g^{\mu\nu}\ \omega_n \\ &+ \int_{\mathscr{M}} \left(-\frac{1}{4}\tilde{\mathcal{H}}^{\sigma\epsilon}\tilde{\mathcal{F}}_{\sigma\epsilon}\right)\vartheta'(\tau)\ \aleph_\mu\aleph_\nu\ \delta g^{\mu\nu}\ \omega_n - \int_{\mathscr{M}} \tilde{\mathcal{H}}^{\mu\nu}\nabla_\mu(\delta\tilde{A}_\nu) \\ &+ \int_{\mathscr{M}} \tilde{\mathcal{H}}^{\mu\nu}\tilde{A}_\nu\delta\aleph_\mu\ \omega_n + \int_{\mathscr{M}} \left(-\frac{1}{4}\tilde{\mathcal{H}}^{\sigma\epsilon}\tilde{\mathcal{F}}_{\sigma\epsilon}\right) 2\ \vartheta'(\tau)\ g^{\mu\nu}\aleph_\nu\delta\aleph_\mu\ \omega_n\end{aligned}$$

where we have again the dual electromagnetic energy-momentum:

$$\tilde{T}^{\text{em}}_{\mu\nu} := \frac{1}{8}\tilde{\mathcal{H}}^{\sigma\epsilon}\tilde{\mathcal{F}}_{\sigma\epsilon}\, g_{\mu\nu} + \frac{1}{4}\tilde{\mathcal{H}}^{\lambda\rho}\left(g_{\mu\lambda}\tilde{\mathcal{F}}_{\rho\nu} + \tilde{\mathcal{F}}_{\mu\lambda}g_{\rho\nu}\right) \tag{7.168}$$

We recall the Lagrangian gravitation with symmetrized electromagnetism:

$$\mathscr{L}_{SGEM} = \mathscr{L}_{EP} + \frac{1}{2}\left(\mathscr{L}_{YM} - \tilde{\mathscr{L}}_{YM}\right)$$

7.5.4.5 *Field Equations*

The variation of Yang-Mills within an Einstein-Cartan spacetime holds:

$$\begin{aligned}\delta\mathscr{S}_{YM} &= \int_{\mathscr{M}} T^{\text{em}}_{\mu\nu}\;\delta g^{\mu\nu}\;\omega_n + \int_{\mathscr{M}} \nabla_\mu \mathcal{H}^{\mu\nu}\;\delta A_\nu\;\omega_n \\ &\quad - \int_{\mathscr{M}} \frac{1}{4}\mathcal{H}^{\sigma\epsilon}\mathcal{F}_{\sigma\epsilon}\;g_{\mu\nu}\;\delta g^{\mu\nu}\omega_n \\ &\quad + \int_{\mathscr{M}} \left[\mathcal{H}^{\mu\nu}A_\nu + \left(-\frac{1}{4}\mathcal{H}^{\sigma\epsilon}\mathcal{F}_{\sigma\epsilon}\right) 2\;\vartheta'(\tau)\;g^{\mu\nu}\aleph_\nu\right]\delta\aleph_\mu\;\omega_n\end{aligned}$$

whereas for the dual:

$$\begin{aligned}-\delta\tilde{\mathscr{S}}_{YM} &= \int_{\mathscr{M}} \tilde{T}^{\text{em}}_{\mu\nu}\;\delta g^{\mu\nu}\;\omega_n + \int_{\mathscr{M}} \nabla_\mu \tilde{\mathcal{H}}^{\mu\nu}\;\delta \tilde{A}_\nu\;\omega_n \\ &\quad - \int_{\mathscr{M}} \frac{1}{4}\tilde{\mathcal{H}}^{\sigma\epsilon}\tilde{\mathcal{F}}_{\sigma\epsilon}\;g_{\mu\nu}\;\delta g^{\mu\nu}\omega_n \\ &\quad + \int_{\mathscr{M}} \left[\tilde{\mathcal{H}}^{\mu\nu}\tilde{A}_\nu + \left(-\frac{1}{4}\tilde{\mathcal{H}}^{\sigma\epsilon}\tilde{\mathcal{F}}_{\sigma\epsilon}\right) 2\;\vartheta'(\tau)\;g^{\mu\nu}\aleph_\nu\right]\delta\aleph_\mu\;\omega_n\end{aligned}$$

and for the Einstein-Palatini action:

$$\begin{aligned}\delta\mathscr{S}_{EP} &= \frac{1}{2\chi}\int_{\mathscr{M}} \left[G_{\mu\nu} + \mathcal{R}\;\vartheta'(\tau)\;\aleph_\mu\aleph_\nu\right]\delta g^{\mu\nu}\omega_n \\ &\quad + \frac{1}{2\chi}\int_{\mathscr{M}} \left(3 + 2\;\mathcal{R}\;\vartheta'(\tau)\right) g^{\mu\nu}\aleph_\nu\;\delta\aleph_\mu\;\omega_n\end{aligned}$$

Due to arbitrariness of δA_ν, $\delta\tilde{A}_\nu$, $\delta g^{\mu\nu}$, and $\delta\aleph_\mu$, we deduce the following basic equations:

Gravito-(symmetrized)-electromagnetism field equations for Einstein-Cartan vacuum. The coupled field equations of the gravito-electromagnetism in Einstein-Cartan vacuum spacetime are:

$$\begin{cases}\nabla_\mu \mathcal{H}^{\mu\nu} = 0 \\ -\nabla_\mu \tilde{\mathcal{H}}^{\mu\nu} = 0 \\ \dfrac{1}{2\chi}G_{\mu\nu} + T^{\text{em}}_{\mu\nu} + \;\mathscr{L}_{SGEM}\;\vartheta'(\tau)\aleph_\mu\aleph_\nu = 0 \\ \dfrac{1}{2\chi}\mathcal{D}^\mu + \dfrac{1}{2}\left(\mathcal{H}^{\mu\nu}A_\nu - \tilde{\mathcal{H}}^{\mu\nu}\tilde{A}_\nu\right) + \mathscr{L}_{SGEM}\;2\;\vartheta'(\tau)g^{\mu\nu}\aleph_\nu = 0\end{cases} \qquad (7.169)$$

in which we have respectively the Einstein tensor (within Riemann-Cartan manifold), the distortion vector, and the Lagrangian:

$$G_{\mu\nu} := \mathcal{R}_{\mu\nu} - \frac{\mathcal{R}}{2} g_{\mu\nu}, \quad \mathcal{D}^{\mu} := 3g^{\mu\nu}\aleph_{\nu}, \tag{7.170}$$

and

$$\mathscr{L}_{SGEM} := \frac{\mathcal{R}}{2\chi} - \frac{1}{2}\left(\frac{1}{4}\mathcal{H}^{\mu\nu}\mathcal{F}_{\mu\nu} - \frac{1}{4}\tilde{\mathcal{H}}^{\mu\nu}\tilde{\mathcal{F}}_{\mu\nu}\right) \tag{7.171}$$

Remark 7.28. The system of equations (7.169) is merely the light version of (7.141) where the third-order torsion tensor is reduced to a torsion vector.

Remark 7.29. It is interesting to note from this system that if we admit that the torsion is zero (that is, that the vacuum is admitted as Einstein spacetime), the third equation reduces to $\mathcal{H}^{\mu\nu}A_{\nu} - \tilde{\mathcal{H}}^{\mu\nu}\tilde{A}_{\nu} \equiv 0$, implying the vanishing of optical helicity. Returning to the three-dimensional formulation, this implies the electromagnetic helicity and the spin of the angular momentum of the electromagnetic field are necessarily equal to zero. This would imply that in an Einstein spacetime, the electromagnetic field is subject to this condition, which is a very drastic limiting condition. In fact, an initial assumption that the torsion is equal to zero does not allow us to obtain the third linking equation since $\delta\aleph^{\gamma}_{\mu\nu} \equiv 0$.

7.5.4.6 *Dark Energy, Dark Matter, and Torsion Field*

To begin with, let us recall that the phenomenological modeling of dark energy has mainly consisted of introducing a cosmological constant Λ which is supposed to be uniform throughout the Universe and which finds its place in the Einstein field equation $\overline{G}_{\mu\nu} + \Lambda\ g_{\mu\nu} = 0$ in a Riemannian spacetime. For relating the present model to dark energy and dark matter, let us define first an extension of the Einstein tensor:

$$\begin{aligned}\tilde{G}_{\mu\nu} &= \overline{G}_{\mu\nu} + \Lambda_{\aleph}\ g_{\mu\nu} - g^{\sigma\epsilon}(\overline{\nabla}_{\sigma}\aleph_{\epsilon} + \overline{\nabla}_{\epsilon}\aleph_{\sigma})\ g_{\mu\nu} \\ &\quad + [2 + \mathscr{L}_{GSEM}\ \vartheta'(\Lambda_{\aleph})]\ \aleph_{\mu}\aleph_{\nu} + (\overline{\nabla}_{\mu}\aleph_{\nu} + \overline{\nabla}_{\nu}\aleph_{\mu})\end{aligned} \tag{7.172}$$

By rewriting this equation with the usual Einstein field equation, we obtain:

$$\tilde{G}_{\mu\nu} = \overline{G}_{\mu\nu} + \Omega^{\mathrm{iso}}_{\mu\nu} + \Omega^{\mathrm{aniso}}_{\mu\nu} \tag{7.173}$$

in which we can distinguish the isotropic and anisotropic contributions.

Field of dark energy. The Einstein-Maxwell field equation (7.172) consistent with these results allows us to distinguish two contributions due essentially to the non-zero torsion field:

(1) the first is an isotropic energy:

$$\Omega^{\text{iso}}_{\mu\nu} := g^{\sigma\epsilon}\left[\aleph_\sigma\aleph_\epsilon - (\overline{\nabla}_\sigma\aleph_\epsilon + \overline{\nabla}_\epsilon\aleph_\sigma)\right] g_{\mu\nu} \tag{7.174}$$

(2) the second is anisotropic:

$$\Omega^{\text{aniso}}_{\mu\nu} := [2 + \mathscr{L}_{GSEM}\ \vartheta'(\Lambda_\aleph)]\,\aleph_\mu\aleph_\nu + (\overline{\nabla}_\mu\aleph_\nu + \overline{\nabla}_\nu\aleph_\mu) \tag{7.175}$$

Both contributions involve the torsion vector field as well as the symmetric part of the gradient using the Levi-Civita connection.

Tensors $\tilde{G}_{\mu\nu}$ and $\mathscr{D}^{\mu\nu}_{\nu}$ are purely deduced from the geometric structure of the spacetime whereas the other terms include both geometry and physics. First, we observe that the term $\Lambda_\aleph := g^{\alpha\beta}\aleph_\alpha\aleph_\beta$ in (7.172) may be negative or positive since the metric is pseudo-Riemannian. *It therefore plays the role of a cosmological constant. For spacelike torsion $\Lambda_\aleph < 0$, it acts as dark energy, whereas for timelike torsion $\Lambda_\aleph > 0$, it acts as dark matter.* Second, this observation can be slightly extended. The most recent experimental results from DESI (Dark Energy Spectroscopic Instrument), at the time of writing this book, have shown that the accelerated expansion of the Universe is due to a non-uniform distribution of the hypothetical dark energy (modeled in standard cosmology by a single cosmological constant Λ) but one that varies depending on the event point in spacetime.

7.5.4.7 *Approximate Relation of Optical Helicity and Torsion Vector*

If we can neglect the term $\vartheta'(\tau) \simeq 0$ and we can assume a nearly homogeneous torsion field $\overline{\nabla}_\sigma\aleph_\epsilon \simeq 0$, meaning that ϑ varies slowly in terms of torsion, we obtain significantly simpler field equations:

$$\begin{cases} \nabla_\mu\mathcal{H}^{\mu\nu} = 0 \\ \nabla_\mu\overline{\mathcal{H}}^{\mu\nu} = 0 \\ \dfrac{1}{2\chi}\left(\overline{G}_{\mu\nu} + \Lambda_\aleph\, g_{\mu\nu} + 2\aleph_\mu\aleph_\nu\right) + \tilde{T}^{\text{em}}_{\mu\nu} = 0 \\ \dfrac{3}{2\chi}\, g^{\mu\nu}\aleph_\nu + \left(\mathcal{H}^{\mu\nu}A_\nu - \tilde{\mathcal{H}}^{\mu\nu}\tilde{A}_\nu\right) = 0 \end{cases} \tag{7.176}$$

We recover the usual form of the electromagnetic energy-momentum:

$$T_{\mu\nu}^{\text{em}} := \frac{1}{8}\mathcal{H}^{\alpha\beta}\mathcal{F}_{\alpha\beta}\, g_{\mu\nu} + \frac{1}{4}\mathcal{H}^{\lambda\rho}\left(g_{\mu\lambda}\mathcal{F}_{\rho\nu} + \mathcal{F}_{\mu\lambda}g_{\rho\nu}\right) \tag{7.177}$$

From the fourth row of (7.176), we deduce the following:

Optical helicity, spin angular momentum, and torsion vector. The torsion vector in terms of the optical helicity and the spin angular momentum [291] (Chern-Simons current) is given by:

$$\aleph_\mu = -\frac{1}{3}\chi\, g_{\mu\alpha}\left(\mathcal{H}^{\alpha\beta}A_\beta - \tilde{\mathcal{H}}^{\alpha\beta}\tilde{A}_\beta\right) \tag{7.178}$$

We observe that the physical constants are only classical ones. There is no need to introduce extra new constants.

This approximated case is apparently simpler because some coupling terms vanish. It nevertheless clearly highlights that the electromagnetic field is a major physical contributor to the acceleration of the expansion. In general, the electromagnetic field breaks the spacetime isotropy due to either directly the energy-momentum tensor $T_{\mu\nu}^{\text{em}}$, or the isotropic breaking from the tensor product $\aleph \otimes \aleph$, except in very rare cases.

7.5.4.8 *Dark Energy and Torsion Field*

The importance of magnetic fields in the acceleration of the expansion is here clearly and explicitly verified. In brief, DESI (https://www.desi.lbl.gov) measures the influence of hypothetical dark energy by studying the distribution of matter in the Universe. Recent DESI results, combined with additional data on the CMB and on gravitational lensing, appear to support a dynamic dark energy model. A single cosmological constant model, essentially based on the Friedmann equations of cosmology, is increasingly being questioned. In the present book, we propose a possible alternative dark energy model by considering the electromagnetic background of the Universe. The accounting of electromagnetism is supported by (7.178) where it is stated that an Einstein-Cartan spacetime is a valuable model in cosmology and where the torsion is merely related to the optical helicity and spin angular momentum of the electromagnetic field. There is not yet a need to introduce a “new physics theory”.

7.5.4.9 *Axion and Electromagnetism*

We limit ourselves to the geometric construction of Einstein-Cartan spacetime in which only gravitation and electromagnetism are accounted for. We have refrained from introducing particle physics which seem to us to be decoupled from our interest although their role is certainly not negligible. It is nevertheless worth giving a few words on the notion of axion which was thought to be related to the dark matter in the Universe e.g. [293, 328]. Without going into details, we mention the possibility of considering other actions including both electromagnetism and axion pseudo-particles e.g. [292]. Suggested action to study the interaction of photons (electromagnetic field) and axions takes the following form e.g. [329]:

$$\begin{aligned}\mathscr{S}_{aem} := &\int_{\mathscr{M}} \frac{1}{2}\left(g^{\mu\nu}\overline{\nabla}_{\mu}\varphi\overline{\nabla}_{\nu}\varphi - m_a^2\varphi^2\right)\overline{\omega}_n \\ &- \int_{\mathscr{M}} \frac{c_{aem}}{4}\varphi\ {}^*\mathcal{H}^{\mu\nu}\mathcal{F}_{\mu\nu}\ \overline{\omega}_n - \int_{\mathscr{M}} \frac{1}{4}\mathcal{H}^{\mu\nu}\mathcal{F}_{\mu\nu}\overline{\omega}_n \qquad (7.179)\end{aligned}$$

where φ is the axion field, m_a its mass, c_{aem} a coupling coefficient between axion and photon, and $\mathcal{H}^{\mu\nu}$ and ${}^*\mathcal{H}^{\mu\nu}$ the electromagnetic excitation and its dual respectively. The first integral defines axion field action, the second the coupling between axion and photon, and the third the photon action e.g. [293]. Einstein spacetime is assumed.

Remark 7.30. The investigation of the mutual influence between electromagnetism and axions constitutes an active area of research aimed at uncovering the physics underlying the dark matter hypothesis. A common assumption is the existence of pseudo-particles filling the Universe, which are expected to interact strongly with magnetic helicity and to play a role in cosmology e.g. [328]. In contrast, our vacuum spacetime model relies on two classical foundations: Einstein-Cartan theory for geometry and coupled gravitation-electromagnetism for physics. In this framework, spacetime is empty of matter but not of energy.

7.5.5 *Relation with Brans-Dicke-Like Cosmology*

On the one hand, numerous research teams are currently looking into the possibility of linking dark energy with the electromagnetic field e.g. [62]. This could be an interesting approach because it would avoid introducing, at least at this stage, "new physics", especially since it could be linked more or less to CMB. The link between CMB anisotropy and dark energy was already suggested soon after the discovery of the accelerated expansion, such as the existence of a scalar field factor of the Yang-Mills Lagrangian:

$$\mathscr{L}_{YM-\phi} := -\frac{1}{4}\,\vartheta(\phi)\;\mathcal{H}^{\mu\nu}\mathcal{F}_{\mu\nu} \tag{7.180}$$

in which $\vartheta(\phi)$ stems from a gauge kinetic scalar function e.g. [324] (see *EUCLID* project). We note that this amounts to weighting the Yang-Mills Lagrangian by a scalar function as if it is considered a new volume-form proportional to $\overline{\omega}_n$. The only difference is that we start in this work from a geometric structure with the sole concern of an *a priori* volume-form compatible with the connection. In this sense, our approach could also be considered "geometrically based" whereas the approach in e.g. [324] is considered as "physically realistic".

On the other hand, numerous theoretical models and experimental studies currently exist to find an alternative to Einstein gravitation in cosmology, in the analysis of gravitational waves and the dynamics of clusters and galaxies. The standard ΛCDM model requires the introduction of a major unknown concept which is dark energy, accounting for nearly 70% of the total mass-energy of the Universe, responsible for the acceleration of its expansion. Historically, among these approaches, the Brans-Dicke (BD) cosmological model was motivated by the desire to extend Einstein gravitation by introducing a gravitational constant $G(t)$ which depends on time. To do this, two new variables were introduced [330]: a scalar field ϕ and an additional constant ϖ_{BD}. The model was supposed to replace, or at least make it possible to study in parallel, the standard model of ΛCDM cosmology.

Regardless of the preferred choice concerning magnetic field genesis, primordial or dynamo, and also its amplification during cosmic evolution, the idea is to start with an action considering a state where the magnetic fields are present.

In a completely different framework, the action (7.128) presents similarities with an action originally proposed by Brans and Dicke [330] and used by numerous authors e.g. [331] for string theory of relative gravitation coupled with electromagnetism of cosmology and which takes the form e.g. [327, 332]:

$$\mathscr{S}_{ST} = \int_{\mathscr{B}} \left[\frac{1}{2\chi}\left(\overline{\mathcal{R}} - \frac{\varpi_{BD}}{\phi^2}\,\overline{\nabla}_\mu\phi\,\overline{\nabla}^\mu\phi\right) - \frac{1}{4}\overline{\mathcal{H}}^{\mu\nu}\overline{\mathcal{F}}_{\mu\nu}\right]\phi\,\overline{\omega}_n \tag{7.181}$$

where $\phi > 0$ is the so-called *dilaton* or Jordan function e.g. [56]. ϖ_{BD} is called the Brans-Dick parameter. The gravitation constant 2χ was introduced in (7.181) for simplicity. Such an action is used in string cosmology based on the scalar tensor gravitation e.g. [333]. The Brans-Dicke theory seems to be the first historical attempt to extend Einstein gravitation by introducing an additional scalar variable ϕ. Originally, the scalar is not a geometrical variable and by extension the spacetime within Brans-Dicke theory is a kind of mixed theory including geometry and physics.

What is also interesting is the introduction (by physical hypothesis) of the Jordan coupling function ϕ, to be matched with the function $e^{\vartheta(\mathcal{T})}$ of the action (6.163) which is on the other hand a mathematical necessity to have compatibility of ω_n with ∇. The additional term $\nabla_\mu\phi\ \nabla^\mu\phi$ resembles terms due to contortion for scalar curvature (6.114). Similarity between the dilaton gravity in the Brans-Dick action and the action based on connection compatible with volume-form was suggested in e.g. [132].

Correlating torsion with dilation is very inspiring for connecting string theory with Einstein-Maxwell spacetime. The present work is more oriented to a geometric Universe based on curved and twisted manifolds, given that magnetic fields are related to its geometric structure. To achieve the comparison, it makes sense to rewrite the action $\mathscr{S}_{GEM}$ (7.128) in an alternative but equivalent form:

$$\mathscr{S}_{GEM} = \int_{\mathscr{M}} \left[\frac{1}{2\chi}\left(\overline{\mathcal{R}} - g^{\beta\lambda}(\mathcal{T}^{\alpha}_{\beta\mu}\mathcal{T}^{\mu}_{\alpha\lambda} - \mathcal{T}^{\alpha}_{\alpha\mu}\mathcal{T}^{\mu}_{\beta\lambda})\right) - \frac{1}{4}\int_{\mathscr{M}} \mathcal{H}^{\mu\nu}\mathcal{F}_{\mu\nu}\right] e^{\vartheta}\ \overline{\omega}_n \tag{7.182}$$

where we have the relation $\mathcal{T}^{\nu}_{\mu\nu} = \nabla_\mu\vartheta$ [132]. The gravitational part of the action is attributed to Brans and Dicke and called BD relative gravitation [330]; when adding a cosmological constant we obtain the $BD{-}\Lambda CDM$ model. Although not exactly the same, (7.182) and (7.181) have many similarities which show that the Einstein-Cartan spacetime action might be considered a more general framework and encompasses the action containing the dilatons in the string theory of relative gravitation. However, a hasty conclusion should not be in order. Namely, the consideration of action involving such a scalar function to model both dark energy and dark matter was already suggested in the past e.g. [334]. In the particular case where torsion is defined by its torsion vector $\aleph_\mu$, we have:

$$g^{\beta\lambda}(\mathcal{T}^{\alpha}_{\beta\mu}\mathcal{T}^{\mu}_{\alpha\lambda} - \mathcal{T}^{\alpha}_{\alpha\mu}\mathcal{T}^{\mu}_{\beta\lambda}) = 6\ g^{\alpha\beta}\aleph_\alpha\aleph_\beta \tag{7.183}$$

which allows us to have a convenient expression of (7.182) as follows:

$$\mathscr{S}_{GEM} = \int_{\mathscr{M}} \left[\frac{1}{2\chi}\left(\overline{\mathcal{R}} - 6\ g^{\mu\nu}\aleph_\mu\aleph_\nu\right) - \frac{1}{4}\int_{\mathscr{M}} \mathcal{H}^{\mu\nu}\mathcal{F}_{\mu\nu}\right] e^{\vartheta}\ \overline{\omega}_n \tag{7.184}$$

where the torsion vector is a gradient of ϑ.

In sum, the action of Einstein-Cartan vacuum with gravitation and electromagnetism may be written as follows:

$$\mathscr{S}_{GEM} = \int_{\mathscr{M}} \left[\frac{1}{2\chi} \left(\overline{\mathcal{R}} - 6\, g^{\mu\nu} \frac{\nabla_\mu \phi \nabla_\nu \phi}{\phi^2} \right) - \frac{1}{4} \int_{\mathscr{M}} \mathcal{H}^{\mu\nu} \mathcal{F}_{\mu\nu} \right] \phi\, \overline{\omega}_n \quad (7.185)$$

with $\phi := e^{\vartheta}$ since $\aleph_\mu = \nabla_\mu \vartheta$. Comparison of the action (7.181) and (7.185) is very astonishing.

Remark 7.31. The use of ∇ compatible with ω_n in an Einstein-Cartan spacetime is perhaps not imposed by a physical law, as far as we know, but it seems to achieve a harmonious formulation of equations, in particular to satisfy the *Minimal Coupling Principle* for developing a relative gravitation model that accounts for spacetime torsion e.g. [134].

Remark 7.32. The present book includes electromagnetism occurring in vacuum spacetime. Other approaches exist, namely by assuming perfect fluid occupying the spacetime, or by assuming an action based on a Lagrangian depending on a curvature function $f(\mathcal{R})$ e.g. [335] as a Brans-Dicke-like model without introducing torsion. In this work, the additional scalar function is reduced to $\phi := f'(\mathcal{R})$.

7.5.6 *Electromagnetic Wave Propagation*

Maxwell electromagnetism constitutes a powerful theory, even if it remains a classical model as with Newtonian gravitation and Einstein gravitation. But unlike the theory of gravitation, electromagnetism could be combined with the theory of quantum physics to create a fully quantum model: quantum electrodynamics. Let's recall the forms of Maxwell's equations in various vacuum spacetimes:

(1) for four-dimensional formulation by using electromagnetic potential A_μ such that $\mathcal{F}_{\mu\nu} = \overline{\nabla}_\mu A_\nu - \overline{\nabla}_\nu A_\mu$:

$$-\, g^{\nu\beta} \overline{\nabla}_\nu \overline{\nabla}_\beta\, A^\mu + g^{\mu\alpha} \overline{\nabla}_\alpha \overline{\nabla}_\nu\, A^\nu + g^{\mu\alpha} \overline{\mathfrak{R}}_{\alpha\nu}\, A^\nu = 0 \quad (7.186)$$

Without magnetic monopoles, the second term vanishes (Lorenz gauge).

(2) for flat spacetime and without magnetic monopoles, we recover the classical D'Alembertian equation:

$$-\, g^{\nu\beta} \overline{\nabla}_\nu \overline{\nabla}_\beta\, A^\mu = 0 \quad (7.187)$$

Explicitly, we use flat metric $g^{\nu\beta} := \mathrm{diag}\,[+1, -1, -1, -1]$ projected onto Cartesian coordinates $(x^0 := ct, x^1, x^2, x^3)$ to obtain classical wave propagation:

$$\partial_0^2 A^\mu - \overline{\Delta} A^\mu = 0 \quad \text{or} \quad \Box \mathbf{A} := \partial_0^2 \mathbf{A} - \overline{\Delta} \mathbf{A} = 0 \tag{7.188}$$

for which the generic solution to the wave equation of the electromagnetic potential takes the form of $A^\mu(x^\alpha) = A_0^\mu \; e^{k_\alpha x^\alpha}$ where k_α is the four-wave vector. Thanks to homogeneity, the wave vector components must verify the condition $k_0^2 - \|\mathbf{k}\|^2 = 0$ with the three-wave vector $\mathbf{k} := (k_1, k_2, k_3)$.

The generic form of Maxwell's equation (first line of (7.141)) seems unaffected by gravitation. This does not mean that there is no interaction between the two fundamental forces. This apparent indifference is related to the fact that the derivation of Maxwell's equations (6.8) and (6.14) did not originally involve the structure of spacetime, other than differentiability of the manifold e.g. [254].

7.5.6.1 *Maxwell's Equations for Curved and Twisted Spacetime*

Electromagnetic wave equation (6.67) was known for a long time e.g. [64] and since then was used to analyze light propagation in the presence of strong gravitation and is useful to explain gravitational lensing. On the one hand, Einstein showed through his theory of gravitation that light bends when traveling near massive bodies. In particular, he demonstrated that light follows null geodesics of spacetime, even though photons are massless. On the other hand, it is reasonable to expect that light itself, being an electromagnetic wave, also curves spacetime in a manner similar to massive bodies, albeit to a much smaller extent. From a broader perspective, electromagnetic energy, like gravitational energy, may influence the geometric structure of vacuum spacetime.

In quantum physics, a *quantum vacuum* is defined as a spacetime where quantum phenomena fluctuations are present with the lowest ground-state energy of the electromagnetic field, and where the fields are quantized e.g. [70, 336]. Under the approximation hypothesis that makes Planck's constant tend towards zero $\ell_P \to 0$, a quantum vacuum transforms into a classical vacuum. In such a case, this does nevertheless mean that vacuum spacetime is empty space but may include electromagnetic energy (no matter but radiation). Classically, it is known that two linear electromagnetic waves in a flat Minkowski vacuum, within the macroscopic Maxwell theory, superpose but do not scatter.

In the present work, the introduction of a Riemann-Cartan geometry imposes a bent and twisted spacetime where torsion, in addition to curvature which takes into account Einstein gravitation, "fills" the vacuum with an electromagnetic field. More specifically, torsion introduces the Chern-Simons current, the optical helicity and spin angular momentum of electromagnetic waves. In other words, the addition of torsion opens the way to a vacuum model that would be expected to allow scattering of two electromagnetic waves.

Accounting for the torsion was shown, in other domains such as wave propagation in elastic defected material, to induce wave dispersion e.g. [211].[5] Let $\mathscr{M}$ be a continuum endowed with a metric $g_{\alpha\beta}$ and a metric-compatible connection $\Gamma^{\gamma}_{\alpha\beta}$. The Maxwell equations (7.141) are now considered:

$$\begin{aligned}\nabla_\nu \mathcal{H}^{\mu\nu} &= \varepsilon_0\ \nabla_\nu \left(g^{\mu\alpha} g^{\alpha\nu} \mathcal{F}_{\alpha\beta}\right) = \varepsilon_0\ \nabla_\nu \left[g^{\mu\alpha}\nabla_\alpha A^\nu - g^{\nu\beta}\nabla_\beta A^\mu\right] \\ &= \varepsilon_0\ g^{\mu\alpha}\left[\nabla_\alpha\nabla_\nu A^\nu - \aleph^{\gamma}_{\nu\alpha}\nabla_\gamma A^\nu + \Re^{\nu}_{\nu\alpha\gamma}A^\gamma\right] - \varepsilon_0\ g^{\nu\beta}\nabla_\nu\nabla_\beta A^\mu = 0\end{aligned}$$

where we have used Schouten's relations e.g. [49]. We then deduce:

Maxwell's equations within Einstein-Cartan vacuum spacetime. The extended electromagnetic wave equations within Einstein-Cartan vacuum spacetime write:

$$-g^{\alpha\beta}\nabla_\alpha\nabla_\beta A^\mu + g^{\mu\alpha}\nabla_\alpha\nabla_\beta A^\beta - g^{\mu\alpha}\aleph^{\gamma}_{\beta\alpha}\nabla_\gamma A^\beta + g^{\mu\alpha}\Re_{\alpha\beta}A^\beta = 0 \tag{7.189}$$

Remark 7.33. It is observed that dual Maxwell's equations are obtained by introducing the dual potential into the second equation of the system (7.169):

$$-g^{\alpha\beta}\nabla_\alpha\nabla_\beta \tilde{A}^\mu + g^{\mu\alpha}\nabla_\alpha\nabla_\beta \tilde{A}^\beta - g^{\mu\alpha}\aleph^{\gamma}_{\beta\alpha}\nabla_\gamma \tilde{A}^\beta + g^{\mu\alpha}\Re_{\alpha\beta}\tilde{A}^\beta = 0 \tag{7.190}$$

[5]Like any continuum model, spacetime endowed with a connection with non-zero twist and curvature consequently presents defects analogous to the fields of dislocations and disclinations e.g. [49]. This in principle influences all physical waves in the environment when they take shape in time and space. Some irreversibility of the waves might be observed accordingly.

We notice that the standard and the dual Maxwell's equations include a classical wave part, a divergence term, and the contribution of torsion and Ricci curvature. The first term expresses a D'Alembertian operator. The second term may be dropped if we assume a null divergence as a gauge condition. The condition $\nabla_\nu A^\nu \equiv 0$ would extend the Lorenz condition in Riemann-Cartan geometry. For non-twisted and non-curved continuum, the electromagnetic wave equation reduces to $\Box\, A^\mu = 0$. The third term introduces a first covariant derivative which leads to a diffusion of the wave (spacetime attenuation), and the last term points out a breathing mode whenever the boundary conditions allow it e.g. [211]. What should be observed too is that the torsion and Ricci curvature influence the wave propagation linearly. Continuum geometry and *in fine* gravitation are in fact tightly linked to electromagnetism. This may not be perceived at first sight.

Since the emergence of relative gravitation, coupling between Einstein gravitation and Maxwell electromagnetism has given rise to many studies. After the verification of the bending of light when crossing a region where the gravitational field is intense by [79], introduction of the Einstein-Cartan spacetime was proposed by some authors to explore this coupling a little further; they had proposed that coupling could only come from constitutive laws linking excitation and electromagnetic strength e.g. [248]. It seems from our results that, by first considering the symmetrized Yang-Mills Lagrangian and the variation calculus with a volume-form compatible with metric but also with connection, a formal coupling between torsion field and Chern-Simons currents (optical helicity and spin angular momentum of light) might be mathematically established (7.178).

Strictly speaking, the wave equation (7.189) cannot be solved in isolation because it is part of a coupled system. The classical approximation method, however, consists of giving (assuming) an approximate solution to spacetime geometry, often a metric tensor associated with a curvature and also a trial torsion field. The Schwarzschild metric is often suggested for light rays passing near a massive spherical object, meaning that the influence of electromagnetism is negligible compared to the gravitational field near these objects e.g. [337]. A rigorous solution must be based on the entire coupled system (7.176) in which the three unknowns are metric, torsion, and electromagnetic potential.

7.5.6.2 *Vacuum Birefringence*

The gravito-electromagnetism equations (7.176) imply that the spacetime geometry (metric, torsion, and curvature) created by the electromagnetic wave — such as light — should influence the propagation of this electromagnetic wave itself. It means a self-gravitation of the electromagnetic wave. What (7.178) explicitly shows is that the influence of the underlying electromagnetic field is essentially due to optical helicity and spin angular momentum. The influence of a magnetic field on light propagation has been known for a long time [338] and thus seems to play a key role in birefringence since the two are ultimately electromagnetic fields but which do not necessarily superimpose linearly e.g. [339]. Accounting for primordial or non-primordial magnetic fields is all the more significant since they exist everywhere, not only in intergalactic voids e.g. [340], but also near the giant and spinning filaments e.g. [290, 341].

The Universe is the theater of a magnetic field that operates everywhere, and this field is intimately linked to the structuring of the Universe at large scale, in addition to the gravitational field. The presence of galaxies, giant filaments and galaxy clusters is none other than the fruit of the interaction of gravitation and electromagnetism. In cosmology, the evolution of primordial magnetic fields would be one of the parameters that led to the current distribution of the universal magnetic field [55]. The initial conditions, unknown and ultimately unattainable, played an important role and the comparative analysis of several scenarios had made it possible to weigh the relative influence of an initial uniform distribution, the presence of initial helical fields (or not), and the distribution of a scalar-invariant field (see e.g. [66] for a complete analysis of initial conditions). The fact remains that this field exists in the present time and fills the vacuum of spacetime, even in the absence of a source (electric and magnetic charges). The vacuum spacetime is not an empty spacetime.

The propagation of an electromagnetic wave in a vacuum is governed by Maxwell's classical equations, which are linear and can be deduced from the Lagrangian:

$$\mathscr{L} = -\frac{1}{4}\mathcal{H}^{\mu\nu}\,\mathcal{F}_{\mu\nu} = \frac{1}{2}\left(\mathbf{D}\cdot\mathbf{E} - \mathbf{B}\cdot\mathbf{H}\right) = \frac{1}{4}\tilde{\mathcal{H}}^{\mu\nu}\,\tilde{\mathcal{F}}_{\mu\nu}$$

Therefore, several electromagnetic waves that are added together simply superimpose on each other, as long as we remain within the framework of classical electrodynamics. In terms of spacetime geometry, this means that we remain in a flat spacetime. However, it is now increasingly accepted that empty spacetime (of matter) is not empty of energy. This energy is reflected, among other things, by the presence of a magnetic field, which at the quantum scale no longer overlaps linearly with a light wave. In other words, the electrodynamics of two photons (one from the spacetime field, and one from the superimposed light wave) can interact through the creation of a virtual electron-positron. This necessarily induces the introduction of an additional term in the Lagrangian above. It was Heisenberg, with his student Euler in 1936, then 1937, who proposed with Hoeckel [80, 81] an additional macroscopic term to extend the Lagrangian. From this was born the vast field of research on light-light scattering, later named as vacuum birefringence by [342] in 1964. Since then, numerous works have been performed, particularly in astrophysics and cosmology e.g. [343] (and references herein with very nice and complete descriptions of experimental devices and methods) [344]. *The propagation of electromagnetic waves is more or less strongly influenced by a pre-existing electromagnetic background field.* Since the torsion vector occurring in (7.189) is mathematically associated with the Chern-Simons current, we also face a kind of birefringence. In some sense, accounting for a Riemann-Cartan geometric background allows us to prospect a continuum description of photon-photon interaction, and then to extend to nonlinear optics e.g. [84].

Light bending and birefringence in the vacuum spacetime. The wave propagation equation (7.189) tells us two essential aspects regarding the influence of the geometric structure of vacuum spacetime on the light ray.

(1) The first, the best known, is that of the deflection of the ray when passing through a gravitational field, concretized by the presence of the Ricci curvature (1919).
(2) The second, just as important but less known than the first, is the birefringence of vacuum spacetime (1936) caused by a background electromagnetic field but concretized, as we have shown in this chapter, by the torsion of spacetime.

In short, the use of Riemann-Cartan geometry seems entirely appropriate to understand empty spacetime in its entirety.

Remark 7.34. Classical electrodynamics in vacuum spacetime without source (concrete by a portion of the Universe without matter or charge), such as in interstellar vacuum, is an apparently linear theory governed by $\nabla_\nu \mathcal{H}^{\mu\nu} = 0$. However, when light passes through an electromagnetic field, the vacuum no longer seems to behave as an isotropic medium and a second-order effect is observed e.g. [84, 85]. Without going into details which go well beyond the objective of this work, an approach in the presence of a weak external magnetic field consists of considering the effective Lagrangian (fields which vary slowly in time and space):

$$\mathscr{L} = -\frac{1}{4}\mathcal{H}^{\mu\nu}\mathcal{F}_{\mu\nu} + A_e\left[4\left(-\frac{1}{4}\mathcal{H}^{\mu\nu}\mathcal{F}_{\mu\nu}\right)^2 + 7\left(-\frac{1}{4}\,{}^{*}\mathcal{H}^{\mu\nu}\mathcal{F}_{\mu\nu}\right)^2\right] \quad (7.191)$$

where the small coefficient $A_e = A_e(\mu_0, \epsilon_0, \hbar, m_e, \alpha)$ e.g. [345] allows us to observe that the last two contributions of the Lagrangian, called Heisenberg-Euler-Weisskopf Lagrangian, is of second order [80]. Rewriting this with vectorial formulation, the expansion of the Lagrangian to quartic order takes the form of e.g. [82]:

$$\mathscr{L} = \frac{1}{2}(\mathbf{D}\cdot\mathbf{E} - \mathbf{B}\cdot\mathbf{H}) + A_e\left[(\mathbf{D}\cdot\mathbf{E} - \mathbf{B}\cdot\mathbf{H})^2 + 7(\mathbf{B}\cdot\mathbf{E})^2\right] \quad (7.192)$$

It is obvious that this Lagrangian induces a nonlinear anisotropic medium for the vacuum [80]. The hope is now to improve experimental measurement of magnetic birefrigence of vacuum due to the electromagnetic fluctuating field of the spacetime vacuum. This approach seems to be confirmed by quantum electrodynamics where experimental measurements still present great challenges either for detecting the vacuum magnetic birefrigence e.g. [345, 346], or for modeling the semi-classical quantum vacuum using computational methods e.g. [336]. Recall that nonlinear electrodynamics, based on the Euler-Heisenberg Lagrangian, allows us to model light birefringence but also was exploited to give some explanation to the existence of dark energy e.g. [87], in relation to the quantum vacuum e.g. [70, 71]. However, the present book is voluntarily limited to the continuum field approach including the curvature and torsion of the spacetime manifold. Linear electrodynamics, based on solely the symmetrized Yang-Mills, is sufficient to give at least some explanation of dark energy.

Remark 7.35. First of all, recall that we use in the present book an action based on the Yang-Mills Lagrangian (for the sake of simplicity, consider only the standard Lagrangian without losing the generality of our purpose):

$$\mathscr{S}_{YM} := \int_{\mathscr{M}} \mathscr{L}_{YM}\,\omega_n, \quad \mathscr{L}_{YM} := -\frac{1}{4}\mathcal{H}^{\mu\nu}\mathcal{F}_{\mu\nu} \quad (7.193)$$

that seems at first sight to be only the first term of the Heisenberg-Euler Lagrangian (7.191). To capture the scattering phenomenon of one photon by another, the accounting of nonlinear optics by means of this second term is necessary [81]. Understanding the influence of strong magnetic fields on light propagation is an important issue, particularly in cosmology and astrophysics, where magnetic fields can reach strengths of up to $10^{12}\,[G]$. With the geometric approach advocated in the present book, the complexity is transferred to the geometric structure of the spacetime, whereas in the Heisenberg-Euler approach the difficulty (complexity from second-order quantum terms) is transferred to the constitutive laws. A direct comparison of these two approaches is very tedious but is certainly not impossible. The refractive indices (corresponding to dielectric and permeability components of the constitutive tensor) obtained with the Heisenberg-Euler Lagrangian method e.g. [344], which is nevertheless very advanced at present, and legitimately the most used, benefit from being compared with nonlinear wave propagation (7.189); one advantage of the latter could be its synthetic aspect which immediately includes gravitation and electromagnetism.

7.5.6.3 *Particular Case of Einstein Vacuum Spacetime*

To better understand the complexity of (7.189), consider the simpler special case in which the torsion field $\aleph$ identically vanishes. In such a case, the field equations are e.g. [6]:

$$\begin{cases} -g^{\alpha\beta}\, \overline{\nabla}_\alpha\, \overline{\nabla}_\beta A^\mu + g^{\mu\alpha}\, \overline{\mathfrak{R}}_{\alpha\beta}\, A^\beta = 0 \\ \dfrac{1}{2\chi}\left(\overline{\mathcal{R}}_{\mu\nu} - \dfrac{\overline{\mathcal{R}}}{2}\, g_{\mu\nu}\right) + \dfrac{1}{8}(\mathcal{H}^{\alpha\beta}\mathcal{F}_{\alpha\beta})g_{\mu\nu} + \dfrac{1}{4}\mathcal{H}^{\alpha\beta}(g_{\mu\alpha}\mathcal{F}_{\beta\nu} + \mathcal{F}_{\mu\alpha}g_{\beta\nu}) = 0 \end{cases}$$

By taking the first of the two equations, it would be tempting to solve it and deduce the solutions of classical waves in a curved spacetime, see e.g. [64] for the FLRW spacetime. However, considering the system as a whole, the solution obtained by limiting itself only to the first equation implicitly neglects the influence of electromagnetism on gravitation (but not vice versa). In a weak electromagnetic field, this kind of solution would be acceptable, but not in the case of a strong field e.g. [260]. The hypotheses of isotropy and homogeneity of spacetime on large scales cannot be assumed without precaution because the spacetime tends to be anisotropic due to the electromagnetic stress-energy tensor $T^{\mathrm{em}}_{\mu\nu}$. A diagonal metric tensor as well as the Ricci curvature field are not allowed in such a case. Maxwell's equations of electromagnetism reduce to wave equations whenever the spacetime is Ricci flat. However, if we neglect anisotropic contributions and implicitly assume that the isotropic part of the electromagnetic field is dominant, say $T^{\mathrm{em}}_{\mu\nu} \simeq T^{\mathrm{em}}_{\mathrm{sp}}\; g_{\mu\nu}$, then the metric and the Ricci curvature tensor could be assumed diagonal. We obtain the system of equations:

$$\begin{cases} g^{\alpha\beta}\overline{\nabla}_\alpha\ \overline{\nabla}_\beta A^0 - g^{00}\ \overline{\mathcal{R}}_{00}\ A^0 = 0 \\ g^{\alpha\beta}\overline{\nabla}_\alpha\ \overline{\nabla}_\beta A^i - g^{(ii)}\ \overline{\mathcal{R}}_{(ii)}\ A^i = 0, \quad i = 1,2,3 \end{cases} \tag{7.194}$$

(no summation on the spatial index $i = 1,2,3$). The set of the Einstein-Maxwell equations write, provided the isotropy assumption for time and space:

$$\begin{cases} \left(\overline{\mathcal{R}}_{00} - \dfrac{\overline{\mathcal{R}}}{2}\ g_{00}\right) - 2\chi\ T^{\text{em}}_{\text{sp}}\ g_{00} = 0 \\ \left(\overline{\mathcal{R}}_{ii} - \dfrac{\overline{\mathcal{R}}}{2}\ g_{ii}\right) - 2\chi\ T^{\text{em}}_{\text{sp}}\ g_{ii} = 0, \quad i = 1,2,3 \end{cases} \tag{7.195}$$

The metric compatibility of the connection $\nabla \mathbf{g} = 0$ allows us to lower the indices of the electromagnetic potential $A_\nu = g_{\nu\mu}A^\mu$. It is obvious that the principle of isotropy may be violated except for very particular cases of electromagnetic fields.

Remark 7.36. Very detailed studies had already been undertaken to analyze the influence of gravitation on the propagation of electromagnetic waves in the past. In particular, results, which we can now consider classic, had been obtained in the 1960s for weak gravitational fields e.g. [260]. Conversely, the analysis of electromagnetism on the metric field seems less obvious at the concrete level even if numerous studies, especially mathematical, on the resolution of the Einstein-Maxwell equations exist for some class of solutions e.g. [347], and also on the mathematical foundations of the existence of global solutions [348]. Anyhow, Einstein-Maxwell solutions are tricky models, and one of the main difficulties lies in the loss of isotropy in the presence of electromagnetic fields. In the case of strong gravitation and electromagnetic fields, solving these equations should account for the coupling.

7.5.6.4 *Minimal Coupling Principle in Electromagnetism*

Gauge invariance and minimal coupling are basic principles for gravitation and electromagnetism theory. Although their applications present no particular difficulties for curved Einstein spacetime, satisfying both of them may be contradictory in the framework of Einstein-Cartan gravito-electromagnetism in the absence of magnetic monopoles e.g. [217, 349]. The Minimal Coupling Principle constitutes a method for extending models from Minkowski, Einstein and Einstein-Cartan spacetimes in the presence of spinning fluids and other physical environments such as electromagnetic fields e.g. [286]. As already warned by past studies, some care should be undertaken when dealing with a manifold endowed with connection with non-zero torsion e.g. [131], namely to use intrinsic operators as exterior differential, co-differential and connection ∇ instead of partial derivative which are associated with a particular coordinate system. For the derivation of Maxwell's equations, we start with the Yang-Mills action:

$$\mathscr{S}_{YM} := -\frac{1}{4}\int_{\mathscr{B}} \mathcal{H}^{\mu\nu}\mathcal{F}_{\mu\nu}\,\omega_n \tag{7.196}$$

in which we can find the Faraday tensor $\mathcal{F}_{\mu\nu}$ that can represent the three cases:

(1) for Minkowski spacetime with respect to an inertial frame

$$\mathcal{F}^{\mathrm{M}}_{\mu\nu} = \partial_\mu A_\nu - \partial_\nu A_\mu \longrightarrow \mathcal{F}_{\mu\nu}$$

(2) for Einstein spacetime using the Levi-Civita connection

$$\mathcal{F}^{\mathrm{E}}_{\mu\nu} = \overline{\nabla}_\mu A_\nu - \overline{\nabla}_\nu A_\mu \longrightarrow \mathcal{F}_{\mu\nu}$$

(3) for Einstein-Cartan spacetime with a metric-compatible connection e.g. [132]

$$\mathcal{F}^{\mathrm{EC}}_{\mu\nu} = \nabla_\mu A_\nu - \nabla_\nu A_\mu \longrightarrow \mathcal{F}_{\mu\nu}$$

The use of this latter definition has often been criticized as not respecting gauge invariance (Div$\mathbf{B}$ = 0 which is the classic three-dimensional formulation) which implicitly admits the absence of magnetic monopoles e.g. [350]. In the present approach, we do not consider *a priori* such a restriction and some comments hereafter are given to illustrate that purpose.

For the three situations, expression of $\mathcal{H}^{\mu\nu}$ is determined using the constitutive laws which should be covariant. In the same way as for the Faraday strength tensor, the volume-form ω_n can also represent all three situations respectively:

(1) for Minkowski spacetime again if using an inertial frame for simplifying

$$\omega_n^{\mathrm{M}} := dx^0 \wedge dx^1 \wedge dx^2 \wedge dx^3 \longrightarrow \omega_n$$

(2) for Einstein spacetime with the Riemann volume-form

$$\omega_n^{\mathrm{R}} := \overline{\omega}_n = \sqrt{|\mathrm{Det}\mathbf{g}|}\; dx^0 \wedge dx^1 \wedge dx^2 \wedge dx^3 \longrightarrow \omega_n$$

(3) for Einstein-Cartan spacetime

$$\omega_n^{\mathrm{RC}} := e^{\vartheta(\aleph)} \sqrt{|\mathrm{Det}\mathbf{g}|}\; dx^0 \wedge dx^1 \wedge dx^2 \wedge dx^3 \longrightarrow \omega_n$$

Using $\overline{\omega}_n$ in an action where ∇ has torsion does not allow us to respect the Minimal Coupling Principle equations e.g. [131] because Maxwell's equations would then include an additional term in (7.125).

7.5.6.5 *Some Comments on the Place of Electromagnetism in Physics*

Axiomatically, the original derivation of Maxwell's equations by means of differential forms does not require geometric structures such as metric or connection but only differentiability of the manifold (region where the electromagnetic field occurs) e.g. [254]. Now let's give some comments about the position of the electromagnetism theory in general physics. Just like gravitation, which is a dynamic component of spacetime, the electromagnetic field of the vacuum is not only a background to the picture in which physical phenomena take place, but constitutes in itself a dynamic field playing a primordial role in the physical evolution of spacetime at the quantum level (fluctuations at zero-point-energy) e.g. [36, 72]. As we have seen and shown throughout this chapter, this major role also has repercussions on the geometry of spacetime at the macroscopic and even galactic level, the field equations of which were derived in the chapter.

First, the general equation (7.189) is not as easy to solve as it appears since the connection involves the torsion and at the end it is an extremely nonlinear equation because the connection also depends on the torsion. Solutions of this equation are very difficult to find except for linearized and simplified cases. Such a simplified case may be achieved by imposing a fixed metric, as in anti-de Sitter spacetime, together with an effective torsion field, as in the previously discussed spherically symmetric example involving an *a priori* magnetic monopole field. Beyond this expected difficulty, (7.189) is a natural extension of the classical electromagnetic wave propagation in curved spacetime e.g. [64, 65, 268]. A major interest of this propagation equation would be to apply it to the behavior of electromagnetic signals around stars, such as the sun, where not only is gravitation intense but also the occurrence of electromagnetic fields is not negligible; see for instance a model based on the three-dimensional approach e.g. [94].

Second, another comment should be made at a more fundamental level. The combined presence of a gravitational field and an electromagnetic field could raise hierarchical questions in terms of length and time scales, even if both of them belong to the class of action-at-a-distance. On the one hand, a consistent theory of gravity should be valid at any length scale and should hopefully take into account the quantum nature of fundamental particles. This would make it possible for gravity to be included within a unified framework with the other three interactions. Some theories such as loop quantum gravity and string theory were introduced to link gravitation with quantum physics, but this goal has not yet been fully achieved. To reconcile gravity with the standard model, in which all forces are mediated by specific particles, it has been hypothesized that a particle called the graviton mediates gravitational interactions. However, in cosmology, the geometric approach based on curvature fields remains the dominant and most successful framework for understanding large-scale gravitational phenomena. On the other hand, while physics appears to be non-classical at very small length scales, classical behavior emerges at larger scales, where quantum effects become far less relevant. Even though quantum physics has provided explanations for many physical phenomena across various scales and has achieved remarkable success, physicists have long recognized that it is still not sufficiently general to explain all observed features of the Universe, notably its accelerated expansion. In this context, electromagnetic interactions occupy an intermediate regime e.g. [75].

7.5.6.6 *Some Comments on Electromagnetism and Quantum Vacuum*

Beyond the important role of electromagnetic fields in cosmology and particularly during the evolution and formation of very large galactic structures [66], we cannot ignore fundamental investigations and research on the local interaction between the gravitational and electromagnetic fields in relation to the quantum vacuum [72, 77].

Nonlinear electrodynamics has been used, for example, to analyze light birefringence phenomena, primarily based on the Lagrangian e.g. [81], but also to provide a possible explanation for the physical modeling of dark energy using the structure of the quantum vacuum e.g. [87, 336]. At the Planck length scale, the quantum approach is appropriate. Since the discovery of the influence of the magnetic field on light propagation [83], numerous studies have been conducted to investigate the electromagnetic properties of a quantum vacuum e.g. [70]. The common link between these various studies revolves around electric helicity e.g. [6] and magnetic helicity [351], which are an integral part of the optical helicity of the electromagnetic wave e.g. [70, 107, 288, 289, 291, 298]. The latter plays an essential role in optics. The interactivity between optical helicity and particle physics, particularly its conservation over time, has been established for a long time e.g. [352, 353] (the latter clearly distinguishing the phenomena of quantum vacuum fluctuations from Casimir forces), with an analogy to the seminal work on the flow of barotropic fluids [300] concerning the discovery of helicity conservation governed by a nonlinear Euler equation of fluid mechanics. In Einstein-Cartan spacetime, and generally in a generalized continuous medium with curvature and torsion, we have seen that this conservation could be violated, particularly with a non-zero torsion field. In Einstein-Cartan spacetime, and more generally in a continuous medium endowed with curvature and torsion, we have seen that conservation laws may be violated, particularly in the presence of a non-zero torsion field. Moreover, helicity plays a key role in the spontaneous generation of magnetic fields through the flow of conducting fluids. Optical helicity is also closely related to the photon wave function, which formally illustrates the gradual transition from Maxwell's classical electromagnetism to the Dirac-Weyl equations formulated in terms of Weber vectors, also known as Riemann-Silberstein vectors (different names are used, but they refer to essentially the same objects, up to proportionality factors) e.g. [69, 74]. In this regard, we found that the symmetrized Yang-Mills Lagrangian used to construct our model closely resembles that employed by [288], with the important distinction that our electromagnetic variables clearly separate the fields E, H from the excitations D, B. This distinction may prove useful if the model developed in this book is extended to media other than vacuum. In short, even without explicitly adopting a quantum description of the vacuum, the spacetime model developed here can be viewed as a modest attempt to provide a multi-scale geometric framework linking the infinitely small (the Planck scale of the quantum vacuum) to the infinitely large (cosmological scales), given the prominent role of linear and nonlinear electromagnetic fields in approaches to quantum mechanics e.g. [72, 77, 87, 336]. For the sake of simplicity, in vacuum spacetime without sources, Maxwell's equations are generally accepted as part of classical physics. However, they can also be reformulated as a Schrödinger-like equation for the photon by using the Weber tensor and the concept of a wave function e.g. [73]. In this sense, Maxwell's equations may be interpreted as arising from a form of "first quantization" of geometrical optics for electromagnetic radiation e.g. [74], placing electromagnetism in an intermediate regime. These considerations highlight the ambiguous status of electromagnetism in its classification as either classical or quantum physics.

In any case, the preliminary results that we have obtained suggest that a coupled approach to relative gravitation with electromagnetism is more than relevant, rather than studying them separately before combining them. This is ultimately one of the main motivations of this chapter. In this approach, we have seen that the role of the torsion field is more than desirable in being able to connect the physics and geometry of spacetime, always within the global framework of the geometrization of physics.

7.5.7 *Magnetic Monopoles and Torsion Field*

The present work supports that the torsion field $\aleph$ is a candidate for dark matter and dark energy and accordingly assumes implicitly the existence of magnetic monopoles, which may be also directly related to dark matter e.g. [89]. It is quite a strong assumption and requires some explanations. See reviews for more detailed developments e.g. [26].

Maxwell's equations and magnetic charge. The existence of magnetic monopoles—that is, isolated magnetic charges—has long fascinated physicists in their quest to symmetrize Maxwell's equations in the presence of matter. Because magnetic monopoles have never been observed, Maxwell did not introduce a magnetic source term analogous to the electric charge term. In macroscopic physics, it is assumed that magnets always exist with both a north and a south pole. At the subatomic level, although the electron carries an isolated electric charge, its intrinsic magnetism—arising from its spin—corresponds to a magnetic dipole. Thus, magnets and ordinary magnetic phenomena are not due to magnetic monopoles; rather, they arise from the interplay of electric charges, electric currents, and the intrinsic quantum-mechanical dipole magnetism of elementary particles.

Magnetic monopoles and quantum electromagnetism. Pierre Curie first proposed the existence of magnetic monopoles in 1894. Forty years later, the idea was revisited following Dirac's proposal of electric charge quantization. Dirac provided a physical and theoretical basis supporting the possible—though not experimentally confirmed—existence of magnetic monopoles e.g. [28]. Using fundamental principles of quantum mechanics, Dirac showed that the existence of a magnetic monopole would provide a natural explanation for the quantization of electric charge. He showed that quantum mechanics constrained the values of the smallest electric charge $q_e := e$ and elementary magnetic charge $q_m := g$ in the Universe by requiring that they satisfy the condition $eg = \hbar/2$, where $\hbar$ is the Planck's constant. Quantization of electric charge can then be understood as arising from the possible existence of a magnetic monopole of strength $q_m := g$ somewhere in the Universe.

Magnetic monopoles and the Standard Model. Grand Unified Theories (GUTs)—which aim to unify electromagnetism, the nuclear forces, and gravity within a single framework—predict the possible existence of magnetic monopoles. However, their actual existence remains questionable despite significant theoretical and experimental efforts e.g. [26]. It is well recognized that, despite its enormous success, the Standard Model is an incomplete theory, as it does not incorporate gravity and fails to account for dark matter, cosmological inflation, and the observed dominance of matter over antimatter in the universe. In such frameworks, magnetic monopoles arise as an unavoidable mathematical consequence of the grand unification of the fundamental forces, appearing in emergent formalisms such as GUTs and superstring theory. Unlike the Dirac monopole—which, like electrons and other elementary particles, is point-like—GUT monopoles are extended objects composed of clouds of quantum fields. They possess a finite size and an internal substructure consisting of many elementary quanta. More importantly, GUT monopoles are extremely massive, with masses on the order of 10^4 trillion times that of the proton e.g. [27]. This enormous mass effectively rules out their production in present-day particle accelerators, as well as in any foreseeable future facilities.

Parker bound and magnetic monopoles. Another important constraint in cosmology is known as the Parker bound. Since the fundamental laws of physics do not prohibit the existence of magnetic monopoles, it is reasonable to expect that they could have been produced during the Big Bang or shortly thereafter in the early Universe. The present-day number density of relic monopoles originating from the Big Bang depends on the monopole mass. According to simple estimates in Big Bang cosmology—excluding the effects of cosmic inflation—the predicted monopole number density surviving today would be far higher than current observations. However, if the Universe underwent an inflationary phase shortly after the Big Bang, as required by modern cosmological theories, the pre-inflationary high monopole density (irrespective of monopole mass) would have rapidly declined to levels well below observational limits, thereby resolving the monopole overabundance problem. Based on theoretical and astrophysical considerations, upper limits have been placed on the monopole flux, or equivalently on their number density in the Universe. The most stringent of these limits is known as the Parker bound. Parker derived an upper limit on the monopole flux of approximately $10^{-15}\,\mathrm{cm}^{-2}\,\mathrm{s}^{-1}\,\mathrm{steradian}^{-1}$ (where a steradian is a unit of solid angle) for monopoles with masses below 1017 [GeV]. The Parker bound therefore excludes all monopole densities predicted by non-inflationary Big Bang cosmology. In inflationary scenarios, as mentioned earlier, the monopole density is diluted to levels below current experimental sensitivity. An additional astrophysical consideration is that if magnetic monopoles existed in the Universe today, they should appear in the cosmic ray flux reaching Earth. To date, however, no cosmic ray experiment has reported a convincing monopole candidate. Numerous experimental searches have been conducted for this purpose, several of which are reviewed in [26].

Magnetic monopoles in spin ice. Recent studies have shown evidence for magnetic monopole-like excitations in certain forms of matter, such as ice. Spin ice materials are magnetic systems in which the orientations of magnetic spins map onto the positions of hydrogen atoms in ice. Their low-temperature magnetic state has been predicted to correspond to a phase that obeys Gauss's law and supports magnetic monopole excitations—commonly referred to as a Coulomb phase. Recent experimental results have confirmed the existence of such phases in magnetic materials and strongly support the magnetic monopole theory of spin ice e.g. [255]. More precisely, no elementary particles carrying a net magnetic charge have been directly observed; instead, the monopoles in spin ice are classified as emergent excitations, in the sense that it is not the particles themselves that are directly observed but rather their effects on neighboring spins—for example during phase transitions analogous to liquid-gas transitions in ice-like systems. Reference [255] conjectured that these emergent magnetic monopoles play a central role in the underlying physics of spin ice materials.

Magnetics monopole and cosmic rays. Some recent studies have investigated the elusive magnetic monopoles in the Universe at large spacetime scales [106]. In these works, the authors searched for magnetic monopoles by modeling their possible creation in the chaotic collisions occurring high in Earth's atmosphere. By simulating the production of magnetic monopoles in the debris of atoms shattered by high-energy cosmic rays, the team was able to place robust lower limits on the amount of energy required to produce a monopole. Although these results do not constitute a direct detection or a definitive measurement of magnetic charge, they provide indirect support for the possible existence of magnetic monopoles. To the best of our knowledge, these studies represent the first results addressing the potential production of magnetic monopoles in atmospheric cosmic ray collisions.

Magnetic monopoles and torsion field. From a more geometric perspective, recent studies have related magnetic monopoles to defects in spacetime e.g. [90]. In such models, spacetime is described as a continuous manifold whose compactness is characterized by a metric $\mathbf{g}$, its connectedness is defined by a connection ∇, and the non-vanishing torsion field $\aleph$ associated with this connection can be interpreted as a measure of defect density e.g. [49]. Similarly, the torsion field has also been related to the electromagnetic Faraday tensor $\mathcal{F}_{\mu\nu}$ as shown explicitly in [6]. The present study may therefore be regarded as an attempt, within a non-quantized theoretical framework, to relate the geometry of spacetime—modeled as an Einstein–Cartan manifold—to the hypothesized existence of magnetic monopoles. Finally, the torsion field may also contribute to dark matter and possibly dark energy, whose phenomenological effects are well established elsewhere.

7.6 General Comments About this Chapter

One of the most fundamental problems in physical cosmology is to answer the question: "*What is the Universe and what does it contain?*" For a long time, the answer was simply matter and radiation. According to the earliest theory of relativistic gravitation, developed more than a hundred years ago, the Universe is expanding, and the manner of this expansion is governed by its matter content, with radiation acting as a background component. The early equations describing this behavior are the Friedmann equations, in which the Universe is modeled as being filled with a perfect fluid. In the 1990s, experimental observations became sufficiently precise to confirm that the Universe indeed contains matter and radiation. Approximately 30% of the Universe consists of matter—not only ordinary (baryonic) matter but also hypothetical dark matter, which is inferred indirectly from observations and is thought to prevent the Universe from expanding too rapidly. Radiation, by contrast, accounts for only about 0.01% of the total energy density. However, the discoveries by Perlmutter and collaborators [19], and by Riess and collaborators [18], demonstrating the accelerated expansion of the Universe, profoundly altered this paradigm. Current estimates indicate that roughly 70% of the universe consists of neither matter nor radiation, but rather an unknown form of energy intrinsic to spacetime itself, now referred to as dark energy. The behavior of dark energy is consistent with that expected from a cosmological constant within general relativity, or equivalently from the vacuum (zero-point) energy in quantum field theory. The density of dark energy is determined cosmologically, that is, phenomenologically—by inferring the value required to reproduce the observed expansion history of the universe. In this context, the role of the Dark Energy Survey collaboration is crucial for increasingly precise measurements of this quantity e.g. [354].

At the time of writing this book, the most recent results from DESI (the Dark Energy Spectroscopic Instrument) raise doubts about the assumption of spatiotemporal homogeneity and, consequently, about the isotropic nature of the cosmological constant Λ, which is the standard approach used to model the unknown dark energy. These significant results, which still require confirmation through further analysis of forthcoming observational data, already prompt us to question whether a single constant Λ is sufficient to account for dark energy.

7.6.1 *Main Contribution of the Chapter*

The search for a physical explanation of the hypothetical dark energy is addressed herein within the framework of an electro-vacuum scenario. The goal is to investigate whether a background electromagnetic field, when combined with gravitation, could contribute to dark energy, for example by producing a repulsive effect on the universe as a whole. The results suggest that electro-vacuum spacetime may indeed contribute to the accelerated expansion of the universe. What we refer to as the "tearing of spacetime" at quantum length scales—leading to its description as a collection of microcosms modeled by a torsion field—can induce a positive acceleration (or deceleration) of the cosmic expansion. This chapter therefore constitutes a central part of this book, if not its very core. To this end, we derive the cosmological field equations for several situations, considering—for the sake of simplicity—the case of a torsion vector $\aleph_\mu$ without loss of generality for the purposes of our analysis.

(1) A homogeneous and isotropic spacetime model, filled with a perfect fluid, leads to the Friedmann equations, and then to FLRW metric.

Cosmological constant Λ is added as an *ad hoc* constant to model the dark energy.

(2) In an Einstein-Cartan spacetime, with a homogeneous torsion field, the model allows us to define a "global cosmological constant" $\overline{\Lambda}_{\aleph} = 3g^{\sigma\epsilon}\aleph_{\sigma}\aleph_{\epsilon}$ deduced from averaging torsion field, which could play the role of dark energy. However, without physical sources such as an electromagnetic field as the background, the torsion vector identically vanishes. This seems to be a problem of model consistency.

(3) In an Einstein-Cartan spacetime, we add gravitational and electromagnetic fields. By using Riemann volume-form $\overline{\omega}_n$, we deduce a "local cosmological constant" $\Lambda_{\aleph} := g^{\sigma\epsilon}\aleph_{\sigma}\aleph_{\epsilon}$ acting as dark matter for timelike vector $\aleph_{\mu}$ and dark energy for spacelike torsion vector $\aleph_{\mu}$. Torsion is a source of spacetime anisotropy $2\aleph_{\mu}\aleph_{\nu}$ occurring in the field equations.

(4) In an Einstein-Cartan spacetime, again with gravitation and electromagnetism, by using the compatible volume-form $\omega_n := e^{\vartheta(\mathbf{g},\aleph)}\overline{\omega}_n$, we deduce a "local cosmological constant" including isotropic energy tensor $\Omega^{\mathrm{iso}}_{\mu\nu}$ and anisotropic energy tensor $\Omega^{\mathrm{aniso}}_{\mu\nu}$. Locally, these tensors can play the roles of dark matter and dark energy, while also acting as a source of spacetime anisotropy.

A brief analysis shows that the first two cases can be used to obtain the Friedmann equations relatively easily. The averaged torsion method appears to offer an interesting approach for developing a purely gravitational formulation and a concise form of the Friedmann equations. However, a closer examination reveals potential inconsistencies, such as the need for physical sources. Including a torsion field poses a fundamental challenge because, in principle, this field is linked to the presence of an electromagnetic field, which exists throughout the universe, even in regions of interstellar vacuum. Once this electromagnetic radiation—and in particular magnetic fields—are taken into account, the field equations necessarily imply spacetime anisotropy. Under these conditions, deriving tractable Friedmann equations becomes difficult. The two remaining models are very similar and even converge into one when the covariant derivative of the torsion vector vanishes. Both of these models also induce spacetime anisotropy, preventing a straightforward derivation of analogous Friedmann equations. Despite the incompleteness of this attempt to model the spacetime background—particularly without solving explicit models—the fact that it does not rely on introducing an *ad hoc* constant to describe dark energy, and allows for a mathematically derived, non-homogeneous, and non-constant form of dark energy, represents a notable theoretical advance that could be further explored in future work.

7.6.2 *Remarks on the Issues of the Extended Spacetime Model*

The standard approach to cosmology assumes that the Universe is a spacetime in which there are fields of matter and energy. The classical model is based on Einstein gravitation, assuming isotropy and homogeneity of spacetime in order to derive the Friedmann equations and its possible finer extensions.

In such a framework, gravitation is the only dominant force to form the structure of the Universe in very large dimensions, namely galaxies and giant filaments as well as intergalactic sidereal voids. On a conceptual level, this chapter represents an attempt to extend the theory of general relativity, generalizing Einstein's field equations—originally limited to gravitation—to include coupling with electromagnetism.

However the extended model is not limited to introducing electromagnetism, which already singularly complicates the equations, because the development goes beyond the limitation of a simply connected spacetime towards a multiply connected manifold by taking into account linear and surface defects and then includes non-zero torsion field in addition to curvature, which inevitably leads to another complication of the field equations. We can no longer find, except in very exceptional cases, equations such as those of Friedmann.

The Universe thus modeled then becomes a contiguous set of microcosms intimately linked to the existence of non-closed loops analogous to the dislocations and disclinations in solid crystalline materials. The principles of isotropy and homogeneity may then be violated, noting however that the obtained field equations are explicit, and can be used to construct alternative theories in the future. Considering both electromagnetism and multiply connected spacetime seems to us to be an interesting avenue of research. It must nevertheless be emphasized that this kind of model extension does not address philosophical questions regarding the existence of a multiverse, nor does it answer fundamental questions about the finitude or infinitude of the universe, which remain open.

Comments on the vacuum spacetime. Last but not least, the main objective of this chapter is not to construct the most sophisticated model, but rather to consider a model that is as simple as possible while still sufficiently rich to explore the combined influence of general relativity and covariant electromagnetism. When coupled with an appropriate geometric structure of spacetime—specifically, an Einstein–Cartan spacetime—this framework allows us to investigate the structure and consistency of a theoretically sound approach to cosmology. In particular, it provides insights into the notion of vacuum spacetime, the accelerated expansion of the universe, and its possible connection with the quantum vacuum. In this context, gravitation induces curvature of spacetime, while the electromagnetic field—through its optical helicity, encompassing both electric and magnetic helicity, and the spin angular momentum of the electromagnetic wave—generates spacetime torsion.

Chapter 8

Some Concluding Remarks

"Quelque variée soit l'imagination de l'homme, la nature est mille fois plus riche encore." (Henri Poincaré, La valeur de la Science, 1905)

In 1929, Hubble's seminal paper demonstrated the expansion of the Universe and marked a decisive turning point in our understanding of cosmic evolution [12]. Hubble found a linear relationship between the radial velocities of galaxies and their distances, where the distances were estimated from the properties of individual stars and the mean luminosities of nebulae in galaxy clusters (the velocity measurements were derived from earlier work by Vesto Melvin Slipher). Although the expansion rate inferred by Hubble was about seven times larger than the currently accepted value, the concept of a uniformly expanding Universe was confirmed and widely accepted for many decades.

Nearly 70 years later, however, two independent teams of astronomers reported that the expansion of the Universe is accelerating [18, 19]. Prior to this discovery, it was generally expected that cosmic expansion would be slowing down and might eventually come to an almost complete halt in the distant future. The discovery of accelerated expansion fundamentally changed this view. A commonly adopted — and perhaps the simplest — approach is to postulate the existence of dark matter, which provides the attractive gravitational component needed to bind cosmic structures, and dark energy, which acts as a repulsive component driving the accelerated expansion. Both effects can be modeled, without explicitly invoking quantum-mechanical particle physics, through the introduction of the so-called cosmological constant: $\Lambda < 0$ for dark matter associated with the

$\mathscr{A}d\mathscr{S}$ spacetime, and $\Lambda > 0$ for dark energy. This acceleration is nowadays accepted by the vast majority of physicists as being due to an unknown dark energy that is thought to constitute more than seventy percent of the matter-energy of the Universe. This lack of knowledge remains nevertheless one of the greatest problems in physics. However, this problem is not entirely new, as it stems from the co-existence of two main streams of thought: the first considers the Universe to be flat and that matter, such as particles and all other solid bodies, is decoupled from flat spacetime; the other school of thought stipulates that the Universe is not flat, that it has a curvature, even a twist, and that these geometric contents are integral actors in the physical interactions of gravitation and electromagnetism in a vacuum.[1] The latter gained increasing importance following Einstein's discovery of general relativity, and became even more relevant with the development of Maxwell's theory of electromagnetism.

A major motivation of the present work, which aligns with this second stream of thought, is to study the accelerated expansion of the Universe using a framework based on generalized continuum media. The main distinction of this book, however, is that it considers fields in vacuum rather than fluid media.

8.1 Actions-at-a-Distance and Tidal Forces

Any discussion on spacetime in theories of gravitation and *a fortiori* in cosmology is necessarily linked to the concept of action-at-a-distance (interaction fields). The physical manifestation of these interactions, whether gravitational or electromagnetic, is described in terms of geodesics and autoparallels. The trajectory of an object solely under the influence of gravity is called a geodesic (Newton and Einstein gravitation). The trajectory under curvature and torsion is called an autoparallel curve. Indeed, the relative motion of two test particles (a test particle does not modify the spacetime in which it evolves) is of most importance for embodying the effects of gravitation and electromagnetism in vacuum spacetime, called tidal forces. The longstanding debate over action-at-a-distance in physical theory therefore remains relevant, even though it now takes a different form, having shifted toward a debate centered on the concept of fields. e.g. [3].

[1]This second school of thought even proposes that matter is an integral part, or even a singularity, of an omnipresent Universe, filled with fluid. But considering only a perfect fluid that would constitute the essential element of empty spacetime, as in the vortex atom theory, has reached its limits e.g. [355].

As announced at the beginning of this book, the approach we adopted was to work not on particles and matter but directly on the structure of spacetime. The discussion then moves to the various spacetime models, given that the unmediated transfer of interaction is possible with the presence of vacuum (not empty) spacetime.

Setting aside the quantum physics approach, we only consider geometric structure at macroscopic length scale of spacetime as a generalized continuum (compact and connected manifold). In this work, gravitation is related to curvature of spacetime conforming to general relativity, whereas torsion is mathematically shown to be related to the electromagnetic field. Emptiness of the vacuum spacetime is filled with curvature and torsion in addition to metric. These three formulae express how spacetime curvature and torsion influence two nearby geodesic/autoparallel curves, making them converge to or diverge from each other. The acceleration of the gap is given by the Jacobi equation:

$$\frac{D^2\xi}{D\tau^2} = \mathscr{K}\left(\mathbf{u}, \xi\right)$$

such that the geometrical structure of the vacuum spacetime model is characterized by the shape of the tidal forces $\mathscr{K}$, which is in some sense the effect of the geometry on the force generated by the vacuum. The right-hand side terms are tidal forces[2]:

$$\begin{aligned}
\mathscr{K}_N &:= \nabla\left(\nabla\mathscr{U}\right)\xi, \\
\mathscr{K}_E &:= \overline{\Re}\left(\mathbf{u}, \xi, \mathbf{u}\right), \\
\mathscr{K}_{EC} &:= \Re\left(\mathbf{u}, \xi, \mathbf{u}\right) + \aleph\left(\frac{D\xi}{D\tau}, \mathbf{u}\right) + \left(\nabla_{\mathbf{u}}\aleph\right)\left(\xi, \mathbf{u}\right)
\end{aligned}$$

with index N, E, and EC for respectively Newton, Einstein and Einstein-Cartan spacetime. The first two tidal forces are essentially due to gravitation and electromagnetism by means of spacetime curvature and torsion. For Newtonian gravitation, the tidal forces do not depend on the velocity u^α conversely to Einstein and Einstein-Cartan gravitation since time is uncoupled from space. For Einstein-Cartan spacetime, we again observe and stress that even non-curved spacetime may induce tidal forces due to the influence of the electromagnetic field. Unlike existing results concerning geodesics and/or autoparallels, neither masses causing gravitation (the current method) nor electric charges and their movements, which are

[2]The geodesic deviation equation is also called the Jacobi equation in the framework of differential geometry.

the sources of magnetic and electric fields, explicitly appear in our Jacobi equation. Instead, vacuum spacetime plays its role as such, with only its geometric properties.

8.2 Gravitation and Electromagnetism

Vacuum spacetime is enriched by the presence of gravitation and electromagnetic as actions-at-a-distance among the fundamental forces of the Universe. By including electromagnetism, our spacetime model considers two hierarchies since electromagnetism has several facets: the classical macroscopic approach but also as a first quantized form of the photon equation. The interweaving between relative gravitation, electromagnetism, with a hypothetical dark energy materialized by the cosmological constant or other more sophisticated variables, has already been highlighted using the Einstein-Maxwell equations on several occasions both in the context of Einstein spacetime e.g. [97] and that of Einstein-Cartan e.g. [38]. Vacuum birefringence is due to interactions of light with vacuum magnetic fluctuations [80], caused by the electromagnetic field but *in fine* concretized by spacetime torsion. This latter phenomenon is of second-order magnitude. Nevertheless, these two geometric ingredients of Einstein-Cartan spacetime seem then entirely appropriate to extend vacuum spacetime models in classical physics, although the accounting for light-magnetic interactions might suggest a (modest) path to semi-classical quantum electrodynamics e.g. [82, 336].

Since cosmological constant Λ is mostly attributed to dark energy which would explain the accelerated expansion of the Universe, it therefore seems logical that the residual electromagnetic field in the Universe would have a link between dark energy and its electromagnetic origins, and particularly on the bouncing effects e.g. [60]. It is worth recalling the equations of fields that support the cosmic accelerated expansion. The complete system of field equations of an Einstein-Cartan spacetime in the presence of gravitation and electromagnetism is:

$$\begin{cases} \nabla_\mu \mathcal{H}^{\mu\nu} = 0 \\ -\nabla_\mu \tilde{\mathcal{H}}^{\mu\nu} = 0 \\ \dfrac{1}{2\chi}\, G_{\mu\nu} + T^{\text{em}}_{\mu\nu} = 0 \\ \dfrac{1}{2\chi}\, \mathscr{D}^{\mu\nu}_{\gamma} + \dfrac{1}{2\chi}\left(\mathcal{H}^{\mu\nu}\, A_\gamma - \tilde{\mathcal{H}}^{\mu\nu}\, \tilde{A}_\gamma\right) - \mathscr{L}_{GSEM}\left(\dfrac{\partial\vartheta}{\partial\aleph^{\gamma}_{\mu\nu}} - \dfrac{\partial\vartheta}{\partial\aleph^{\gamma}_{\nu\mu}}\right) = 0 \end{cases}$$

where a symmetrized Yang-Mills Lagrangian allows us to derive simultaneously the two pairs of Maxwell's equations. The present work shows that the tensor $\mathscr{D}_{\gamma}^{\mu\nu}$ quantifying spacetime distortion is explicitly related among others to Chern-Simons current and gives new interesting insights for analyzing optical helicity and spin angular momentum of light e.g. [108, 222, 325]. Under light assumptions, we interestingly relate the torsion-vector with Cherns-Simons current which physically includes optical helicity and spin angular momentum (7.178):

$$\aleph_{\mu} = -\frac{1}{3}\chi\, g_{\mu\alpha}\left(\mathcal{H}^{\alpha\beta} A_{\beta} - \tilde{\mathcal{H}}^{\alpha\beta} \tilde{A}_{\beta}\right)$$

The essential role of light propagation and light-matter duality, through its helicity, is thus pointed out. While the old Kelvin's vortex theory of atoms of the 1860s posited that spacetime was filled with a perfect fluid, the present work suggests that it is rather the helicity of light that is directly linked to the torsion of vacuum spacetime (and *vice versa*), which would replace a hypothetical omnipresent fluid (as for Friedman equations). This naturally brings to mind wave-particle or light-matter duality, with optical helicity playing the significant role. Mathematically, this would confirm by extension the importance of knot theory in physics, given that optical helicity is the source of twisting, winding, and even linking of electromagnetic tubes within vacuum spacetime.

8.3 Accelerating Expansion of the Universe and Torsion Field

Staying in the context of general relativity in classical physics, this work points out the unavoidable role of torsion. Since the evidence of dark matter and dark energy comes from gravitational phenomenon and also from electromagnetism, it is quite natural to consider the geometrical properties of spacetime associated with these two phenomena. Most recent results by DESI[3] on dark matter and dark energy state that a single cosmological constant would not be appropriate due to anisotropy and heterogeneity of the Universe even at very large scales. The acceleration or deceleration of the expansion is shown to be controlled by two *cosmological fields essentially due to torsion* instead of a single cosmological constant. The first is an isotropic energy $\Omega_{\mu\nu}^{\text{iso}} := g^{\sigma\epsilon}\left[\aleph_{\sigma}\aleph_{\epsilon} - (\overline{\nabla}_{\sigma}\aleph_{\epsilon} + \overline{\nabla}_{\epsilon}\aleph_{\sigma})\right] g_{\mu\nu}$. The second is anisotropic $\Omega_{\mu\nu}^{\text{aniso}} := \left[2 + \mathcal{R}\,\vartheta'(\tau)\right]\aleph_{\mu}\aleph_{\nu} + (\overline{\nabla}_{\mu}\aleph_{\nu} + \overline{\nabla}_{\nu}\aleph_{\mu})$ where $\tau := g^{\sigma\epsilon}\aleph_{\sigma}\aleph_{\epsilon}$ and $\omega_n = e^{\vartheta(\tau)}\overline{\omega}_n$, and ϑ may be interpreted as an *axion-like*

[3] Dark Energy Spectroscopic Instrument

particle or a kind of *dilaton* linked to torsion. On the one hand, torsion is a variable measuring the defects of spacetime probably due to the cooling and inflation during the birth of the Universe; on the other hand it is also a variable which is closely linked to the electromagnetic field. In a way this geometrization of the electromagnetic field by torsion and the link which unites torsion with the cosmological constant is in conformity with recent studies which place torsion as a probable element, at least in part, to quantify dark matter and energy. However, aside from approaches based on averaged torsion (where torsion is assumed to form a homogeneous field), a major drawback of incorporating torsion is the nearly unavoidable introduction of spacetime anisotropy. In the present study, torsion is an ambient vacuum field which is in no way linked to the presence of fluids but to a background electromagnetic field. The standard model of cosmology ΛCMB is essentially based on isotropy and homogeneity assumptions of the Universe at very large scales. They made it possible to develop Friedman's classic equations, and thus to suggest phenomenological models on dark energy. This work seems to indicate that, except by making drastic hypotheses, spacetime must logically present non-homogeneity and anisotropy properties. One of the questions is whether this anisotropy is so pronounced as not to be negligible.

8.4 General Final Remarks

Historical context for this work. Reflecting the history of scientific thought in mechanics, the study of the motion of bodies and particles—and indeed of physics as a whole—has always relied on the concepts of space and time, which remain the fundamental measuring tools used by humankind in its quest for knowledge. Two distinct lines of thought form the underlying foundations that have driven the advancement of physics, including cosmology: the first posits an immutable spacetime, while the second regards spacetime as dynamic and as a constituent element of physical phenomena.

The first line of thought posits that spacetime is an immutable object and remains so regardless of the physical phenomena occurring in our Universe. Broadly speaking, this view was predominant until the advent of general relativity in 1915. Classical mechanics and even the early developments of quantum mechanics are based on this assumption. In the long history of scientific evolution, two advances—Newton's theory of mechanics and Maxwell's theory of electromagnetism—had fostered the gradual shift from this *immutable space and time* approach to a *dynamic coupled spacetime* approach with a transition. The culmination of this transition period was Einstein's discovery of special relativity in 1905, along with Minkowski's four-dimensional spacetime, not forgetting Poincaré's essential contribution to this field. With the discovery of general relativity in 1915, spacetime became a dynamic object. More precisely, vacuum spacetime is no longer just an independent playground but constitutes an integral component of physical phenomena that occur therein. This is all the more remarkable since it allows for the mathematical inclusion of the possibility of remote unmediated interactions such as gravitation and electromagnetism. Action-at-a-distance had struggled to be accepted by the greatest scientists before Newton. The advent of Maxwell's theory of electromagnetism had shaken up the thinking of the time. In the context of relative gravitation, it is now well established that empty spacetime is no longer simply a medium for mediating physical phenomena but constitutes a dynamic element of the theory with inherent energy and even a potential source of matter, as described by quantum mechanics. In the theory of classical physics, it is now accepted that two fields exist in a vacuum: gravitation and the electromagnetic field, without any material support. The latter even constitutes a fundamental component of what is called the quantum vacuum, because the vacuum energy in this theory is due to the zero-point fluctuations of electric and magnetic fields. This book is an attempt to generalize the spacetime model a little further by attributing a geometric, rather than a physical, meaning to the construction of spacetime with electromagnetism. More precisely, optical helicity and spin angular momentum are then directly related to the torsion vector. One of the main contributions of this book would therefore be the derivation field equations linking mathematically the geometrical structure of spacetime with action-at-a-distance physics.

Geometry and physics. Translating Poincaré, we may read "Every generalization of a certain theory is a hypothesis; the hypothesis therefore has a necessary role that no one has ever contested. Only it must always be, as soon as possible and as often as possible, subject to verification"[4]. The historical and formal link between geometry and physics has long been a source of motivation for our understanding of the Universe and, by extension, the spacetime in which we live, as a whole. The passage from absolute time and space to spacetime as a dynamical part of the theory is enlightening. We have limited ourselves to classical physics in this book and focused solely on

[4]La Science et l'Hypothèse, Poincaré, Edition Champs Flammarion, Paris 1968.

long-range interactions with infinite radii of action. We note, however, that consideration of spacetime microcosms (the Universe as grains of spacetime), which more or less directly constitute the basis of Riemann-Cartan manifolds and also of the so-called Planck-Kleinert crystals, seems to be a source of inspiration for building not only classical physics but also eventually a theory of loop quantum gravitation. In the meantime, limitation to classical geometry has allowed us to work within differential geometry and a generalized continuum hypothesis for putting gravitation and electromagnetism in the same geometric basket. It seems a reasonable generalization because it provides a certain continuity on the approach specific to relative gravitation based on Einstein-(anti)-de Sitter spacetime towards the dominant role of torsion fields to further study dark matter and dark energy.

Spacetime and length scales. The present book deliberately omitted quantum physics, and instead focused on the geometric structure of spacetime as an infinite set of spacetime grains (microcosms). In this way, torsion is indeed associated with the relative translation of microcosms while curvature is due to relative rotation of microcosms of the spacetime. This physical interpretation conforms to Schouten's theorem for Riemann-Cartan manifolds, which models the Einstein-Cartan spacetime. Implicitly, this geometric approach is the fundamental ingredient to construct the generalized continuum underlying the spacetime model we developed in this work. It is known that the non-compatibility of quantum physics and general relativity arises in part from the fact that metric tensor and connection are physical and dynamic variables. Spacetime is curved and maybe twisted. This shows that spacetime geometry stands apart from other aspects of quantum physics, whose definitions of variables are based on the existence of classical non-relativistic gravitational spacetime. The unification of all physical theories is thus dependent on the possibility of making the geometry of spacetime and physical particles consistent down to the smallest scales. It should be remembered that various characteristic length scales used for classical physics are for relative gravitation lower than $\ell_{RG} \simeq 10^{11}$ $[m]$, for cosmology nearly $\ell_{COS} \simeq 10^{26}$ $[m]$, whereas for quantum physics are at nearly Planck length $\ell_P \simeq 1.6 \times 10^{-35}$ $[m]$, given the radii of action for strong nuclear $\ell_{Str} \simeq 10^{-15}$ $[m]$ and weak $\ell_{Weak} \simeq 10^{-18}$ $[m]$ interactions, and also the observed cosmological constant $\Lambda_{obs} \simeq 10^{-52}$ $[m^2]$. In view of discrepancies of these characteristic length scales, it is understandable that unifying such theories remains complicated. A critical question, already posed by Ehlers for example, is to know up to what length scale we can consider spacetime as a continuum (compacted and connected). And consequently, what is the geometric structure of possibly discrete spacetime for smaller scales which allows a continuous and valid transition to be made with continuous spacetime (differentiable manifold with connection)? The present book shows that taking into account curvature and torsion seems to provide some (fragmentary) answers between quantum and galactic distances.

Some thoughts on spacetime. To conclude, we must indicate as much as possible the place of this book, however modest it may be, in a set of reflections on space and time. This work is neither a mathematics book nor a physics book. It lies in the space between these two areas of science. Mathematics is not there to provide equations or solve them for physics and conversely physics cannot ask mathematics to reveal a new discovery. But the latter can on the other hand provide elements to intuit eventually some new scientific truths, if there are any. In simplified terms, all the laws of physics are drawn from experiment, but these experiments need an appropriate language to be formulated in a scientifically rigorous way. These laws come from experience but they do not come directly from it, and their formulation may take years, decades, or even centuries. Moreover, these laws are continually processed, refined, and generalized over time.

Beyond the mutual influences between physics and mathematics in the search for models that we hope can reliably describe and predict the genesis and evolution of the Universe, cosmology itself is deeply connected to philosophy and, to some extent, theology. Timeless questions naturally arise: Which experimentally observed facts do we seek to explain through cosmology? Which of these facts can truly be explained by physical models of the Universe's evolution? To what extent can explanations calibrated by observations and measurements—even those obtained using the most sophisticated instruments available today—remain valid in the long term? Should we go further and ask about the meaning of the Universe's evolution, a question that currently remains beyond our reach?

Indeed, one may even ask whether such questions can ever be fully addressed, given that the phenomena under observation unfold over billions of years, while the time available to humanity for observation spans only a few hundred, or at most a few thousand, years. The survival of the human species thus becomes an essential condition for transmitting, from generation to generation, the data and inevitably incomplete concepts developed over time by scientists and philosophers. Otherwise, this endeavor in cosmological research would remain a beautiful—but brief—chapter in the history of humankind.

Annexes

Spherically symmetric spacetime and Birkhoff's theorem

Problems with spherical symmetry constitute very instructive examples in the study of spacetime in relative gravitation and in cosmology. Moreover, Schwarzschild's spacetime example is the first of its kind to be solved analytically in 1916. We consider in this Appendix a spherical symmetric massive body and focus on the gravitational field it generates outside the body.

Spherically symmetric spacetime

In the three-dimensional space $\mathscr{E}$, we define an origin O. Any point of this space is defined by the vector-position $\mathbf{OM}$. First of all, the most general shape of an isotropic scalar function of the following vector as arguments reduce to:

$$F : (\mathbf{OM}, d\mathbf{OM}) \in \mathscr{E}^2 \rightarrow F(\mathbf{OM} \cdot \mathbf{OM}, \mathbf{OM} \cdot d\mathbf{OM}, d\mathbf{OM} \cdot d\mathbf{OM}) \in \mathbb{R}$$

thanks to Cauchy's theorem on isotropic functions e.g. [49]. We recall that the usual decomposition of the vector position in a three-dimensional space $\mathscr{D}$ is given by:

$$\mathbf{OM} = r\ \mathbf{e}_r, \qquad d\mathbf{OM} = dr\ \mathbf{e}_r + rd\theta\ \mathbf{e}_\theta + r\sin\theta d\varphi\ \mathbf{e}_\varphi \tag{A.1}$$

in the local orthonormal spherical base $(\mathbf{e}_r, \mathbf{e}_\theta, \mathbf{e}_\varphi)$. Therefore, it is obvious to write the most general spatially isotropic metric that can be defined with these arguments by introducing the time τ and its differential $d\tau$. Using spherical coordinates, and by considering only quadratic terms, we have the line element:

$$ds^2 = a\ d\tau^2 + 2b\ rd\tau dr - c\ r^2 dr^2 - d\left(dr^2 + r^2 d\theta^2 + r^2 \sin^2\theta d\varphi^2\right)$$

since we have in a spherical coordinate system:

$$\mathbf{OM} \cdot \mathbf{OM} = r^2, \quad \mathbf{OM} \cdot d\mathbf{OM} = rdr, \quad d\mathbf{OM} \cdot d\mathbf{OM} = dr^2$$

and where the arbitrary coefficients are only a function of the time τ and the radius r due to symmetry:

$$a = a(\tau, r), \quad b = b(\tau, r), \quad c = c(\tau, r), \quad d = d(\tau, r)$$

Remark A.1. Instead of x^0 as in the body of the book, we utilize the notation τ for the time variable in order to simplify the notation.

For the needs of the cause, we rewrite the line element as:

$$ds^2 = A(\tau, r)\ d\tau^2 + 2B(\tau, r)\ d\tau dr - C(\tau, r)\ dr^2 - D(\tau, r)\ d\Omega^2 \qquad \text{(A.2)}$$

in which we define the still new arbitrary functions:

$$A := a(\tau, r), \quad B := r\ b(\tau, r), \quad C := r^2 c(\tau, r) + d(\tau, r), \quad D := r^2\ d(\tau, r)$$

and the solid angle:

$$d\Omega^2 := d\theta^2 + \sin^2\theta\ d\varphi^2$$

The idea is now to simplify the line element with a change of variables.

(1) Choice of a new radius coordinate. This is obvious by taking the new radius:

$$\overline{r} := D(\tau, r) = r^2\ d(\tau, r) \qquad \text{(A.3)}$$

(2) Choice of new time. To eliminate the crossed term $d\tau dr$, let's rewrite the line element as:

$$ds^2 := \overline{A}(\tau, r)\ d\tau^2 + 2\overline{B}(\tau, r)\ d\tau d\overline{r} - \overline{C}(\tau, r)\ d\overline{r}^2 - \overline{r}^2\ d\Omega^2 \qquad \text{(A.4)}$$

where the functions $\overline{A}(\tau, r)$, $\overline{B}(\tau, r)$, and $\overline{C}(\tau, r)$ are still arbitrary new ones, and let's define a new time $\overline{\tau}$ such as:

$$d\overline{\tau} := \mu(\tau, \overline{r})\left[\overline{A}(\tau, r)\ d\tau + \overline{B}(\tau, r)\ d\overline{r}\right] = d\mathfrak{F} \qquad \text{(A.5)}$$

with the integrating factor $\mu(\tau, \overline{r})$ calculated to make this differential a total differential of a scalar function $\mathcal{F}$. The square of the time gives:

$$d\overline{\tau}^2 = \mu^2 \left[\overline{A}^2 \, d\tau^2 + 2\overline{AB} \, d\tau d\overline{r} + \overline{B}^2 \, d\overline{r}^2 \right] \tag{A.6}$$

By introducing the expression of the term $d\tau^2$ into (A.4), we eliminate the crossed term $d\tau d\overline{r}$ and obtain the diagonal line element:

$$ds^2 = \frac{1}{\mu^2 \overline{A}} \, d\overline{\tau}^2 - \left(\overline{C} + \frac{\overline{B}}{\overline{A}^2} \right) d\overline{r}^2 - \overline{r}^2 \, d\Omega^2 \tag{A.7}$$

We define the two functions:

$$e^{2\mu(\overline{\tau}, \overline{r})} := \frac{1}{\mu^2 \overline{A}}, \quad e^{2\nu(\overline{\tau}, \overline{r})} := \overline{C}(\overline{\tau}, \overline{r}) + \frac{\overline{B}(\overline{\tau}, \overline{r})}{\overline{A}^2}(\overline{\tau}, \overline{r}) \tag{A.8}$$

Dropping the overline bar over variables to avoid cumbersome notation, we obtain the general form of the spherically symmetric line element:

$$\boxed{ds^2 = e^{2\mu} \, d\tau^2 - e^{2\nu} \, dr^2 - r^2 \left(d\theta^2 + \sin^2 \theta \, d\varphi^2 \right)} \tag{A.9}$$

where the time and the radius are not necessarily the usual time and the classical radius, respectively, of the spherical coordinate system.

Birkhoff's theorem

Theorem A.1 (label=thm1). *(Birkhoff,* 1923*) All vacuum spacetimes of Einstein's field equations satisfying spherical symmetry ($SO(3)$ symmetry) are necessarily static.*

Proof. The proof is done in two steps.

(1) Levi-Civita connection (for the sake of simplicity, we drop the overline bar). The connection coefficients are derived from the spherically symmetric metric:

$$\overline{\Gamma}^{\gamma}_{\alpha\beta} = g^{\gamma\lambda} \left(\frac{\partial g_{\alpha\lambda}}{\partial x^{\beta}} + \frac{\partial g_{\lambda\beta}}{\partial x^{\alpha}} - \frac{\partial g_{\alpha\beta}}{\partial x^{\lambda}} \right)$$

where $x^{\mu} = \tau, r, \theta, \varphi$. We do not go into details of the calculus.

(2) Ricci curvature tensor:

$$\Re_{\alpha\beta} = \Re^{\lambda}_{\lambda\alpha\beta}$$

The components of the Ricci curvature tensor are the following:

$$\begin{aligned}
\Re_{\tau\tau} &= \left[\partial_\tau \nu + (\partial_\tau \nu)^2 + \partial_\tau \mu \; \partial_\tau \nu\right] \\
&\quad + \left[\partial_r^2 \mu + (\partial_r \mu)^2 - \partial_r \mu \; \partial_r \nu + \frac{2}{r}\partial_r \mu\right] e^{2(\mu-\nu)} \\
\Re_{rr} &= \left[\partial_\tau \nu + (\partial_\tau \nu)^2 - \partial_\tau \mu \; \partial_\tau \nu\right] e^{2(\nu-\mu)} \\
&\quad - \left[\partial_r^2 \mu + (\partial_r \mu)^2 - \partial_r \mu \; \partial_r \nu - \frac{2}{r}\partial_r \nu\right] \\
\Re_{\theta\theta} &= 1 + e^{-2\nu}\left[r\left(\partial_r \nu - \partial_r \mu\right) - 1\right] \\
\Re_{\varphi\varphi} &= \left\{1 + e^{-2\nu}\left[r\left(\partial_r \nu - \partial_r \mu\right) - 1\right]\right\} \sin^2\theta
\end{aligned}$$

The only off-diagonal term is:

$$\Re_{\tau r} = \frac{2}{r}\partial_\tau \nu$$

We directly deduce that the function $\nu(\tau, r)$ does not depend on the time τ. One obtains the expressions:

$$\begin{aligned}
\Re_{\tau\tau} &= \left[\partial_r^2 \mu + (\partial_r \mu)^2 - \partial_r \mu \; \partial_r \nu + \frac{2}{r}\partial_r \mu\right] e^{2(\mu-\nu)} \\
\Re_{rr} &= -\left[\partial_r^2 \mu + (\partial_r \mu)^2 - \partial_r \mu \; \partial_r \nu - \frac{2}{r}\partial_r \nu\right] \\
\Re_{\theta\theta} &= 1 + e^{-2\nu}\left[r\left(\partial_r \nu - \partial_r \mu\right) - 1\right] \\
\Re_{\varphi\varphi} &= \left\{1 + e^{-2\nu}\left[r\left(\partial_r \nu - \partial_r \mu\right) - 1\right]\right\} \sin^2\theta
\end{aligned}$$

Partial derivative of the component $\Re_{\theta\theta}$ by the time τ implies that:

$$\partial_\tau \partial_r \mu = 0 \quad \Longrightarrow \quad \mu(\tau, r) = \alpha(r) + \beta(r)$$

Using the exponential properties, we can consider a new time by observing that:

$$e^{2\mu(\tau,r)} = e^{2\alpha(\tau)+2\beta(r)} = e^{2\alpha(\tau)} \; e^{\{2\beta(r)}$$

and introduce it into the line element:

$$ds^2 = e^{2\alpha(\tau)} \; e^{\{2\beta(r)} \; d\tau^2 - e^{2\nu(r)} \; dr^2 - r^2 d\theta^2 - r^2 \sin^2\theta \; d\varphi^2 \tag{A.10}$$

Making a change of variable for the time $e^{2\alpha(\tau)}\, d\tau^2 \rightarrow d\tau^2$ means that the line element can be rewritten as:

$$\boxed{ds^2 = e^{2\mu(r)}\, d\tau^2 - e^{2\nu(r)}\, dr^2 - r^2 d\theta^2 - r^2 \sin^2\theta\, d\varphi^2} \tag{A.11}$$

The metric associated with this line element is necessarily static because none of its components depend on the time τ. This remains valid even if we introduce a non-stationary source with a spherical symmetry. □

Remark A.2. It should be noticed that the variable radius is in fact a function of the usual radius of a spherical coordinate system. The term r does not necessarily mean the usual geometric radius e.g. [243].

Remark A.3. In other words, in the framework of general relativity, the gravitational field of any spherically symmetric vacuum spacetime is therefore necessarily static according to Birkhoff's theorem. Practically, this means that the spacetime outside of a spherical (nevertheless non-rotating) massive body must be given by the Schwarzschild metric as we will see in the chapter on Einstein's relative gravitation.

Bibliography

[1] Ehlers J. The nature and structure of spacetime. In: Mehra J, editor. *The Physicist's Conception of Nature.* Dordrecht: Holland; 1973. p. 71–91.

[2] Gonseth F. *Les fondements des mathématiques: De la géométrie d'Euclide à la relativité générale et à l'intuitionisme.* Paris: Albert Blanchard; 1926.

[3] McMullin E. The origins of the field concept in physics. *Phys Perspect.* 2002;4:13–39.

[4] Medina SB, Nowakowski M, Batic D. Einstein-Cartan cosmologies. *Ann Phys.* 2019;400:64–108.

[5] Nakahara M. *Geometry, topology and physics.* Bristol: Institute of Physics Publishing; 1996.

[6] Rakotomanana RL. *Covariance and gauge invariance in continuum physics: Application to mechanics, gravitation, and electromagnetism.* Cham: Birkhäuser; 2018.

[7] Misner CW, Wheeler JA. Classical physics as geometry: Gravitation, electromagnetism, unquantized charge, and mass as properties of curved empty space. *Ann Phys.* 1957;2:525–603.

[8] Rainich GY. Electrodynamics in the general relativity theory. *Trans Am Math Soc.* 1925;27:106–36.

[9] Maugin GA. *Continuum mechanics through the twentieth century: A concise historical perspective.* Berlin: Springer; 2013.

[10] Kleinert H. Emerging gravity from defects in world crystal. *Braz J Phys.* 2005;35:359–61.

[11] Rakotomanana RL. Contribution à la modélisation géométrique et thermodynamique d'une classe de milieux faiblement continus. *Arch Ration Mech Anal.* 1997;141:199–236.

[12] Hubble E. A relation between distance and radial velocity among extragalactic nebulae. *Proc Natl Acad Sci USA.* 1929;15:168–73.

[13] Lindel F, Bennett R, Buhmann SY. Theory of polaritonic quantum-vacuum detection. *Phys Rev A.* 2020;102:041701.

[14] Riek C, Seletskiy DV, Moskalenko AS, Schmidt JF, Krauspe P, Eckart S, *et al.* Direct sampling of electric-field vacuum fluctuations. *Science.* 2015;350(6259):420–3.

[15] Abdalla E, *et al.* Cosmology intertwined: A review of the particle physics, astrophysics, and cosmology associated with the cosmological tensions and anomalies. *J High Energy Astrophys.* 2022;34:49–211.

[16] Padmanabhan T. Cosmological constant—the weight of the vacuum. *Phys Rep.* 2003;380:235–320.

[17] Lindel F, Bennett R, Buhmann SY. Macroscopic quantum electrodynamics approach to nonlinear optics and application for polaritonic quantum-vacuum detection. *Phys Rev A.* 2021;103:033705.

[18] Riess AG, Filippenko AV, *et al.* Observational evidence from supernovae for an accelerating universe and cosmological constant. *Astron J.* 1998;116:1009–38.

[19] Perlmutter S, Aldering A, *et al.* Measurements of Ω and Λ from 42 high-redshift supernovae. *Astrophys J.* 1999;517:565–86.

[20] Amendola L, *et al.* Euclid Theory Working Group. Cosmology and fundamental physics with the Euclid satellite. *Liv Rev Relativ.* 2018;21:12.

[21] Frieman JA, Turner MS, Huterer D. Dark energy and the accelerating universe. *Annu Rev Astron Astrophys.* 2008;46:385–432.

[22] Turner MS. Dark matter and dark energy in the universe. In: The Third Stromlo Symposium: The Galactic Halo. *ASP Conf Ser.* 1999;666:1–22.

[23] Peebles PJE, Ratra B. The cosmological constant and dark energy. *Rev Mod Phys.* 2003;75:559–606.

[24] Carroll SM. The cosmological constant. *Living Rev Relativ.* 2001;4:1. Available from: http://www.livingreviews.org/lrr-2001-1.

[25] Biasi A, Evnin O, Syspas S. De Sitter bubbles from anti-de Sitter fluctuations. *Phys Rev Lett.* 2022;129:251104.

[26] Mavromatos NE, Mitsou VA. Magnetic monopoles revisited: Models and searches at colliders and in the cosmos. *Int J Mod Phys A.* 2020;35:2030012.

[27] Preskill JP. Cosmological production of superheavy magnetic monopoles. *Phys Rev Lett.* 1979;43:1365–8.

[28] Dirac PAM. Quantised singularities in the electromagnetic field. *Proc R Soc A.* 1931;133:60–72.

[29] Zel'Dovich YB. The cosmological constant and the theory of elementary particles. In: Sahni V, Krasinski A, editors. *Gen Relativ Gravit.* 2008;40:1557–91.

[30] Weinberg S. The cosmological constant problem. *Rev Mod Phys.* 1989;61: 1–23.

[31] Gupta RP. Testing CCC + TL cosmology with observed baryon acoustic oscillation features. *Astrophys J.* 2024;964:55.

[32] Synge JL. On the deviation of geodesics and null-geodesics, particularly in relation to the properties of spaces of constant curvatures and indefinite line-element. *Ann Math.* 1934;35:705–13.

[33] Levi-Civita T. Sur l'écart géodésique. *Math Ann.* 1927;97:291–320.

[34] Balakin A, van Holten JW, Kerner R. Motions and worldline deviations in Einstein-Maxwell theory. *Class Quantum Grav.* 2000;17:5009–23.

[35] Podolosky J, Svarc R. Interpreting spacetimes of any dimension using geodesic deviation. *Phys Rev D.* 2012;85:044057.

[36] Rugh SE, Zinkernagel H. The quantum vacuum and the cosmological constant. *Stud Hist Philos Mod Phys.* 2002;33:663–705.

[37] Hehl FW, von der Heyde P. Spin and the structure of spacetime. *Ann Inst Henri Poincaré A.* 1973;19:179–96.

[38] Ivanov AN, Wellenzohn M. Einstein-Cartan gravity with torsion field serving as origin for the cosmological constant or dark energy density. *Astrophys J.* 2016;829:47.

[39] Rakotomanana RL. Second gradient continuum: Role of electromagnetism interacting with gravitation on the presence of torsion and curvature. In: *Generalized Models and Non-Classical Approaches in Complex Materials.* 1. Springer; 2018. p. 675–94.

[40] Acedo L. Autoparallel vs geodesic trajectories in a model of torsion gravity. *Universe.* 2015;1:422–45.

[41] Cruz N, Olivares M, Villanueva JR. The geodesic structure of the Schwarzschild anti-de Sitter black hole. *Class Quantum Grav.* 2005;22: 1167–90.

[42] Grunau S, Kagramanova V. Geodesics of electrically and magnetically charged particles in the Reissner-Nordström space-time: Analytical solutions. *Phys Rev D.* 2011;83:044009.

[43] Greenberg PJ. The equation of geodesic deviation in Newtonian theory and the oblateness of the Earth. *Il Nuovo Cimento.* 1974;24:272–86.

[44] Hackmann E, Lämmerzahl C. Geodesic equation in Schwarzschild-(anti-)de Sitter space-times. *Phys Rev D.* 2008;78:024035.

[45] Philipp D, Perlick V, Lammerzahl C, Deshpande K. On geodesic deviation in Schwarzschild spacetime. *Metrol Aerospace IEEE.* 2015:198–203.

[46] Ashby N. Relativity in the Global Positioning System. *Living Rev Relativ.* 2003;6:1.

[47] Cartan E. *On manifolds with an affine connection and the theory of general relativity.* Napoli: Bibliopolis; 1986.

[48] Mavromatos NE, Pais P, Iorio A. Review: Torsion at different scales: From materials to universe. *Universe.* 2023;9:1–45.

[49] Rakotomanana RL. A geometric approach to thermomechanics of dissipating continua. In: *Progress in Mathematical Physics Series.* Boston: Birkhäuser; 2003.

[50] de Andrade VC, Pereira JG. Torsion and the electromagnetic field. *Int J Mod Phys D.* 1999;8:141–51.

[51] Bose A, Chakraborty S. Homogeneous and isotropic spacetime, modified torsion field and complete cosmic scenario. *Eur Phys J C.* 2020;80:205.

[52] Capozziello S, Cianci R, Stornaiolo C, Vignolo S. f(R) cosmology with torsion. *Phys Scr.* 2008;78:065010.

[53] Shapiro IL. Physical aspects of spacetime torsion. *Phys Rep.* 2002;357: 113–213.

[54] Milton GW. A possible explanation of dark matter and dark energy involving a vector torsion field. *Universe.* 2022;8:298.

[55] Durrer R, Neronov A. Cosmological magnetic fields: Their generation, evolution and observation. *Astron Astrophys Rev.* 2023;31:62.

[56] Grasso D, Rubinstein HR. Magnetic fields in the early universe. *Phys Rep.* 2001;348:163–266.

[57] Giovanelli M. "...But I still can't get rid of a sense of artificiality": The Reichenbach-Einstein debate on the geometrization of the electromagnetic field. *Stud Hist Philos Mod Phys.* 2016;54:35–51.

[58] Reichenbach H. Zur Einordnung des neuen Einsteinschen Ansatzes über Gravitation und Electrizität. *Z Phys.* 1929;53:683–98.

[59] Weyl H. *Raum, Zeit, Materie.* Berlin: Springer; 1921.

[60] Abbott B, *et al.* Observation of gravitational waves from binary black hole merger. *Phys Rev Lett.* 2016;116:061102.

[61] Prasanna AR. Static fluid spheres in Einstein-Cartan theory. *Phys Rev D.* 1975;11:2076–82.

[62] Jimenez JB, Maroto AL. Cosmological electromagnetic fields and dark energy. *J Cosmol Astropart Phys.* 2009;03:016.

[63] Bunney CRD, Gradoni G. Electromagnetism in curved spacetimes: Coupling of the Doppler and gravitational redshifts. *IEEE Antennas Propag Mag.* 2022;64:1–11.

[64] Staicova D, Stoilov M. Electromagnetic waves in cosmological spacetime. *Universe.* 2023;29:292.

[65] Mashhoon B. Wave propagation in gravitational field. *Phys Lett A.* 1986;122:299–304.

[66] Mtchedlidze S, Dominguez-Fernandez P, Du X, Brandenburg A, Kahniashvili T, O'Sullivan S, *et al.* Evolution of primordial magnetic fields during large-scale structure formation. *Astrophys J.* 2022;929:127.

[67] Gasperini G, Giovannini M, Veneziano G. Primordial magnetic fields from string cosmology. *Phys Rev Lett.* 1995;75:3796–9.

[68] Cornwall JM. Speculations on primordial magnetic helicity. *Phys Rev D.* 1997;56:6146–54.

[69] Elbistan M, Horvathy PA, Zhang PM. Duality and helicity: The photon wave approach. *Phys Lett A.* 2017;381:2375–9.

[70] Battesti R, Rizzo C. Magnetic and electric properties of a quantum vacuum. *Rep Prog Phys.* 2013;76:016401.

[71] Rafelski J, Labun L, Hadad Y, Chen P. Quantum vacuum structure and cosmology. Lecture at the Tenth Workshop on Non-Perturbative Quantum Chromodynamics; 2009. Available from: https://doi.org/10.48550/arXiv.0909.2989.

[72] Ru W. The electromagnetic vacuum field: A fundamental component of quantum-mechanical ontology. *J Phys Math.* 2024;510:1–2.

[73] Bialynicki-Birula I. On the wave function of the photon. *Acta Phys Pol A.* 1994;86:97–116.

[74] Kobe DH. A relativistic Schrödinger-like equation for a photon and its second quantization. *Found Phys.* 1999;29:1203–31.

[75] Sebens CT. Electromagnetism as quantum physics. *Found Phys.* 2019;49:365–89.

[76] Ruggiero ML, Tartaglia A. Einstein-Cartan as theory of defects in space-time. *Am J Phys.* 2003;71:1303–13.

[77] Cetto AM, de la Pena L. The electromagnetic vacuum field as an essential hidden ingredient of the quantum-mechanical ontology. *Entropy.* 2022;24:1717.

[78] Hehl FW, von der Heyde P, Kerlick GD. General relativity with spin and torsion and its deviation from Einstein's theory. *Phys Rev D.* 1974;16: 1066–9.

[79] Dyson FW, Eddington AS, Davidson C. A determination of the deflection of light by the sun's gravitational field, from observations made at the total eclipse of May 29, 1919. *Phil Trans R Soc A.* 1920;220:291–333.

[80] Euler H. Über die Streuung von Licht an Licht nach der Diracschen Theorie. *Ann Phys.* 1936;418:398–448.

[81] Heisenberg W, Euler HH. Folgerungen aus der Diracschen Theorie des Positrons. *Z Phys.* 1936;98:714–32.

[82] Dunne GV. The Heisenberg-Euler effective action: 75 years on. *Int J Mod Phys Conf Ser.* 2012;14:42–56.

[83] Watson WH. The effect of a transverse magnetic field on the propagation of light in vacuo. *Proc R Soc Lond A.* 1929;123:345–51.

[84] Heinzl T, Liesfeld B, Amthor KU, Schwoerer H, Sauerbrey R, Wipf A. On the observation of vacuum birefringence. *Opt Commun.* 2006;267:318–21.

[85] Valluri SR, Mielniczuk JW, Chishtie F, Lamm D, Auddy S. Vacuum birefringence, the photon anomalous magnetic moment and the neutron star RX J1856.5-3754. *Mon Not R Astron Soc.* 2017;472:2398–402.

[86] Labun L, Rafelski J. Vacuum structure and dark energy. *Int J Mod Phys D.* 2010;29:2299–2304.

[87] Labun L, Rafelski J. Dark energy simulacrum in nonlinear dynamics. *Phys Rev D.* 2010;81:065026.

[88] Cabral F, Lobo FSN, Rubiera-Garcia D. Einstein-Cartan-Dirac gravity with U(1) symmetry breaking. *Eur Phys J C.* 2019;79:1023.

[89] Pereira SH, Vicente AM, Jesus JF, Holanda RFL. Dark matter from torsion in Friedmann cosmology. *Eur Phys J C.* 2022;82:356.

[90] Gera S, Sengupta S. Magnetic monopole as a spacetime defect. *Phys Rev D.* 2021;104:044038.

[91] Katanaev MO. Spin distribution for the 't Hooft-Polyakov monopole in the geometry theory of defects. *Universe.* 2021;7:256.

[92] Rahaman F, Mondal R. A study of global monopole in Einstein-Cartan theory. *Astrophys Space Sci.* 2005;299:167–74.

[93] Jedamzik K, Pogosian L. Relieving the Hubble tension with primordial magnetic fields. *Phys Rev Lett.* 2020;125:181302.

[94] Turyshev SG, Toth V. Diffraction of electromagnetic waves in the gravitational field of the Sun. *Phys Rev D.* 2017;96:024008.

[95] Anjan AS, Shahnawaz AA, Somarsi S. Do cosmological constant observations allow a negative Λ? *Mon Not R Astron Soc.* 2023;518:1098–105.

[96] Marongwe S. On the electromagnetic vacuum origins of dark energy, dark matter, and astrophysical jets. *Adv High Energy Phys.* 2021;2021:5588662.

[97] Posada-Aguirre NC, Batic D. Electrovac universes with a cosmological constant. *Cent Eur J Phys.* 2014;12:297–304.

[98] Leonhardt U. Lifschitz theory of the cosmological constant. *Ann Phys.* 2019;411:167913.

[99] Fernando J, Giglio T, Rodrigues WA Jr. Gravitation and electromagnetism as geometrical objects of Riemann-Cartan spacetime structure. *Adv Appl Clifford Algebras.* 2012;22:640–64.

[100] Hammond RT. Einstein-Maxwell theory from torsion. *Class Quantum Grav.* 1989;6:195–8.

[101] Nieto JA, Saucedo J, Villanueva VM. Geodesic deviation equation for relativistic tops and the detection of gravitational waves. *Rev Mex Fis S.* 2007;53:141–5.

[102] Darabi F, Mousavi M, Atazadeh K. Geodesic deviation equation in f(T) gravity. *Phys Rev D.* 2015;91:084023.

[103] Swaminarayan NS, Safko JL. A coordinate-free derivation of a generalized geodesic deviation equation. *J Math Phys.* 1983;24:883–5.

[104] Manoff S. Deviation operator and deviation equations over spaces with affine connections and metrics. *J Geom Phys.* 2001;39:337–50.

[105] Kerner R. Generalized geodesic deviations: A Lagrangean approach. *Banach Center Publ.* 2003;59:173–88.

[106] Iguro S, Plestid R, Takhistov V. Monopoles from an atmospheric fixed target experiment. *Phys Rev Lett.* 2022;128:201101.

[107] Afanasiev GN, Stepanovsky YP. The helicity of the free electromagnetic field and its physical meaning. *Il Nuovo Cimento.* 1996;109:271–9.

[108] Allen L, van Beijbergen MW, Spreeuw RJC, Woerdman JP. Orbital angular momentum of light and the transformation of Laguerre-Gaussian laser modes. *Phys Rev A.* 1992;45:8185–9.

[109] Barnett SM. Optical angular-momentum flux. *J Opt B Quantum Semiclass Opt.* 2002;4:7–16.

[110] Lovelock D, Rund H. *Tensors, Differential Forms, and Variational Principles.* New York: Dover Publications; 1975.

[111] Bain J. Theories of Newtonian gravity and empirical indistinguishability. *Stud Hist Philos Mod Phys.* 2004;35:345–76.

[112] Defrise P. Analyse géométrique de la cinématique des milieux continus. *Inst Roy Meteorol Belg Publ Serie B.* 1953;6:5–63.

[113] Kadianakis ND. The kinematics of continua and the concept of connection on classical spacetime. *Int J Eng Sci.* 1996;34:289–98.

[114] Bernal AN, Sánchez M. Leibnizian, Galilean, and Newtonian structures of spacetime. *J Math Phys.* 2003;44:1129–49.

[115] Rosen G. Galilean invariance and the general covariance of nonrelativistic laws. *Am J Phys.* 1972;40:683–7.

[116] Krause J. Christoffel symbols and inertia in flat spacetime theory. *Int J Theor Phys.* 1976;15:801–7.

[117] Dixon WG. On the uniqueness of the Newtonian theory as a geometric theory of gravitation. *Commun Math Phys.* 1975;45:167–82.

[118] Kibble TWB. Lorentz invariance and gravitational field. *J Math Phys.* 1961;3:212–21.

[119] Ricci MMG, Levi-Civita T. Méthodes de calcul différentiel absolu et leurs applications. *Math Ann.* 1900;54:125–201.

[120] Frankel T. *The Geometry of Physics: An Introduction.* Cambridge: Cambridge University Press; 1997.

[121] Ross DK. Planck's constant, torsion, and space-time defects. *Int J Theor Phys.* 1989;28:1333–40.

[122] Yang G, Duan Y, Huang Y. Topological invariant in Riemann-Cartan manifold and spacetime defects. *Int J Theor Phys.* 1998;37:2953–64.

[123] Havas P. Four-dimensional formulations of Newtonian mechanics and their relation to the special and general theory of relativity. *Rev Mod Phys.* 1964;36:938–65.

[124] Bianchi L. Sui simboli a quatro indici e sulla curvature di Riemann. *Rend Accad Lincei V.* 1902;11:3–7.

[125] Ryder L. *Introduction to General Relativity.* New York: Cambridge University Press; 2009.

[126] Steeb WH. Invariance conditions, and physical laws. *Z Naturforsch A.* 2023;33:742–8.

[127] McKellar RJ. The uniqueness of gravity as a Poincaré or Lorentz gauge theory. *J Math Phys.* 1981;22:2934–42.

[128] Radofilao J. *Initiation à la géométrie différentielle.* Antananarivo: Edition FTM; 1986.

[129] Duval Ch, Kunzle HP. Dynamics of continua and particles from general covariance of Newtonian gravitation theory. *Rep Math Phys.* 1978;13: 351–68.

[130] Souriau JM. Modèle de particule à spin dans le champ électromagnétique et gravitationnel. *Ann IHP Phys Théor.* 1974;20:315–64.

[131] Delhom A. Minimal coupling in presence of non-metricity and torsion. *Eur Phys J C.* 2020;80:728.

[132] Saa A. Volume-forms and minimal action principles in affine manifolds. *J Geom Phys.* 1995;15:102–8.

[133] Rakotomanana RL. *Introduction à la mécanique des milieux continus.* Paris: Ellipses; 2022.

[134] Mosna RA, Saa R. Volume elements and torsion. *J Math Phys.* 2005;46:112502.

[135] Saa A. Gauge fields on Riemann-Cartan space-times. *Mod Phys Lett A.* 1994;9:971–81.

[136] Whyburn GT. On the structure of continua. Invited lecture, Ann Arbor, USA, September 13, 1935.

[137] Sotiriou TP, Liberati S. Metric-affine f(R) theories of gravity. *Ann Phys.* 2007;322:935–66.

[138] Cajori F. Origins of fourth dimension concepts. *Am Math Soc Bull.* 1926;33:397–406.

[139] Knox E. *Philosophy of Spacetime Physics.* Oxford: Oxford Research Encyclopedia of Physics; 2025.

[140] Earman J. Covariance, invariance, and the equivalence of frames. *Found Phys.* 1974;4:267–89.

[141] Truesdell C, Noll W. *The Non-linear Field Theories of Mechanics.* 2nd ed. Berlin: Springer-Verlag; 1991.

[142] Hartong J, Obers NA, Oling G. Review on non-relativistic gravity. *Front Phys.* 2023;11:1–23.

[143] Appleby PG. Inertial frames in classical mechanics. *Arch Ration Mech Anal.* 1977;67:337–50.

[144] Rüdde C, Straumann N. On Newton-Cartan cosmology. *Helv Phys Acta.* 1997;71:318–35.

[145] Goenner HFM. A variational principle for Newton-Cartan theory. *Gen Relat Gravit.* 1984;16:513–26.

[146] Brading KA, Ryckman TA. Hilbert's "Foundations of Physics": Gravitation and electromagnetism within the axiomatic method. *Stud Hist Philos Mod Phys.* 2008;39:102–53.

[147] Logunov AA, Mestvirishvili MA. Hilbert's causality principle and equations of general relativity exclude the possibility of black hole formation. *Theor Math Phys.* 2012;170:413–19.

[148] Zeeman EC. Causality implies the Lorentz group. *J Math Phys.* 1964;5: 490–3.

[149] Williams G. A discussion of causality and the Lorentz group. *Int J Theor Phys.* 1973;7:415–21.

[150] Pettey D. One-one-mappings onto locally connected generalized continua. *Pacific J Math.* 1971;50:573–82.

[151] Hayashi K, Shirafuji T. Energy, momentum and angular momentum in Poincaré gauge theory. *Prog Theor Phys.* 1985;73:54–74.

[152] Verçyn A. Metric-torsion gauge theory of continuum line defects. *Int J Theor Phys.* 1990;29:7–21.

[153] Marsden JE, Hughes TJR. *Mathematical Foundations of Elasticity.* Englewood Cliffs: Prentice-Hall; 1983.

[154] Papapetrou A. Spinning test-particles in general relativity I. *Proc R Soc Lond A.* 1951;209:248–58.

[155] Puetzfeld D, Obukhov YN. Probing non-Riemannian spacetime geometry. *Phys Lett A.* 2008;372:6711–6.

[156] Garcia de Andrade LC. On the necessity of non-Riemannian acoustic spacetime in fluids with vorticity. *Phys Lett A.* 2005;346:327–9.

[157] Mao Y, Tegmark M, Guth AH, Cabi S. Constraining torsion with Gravity Probe B. *Phys Rev D.* 2007;76:104029.

[158] Lichnerowicz A. *Théories relativistes de la gravitation et de l'électromagnétisme.* Paris: Masson & Cie; 1955.

[159] Manoff S. Frames of reference in spaces with affine connections and metrics. *Class Quantum Grav.* 2001;18:1111–25.

[160] Norton JD. General covariance and the foundations of general relativity: eight decades of disputes. *Rep Prog Phys.* 1993;56:791–858.

[161] Kempers LJTM. The principle of material frame indifference and the covariance principle. *Il Nuovo Cimento B.* 1989;103:227–36.
[162] Frewer M. More clarity on the concept of material frame-indifference in classical mechanics. *Acta Mech.* 2009;202:213–46.
[163] Pooley O. Background independence, diffeomorphism invariance, and the meaning of coordinates. In: Lehmkuhl D, Schiemann G, Scholz E, editors. *Towards a Theory of Spacetime Theories.* New York: Birkhäuser; 2016. p. 1–39.
[164] Bargmann V. Relativity. *Rev Mod Phys.* 1957;29:161–74.
[165] Ryskin G. Misconception which led to the "material frame-indifference" controversy. *Phys Rev A.* 1985;32:1239–40.
[166] Earman J. Chapter 1: The implications of general covariance for the ontology and ideology of spacetime. In: Dicks D, editor. *Philosophy and Foundations of Physics: The Ontology of Spacetime.* 2006. p. 3–23.
[167] Lompay RR, Petrov A. Covariantized differential identities and conservation laws in metric-torsion theories of gravitation. I. General consideration. *J Math Phys.* 2013;54:062504.
[168] Cartan E. Sur les équations de la gravitation d'Einstein. *J Math Pures Appl.* 1922;1:141–203.
[169] Tamarasselvame NA, Rakotomanana L. On the form-invariance of Lagrangean function for higher gradient continuum. In: Altenbach H, Maugin G, Erofeev V, editors. *Mechanics of Generalized Continua.* Berlin: Springer-Verlag; 2011. p. 291–322.
[170] Schmidt H-J. Fourth-order gravity: equations, history, and applications to cosmology. *Int J Geom Methods Mod Phys.* 2007;4:209–48.
[171] Anderson IM. On the structure of divergence-free tensors. *J Math Phys.* 1978;19:2570–5.
[172] Lovelock D. The uniqueness of the Einstein field equations in a four-dimensional space. *Arch Ration Mech Anal.* 1969;33:54–70.
[173] Lovelock D. The uniqueness of the Einstein-Maxwell field equations. *Gen Relat Gravit.* 1974;5:399–408.
[174] Anderson IM. The principle of minimal gravitational coupling. *Arch Ration Mech Anal.* 1981;75:349–72.
[175] Sotiriou TP. The viability of theories with matter coupled to the Ricci scalar. *Phys Lett B.* 2008;664:225–8.
[176] Capozziello S, De Laurentis D. Gravity from local Poincaré gauge invariance. *Int J Geom Methods Mod Phys.* 2009;6:1–24.
[177] Bruzzo U. The global Utiyama theorem in Einstein-Cartan theory. *J Math Phys.* 1987;28:2074–7.
[178] Utiyama R. Invariant theoretical interpretation of interaction. *Phys Rev.* 1956;101:1597–607.
[179] Cho YM. Gauge theory of Poincaré symmetry. *Phys Rev D.* 1976;14: 3335–40.
[180] Cho YM. Gauge theory, gravitation, and symmetry. *Phys Rev D.* 1976;14:3341–4.

[181] Utiyama R, Fukuyama T. Gravitational field as a generalized gauge field. *Prog Theor Phys.* 1971;45:612–27.

[182] Ali SA, Cafaro C, Capozziello S, Corda C. On the Poincaré gauge theory of gravitation. *Int J Theor Phys.* 2009;48:3426–48.

[183] Banerjee R, Roy D. Poincaré gauge, Hamiltonian symmetries, and trivial gauge transformations. *Phys Rev D.* 2011;84:124034.

[184] Hehl FW, Obukhov YN, Puetzfeld D. On Poincaré gauge theory of gravity, its equation of motions, and Gravity Probe B. *Phys Lett A.* 2013;377: 1775–81.

[185] Rakotomanana RL. Some thoughts concerning spacetime models in cosmology: role of gravitation and electromagnetism. *17th Marcel Grossmann Meetings*, Pescara, Italy, Jul 2024. [cited 2024]. Available from: https://hal.archives-ouvertes.fr/hal-04917087. p. 1–20.

[186] Noether E. Invariante Variationsprobleme. *Nachr Ges Wiss Göttingen Math-Phys Kl.* 1918:235–57.

[187] Eringen AC, Suhubi ES. Nonlinear theory of simple microelastic solids I. *Int J Eng Sci.* 1964;2:189–203.

[188] Eringen AC, Suhubi ES. Nonlinear theory of simple microelastic solids II. *Int J Eng Sci.* 1964;2:389–404.

[189] Kondo K. Non-Riemannian geometry of imperfect crystals from a macroscopic point of view. In: Kondo K, editor. *RAAG Memoirs of the Unifying Study of Basic Problems in Engineering and Physical Sciences by Means of Geometry.* Vol. 1. Tokyo: Gakajutsu Bunken Fukyukai; 1955. p. 459–80.

[190] Voyiadjis GZ, Song Y. Strain gradient continuum plasticity theories: theoretical, numerical and experimental investigations. *Int J Plast.* 2019;121:21–75.

[191] Noll W. Materially uniform simple bodies with inhomogeneities. *Arch Ration Mech Anal.* 1967;27:1–32.

[192] Rakotomanana RL. More on the geometric approach for gravitation coupled with electromagnetism within a Riemann-Cartan vacuum spacetime. [cited 2021]. Available from: https://hal.archives-ouvertes.fr/hal-03216920. p. 1–19.

[193] Ramaniraka NA, Rakotomanana LR. Models of continuum with microcrack distribution. *Math Mech Solids.* 2000;5:301–36.

[194] Lazar M, Anastassiadis C. The gauge theory of dislocations: conservation and balance laws. *Philos Mag.* 2008;88:1673–99.

[195] Carter B. Elastic perturbation theory in general relativity and a variation principle for a rotating solid star. *Commun Math Phys.* 1973;30:261–86.

[196] Manoff S. Lagrangian theory of tensor fields over spaces with contravariant and covariant affine connections and metrics and its application to Einstein's theory of gravitation in V4 spaces. *Acta Appl Math.* 1999;55: 51–125.

[197] Petrov AN, Lompay RR. Covariantized Noether identities and conservation laws for perturbations in metric theories of gravity. *Gen Relat Gravit.* 2013;45:545–79.

[198] Lovelock D. The Einstein tensor and its generalizations. *J Math Phys.* 1971;12:498–501.

[199] Abedi H, Capozziello S, Capriolo M, Abbassi AM. Gravitational energy-momentum pseudo-tensor in Palatini and metric f(R) gravity. *Ann Phys.* 2022;439:168796.

[200] Hehl FW, Kerlick GD, von der Heyde P. On hypermomentum in general relativity I. The notion of hypermomentum. *Z Naturforsch A.* 1976;31: 111–4.

[201] Lompay RR. On the energy-momentum and spin tensors in the Riemann-Cartan space. *Gen Relat Gravit.* 2014;46:1692.

[202] Gurtin ME, Anand L. A theory of strain-gradient plasticity for isotropic, plastically irrotational materials. Part I: Small deformations. *J Mech Phys Solids.* 2005;53:1624–49.

[203] Gurtin ME, Anand L. A theory of strain-gradient plasticity for isotropic, plastically irrotational materials. Part II: Finite deformations. *J Mech Phys Solids.* 2005;53:2297–318.

[204] Blakojevic M. *Gravitation and Gauge Symmetries.* Bristol: Institute of Physics Publishing; 2002.

[205] Kleinert H. Nonholonomic mapping principle for classical and quantum mechanics in spaces with curvature and torsion. *Gen Relat Gravit.* 2000;32:769–839.

[206] Cabral F, Lobo FSN, Rubiera-Garcia D. Fundamental symmetries and spacetime geometries in gauge theories of gravity: prospects for unified field theories. *Universe.* 2020;6:238.

[207] Obukhov YN, Puetzfeld D. Conservation laws in gravity: a unified framework. *Phys Rev D.* 2014;90:020404.

[208] Puetzfeld D, Obukhov YN. Equations of motion in gravity theories with nonminimal coupling: a loophole to detect torsion macroscopically. *Phys Lett D.* 2013;88:064025.

[209] Coleman BD, Noll W. The thermodynamics of elastic materials with heat conduction and viscosity. *Arch Ration Mech Anal.* 1963;13:167–70.

[210] Curnier A. *Mécanique des solides déformables.* Lausanne: Presses Polytechniques Universitaires Romandes; 2004.

[211] Futhazar G, Le Marrec L, Rakotomanana LR. Covariant gradient continua applied to wave propagation within defected material. *Arch Appl Mech.* 2014.

[212] Maugin G. *Material Inhomogeneities in Elasticity.* London: Chapman and Hall; 1993.

[213] Rakotomanana RL. *Élements de dynamiques des structures et solides déformables.* Lausanne: Presses Polytechniques et Universitaires Romandes; 2009.

[214] Samavaki M, Tuomela J. Navier-Stokes equations on Riemannian manifolds. *J Geom Phys.* 2020;148:1–15.

[215] Dalla Torre F, Lapovok R, Sandlin J, Thomson PF, Davis CHJ, Pereloma EV. Microstructures and properties of copper processed by equal channel angular extrusion for 1–16 passes. *Acta Mater.* 2004;52:4819–32.

[216] Kleinert H. *Multivalued Fields: in Condensed Matter, Electromagnetism, and Gravitation.* Singapore: World Scientific; 2008.

[217] Hehl FW, von der Heyde P, Kerlick GD, Nester JM. General relativity with spin and torsion: foundations and prospects. *Rev Mod Phys.* 1976;48: 393–416.

[218] Hehl FW. How does one measure torsion of space-time? *Phys Lett A.* 1971;36:225–6.

[219] Maugin GA. Towards an analytical mechanics of dissipative materials. *Rend Sem Mat Univ Pol Torino.* 2000;58:171–80.

[220] Cermelli P, Gurtin ME. On the characterization of geometrically necessary dislocations in finite plasticity. *J Mech Phys Solids.* 2001;49:1539–68.

[221] Malvern LE. *Introduction to the Mechanics of a Continuous Medium.* Englewood Cliffs: Prentice-Hall; 1969.

[222] Hammond RT. Electromagnetic spin creates torsion. *Int J Mod Phys D.* 2018;17:1847005.

[223] dell'Isola F, Madeo A, Seppecher P. Cauchy tetrahedron argument applied to higher contact interactions. *Arch Ration Mech Anal.* 2016;219:1305–41.

[224] Germain P. The method of virtual power in continuum mechanics Part 2: Microstructure. *SIAM J Appl Math.* 1973;25:556–75.

[225] Dixon WG. Dynamics of extended bodies in general relativity, I. Momentum and angular momentum. *Proc R Soc Lond A.* 1970;314:499–527.

[226] Coleman BD, Gurtin ME. The thermodynamics with internal state variables. *J Chem Phys.* 1967;47:597–613.

[227] Moreau JJ. Sur les lois de frottement, de plasticité et de viscosité. *CR Acad Sc Paris A.* 1970;271:608–11.

[228] Hackl K, Fischer FD. On the relation between the principle of maximum dissipation and inelastic evolution given by dissipation potentials. *Proc R Soc A.* 2008;464:117–32.

[229] Lychev S, Koifman K. *Geometry of incompatible deformations: differential geometry in continuum mechanics.* Berlin: De Gruyter; 2019.

[230] Maugin GA. Basics of the material mechanics of Materials (M3) for arbitrary continua. *Rend Sem Mat Univ Pol Torino.* 2007;65:205–15.

[231] Obukhov YN, Portamles-Oliva F, Puetzfeld D. Invariant conserved currents in general relativity. *Phys Rev D.* 2015;92:104010.

[232] Philipp D, Putzfeld D, Lämmerzahl C. On the applicability of the geodesic deviation equation in general relativity. In: *Relativistic Geodesy.* Vol. 196. Springer; 2019. p. 419–51.

[233] Kadianakis ND. Affine connections and frames of reference in classical mechanics. *Rep Math Phys.* 1991;30:21–32.

[234] Whyburn GT. On the compactness of mappings. *Proc Natl Acad Sci USA.* 1964;52:1426–51.

[235] Ehlers J, Geroch R. Equation of motion of small bodies in relativity. *Ann Phys.* 2004;309:232–6.

[236] McMullin E. The origins of the field concept in physics. *Phys Perspect.* 2002;4:13–39.

[237] Overduin J, Fahr HJ. Matter, spacetime and the vacuum. *Naturwissenschaften.* 2001;88:491–503.

[238] Leclerc M. Mathisson-Papapetrou equations in metric and gauge theories of gravity in a Lagrangian formulation. *Class Quantum Grav.* 2005;22: 3203–21.

[239] Shen W, Moritz H. On the separation of gravitation and inertia and the determination of the relativistic gravity field in the case of free motion. *J Geod.* 1996;70:633–44.

[240] Cariglia M, Houri T, Krtous P, Kubiznak D. On integrability of the geodesic deviation equation. *Eur Phys J C.* 2018;78:661.

[241] Griffiths JB, Podolsky J. *Exact Space-Times in Einstein's General Relativity.* Cambridge: Cambridge University Press; 2009.

[242] Synge JL. *The General Theory.* Amsterdam: North-Holland; 1960.

[243] Eisenstaedt J. Histoire et singularités de la solution de Schwarzschild. *Arch Hist Exact Sci.* 1982;7:157–98.

[244] Fuchs H. Solutions of the equations of geodesic deviation for static spherical symmetric space-times. *Ann Phys.* 1983;7:231–3.

[245] Fuchs H. Deviation of circular geodesics in static spherically symmetric space-times. *Astron Nachr.* 1990;5:271–6.

[246] Maldacena J. The large N limit of superconformal field theories and supergravity. *Adv Theor Math Phys.* 1998;2:231–52.

[247] Cruz N, Olivares M, Saavedra J, Villanueva JR. Null geodesics in the Reissner-Nordström Anti-de Sitter black holes. arXiv:1111.0924 [gr-qc]. 2011;1–13.

[248] Hehl FW, Obukhov YN. How does the electromagnetic field couple to gravity, in particular to metric, nonmetricity, torsion, and curvature. In: *Lecture Notes in Physics.* Vol. 562. 2001. p. 479–504.

[249] Gersten A, Moalem A. Maxwell's equations, quantum physics and the quantum graviton. *J Phys Conf Ser.* 2010;330:044038.

[250] Edelen DGB. The affine theory of electricity and gravitation. *Arch Ration Mech Anal.* 1960;6:1–19.

[251] Einstein A. Zur Elektrodynamik bewegter Körper. *Ann Phys.* 1905;17: 891–921.

[252] Kovetz A. *Electromagnetic Theory.* Oxford: Oxford University Press; 2000.

[253] Hehl FW. Maxwell's equations in Minkowski's world: their premetric generalization and the electromagnetic energy-momentum tensor. *Ann Phys.* 2018;23:1–14.

[254] Hehl FW, Obukhov YN. *Foundations of Classical Electrodynamics: Charge, Flux, and Metric.* Boston: Birkhäuser; 2003.

[255] Castelnovo C, Moessner R, Sondhi SL. Magnetic monopoles in spin ice. *Nature.* 2007;451:42–5.

[256] 't Hooft G. Magnetic monopoles in unified gauge theories. *Nucl Phys B.* 1974;79:276–90.

[257] Preskill JP. Magnetic monopoles. *Annu Rev Nucl Part Sci.* 1979;34: 461–530.

[258] Romero JM, Bellini M, Aguilar JEM. Gravitational waves and magnetic monopoles during inflation with Weitzenböck torsion. *Phys Dark Univ.* 2016;13:121–5.

[259] Yang CN, Mills RL. Conservation of isotopic spin and isotopic gauge invariance. *Phys Rev.* 1954;96:191–5.

[260] Plebanski J. Electromagnetic waves in gravitational fields. *Phys Rev.* 1960;118:1396–408.

[261] Charap JM, Duff MJ. Gravitational effects on Yang-Mills topology. *Phys Lett B.* 1977;69:445–7.

[262] Waite TR. Electromagnetic sources in general relativity theory. *Phys Rev.* 1961;123:1888–91.

[263] Hammond RT. Gravitation, torsion, and electromagnetism. *Gen Relat Gravit.* 1987;20:813–27.

[264] Obukhov YN. Electromagnetic energy and momentum in moving media. *Ann Phys Berlin.* 2008;17:830–51.

[265] Anghinoni B, Flizikowski GAS, Malacarne LC, Partanen M, Bialkowski SE, Astrath NGC. On the formulations of the electromagnetic stress-energy tensor. *Ann Phys.* 2022;443:169004.

[266] Chrusciel PT. On the unified affine electromagnetism and gravitation theories. *Acta Phys Pol B.* 1984;15:35–51.

[267] Witten L. Geometry of gravitation and electromagnetism. *Phys Rev.* 1959;115:206–14.

[268] Lichnerowicz A. Magnetohydrodynamics: Waves and shock waves in curved spacetime. In: *Mathematical Physics Studies.* Vol. 14. Springer; 1994. p. 18–32.

[269] Nordström G. On the energy of gravitational field in Einstein's theory. *Proc Kon Ned Akad Wet.* 1918;20:1238–45.

[270] Reissner H. Über die Eigengravitation des elektrischen Feldes nach der Einsteinschen Theorie. *Ann Phys.* 1916;50:106–20.

[271] Pugliese D, Quevedo H, Ruffini R. Motion of charged test particles in Reissner-Nordström spacetime. *Phys Rev D.* 2011;104052.

[272] Chatterjee AK, Flathmann K, Nandan H, Rudra A. Analytical solutions of the geodesic equation for Reissner-Nordström-(anti-)de Sitter black holes surrounded by different kinds of regular and exotic matter fields. *Phys Rev D.* 2019;100:024044.

[273] Hawking SW, Ellis GFR. *The Large Scale Structure of Space-time.* Cambridge: Cambridge University Press; 1973.

[274] Zhou M, Zhdankin V, Kunz MW, Loureiro NF, Uzdensky DA. Spontaneous magnetization of collisionless plasma. *Proc Natl Acad Sci USA.* 2022;119:e2119831119.

[275] Barrow JD, Tsagas CG, Fanaras G. Friedmann-like universes with weak torsion: a dynamical system approach. *Eur Phys J C.* 2019;79:764.

[276] Kranas D, Tsagas CG, Barrow JD, Iosifidis D. Friedmann-like universes with torsion. *Eur Phys J C.* 2019;79:341.

[277] Ferraris M, Francaviglia M. Variation formulation of general relativity from 1915 to 1925: "Palatini's method" discovered by Einstein in 1925. *Gen Relat Gravit.* 1982;14:243–54.
[278] Hehl FW, Kerlick GD. Metric-affine variational principles in general relativity. I. Riemannian spacetime. *Gen Relat Gravit.* 1978;9:691–710.
[279] Capozziello S, Cardone VF, Piedipalumbo E, Sereno M, Troisi A. Matching torsion Lambda-term with observations. *Int J Mod Phys D.* 2003;12: 381–94.
[280] Schücker T, Tilquin A. Torsion, an alternative to the cosmological constant? *Int J Mod Phys D.* 2012;21:1250089.
[281] Bull PH, Akrami Y, *et al.* Beyond ΛCDM: problems, solutions, and the road ahead. *Phys Dark Univ.* 2016;12:56–99.
[282] Dadhich N, Pons JM. On the equivalence of the Einstein-Hilbert and the Einstein-Palatini formulations of general relativity for an arbitrary connection. *Gen Relat Gravit.* 2012;44:2337–52.
[283] Vandyck MA. Maxwell's equations in spaces with non-metricity and torsion. *J Phys A Math Gen.* 1996;29:2245–55.
[284] Prasanna AR. Maxwell's equations in Riemann-Cartan space U4. *Phys Lett A.* 1975;54:17–8.
[285] Smalley LL. On the extension of geometric optics from Riemannian to Riemann-Cartan spacetime. *Phys Lett A.* 1986;117:267–9.
[286] Smalley LL, Krisch JP. Minimal coupling of electromagnetic fields in Riemann-Cartan spacetimes for perfect fluids with spin density. *J Math Phys.* 1992;33:1073–81.
[287] Puntigam RA, Lämmerzahl C, Hehl FW. Maxwell's theory on a post-Riemannian spacetime and the equivalence principle. *Class Quantum Grav.* 1997;14:1347–56.
[288] Bliokh KY, Bekshaev AY, Nori F. Dual electromagnetism: helicity, spin, momentum and angular momentum. *New J Phys.* 2013;15:033026. Corrigendum *New J Phys.* 2016;18:089503.
[289] Trueba JL, Ranada AF. The electromagnetic helicity. *Eur J Phys.* 1996;17:141–4.
[290] Vernstrom T, Heald G, Vazza F, Galvin TJ, West JL, Locatelli N, *et al.* Discovery of magnetic fields along stacked cosmic filaments as revealed by radio X-ray emission. *Mon Not R Astron Soc.* 2021;505:4178–96.
[291] Arrayas M, Trueba JL. Spin-orbital momentum decomposition and helicity exchange in a set of non-null knotted electromagnetic fields. *Symmetry.* 2018;10:88.
[292] Peccei RD, Quinn HR. CP conservation in the presence of pseudoparticles. *Phys Rev Lett.* 1977;38:1440–3.
[293] Campanelli L, Giannotti M. Production of axions by cosmic magnetic helicity. *Phys Rev Lett.* 2006;96:161302.
[294] Hammond RT. Torsion gravity. *Rep Prog Phys.* 2002;65:599–649.

[295] Tiwari RN, Ray S. Static spherical charged dust electromagnetic mass models in Einstein-Cartan theory. *Gen Relat Gravit.* 1997;29:683–90.
[296] Obukhov YN, Hehl FW. Electromagnetic energy-momentum and forces in matter. *Phys Lett A.* 2003;311:277–84.
[297] Beck R. Magnetic fields in spiral galaxies. *Astron Astrophys Rev.* 2016;24: 1–57.
[298] Cameron RP, Barnett SM, Yao AM. Optical helicity of interfering waves. *J Mod Opt.* 2014;61:25–31.
[299] Aiello A. Helicity, chirality, and spin of optical fields without vector potentials. *Phys Rev A.* 2022;106:043519.
[300] Moreau JJ. Constantes d'un îlot tourbillonnaire en fluide parfait barotrope. *CR Acad Sci Paris.* 1961;252:2810–2.
[301] Edtmair O, Seyffadini S. A universal extension of helicity to topological flows. *Arch Ration Mech Anal.* 2025;arXiv:2508.10609 [math.SG].
[302] Puetzfeld D, Obukhov YN. Deviation equation in Riemann-Cartan spacetime. *Phys Rev D.* 2018;97:104069.
[303] Kleinert H. Universality principle for orbital angular momentum and spin in gravity with torsion. *Gen Relat Gravit.* 2000;32:1271–80.
[304] Oprisan CD, Zet G. Gauge theory on a space-time with torsion. *Rom J Phys.* 2006;51:531–40.
[305] Bamba K, Capozziello S, De Laurentis M, Nojiri S, Sáez-Gómez D. No further gravitational modes in F(T) gravity. *Phys Lett B.* 2013;727:194–8.
[306] Friedmann A. Über die Krümmung des Raumes. *Z Phys.* 1922;10:377–86.
[307] Ellis GFR. 83 years of general relativity and cosmology: progress and problems. *Class Quantum Grav.* 1999;16:A37–A75.
[308] Lemaître G. Un Univers homogène de masse constante et de rayon croissant rendant compte de la vitesse radiale des nébuleuses extragalactiques. *Ann Soc Sci Bruxelles A.* 1927;47:19–59.
[309] Robertson HP. On the foundations of relativistic cosmology. *Proc Natl Acad Sci USA.* 1929;15:822–9.
[310] Walker AG. On Riemannian spaces with spherical symmetry about a line, and the conditions for isotropy in general relativity. *J Math.* 1935;6:81–93.
[311] Israel W. Event horizons in static vacuum space-times. *Phys Rev.* 1967;164:1776–9.
[312] Taub AH. General relativistic variational principle for perfect fluids. *Phys Rev.* 1954;94:1468–70.
[313] Minazzoli O, Harko T. New derivation of the Lagrangian of a perfect fluid with a barotropic equation of state. *Phys Rev D.* 2012;86:087502.
[314] Ray JR. Lagrangian density for perfect fluids in general relativity. *J Math Phys.* 1972;13:1451–3.
[315] Schutz BF. Perfect fluids in general relativity: velocity potentials and a variational principle. *Phys Rev D.* 1970;2:2762–73.
[316] Mannheim PD, O'Brien JG, Cox DE. Limitations of the standard gravitational perfect fluid paradigm. *Gen Relat Gravit.* 2011;43:1–23.
[317] Minkowski P. On the cosmological equations in the presence of spatially homogeneous torsion field. *Phys Lett B.* 1986;173:247–50.

[318] Pereira SH, Lima RdC, Jesus JF, Holanda RFL. Acceleration in Friedmann cosmology with torsion. *Eur Phys J C.* 2019;79:950.
[319] Jackiw R, Pi SY. Chern-Simons modification of general relativity. *Phys Rev D.* 2003;68:104012.
[320] Wang J, Yu H, Wu P. Revisiting cosmic acceleration with DESI BAO. *Eur Phys J C.* 2025;83:853.
[321] Iosofidis D. Cosmic acceleration with torsion and non-metricity in Friedmann-like universes. *Class Quantum Grav.* 2020;38:015015.
[322] Rudolph P. Fundamentals and engineering of defects. *Prog Cryst Growth Charact Mater.* 2016;62:89–110.
[323] Weyl H. Gravitation and the electron. *Proc Natl Acad Sci USA.* 1929;15:323–34.
[324] Martinelli M, *et al.* (Euclid Project) Euclid: constraining dark energy coupled to electromagnetism using astrophysical and laboratory data. *Astron Astrophys.* 2021;654:A148.
[325] Barnett SM, Allen L, Cameron RP, Gilson CR, Padgett MJ, Speirits FC, Yao AM. On the natures of the spin and orbital parts of optical angular momentum. *J Opt.* 2016;18:064004.
[326] Fixsen DJ. The temperature of the cosmic microwave background. *Astrophys J.* 2009;707:916–20.
[327] Peracaula JS, Gómez-Valent A, Pérez JdC, Moreno-Pulido C. Brans-Dicke gravity with a cosmological constant smoothes out ΛCDM tensions. *Astrophys J Lett.* 2019;886:L6.
[328] Mirizzi A, Raffelt GG, Serpico PD. Photon-axion conversion in intergalactic magnetic fields and cosmological consequences. In: Kuster M, Raffelt G, Beltrán B, editors. *Lecture Notes in Physics: Axions, Theory, Cosmology, and Experimental Searches.* Vol. 741. Springer; 2008. p. 115–34.
[329] Beutter M, Pargner A, Schwetz T, Todarello E. Axion-electrodynamics: a quantum field calculation. *J Cosmol Astropart Phys.* 2019;2019:1–13.
[330] Brans C, Dicke R. Mach's principle and a relativistic theory of gravitation. *Phys Rev.* 1961;124:925–35.
[331] Angelantonj C, Amendola L, Litterio M, Occhiniero F. String cosmology and inflation. *Phys Rev D.* 1995;51:1607–16.
[332] Khatri V, Singh CP. Brans-Dicke cosmology with configuration term $\Lambda(\mathrm{H}) = \mathrm{c}0 + 3\nu\mathrm{H}2$. *Phys Dark Energy.* 2023;42:101300.
[333] Lempine D, Lemoine M. Primordial magnetic fields in string cosmology. *Phys Rev D.* 1995;52:1–8.
[334] Benisty D, Guendelman EI. Unified dark energy and dark matter from dynamical spacetime. *Phys Rev D.* 2018;98:023506.
[335] Çiftci DK, Faraoni V. Perfect fluid solutions of Brans-Dicke and f(R) cosmology. *Ann Phys.* 2018;391:65–82.
[336] Zhang Z, Aboushelbaya R, Quatu I, Denis E, James A, Timmis RJL, Von Der Leyen MW, Norreys PA, Torres R, Grismayer T, Silva LO. Computational modelling of the semi-classical quantum vacuum in 3D. *Commun Phys.* 2025;8:224.

[337] Wheeler JT. Gravitationally squeezed light. *Gen Relat Gravit.* 1989;21: 293–305.

[338] Erber T. Velocity of light in a magnetic field. *Nature.* 1961;191:25–7.

[339] Haïssinski J, Dagoret S, Urban M, Zomer F. About the connection between vacuum birefringence and the light-light scattering amplitude. *Phys Scr.* 2006;74:678–81.

[340] O'Sullivan SP, Machalski J, VanEck CL, Heald G, *et al.* The intergalactic magnetic field probed by a giant radio galaxy. *Astron Astrophys.* 2019;623:A16.

[341] Carretti E, Vacca V, O'Sullivan SP, Heald GH, Horellou C, Röttgering HJA, Scaife AMM, Shimwell TW, Shulevski A, Stuardi C, Vernstrom T. Magnetic field strength in cosmic web filaments. *Mon Not R Astron Soc.* 2022;512:945–59.

[342] Klein JJ, Nigam BP. Birefringence of the vacuum. *Phys Rev.* 1964; 135:B1279–80.

[343] Agil J, Battesti R, Rizzo C. On the speed of light in a vacuum in the presence of a magnetic field. *Eur Phys J H.* 2023;48:2.

[344] Ahmadiniaz N, Bähtz C, Benediktovitch A, Bömer C, Bocklage L, *et al.* Towards a vacuum birefringence experiment at the Helmholtz International Beamline for Extreme Fields (letter of intent of the BIREF@HIBEF collaboration). *High Power Laser Sci Eng.* 2025;37:e7.

[345] Zavattini G, Della Valle F, Soflau AM, Formaggio L, Crapulli G, Messineo G, Mariotti E, Kunc S, Ejlli A, Ruoso G, Marinelli A, Andreotti M. Polarimetry for measuring the vacuum magnetic birefringence with quasi-static fields: a systematics study for the VMB at CERN experiment. *Eur Phys J C.* 2022;82:159.

[346] Della Valle F, Gastaldi U, Messineo G, Milotti E, Pengo R, Piemontese L, Ruoso G, Zavattini G. Measurements of vacuum magnetic birefringence using permanent dipole magnets: the PVLAS experiment. *New J Phys.* 2013;15:053026.

[347] Plebanski J. A class of solutions of Einstein-Maxwell equations. *Ann Phys.* 1975;90:196–255.

[348] Loizelet J. Solutions globales des équations d'Einstein-Maxwell. *Ann Fac Sci Toulouse.* 2009;18:495–540.

[349] Hojman S, Rosenbaum M, Ryan MP. Gauge invariance, minimal coupling, and torsion. *Phys Rev D.* 1978;17:3141–6.

[350] Stachel J. On the interpretation of the Einstein-Cartan formalism. In: Harvey A, editor. *On Einstein's Path.* New York: Springer; 1990. p. 475–85.

[351] Woltjer L. A theorem on force-free magnetic fields. *Proc Natl Acad Sci USA.* 1958;44:489–91.

[352] Barnett SM, Cameron RP, Yao AM. Duplex symmetry and its relation to the conservation of optical helicity. *Phys Rev A.* 2012;86:013845.

[353] Jaffe RL. Casimir effect and the quantum vacuum. *Phys Rev D.* 2005;72:021301.

[354] Gatti F, *et al.* Dark Energy Survey Year 3 results: simulation-based cosmological inference with wavelet harmonics, scattering transforms, and moments of weak lensing mass maps I—validation on simulations. *FERMILAB-PUB-23-634-PPD.* 2023;arXiv:2310.17557.

[355] Kragh H. The vortex atom: a Victorian theory of everything. *Centaurus.* 2002;44:32–114.

Index

www.ingramcontent.com/pod-product-compliance
Ingram Content Group UK Ltd.
Pitfield, Milton Keynes, MK11 3LW, UK
UKHW022319010626
6152IPUK00004B/18

9 789819 828616